FERTILIZATION

Comparative Morphology,
Biochemistry,
and Immunology

Volume II

ALBERT TYLER
(1906–1968)

FERTILIZATION

Comparative Morphology,
Biochemistry,
and Immunology

edited by

CHARLES B. METZ

Institute of Molecular Evolution
School of Environmental and Planetary Sciences
University of Miami
Coral Gables, Florida

and

ALBERTO MONROY

Institute of Comparative Anatomy
University of Palermo, and
Research Unit for Molecular Embryology
National Council of Research
Palermo, Italy

Volume II

ACADEMIC PRESS

NEW YORK ● LONDON   1969

ACADEMIC PRESS, INC.
111 Fifth Avenue, New York, New York 10003

United Kingdom Edition published by
ACADEMIC PRESS, INC. (LONDON) LTD.
Berkeley Square House, London W1X 6BA

LIBRARY OF CONGRESS CATALOG CARD NUMBER: 66-29671

PRINTED IN THE UNITED STATES OF AMERICA

List of Contributors

Numbers in parentheses indicate the pages on which the authors' contributions begin.

Thomas F. Anderson, The Institute for Cancer Research, Philadelphia, Pennsylvania (1)

C. R. Austin, Physiological Laboratory, Cambridge, England (437)

E. C. Cantino, Department of Botany and Plant Pathology, Michigan State University, East Lansing, Michigan (95)

Koichi Hiwatashi, Biological Institute, Tohoku University, Sendai, Japan (255)

E. A. Horenstein, Division of Research Grants, National Institute of Health, Bethesda, Maryland (95)

Harold Jackson, Paterson Laboratories, Christie Hospital, Manchester, England (467)

H. F. Linskens, Department of Botany, University of Nijmegen, Nijmegen, The Netherlands (189)

Eizo Nakano, Biological Institute, Nagoya University, Nagoya, Japan (295)

*Lajos Pikó,** Division of Biology, California Institute of Technology, Pasadena, California (325)

Giuseppe Sermonti, Institute of Genetics, The University, Palermo, Italy (47)

Lee D. Simon, The Institute for Cancer Research, Philadelphia, Pennsylvania (1)

C. Thibault, Départment de Physiologie Animale, Institut National de la Recherche Agronomique, Jouy-en-Josas, France (405)

L. Wiese, Department of Biological Sciences, Florida State University, Tallahassee, Florida (135)

---

* *Present address:* Developmental Biology Laboratory, Veterans Administration Hospital, Sepulveda, California.

Albert Tyler

Albert Tyler would certainly play a prominent role in any treatise on comparative aspects of fertilization at the morphological, biochemical, and immunological levels. But his relationship to this work was a very special one. He not only wrote the introduction and co-authored a chapter in Volume I with one of us (A.M.), but he contributed much to the planning of both volumes. Furthermore, he served as a most valuable advisor and consultant to the Fertilization and Gamete Physiology Training Program at the Marine Biological Laboratory from which this treatise evolved. His sudden death on November 9, 1968 is a great loss to reproductive and developmental biology, and a personal grief to his multitude of friends and admirers, which includes many of the contributors to these two volumes. We dedicate this second volume to him.

Albert Tyler was born in Brooklyn, New York on June 26, 1906. He began his higher education at Columbia University as a major in chemistry. However, early in his graduate program, he came under the influence of T. H. Morgan. At this time, Morgan's research interests had shifted from genetics back to developmental and reproductive biology. Albert Tyler moved to the California Institute of Technology with Morgan in 1928 and received his Ph.D. there in 1929 under Morgan's supervision—the first Ph.D. in Biology at the Institute. He then joined the faculty of the Institute, and rose through the ranks to the professorship which he held throughout his career.

Tyler and Morgan remained very close friends and colleagues until Morgan's death. The two men shared a common interest in problems of fertilization and development and a strong attraction to marine invertebrates and marine biology in general. The latter interest was generated during their respective student days at the Marine Biological Laboratory and is reflected in the Kerckhoff Marine Laboratory of the California Institute of Technology which they founded at Corona del Mar, California. This laboratory served as a base for collecting and storing living material and as a haven for research for both men, and their students, for the remainder of their lives. Tyler had a deep attachment for this laboratory and the adjacent area. He was thoroughly familiar with the organisms in the local waters and with their distribution, behavior, and ecology. Indeed, his knowledge and insight into the area and marine biology in general was so detailed that he was engaged as a consultant by Marineland of the Pacific at Palo Verde, California. Not surprisingly, then, most of Albert Tyler's research involved marine invertebrate material.

As indicated above, Albert Tyler received formal training in the physicochemical sciences and, from T. H. Morgan, E. B. Wilson, and others, the very best possible instruction in classic cytology, genetics, and developmental biology at the time when these subjects were at their peak. Thus he was prepared as were few others to lead the transition from classic experimental embryology through developmental physiology and into the present decade of molecular biology. His first publications were in the classic tradition, and concerned studies on determination and differentiation in the "determinate" eggs of annelids. Shortly, however, his interests turned to problems of the energetics of development and the respiratory changes in the sea urchin egg following fertilization. He received reinforcement in this direction during 1932–1933 as a National Research Council postdoctoral Fellow in Warburg's laboratory and at the Zoological Station in Naples. The most significant outcome of this period was the series of papers entitled "On the Energetics of Differentiation" in which he sought to introduce the concepts of physical chemistry to the study of the processes of embryonic development.

In the late 1930's, Albert Tyler and some of his associates, including S. H. Emerson, L. C. Pauling, and A. H. Sturtevant, realized that modern immunochemistry, then emerging, was a subject of fundamental interest to several biological disciplines and, in addition, provided tools and models for the study of a variety of other basic biological problems at the physicochemical level. He recognized this as an opportunity to investigate the events of fertilization and development in a more modern and rigorous fashion. Accordingly, he began a reexamination, extension, and modernization of F. R. Lillie's "Fertilizin Theory" of fertilization with the realization that fertilizin and other specific interacting sperm and egg substances were the only systems readily available among metazoa in which biological substances of obvious significance in fertilization and development could be isolated, characterized, and their functions studied. The resulting studies and their offshoots commanded his interest for the next twenty years. He investigated specific egg membrane lysins obtained from sperm. These agents evidently can contribute to the specificity of fertilization, and certainly are required to provide a passage for the spermatozoan to the egg surface. The studies on fertilizin and antifertilizin modernized Lillie's original antigen–antibody analogy, and extended the understanding of fertilizin and antifertilizin chemistry—mode of interaction and role in fertilization. He examined sperm–egg interaction at the ultrastructural level, and presented provocative theories to relate the ultrastructure to the role of fertilizin and antifertilizin in fertilization. Characteristically, he introduced conceptual and experimental novelties into his studies. For example, he prepared nonagglutinating and nonprecipitating antibody by photooxidation and used this in fertilization studies some fourteen years before the enzymatic digestion methods for preparation of univalent antibody were discovered.

Albert Tyler was always alert to the possibility of extending his findings into the widest possible contexts. His studies on sea urchins led him to formulate an autoantibody concept of cell structure which he proceeded to test by laboratory experiments. These tests included efforts to obtain specific complementary substances from pneumococci and an examination for antivenom production in Gila monsters. Later he proposed a theory of cancer, and undertook extensive experiments to prove it using mice. But his major interest during this period concerned fertilization and the physiology of reproduction. The precision of his thinking and laboratory skill set the standard for all who worked in this field.

During his last ten years, Albert Tyler's interests were concentrated on the molecular biology of fertilization and development. His experiments were among the first to give compelling evidence that mRNA must be present in some inactive condition in the unfertilized sea urchin egg, thus contributing to the establishment of the provocative idea of "masked messenger" RNA. This was the starting point for an intensive effort by Tyler's group to solve the problem of the mechanisms of activation of protein synthesis which follows fertilization. Another important contribution was the demonstration of cytoplasmic DNA and its significance in the echinoderm egg.

Albert Tyler was deeply interested in areas of direct benefit to mankind, including conservation and birth control. In the latter context he conducted studies on infertility in relation to autoimmunity in man using clinical material supplied by his brother, Dr. Edward T. Tyler of Los Angeles.

Although the impact of Albert Tyler's published work will continue to be felt, his influence has and will continue to be equally great through the training, guidance, and inspiration he has given to the many students, postdoctorals, and friends who have worked with him in his own laboratory, at the Marine Biological Laboratory, and at other institutions during visits. Such visits, even when brief, had a lasting effect on students. This resulted from an extraordinary combination of energy, personality, imagination, and immediate transfer of ideas to laboratory tests. For example, he visited the Fertilization and Gamete Physiology Training Program for a few days almost every summer from 1962 to 1967. Characteristically, these visits began with a few hours of conversation followed by intensive laboratory work with trainees and staff during the remaining days of the visit.

Albert Tyler was sought after to organize and chair many symposia, he functioned as a consultant to several agencies and foundations including the NSF, NIH, Ford Foundation, Population Council, World Health Organization, and the Panel on Population Problems of the National Academy of Sciences. For fourteen years he served on the Board of Trustees of the Marine Biological Laboratory and as President of the American Society of Naturalists and the Society of General Physiologists. Throughout his life Albert Tyler was always an informal man who abhorred pretense in any form. He especially enjoyed small, intimate gatherings where he shared his enthusiasm for science through discussion with friends. We, his friends, will long remember such gatherings and will miss the excitement and challenge of Albert's subtle wit and penetrating mind.

CHARLES B. METZ
ALBERTO MONROY

Preface

The objectives of this work as outlined in the Preface to Volume I include assembling the available information on gamete physiology and fertilization mechanisms in plants, animals, and microorganisms in the form of a comparative treatise. Each contributor to Volume I has treated one or a few of the recognized fundamental problems. These problems have been examined for the most part without regard to the inter-relationships of the organisms considered. Nevertheless, a comparative approach is implicit for, as emphasized by our late friend Albert Tyler in the first chapter of Volume I, "It is clear that the most important features of any biological process are those which are common to diverse organisms."

On the other hand, much is gained by a comparison of the sequence of the several steps in the total fertilization process. Accordingly, Volume II is largely devoted to the examination of fertilization mechanisms or their analogues in specific groups of organisms. Each of these is treated in depth by a specialist to give as complete a picture of the physiology of the reproductive process as possible. Here, as in any selection, the choice of subjects was necessarily somewhat arbitrary. In addition to an attempt at a comparative coverage, we were guided in our choice by the extent of information available (or in some cases the lack of it), the value of the system as a model for one or more aspects of fertilization, and the uniqueness of the problems and their solutions as revealed by morphological, biochemical, or immunological analysis. In attempting to delimit the scope of the comparative treatment, we recognized that much exciting information is compartmentalized in the literatures of virology, microbiology, protozoology, and the plant, animal, veterinary, and medical sciences as a consequence of scientific specialization. Here we attempt a cross-sectional exposure to these exciting studies.

A final objective of these two volumes is to present provocative aspects and leading questions in the hope that this will stimulate additional research in this subject area which is not only of basic biological interest in its own right but of increasing importance to human welfare in the areas of fisheries, agriculture, and medicine.

Woods Hole, Massachusetts CHARLES B. METZ
September, 1969 ALBERTO MONROY

Contents

CHAPTER 1. The Attachment of Bacteriophages and the
Transfer of Their Genetic Material to Host Cells

Lee D. Simon and Thomas F. Anderson

CHAPTER 2. Bacteria

Giuseppe Sermonti

CHAPTER 3. Fungi

E. A. Horenstein and E. C. Cantino

CHAPTER 9. *In Vitro* **Fertilization of the Mammalian Egg**

C. *Thibault*

CHAPTER 10. **Variations and Anomalies in Fertilization**

C. R. *Austin*

CHAPTER 11. **Control of Fertility Mechanisms Affecting
            Gametogenesis**

Harold *Jackson*

Contents of Volume I

FERTILIZATION

*Comparative Morphology,
Biochemistry,
and Immunology*

Volume II

CHAPTER 1

The Attachment of Bacteriophages and the Transfer of Their Genetic Material to Host Cells*

Lee D. Simon and Thomas F. Anderson

THE INSTITUTE FOR CANCER RESEARCH, PHILADELPHIA, PENNSYLVANIA

I. Introduction

The word "fertilization" in its broad sense may be taken to mean the introduction of external genetic information into a cell. Nature has invented many kinds of devices for this purpose ranging in size from relatively huge sperm cells of which the dimensions are measured in hundreds

* This work was supported by Grants NIH-5-T-1-GM-000658 (to the University of Rochester) and CA-06927 from the National Institutes of Health, U. S. Public Health Service, Grants GB-4640 and GB-982 from the National Science Foundation and an appropriation from the Commonwealth of Pennsylvania.

of thousands of angstroms to relatively tiny virus particles (virions) that measure only a few hundreds of angstroms. In this chapter we shall describe the molecular biology of viral attachment to host cells with particular reference to T-even bacteriophage systems where most is known. The various *molecular organelles* of these virions play sequential roles in recognizing the host cell, in adsorbing to it, and in injecting the viral nucleic acid into it. Other bacteriophages have their own characteristic mechanisms for recognizing host cells and introducing their nucleic acid messages into them. There will be a review of the various types of bacterial receptors for these bacteriophages and the effects phage adsorption has on them. Finally, in the discussion, we shall attempt to compare the bacteriophage–bacterium systems with sperm–egg systems and see that both morphologically (for many phages are shaped like sperm) and functionally (for some phages carry genetic messages from one bacterium to another) the two systems have many features in common.

II. Structure of T-Even Bacteriophages

The first electron micrographs of bacteriophages were taken by Ruska in 1941. In 1942, Luria and Anderson obtained electron micrographs which provided the first direct evidence that phage T2 was not a simple sphere, but rather a complex structure with a head and a long tail. The significance of this morphological complexity of T2 bacteriophage was not understood until Anderson (1952, 1953) using the critical-point technique (Anderson, 1951) to preserve specimen structure, demonstrated with electron micrographs that the tail was the "organelle" by which bacteriophage attach to sensitive cells. This observation was subsequently confirmed for phage T2 (Williams and Fraser, 1956) and for phage T5 (Weidel and Kellenberger, 1955) using another technique, freeze-drying, to eliminate artifacts in the preparation of specimens. The bacteriophage tail, therefore, is a significant structure; it mediates the attachment of phage to bacterial cells (Anderson, 1953) and is involved in the introduction of the bacteriophage's nucleic acid into its host (Hershey and Chase, 1952).

Bacteriophage tails, in particular those of the T-even phages (T2, T4, and T6), are quite intricate, possessing several morphologically, chemically, and serologically different types of structural components (Brenner et al., 1959; Edgar and Lielausis, 1965). The T-even phage tail (Fig. 1) consists of two coaxial hollow tubes, the inner being referred to as the needle and the outer as the sheath; the needle and the sheath terminate at their distal tips in the baseplate which, when viewed from the side, sometimes resembles a crown and which, when viewed along its major

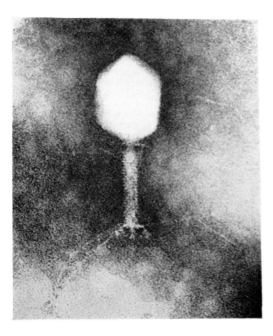

Fig. 1. Top: An electron micrograph of a T4 bacteriophage embedded in sodium silicotungstate. In this and in subsequent figures the magnification bar represents 1000 Å. Bottom: A schematic drawing of the T4 phage shown above.

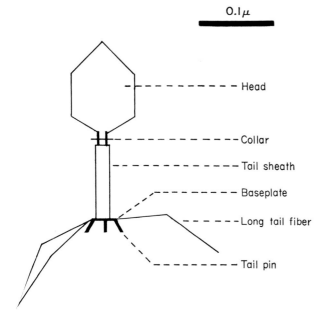

axis, appears as a six-pointed star with a long fiber radiating from each point (Fernández-Morán, 1962; Anderson and Stephens, 1964). These tail fibers seem to play an important role in the specific attachment of T-even phage to sensitive bacteria (Williams and Fraser, 1956; Kellenberger and Séchaud, 1957; Wildy and Anderson, 1964; Simon and Anderson, 1967a). At the opposite end of the tail, nearly touching the head membrane is a thin disc called a "collar." The overall dimensions of the tail are about 200 × 1000 Å, with the long tail fibers extending another 1300–1500 Å.

Each component of a phage tail may itself be a complex structure. This fact is exemplified by the long tail fibers of phage T4. Genetic, serological, and morphological evidence indicates that at least six genes (34, 35, 36, 37, 38, and 57) of bacteriophage T4 are involved in the construction of the phage's tail fibers (Epstein et al., 1963; Edgar and Lielausis, 1965). Edgar and Lielausis have shown that there are at least three antigenic components, called A, B, and C, in the tail fibers. Antigens A, B, and C appear to be controlled by genes 34, 36, and 37, respectively; gene 38 may also affect the synthesis of antigen C. Edgar and Lielausis have speculated that gene 35 may be important in the "association of the tail fiber components with the phage particle." It is interesting that gene 57, unlinked to the other tail fiber genes, seems to affect the production of all three tail fiber antigens. The possibility that gene 57 has a regulatory role in the synthesis of tail fiber components has been suggested.

One of the most remarkable properties of the T-even phage tail is the ability of its sheath to contract, making visible the inner needle (Kozloff and Henderson, 1955; Kellenberger and Arber, 1955). Available evidence suggests that the contracted sheath is composed of six or twelve helically intertwined chains of identical subunits (Brenner et al., 1959; Anderson and Stephens, 1964; Kellenberger and Boy de la Tour, 1964; Anderson and Krimm, 1966; Moody, 1967a, and Krimm and Anderson, 1967). Extended sheaths often show twenty-four cross striations or annuli, each annulus containing six morphological subunits. These subunits have a molecular weight of about 55,000 (Brenner et al., 1959; Sarkar et al., 1964) and they are visible in electron micrographs of negatively stained sheaths (Anderson, 1963). Electron micrographs of extended T-even sheaths have also been interpreted as showing that the sheath subunits are arranged to form six coarse helices (Moody, 1967a; Krimm and Anderson, 1967). After the sheath contracts the cross striations are more difficult to resolve but when they are visible, their number is seen to be twelve. Some authors (Bradley, 1963; Krimm and Anderson, 1967) have suggested that during contraction the number of subunits per annulus is doubled; furthermore, the volumes of extended and contracted sheaths are about the same. Therefore, there does not seem to be a loss of structural units from the T-even sheath during its contraction process (Brenner *et al.*, 1959).

III. Adsorption Cofactors for T-Even Bacteriophages

Certain strains of T-even bacteriophages require tryptophan in order to attach to sensitive bacteria (Anderson, 1945). This fact was determined by comparing the number of plaques produced by virus particles spread with sensitive bacteria on agar medium with and without tryptophan. Without this amino acid some strains of T4 and T6 bacteriophages formed significantly fewer plaques than the control phages plated with tryptophan; the number of plaques formed in the presence of tryptophan was up to 10^5 times greater than the corresponding number without tryptophan. It seems that tryptophan acts as a cofactor, facilitating the adsorption of these strains of bacteriophages.

Anderson (1948) showed that tryptophan reacts with the virus particles rather than with the bacteria. If tryptophan is added to a cofactor-requiring phage strain and then the phages are diluted into a large volume of bacteria without cofactor, the phages will attach to and will lyse the cells. Conversely, if the tryptophan is added to a bacterial suspension which is subsequently diluted into a large volume of phages without tryptophan, the phages will not adsorb to the host cells. About five or six molecules of tryptophan are required to activate one phage (Anderson, 1945; Wollman and Stent, 1950).

Ultracentrifugal studies have shown that the presence of tryptophan decreases the sedimentation velocity of cofactor-requiring T-even phages (Cummings, 1964; Kellenberger *et al.*, 1965). In the presence of tryptophan, T4 particles exhibit an $s_{20,w}$ sedimentation coefficient of about 900, whereas in the absence of tryptophan the $s_{20,w}$ value increases to about 1000. The factor responsible for this change in sedimentation rate appears to be the phages' tail fibers. Both Cummings (1964) and Kellenberger *et al.* (1965), using the electron microscope, found that in the absence of tryptophan the long tail fibers of cofactor-requiring particles are wrapped around their tail sheaths; the addition of tryptophan causes the release of the long fibers so that they extend away from the sheaths. In the presence of tryptophan the tail fibers thus project from the baseplate into the surrounding fluid thereby increasing the frictional resistance of the phage to movement in the centrifugal field of the ultracentrifuge; on the other hand, in the absence of tryptophan the tail fibers are wrapped closely around the tail sheath, reducing the frictional drag, and permitting the phage to sediment more rapidly.

The observation that tryptophan causes the extension of the long tail fibers explains, at the morphological level, the activation of cofactor-requiring T-even phages. The attachment of these virus particles to the bacterial wall involves the long tail fibers; unless these fibers are in the extended configuration, they are not free to interact with the bacterial surface. Cofactor-requiring phages may be stabilized in an active state by treating them with tryptophan and then with an antiserum (Jerne, 1956) which is specific for tail sheaths (Brenner et al., 1962).

An interesting problem was raised by the observation that bacteria infected with cofactor-requiring phages form plaques on agar medium without tryptophan, whereas free cofactor-requiring phages plated on similar medium with sensitive bacteria do not produce plaques. The formation of a visible plaque requires many rounds of bacteriophage adsorption and bacterial lysis. Therefore, when the initially infected bacteria lysed, the phages which were released must have been able to infect other bacteria even though no tryptophan was present. This ability of newly formed cofactor-requiring phages to infect sensitive bacteria in the absence of tryptophan has been called "nascent activity" (Wollman and Stent, 1952). Nascent activity is not due simply to the localized release of tryptophan by lysed bacteria. Cofactor-requiring phages lose their tryptophan-induced activity very quickly when diluted into tryptophan-free media; however, nascently active particles lose their infectivity relatively slowly. Furthermore, bacteria lysed by T1 phages do not facilitate infection by cofactor-requiring T4 viruses (Wollman and Stent, 1952). The reader is referred to Wollman and Stent (1952) for further discussion of nascent activity.

Indole and other substances such as iodobenzene act as negative cofactors in that they inhibit the adsorption of some strains of T2 and T4 phages (Delbrück, 1948). Thus there are at least four types of T-even bacteriophages with respect to cofactor sensitivity: (1) those that are insensitive to both indole and tryptophan; (2) those that respond only to positive cofactors such as tryptophan; (3) those that respond only to negative cofactors such as indole; and (4) those that respond to both tryptophan and indole. It seems that the effects of indole on bacteriophages are nearly the opposite of those discussed for tryptophan. For example, at pH 7.5 the presence of indole causes sensitive T-even particles to sediment rapidly in the ultracentrifuge (Kanner and Kozloff, 1964), analogous to cofactor-requiring phages in the absence of tryptophan. The reaction of phage with antisheath serum prevents sensitivity to indole; the same reaction eliminates the requirement for tryptophan (Jerne, 1956; Brenner et al., 1962). The interaction of indole with sensitive T2 and T4 particles seems to cause the long tail fibers to become bound in the retracted position thereby temporarily inactivating the phages (Brenner *et al.*, 1962) and increasing their sedimentation rate (Kanner and Kozloff, 1964).

The prevalence of bacteriophages that react with cofactors may be related to the increased resistance of the phage tail to physical and chemical damage when the long tail fibers are retracted and wrapped around the sheath. Cadmium cyanide treatment inactivates T-even phages and causes their tail sheaths to contract (Kozloff *et al.*, 1957). However, when a phage's long tail fibers are wrapped closely around its sheath, the phage is insensitive to inactivation by cadmium cyanide; thus cofactor-requiring particles are protected in the absence of tryptophan, whereas indole-sensitive bacteriophages are protected in the presence of indole (Brenner *et al.*, 1962; Kanner and Kozloff, 1964). While their tail fibers are retracted these phages do not respond to antisheath serum (Jerne, 1956; Brenner *et al.*, 1962).

IV. The Infection of *Escherichia coli* by T-Even Bacteriophages

A. THE INFECTIOUS PROCESS AS SEEN IN THE ELECTRON MICROSCOPE

Simon and Anderson (1967a) used the electron microscope to study the adsorption of T-even bacteriophages to *Escherichia coli* cells. They found no association between cofactor-requiring T4 phages and *E. coli* in the absence of tryptophan, but within 1 minute after adding tryptophan, T4 particles are strikingly oriented with respect to the bacterial wall (Fig. 2a,b,c). Many phages are oriented with their tails pointing toward the wall; the distance from their baseplates to the cell wall is about 1000 Å. In Fig. 2b,c, long fibers are seen linking the phage tails to the wall. These long tail fibers are characteristically bent near their centers so that their distal tips touch the cell wall at points some 700–900 Å from the point where a line drawn perpendicular to the baseplate would intersect the cell wall.

Subsequently, the main body of the phage seems to move closer to the bacterial surface. Figure 3 is an electron micrograph of a section showing a T4 particle apparently attached to the cell wall by the tips of the 100-Å long pins on its baseplate. The factors responsible for the phage's movement toward the bacterial surface have not been experimentally determined. Perhaps Brownian motion could randomly agitate the attached virus until by chance it moves close enough to the cell wall for the pins to attach. A more interesting possibility is that the attachment of the long tail fibers triggers the complex sequence of events which follows. The first of these events would be the *active* drawing of the phage particle to the bacterial surface by the flexing of the long fibers.

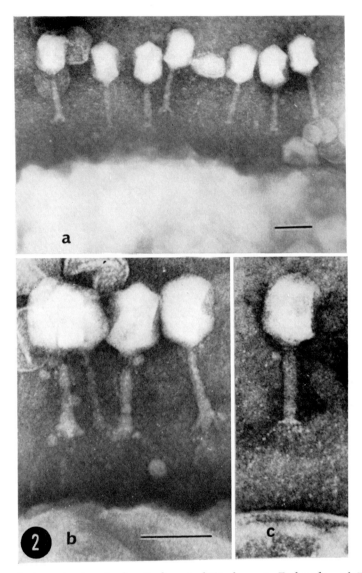

Fig. 2. An early stage in the attachment of T4 phages to *Escherichia coli* B cells as seen in negatively stained preparations. Cofactor-requiring T4 phages were mixed on the surface of a grid with *E. coli* B cells in F medium. Tryptophan was then added, and about 15 seconds later the preparation was negatively stained; within 1 minute it was placed in the column of the electron microscope and dried. (a) The phages are oriented with their tails toward the *E. coli* cell wall. The baseplates of the viruses are about 1000 Å from the bacterial surface. (b) Long tail fibers are visible connecting the baseplates to the *E. coli* wall. (c) The long tail fibers extending from the phage's baseplate to the cell wall are bent meeting the wall about 900 Å from the point where a perpendicular from the baseplate would intersect the cell wall.

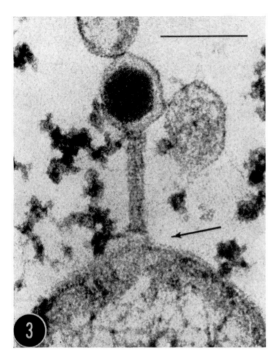

FIG. 3. A section through a T4 phage with its base-plate about 100 Å from an *Escherichia coli* CR63 wall fragment. The phage appears to be attached to the cell wall by its tail pins (arrow).

The next steps after the attachment of the phage's pins involve the contraction of the tail sheath. The baseplate remains attached to the distal end of the sheath but releases itself from the tip of the needle so that as the sheath contracts it slides up the needle and away from the cell wall. This stage can be seen in Fig. 4, which shows many T4 bacteriophages attached to the wall of an *Escherichia coli* B cell. The distance from the baseplate to the bacterial surface after contraction of the sheath is about 370 Å.

At first it seemed possible that perhaps the baseplates might be in intimate contact with an invisible 370-Å thick surface layer which forms

FIG. 4. T4 phages adsorbed in *Escherichia coli* B cells. Cofactor-requiring phages were suspended with bacteria in F medium containing tryptophan. After 10 minutes at 37°C, a droplet of this suspension was negatively stained and examined in the electron microscope. The phages' sheaths are contracted and their baseplates are 300–400 Å from the cell wall. Several hollow needles are visible extending from the baseplates to the cell wall. (Figure on following page.)

FIG. 5. T4 phages adsorbed on an *Escherichia coli* B cell wall as seen in thin sections. The phages are bound to the bacterial surface by short tail fibers extending directly from their baseplates to the cell walls. The needle of one of the phages can be seen to penetrate just through the cell wall (arrow); 30-Å fibrils, probably of phage DNA, extend on the inner side of the cell wall from the distal tips of the T4 needles. (Figure on following page.)

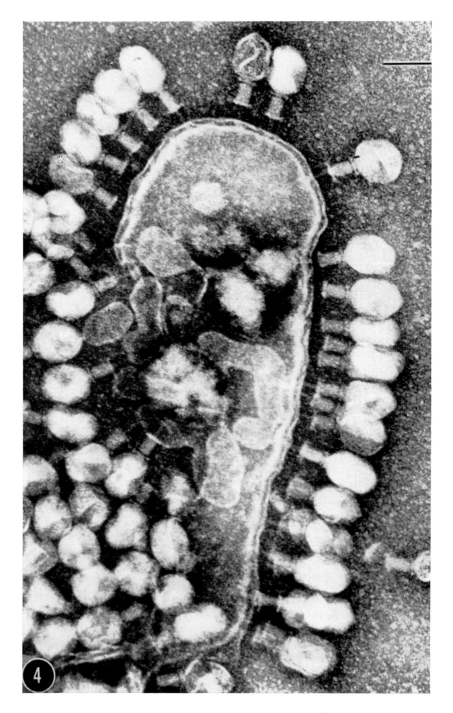

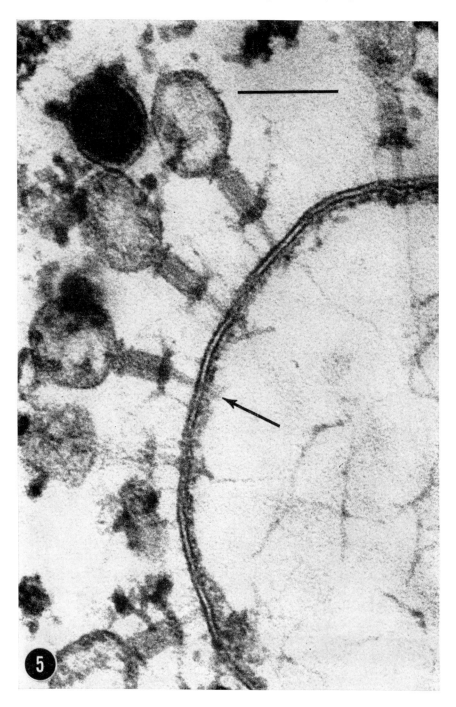

around each bacterium after infection. All attempts, though, to observe such a layer have failed.

In thin sections through T4 particles adsorbed on *E. coli* cells, which the phage have lysed from without (Fig. 5), it is clear that the phages are attached to the bacterial wall by short tail fibers. These short fibers extend directly from the baseplates to the wall. In serial sections (Fig. 6) one can discern at least four short fibers emanating from points about 420 Å apart on the baseplate. It seems that when the sheath contracts, it pulls the baseplate away from the wall, causing the tail pins to uncoil (Simon and Anderson, 1967a) to form the short tail fibers. T2 phages appear to have tail fiber arrangements similar to those observed on T4.

The distal ends of the phage's long tail fibers may remain attached to the cell wall after contraction of the sheath. In micrographs of sections containing adsorbed T2 phages and their long tail fibers (Fig. 7), the long tail fibers connect the baseplates to the cell walls. The proximal parts of these fibers extend from the baseplates in directions away from the bacteria; near their centers, the long fibers bend sharply so that their distal halves are almost perpendicular to the bacterial surfaces. The long fibers usually meet the cell walls about 700 to 900 Å from the points of intersection of the phages' needles with the walls.

The results of Simon and Anderson (1967a) indicate that the adsorption of T2 and T4 phages to *E. coli* involves binding by at least two different phage components—an initial attachment by long tail fibers, and a subsequent attachment by short tail fibers.

This dichotomy may be relevant regarding the observations of Franklin (1961) and of Edgar and Lielausis (1965) that some of the neutralizing activity of anti-T4 serum is directed against phage components other than the long tail fibers. For example, Franklin reported that about 10% of the activity is directed against sheaths. Furthermore, since tail pins or short tail fibers are involved in the attachment process, one might expect that some neutralizing activity would be directed against these pins or short fibers.

Electron micrographs of thin sections [Figs. 5 (arrow) and 6] and statistical data indicate that the needles of infecting T2 and T4 bacteriophages penetrate the *E. coli* cell wall about 120 Å. The thickness of the intact cell wall is about 120 Å. Therefore, it appears that the needles of adsorbed T2 and T4 phages are able to penetrate just through the *E. coli* cell wall and not through the 60–70 Å thick protoplasmic membrane immediately surrounding the bacterial cell (Simon and Anderson, 1967a). It would seem that the phage deoxyribonucleic acid (DNA) is injected at the surface of the cytoplasmic membrane. From there it might be expected that the phage DNA infects the cell in a manner analogous to

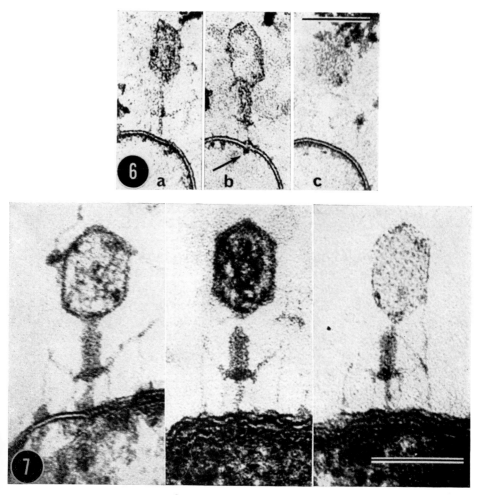

Fig. 6. Consecutive serial sections through an adsorbed T4 phage. The section thickness is under 300 Å. (a) Two short tail fibers extend from the baseplate to the cell wall. The distal tip of the needle is not visible in the plane of this section. The outer layers of the cell wall in the immediate vicinity of the phage's needle are continuous and unbroken. (b) In this section, two short tail fibers again connect the baseplate to the wall. The arrow indicates where the needle appears to have penetrated the broken cell wall. (c) The cell wall in this section is unbroken; only part of the phage's head is visible.

Fig. 7. Thin sections through three adsorbed T2 phages attached by both their long and short tail fibers to intact *Escherichia coli* B cells or large wall fragments. *Escherichia coli* B cells were infected with T2 phages. After 1 hour they were fixed and then embedded. Rather thick, silvery gold-colored sections were cut and, subsequently, examined in the electron microscope. The phages' long tail fibers extend laterally from their baseplates; they flex near their centers and their distal tips are attached to the cell walls. The long fibers attach to the walls at points over 800 Å from the intersections of the needles with the walls. Short tail fibers go directly from the baseplates to the walls.

that in which pure DNA extracted from λ phages (Meyer *et al.*, 1961) and from ΦX174 phages (Guthrie and Sinsheimer, 1960) infects protoplasts. Alternatively, the membrane and wall may be in intimate contact at the site of injection, so that the transfer of the phage DNA into the cytoplasm is facilitated (Bayer, 1967).

A tentative model has been proposed by Kellenberger and Boy de la Tour (1964) to explain the transition of the sheath from the extended to the contracted state. According to their hypothesis, in its extended form the sheath's subunits are bound tightly to the central needle, but only weakly to each other. During contraction, the subunits are released from the needle; they undergo a structural change, bringing new chemical groups into active positions and, thus, permitting the subunits to "self-assemble" into helical bands with a pitch of 30°—the contracted sheath. Two more detailed models of the contraction process have been suggested recently. One model proposes that the contraction of the tail sheath is due to a reduction in the pitch of the six helices which constitute the extended sheath (Moody, 1967b). According to this scheme, the axial rotational symmetry of the contracted sheath is the same (sixfold) as that of the extended sheath. The other model proposes that contraction of the sheath occurs by the merging of adjacent pairs of annuli in the extended sheath (Krimm and Anderson, 1967). Thus, contracted sheaths would have half the number of annuli found on extended sheaths, the number of subunits per contracted sheath annulus being twice the number per extended sheath annulus. According to this second model, the contracted sheath would have twelvefold axial rotational symmetry as opposed to the sixfold symmetry of the extended sheath. In support of Krimm and Anderson's model, electron micrographs of contracted sheaths in axial view consistently indicate that the contracted sheath seems to have twelvefold axial rotational symmetry. However, the resolution of the electron microscope may not be sufficient to rule out the model with sixfold symmetry that Moody proposes.

The baseplates of T-even phages increase about 50% in diameter in association with the contraction of the phages' tail sheaths (Simon and Anderson, 1967b). A change in the morphology of the baseplate associated with the contraction of the tail sheath (Figs. 8 and 9) is not restricted to the T-even coliphages. Several apparently unrelated bacteriophages with contractile sheaths, such as the *Staphylococcus* phage Twort (Vieu *et al.*, 1964), the *Bacillus subtilis* phage SP3 (Eiserling and Romig, 1962), and the *Lactobacillus fermenti* phages 222a and 300 (de Klerk *et al.*, 1965), have baseplates which exhibit striking changes in their molecular configurations associated with the contraction of their tail sheaths. The micrographs of the Twort and SP3 phages indicate that their

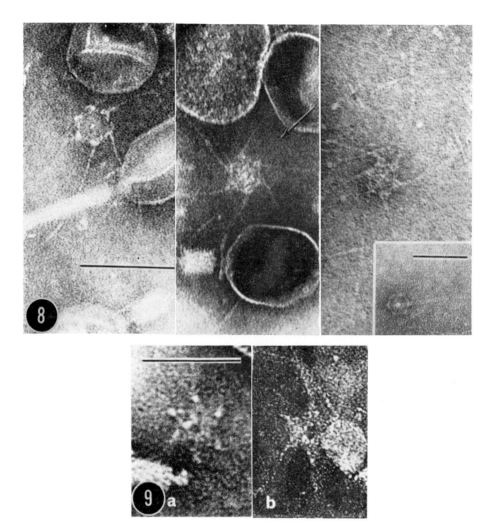

Fig. 8. Axial views of hexagonal T2 and T4 baseplates. These baseplates are shaped like hexagons with vertex-to-vertex diameters of about 400 Å. They have long tail fibers extending from their vertices and plugs in their centers. The baseplate indicated by the arrow has two internal concentric rings. These rings are probably sheath and needle subunits which have remained attached to the baseplate. The larger ring has a diameter of about 180 Å which is the diameter of the extended sheath. Inset (right) shows a 400-Å hexagon apparently bound to the tip of a needle.

Fig. 9. Axial views of isolated T4 baseplates seen by negative staining. (a) T4 phages were disrupted by treatment with 3% H_2O_2 (Kellenberger and Arber, 1955). Droplets of the suspension were then withdrawn, placed on grids, negatively stained, and examined under the electron microscope. The baseplate is shaped like a six-pointed star with a hole through its center and white dots near its tips. The diameter of the baseplates is about 600 Å. (b) This baseplate has a morphology similar to that of the baseplate in (a); however, it was found in the preparation containing some of the smaller hexagonal baseplates shown in Fig. 8.

baseplates increase in diameter; the pictures of the 222a and 300 *L. fermenti* phages show that their baseplates undergo obvious structural rearrangements, although the micrographs do not permit accurate measurement of the diameters of these baseplates. It seems, therefore, that a change in the molecular configuration of the baseplate associated with the contraction of the tail sheath is to some extent, at least, a general phenomenon among bacteriophages.

When the diameter of the T-even baseplate increases by 50% (from 400 to 600 Å) the diameter of the distal ring of sheath subunits which is attached to the baseplate would also be expanded by 50% (Fig. 10). The diameter of the extended sheath is 180 Å; an increase of 50% would change the diameter of the distal ring of sheath subunits to 270 Å which is the diameter of the contracted sheath. This change in the diameter of the distal ring of sheath subunits might cause a cascading effect among the other sheath subunits resulting in the contraction of the entire sheath.

The change in the molecular configuration of the baseplate may also be of significance in releasing the phage's DNA at the appropriate time during the adsorption process. The needle of an adsorbed T-even particle appears to penetrate the *Escherichia coli* cell wall (Simon and Anderson, 1967a), and the phage's DNA passes through the needle into the bacterium. Associated with the contraction of the sheath, the T-even baseplate increases in diameter, and, also, the plug at its center is replaced by

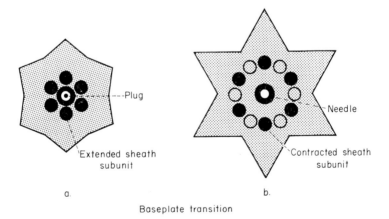

Baseplate transition

Fig. 10. Schematic illustration of the increase in the diameter of the baseplate and sheath associated with contraction of the tail sheath. The diameter of the baseplate increases by about 50%. The ring of sheath subunits attached to the baseplate is presumably also expanded by a similar factor, thereby allowing the intercalation of an adjacent ring of sheath subunits (cf. Anderson and Krimm, 1966). The central plug seen on the smaller baseplate is replaced by a central hole on the larger star-shaped baseplate.

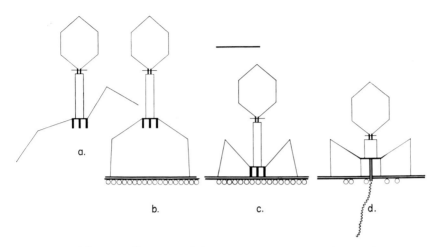

Fig. 11. Schematic illustration of the major steps in the attachment of T2 and T4 phages to the *Escherichia coli* cell wall. (a) An unattached phage is depicted. A few of its long tail fibers and tail pins are seen. (b) The phage's long tail fibers have attached to the cell wall. Opposite long fibers attach to the wall about 1800 Å apart; they are bent slightly near their centers. The baseplate of the phage is over 1000 Å from the cell wall. (c) The phage moves closer to the bacterium until its tail pins are in contact with the bacterial wall. (d) The tail sheath has contracted. Short tail fibers and long tail fibers link the baseplate to the cell wall. The needle through which the phage's DNA is ejected has penetrated the wall. The innermost component of the wall has been disrupted, as will be described in detail in Section VII.

an 80-Å hole through which the needle passes. The plug in the baseplate associated with the extended sheath may be responsible for keeping the phage's DNA from prematurely flowing out through the needle. According to this suggestion, when the baseplate expands and the 80-Å hole forms in its center, the baseplate can slide up the needle; when the plug is removed, the needle is no longer obstructed and the phage's DNA may pass through it into the bacterium.

The following hypothesis has been suggested (Simon and Anderson, 1967b) to explain the ordered sequence of events characterizing the adsorption of T-even bacteriophages to *E. coli* cells (Fig. 11):

1. The long tail fibers attach at their distal ends to an *E. coli* cell wall.

2. The short tail pins are brought into intimate contact with the cell wall as the phage is repositioned either by Brownian motion or by a mechanism built into the phage's tail fibers and/or baseplate.

3. The short tail pins attach to the cell wall.

4. The molecular organization of the baseplate changes from a hexagon with a plug at its center to a six-pointed star with a hole through

its center. This change may be caused by the release of specific materials from the cell wall or by alterations of the tail pins which occur when they attach to the wall.

5. The baseplate is released from the distal tip of the needle. The change in the molecular organization of the baseplate may cause this release.

6. The tail sheath contracts. The change in the structure of the baseplate probably causes the distal subunits of the sheath to undergo changes in their spatial arrangement and these changes may trigger the contraction of the rest of the sheath.

7. The contraction of the sheath draws the baseplate away from the cell wall. As the baseplate moves away from the wall it may pull on the tail pins, stretching them to form the short tail fibers.

8. The phage's needle penetrates just through the cell wall. Since the short tail fibers firmly anchor the baseplate to the wall, the contraction of the sheath apparently drives the needle into the wall.

9. The DNA which had been retained in the phage's head passes out through the hollow needle which has penetrated the wall. The removal of the central plug during the molecular reorganization of the baseplate may permit the phage's DNA to flow out through the distal end of the needle.

B. THE KINETICS OF INFECTION

1. *Adsorption Rates*

The rate of adsorption of phages on bacteria may be described by the following equations:

$$-\frac{dP}{dt} = kBP \tag{1}$$

where dP/dt is the adsorption rate, P is the concentration of free phage, B is the bacterial concentration, and k is the adsorption rate constant (usually expressed as milliliters per minute). Integration of Eq. (1) leads to

$$\frac{P}{P_0} = e^{-kBt} \tag{2}$$

or

$$k = \frac{2.3}{Bt} \log_{10} \frac{P_0}{P} \tag{3}$$

where P_0 is the initial concentration of phage.

Early studies on the attachment kinetics of bacteriophage were reported by M. Schlesinger. After mixing a WLL phage (related to T2) suspension with *Escherichia coli* cells in broth, Schlesinger periodically measured the number of unadsorbed phage in the medium; he repeated this procedure with different concentrations of bacteria and of bacteriophage. These experiments indicated to Schlesinger that the adsorption rate constant, k, is independent of bacterial concentration over the range from 3×10^6 bacteria/ml to 7×10^8 bacteria/ml, is independent of phage concentration from 6×10^3 particles/ml to 6×10^5 particles/ml, and is independent of time from 2 minutes to 12 hours (in Adams, 1959). Schlesinger also observed that the number of free virus particles in the medium decreased exponentially as a function of time.

Schlesinger adapted the von Smoluchowski equation for the rate of coagulation of colloidal particles to estimate the rate at which the relatively small phage particles come close enough to collide with the much larger, almost stationary bacteria. Assuming that each such interaction leads to adsorption, he obtained the equation:

$$- \frac{dP}{dt} = 4\pi DrBP \tag{4}$$

where r is the radius of a sphere approximating the surface area of a bacterium, D is the diffusion constant for a free phage particle, and B is the bacterial concentration. By combining Eqs. (1) and (4) the following relationship regarding the theoretical adsorption rate constant, k_t, may be derived:

$$k_t BP = 4\pi DrBP$$

or

$$k_t = 4\pi Dr \tag{5}$$

Since the diffusion constant for a phage may be measured directly (2.4×10^{-6} cm^2 per minute for T2 phage, according to Putnam, 1954) and a value for r may be estimated (about 10^{-4} cm), Eq. (5) permits the calculation of k_t:

$$k_t = 4\pi \times 2.4 \times 10^{-6} \text{ cm}^2 \text{ per minute} \times 10^{-4} \text{ cm}$$

or

$$k_t = 3 \times 10^{-9} \text{ cm}^3 \text{ per minute}$$

For T4 bacteriophage the experimentally determined adsorption rate constant k is about 2.5×10^{-9} cm^3 per minute at the optimal temperature of 37°C. The observation that the value for k_t, calculated under the assumption that every close interaction between a phage and a bacterium leads to adsorption, is nearly the same as the experimentally determined values for k suggests that a T4 particle is adsorbed on its host almost every time it aproaches the bacterial surface (Delbrück, 1942; Anderson, 1949; Stent and Wollman, 1952). This raises some questions.

It is known from the electron microscopic observations already presented and from serological studies that the binding of specific phage components to specific sites on the bacterial cell wall is required to effect adsorption. With constraints like these, how can each random approach lead to adsorption? It is conceivable that a random collision between any part of a phage and the surface of a sensitive bacterium might lead to the attachment of the phage by a mechanism similar to that proposed by Setlow and Pollard (1962) for the formation of enzyme–substrate complexes: nonspecific London forces might keep a colliding phage and bacterium together for a brief time during which the phage could move by rotational Brownian motion until a steric fit is achieved. This type of interaction might significantly increase the probability of the binding sites on the phage's tail fibers interacting specifically with the receptor sites on the cell wall.

2. The Multistep Nature of Bacteriophage Adsorption

There are conditions for which the preceding considerations of bacteriophage adsorption rates do not seem to provide a satisfactory explanation. For example, Stent and Wollman (1952) noted that the rate of adsorption remained proportional to the bacterial concentration until the bacterial concentration rose above about 2×10^8 cells/ml; at bacterial concentrations above this level the rate of adsorption approached a maximum. Furthermore, between 25° and 5°C the fraction of collisions between T4 and *Escherichia coli* which led to adsorption decreased with temperature. These observations were not accounted for by the simple two-body collision model of bacteriophage adsorption; the fact that above a certain value the bacterial concentration was not rate-limiting caused Stent and Wollman (1952) to postulate the addition of a second, temperature-sensitive step to the simple two-body collision model. They discussed three possible alternatives for this second step.

1. "Activity-inactivity theory." Phage may be in one of two states: active or inactive. At very high bacterial concentrations, all the active phages are adsorbed almost immediately, while the inactive phages are adsorbed at the limiting rate which is dependent on the rate at which they become active.

2. "Alternative collision theory." Phages attach to bacteria in one of two ways: a "good way" which always leads to adsorption, or a "bad way" which is reversible and does not lead to adsorption. At high bac-

terial concentrations the adsorption rate is limited by the rate of "desorption" of the reversibly bound phage.

3. "Surface reaction theory." Phage and bacteria become reversibly attached. After this first step they are either irreversibly adsorbed or they are desorbed. At high bacterial concentrations all phage are reversibly attached so that the rate of irreversible adsorption is determined only by the rate at which reversible attachment leads to irreversible adsorption.

Reversible attachment has been reported for bacteriophages T1, T2, and T4 (Puck *et al.*, 1951; Garen and Puck, 1951; Christensen and Tolmach, 1955). Phage T4 appeared to bind reversibly to glass filters. Cofactor-requiring T4 particles attached to glass filters only in the presence of tryptophan, whereas nonrequiring mutants attached without tryptophan. This observation of cofactor requirement indicated that the reversible binding of T4 to glass filters might be indicative of the nature of adsorption of T4 to bacterial cells. Bacteriophage T2, mixed with *Escherichia coli* B in a medium containing 0.02 M monovalent salt, appeared to attach to the cells reversibly; active phage T2 could be eluted from the *E. coli* B by washing with distilled water. Another experiment with T2 virus showed that their attachment to *E. coli* B at 37°C in the presence of 0.02 M NaCl did not result in cell death (Garen and Puck, 1951). This observation agrees with earlier experiments (Hershey *et al.*, 1944) showing that when Na$^+$ concentrations are below 0.2 M the efficiency of plating of T2 particles decreases. Therefore, a low electrolyte concentration causes most T2 phages to lose their infectivity.

Garen and Puck (1951) concluded that bacteriophage infection involved a temperature-independent, electrostatic, reversible primary attachment reaction followed by a temperature-dependent, irreversible second step which may be enzymic in nature.

The basis for comparison of the interactions between phages and nonbiological objects such as glass filters and ion exchange resins with the interactions between phages and bacteria has been questioned (Hershey, 1957). Some evidence such as the tryptophan requirement for the attachment of T4 viruses to glass filters and to bacteria indicates that such a comparison is valid. There are observations, however, which suggest that phages can interact differently with nonbiological objects than with bacteria. For example, either monovalent or divalent salts permit the attachment of T2 particles to cation exchangers, whereas only monovalent salts facilitate the adsorption of T2 viruses to sensitive *E. coli* cells. Hershey (1957) has argued that the experiments described by Puck *et al.* (1951), Garen and Puck (1951), Puck (1953), and others do not necessitate the conclusion reached by Garen and Puck that reversible attachment of phage to bacteria is an integral part of the infectious process. An alternative explanation, according to Hershey, is that "reversible and irreversible attachments involve different bacterial receptors, or different parts of the phage particle, in which case the two kinds of attachment are not steps in a single process, but competing processes . . ." (Hershey, 1957). These two alternative explanations by Garen and Puck and by Hershey correspond, respectively, to the "surface reaction" and "alternative collision" theories of Stent and Wollman (1952).

Evidence has recently been presented (Christensen, 1965) which indicates that the "surface reaction" theory of attachment is correct in the case of bacteriophage T1. Phage T1 was allowed to attach to *E. coli* at 0°C, where all the phages were found to be reversibly attached. Samples were then taken and diluted by a factor of 1000 into nutrient broth at 37°C where the bacteria were so dilute that adsorption of free phage was essentially stopped. When the sample was centrifuged and assayed it was found that 95% of the phage had rapidly become irreversibly attached. This result indicates that the transition from the reversibly bound to the irreversibly bound state does not require desorption and readsorption of the phage as predicted by the "alternative collision" theory.

It is interesting to note in this context that the adsorption of T-even phages, as observed by electron microscopy, involves two distinct binding reactions with the bacterial cell wall: first the attachment of the long tail fibers and, second, the attachment of the short tail fibers or tail pins (Simon and Anderson, 1967a). Occasionally, the long tail fibers *of the adsorbed phages* are found detached from the bacterial wall. This observation may indicate that the initial long fiber attachment of T-even phage to the bacterial surface is reversible; this inference suggests that the "surface reaction theory" of phage adsorption (Stent and Wollman, 1952) may be correct for the T-even phages.

V. Attachment of Bacteriophages with Noncontractile Tails

There are many phages such as T1, T5 (Fig. 12), and λ (Fig. 13) of which the tails or tail sheaths do not normally contract during the adsorption process. These types of viruses frequently have spikes or other structures probably necessary for attachment and/or penetration at the distal ends of their tails. Coliphage λ is an example of such a particle. Figure 13 shows negatively stained λ phages; their heads have vertex-to-vertex diameters of about 600 Å and their tails are about 1500 Å long and 80 Å wide with a thin spike extending another 230 Å from the distal end of the tail. From electron micrographs of adsorbed λ phages (Fig. 14)

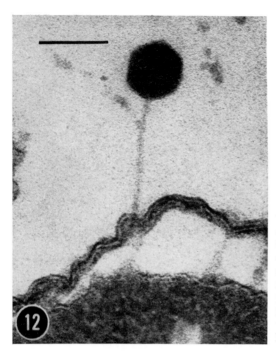

FIG. 12. A T5 phage attached to a plasmolyzed *Escherichia coli* cell. This T5 particle, seen in thin section, has attached to the bacterial wall at a point where the wall has remained in contact with the protoplastic membrane after plasmolysis. (This micrograph was kindly made available by Dr. M. Bayer.)

it is clear that the tails have not contracted. The spike is not visible outside the bacterium, thereby indicating that the spike may have pierced the cell wall. However, this conclusion is preliminary as very little is known at the present time about the attachment mechanism(s) of λ and other noncontractile particles such as the bacteriophage that infects the marine bacterium *Cytophaga marinoflava* (A. F. Valentine and Chapman, 1966).

VI. Receptor Sites for Bacteriophages

A. Receptors on the Cell Wall

Several studies have involved attempts to learn about phage-specific receptor sites on bacterial cell walls by chemically and mechanically fractionating bacterial cells. These experiments have generally used irreversible inactivation of bacteriophage as the test for receptor activity.

Burnet (1934) was one of the first investigators to study the interactions of bacterial extracts with phage suspensions. He showed that a bacterial extract inactivated only those bacteriophage to which the bacteria were sensitive; in other words, the phage-inhibiting agents in the extracts possessed the same specificity as the bacteria from which

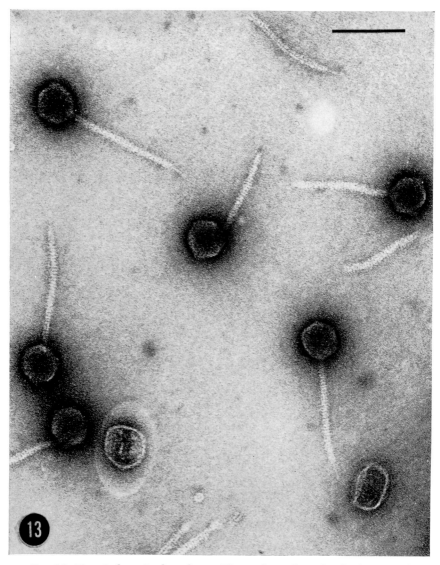

Fig. 13. Negatively stained λ phages. These phages have heads about 600 Å in diameter with hexagonal outlines and 1500-Å long flexible noncontractile tails. A thin spike or fiber extends from the distal tip of most tails.

the extracts were derived. This observation led Burnet to assume that the phage-inhibiting agents in the extracts were the surface components of the bacteria which determined the specificity of bacteriophage adsorption. Burnet suggested that the specificity of the phage attachment

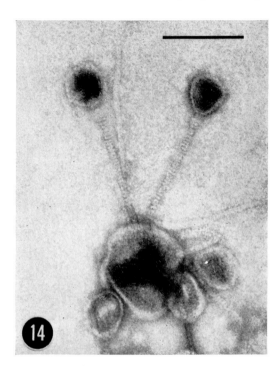

FIG. 14. λ Phages attached to an *Escherichia coli* wall fragment. The thin spikes at the distal ends of the tails are not visible.

reaction was analogous to an antigen–antibody reaction; only those bacteriophage which "fit" the stereospecific antigenic pattern of the bacterial surface could adsorb to the bacteria.

Proof that phage-specific receptor sites are located on the bacterial cell wall came from the demonstration that cell walls isolated from *Escherichia coli* adsorbed and inactivated bacteriophages as specifically and as readily as intact bacteria (Weidel, 1958). One exception to the above statement should be noted: bacteriophage T1 adsorbed only reversibly to isolated bacterial cell walls (Weidel, 1958). It seems that T1 will attach irreversibly only to metabolically active cells, perhaps due to an energy requirement for irreversible attachment.

Investigations (Jesaitis and Goebel, 1952) of the chemical nature of the cell wall's receptor sites showed that an antigenically active lipomucoprotein extracted from Phase II *Shigella sonnei* (sensitive to T2, T3, T4, T5, T6, and T7) inactivated *in vitro* all the T phages to which the bacillus was sensitive. Upon removal of the protein component of the antigen, a lipocarbohydrate remained which was able to inactivate phages T3, T4, and T7. Bacteria can mutate from phage sensitivity to phage resistance. The S. *sonnei* mutant II/3,4,7 (lacking receptor activity for bacteriophages T3, T4, and T7) possessed an altered carbohydrate component of the antigen (Goebel and Jesaitis, 1952). This mutant antigen had no activity against phages T3, T4, and T7 but was able to inactivate phages T2 and T6. The normal antigen contained a heptose which was missing from the mutant. Therefore, it seems that the receptor sites for phages T3, T4, and T7 have been chemically modified in the S. *sonnei* mutant II/3,4,7 thereby preventing the attachment of these phages.

When a suspension of bacteriophage T4 was mixed with the phage-inhibiting lipocarbohydrate from sensitive S. *sonnei,* DNA was found to be released into the medium simultaneously with inactivation of the T4. The phage were also observed in the electron microscope to have empty heads. However, if lipid was extracted from the lipocarbohydrate with 70% ethanol, neither the remaining lipocarbohydrate nor the extracted lipid could inactivate phage T4. Upon addition of the extracted lipid or certain saturated fatty acids to the ethanol-treated lipocarbohydrate, phage inactivating ability was restored, the lipid apparently serving as a cofactor in the reaction (Jesaitis and Goebel, 1953).

Escherichia coli has yielded results similar to those obtained with *Shigella sonnei* regarding the chemical characterization of receptor complexes. Treatment of isolated cell walls with phenol removed most of the protein and all of the activity against phages T2 and T6. The phenol-insoluble portion, including the lipocarbohydrate, of the cell wall retained its shape and its activity against T3, T4, and T7. The lipocarbohydrate was easily hydrolyzed with a concomitant loss of receptor activity for T3, T4, and T7 (W. Weidel *et al.,* unpublished results).

The specific receptors for bacteriophage T5 were obtained by treating *Escherichia coli* walls with dilute alkali (Weidel *et al.,* 1954). These receptor particles appeared to be spheres 300 Å in diameter (Weidel and Kellenberger, 1955). Kinetic studies (Weidel, 1958) showed that one sphere inactivated one T5 virus. The T5 receptor particles had a surface cover of lipoprotein and a core of lipopolysaccharide. The lipopolysaccharide was found to be the receptor material for phages T3, T4, and T7, for when the protein component was removed from the T5 receptor particles they acquired a new receptor activity toward T3, T4, and T7. Bayer (1968) has recently obtained preliminary evidence that the receptor sites for T phages on *E. coli* may be located at points of active wall synthesis. This suggestion is based on the following observations: (1) the protoplasmic membranes of plasmolyzed *E. coli* B cells may adhere closely to the cell walls at the points of wall synthesis (Bayer, 1968); and (2) T phages selectively attach to plasmolyzed *E. coli* B cells at points where the cell walls are in contact with the protoplasmic membranes. These may be the sites of wall synthesis (Fig. 12) (Bayer, 1967), but further experiments are needed to show that this is so.

Modifications in the chemical composition of the cell wall may be quite subtle and still affect phage attachment. For example, the receptor sites for coliphage C21 appear blocked in *E. coli* K-12 mutants of which the galactosyl units have been donated specifically by uridine diphosphogalactose to the lipopolysaccharide of the cell wall. Other *E. coli* K-12 mutants which obtain their galactosyl units from other sources can adsorb phage C21 (Kalckar *et al.*, 1966). No hypothesis has as yet been put forth to explain these observations.

In summarizing our present knowledge of receptor sites for T phages as related to the structure of the bacterial cell wall (Weidel *et al.*, 1960; Bayer and Anderson, 1965; Murray *et al.*, 1965), the following model may be proposed: surrounding the protoplasmic membrane is a rigid mucopolymer layer which is sensitive to lysozyme; partially covering this rigid layer is a layer of globular protein elements; further from the protoplasmic membrane is a layer of lipopolysaccharide which may contain the receptor sites for bacteriophages T3, T4, and T7; the outermost layer of the cell wall is a discontinuous lipoprotein layer which contains the receptor sites for phages T2 and T6.

B. EPISOMALLY CONTROLLED PILI AS RECEPTORS

Besides the adsorption of the relatively large phages described so far, the attachment processes of a few small tailless coliphages have been studied. These phages are "male-specific" which means that only bacteria carrying episomes which cause them to act as donors during conjugation are sensitive to them. An episome may be defined, according to Watson (1965), as "a genetic particle that can exist either free or as a part of a normal cellular chromosome." Crawford and Gesteland (1964) used the electron microscope to examine the adsorption of the polyhedral, tailless, ribonucleic acid (RNA)-containing bacteriophage R-17 which infects only male strains of *Escherichia coli* (Paranchych and Graham, 1962). The *E. coli* strain used by Crawford and Gesteland carried the episome called F (fertility). Cells carrying F are referred to as F$^+$ or Hfr, depending on the state of the episome. When mixed with F$^+$ or Hfr cells, R-17 particles were found attached along the lateral surface area of a few pili (pili are thin tubules, about 80 Å thick, which extend from the bacterial surface) on most Hfr and F$^+$ cells, but no phages would adhere to the pili of F$^-$ cells (Crawford and Gesteland, 1964; R. C. Valentine and Strand, 1965) (Fig. 15). Brinton *et al.* (1964) have suggested that the presence of the F factor in a cell may determine the synthesis of special "F pili" to which male-specific phages can attach. This hypothesis is supported by the following experimental results: (*1*) when F$^+$ cells are grown in the presence of acridine orange they are converted to F$^-$ cells (Hirota, 1960) and at the same time they become resistant to R-17-type

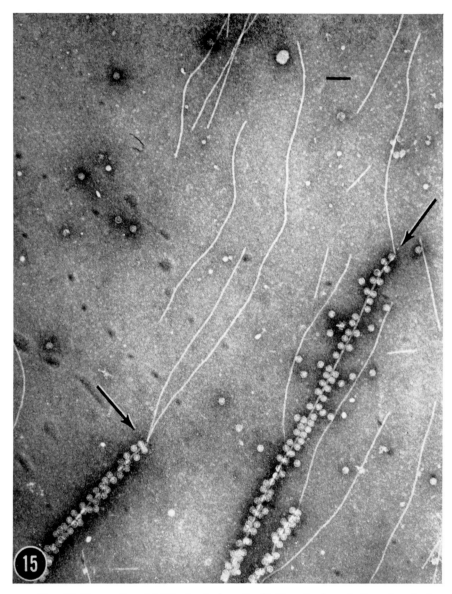

Fɪɢ. 15. Phages f1 and MS2 adsorbed on F pili. The attachment of nearly spherical MS2 phages identifies the lateral surfaces of the F pili; the tips of long, rod-shaped f1 particles attach only to the ends of the F pili (arrow). Usually only one or two f1 phages can attach to an F pilus. (This micrograph was kindly made available by Dr. L. Caro.)

phages (Brinton *et al.*, 1964) although the bacteria generally remain piliated; (2) conversely, when F⁻ cells are changed into F⁺ cells by contact with cells containing the F episome they become sensitive to these phages; and (3) cells carrying the F factor usually exhibit an F⁺ antigen not found on F⁻ cells (Hirota *et al.*, 1964). Thus it seems that the F factor, consisting of DNA (Driskell-Zamenhof and Adelberg, 1963) may direct the synthesis of an antigenically distinct type of pilus to which the male-specific phages can attach. After adsorption, the RNA of these phages might be transferred to the bacterial cytoplasm through the 25 Å central channel observed in some types of pili (Brinton, 1965).

The F pili, induced by the F episome, have a slightly different morphology than common pili (Lawn, 1966); they are thicker and usually longer than the more numerous common pili. Sneath and Lederberg (1961) found that treatment of F⁺ and Hfr cells with periodate prevents their conjugation with F⁻ cells. They suggest that male cells carrying the F factor have a carbohydrate-containing conjugal substance on their surface and that the oxidation of this carbohydrate by periodate may prevent conjugation. Whether the F⁺ antigen (Ørskov and Ørskov, 1960) is related to the carbohydrate conjugal substance has not yet been determined. However, it has been shown that a variety of conditions which cause the F⁺ antigen to appear or disappear simultaneously cause the receptor sites for male-specific phages to appear or disappear, thereby implying the identity of the F⁺ antigen and the phage receptor site (Knolle and Ørskov, 1967).

Another variety of male-specific bacteriophage appears as long filaments, usually 60 Å wide and 8000 Å long. Occasionally, narrow grooves parallel to the long axes are visible on the filaments. These phages, named f1, fd, and M13 contain single-stranded DNA instead of RNA (Marvin and Schaller, 1966; Salivar *et al.*, 1964). There are at least two distinct attachment sites on an F pilus for the binding of the two kinds of male-specific phages since the adsorption of polyhedral phages does not alter the cell's ability to be infected with filamentous phages (Tzagoloff and Pratt, 1964). Furthermore, rarely more than three M13 phage particles are adsorbed per bacterial cell which corresponds to the fact that most male strains of *Escherichia coli* contain only about three F pili (Brinton, 1965). Thus, one might suspect that only one f1 phage adsorbs per F pilus. The observations of Caro and Schnös (1966) confirm this suspicion and, indeed, show that the tips of f1 particles adsorb specifically to the distal ends of F pili, usually one f1 phage per pilus (Fig. 15).

The episomal factor called "R" is also known to affect the production of phage receptor sites in *E. coli* cells. The R factors promote drug resistance in bacteria which they infect (Watanabe *et al.*, 1964). They also confer a sensitivity to male-specific phages and the ability to conjugate. Like the F factor, R factors seem to cause the production of special pili to which male-specific phages can attach and through which bacterial conjugation can occur (Datta *et al.*, 1966).

From these studies on the adsorption of male-specific phages, it seems that not only can the bacterial receptor sites for bacteriophages be changed by bacterial mutations but also by the presence or absence of various episomes such as the F factor. Another type of episome is a prophage, the genome of a phage which has lysogenized a bacterium by integration into the bacterial chromosome. Uetake *et al.* (1958) showed that some prophages could influence the attachment of bacteriophages. They observed that phage E15 can adsorb only on *Salmonella anatum* cells which are not lysogenic for E15, whereas another phage, E34, can adsorb only to *S. anatum* cells that are lysogenic for E15. Serological evidence showed that cells lysogenic for E15 possess a distinct surface antigen not found on nonlysogenic cells which may be the attachment site for E34. Another surface antigen, the receptor for E15, is present on nonlysogenic *S. anatum* and is absent from lysogenic cells. Therefore, episomes such as F, R, and prophages can alter the phage sensitivity of bacteria. This conclusion is not surprising since episomes are known to initiate the production of characteristic bacterial surface antigens and the development of distinct morphological features (Stirm *et al.*, 1967a,b).

C. FLAGELLA AS RECEPTORS

Two bacteriophages have been reported which infect only motile bacteria; these phages have adsorption specificities for the flagella of sensitive cells. One of these viruses, called χ, is specific for various types of *Salmonella* (Meynell, 1961) and the other phage, called PBS1, infects strains of *Bacillus subtilis* (Joys, 1965). The χ particles (Meynell, 1961) have regular polyhedral heads about 675 Å in diameter and noncontrac-tile tails 2300 Å long and 125 Å wide. A single flexible tail fiber about 20 Å thick and 2000 Å long extends from the tip of the tail (Schade *et al.*, 1967). Careful electron microscopic examination has shown that phage χ probably attaches to bacterial flagella by means of its long tail fiber; this fiber appears to wind loosely around a flagellum. There may be many receptor sites for the fiber on the surface of a flagellum since many χ particles will adsorb to a single flagellum. After this initial attachment step, the phage apparently slides along to the base of the flagellum. De-oxyribonucleic acid injection probably occurs at the base of the flagellum, since it is only there that empty headed χ particles are found. Phage PBS1 also attaches to flagella. It has a regular polyhedral head about

1200 Å in diameter; its tail, about 2000 Å long, has a contractile sheath terminated by a baseplate from which long, helically coiled fibers extend. In addition, shorter straight fibers project from the distal halves of contracted sheaths (Eiserling, 1967). The helical nature of the long fibers may permit them to coil around sensitive flagella during the adsorption process. Both χ and PBS1 phages attach only to active flagella possessing specific antigenic patterns. These viruses are not inactivated by isolated flagella, nor are they adsorbed on bacteria of which the flagella have been "paralyzed" (Meynell, 1961; Frankel and Joys, 1966). Diffraction studies have shown that bacterial flagellin polypeptides give rise to both cross-α- and cross-β-reflections, thereby suggesting a change in the molecular configuration of flagellar proteins associated with flagellar activity (Doetsch, 1966). Perhaps there is a change in protein configuration, occurring during flagellar activity that produces suitable phage receptor sites. The movement of flagella may be of further importance in transferring phage from their initial sites of attachment on the flagellar surfaces to the proximal ends or bases of the flagella; motility could also serve to expose otherwise hidden receptor sites at the bases of flagella.

VII. Changes in the Bacterial Wall Resulting from T-Even Bacteriophage Infection

The irreversible attachment of T-even bacteriophages to purified cell walls is accompanied by the release of nitrogen-containing material from the walls (Koch and Weidel, 1956; Barrington and Kozloff, 1956; Weidel and Primosigh, 1958). The amount of nitrogen released is directly proportional to the number of particles adsorbed (Barrington and Kozloff, 1956). These facts suggest that the parts of the phage which interact with the cell wall contain lytic properties. Koch and Dreyer (1958) and Maass and Weidel (1963) have characterized an enzyme found in T2 lysates as a lysozyme. It is the same enzyme found in lysates after removal of phages by acid precipitation and high-speed centrifugation (Koch and Jordan, 1957). Brown and Kozloff (1957) have suggested that lysozyme is present in the T-even phage tail. The observation that the central plug, which is probably associated with the distal tip of the needle, disappears from the baseplate during the course of phage adsorption (Simon and Anderson, 1967b) makes it plausible to speculate that the plug is, in fact, a lysozyme complex. The complex could remain attached to the tip of the needle during the change in the molecular organization of the baseplate. Then the needle, driven by the contracting sheath through the soft outer lipoid layers of the bacterial wall, would carry the lysozyme complex to its substrate—the rigid mucopolymer layer at the inner surface of the wall. As the lysozyme reacts with the mucopolymer, its binding affinities might change so that it would cease to serve as a plug for the needle and would permit the outflow of the phage's DNA. This model may account for several heretofore unanswered questions such as (1) "Where is the lysozyme located in the phage tail?" (2) "Since its outer lipoid layers protect the rigid layer of the wall from lysozyme, how does the lysozyme of infecting bacteriophages reach the mucopolymer on the inner surface of the wall?" (3) "What causes the phage's DNA to be released after the needle has penetrated the wall?"

The loss of material from walls after T-even adsorption has been studied with various chemical and electron-microscopic techniques. The results suggest that the loss of material reflects a weakening or dissolution of the rigid components of the cell wall. Weidel *et al.* (1960) observed in the electron microscope that treatment of isolated rigid layers either with purified T2 lysozyme or with egg-white lysozyme "results in a rapid and thorough disintegration of the structure . . . by disengaging one from another the tiny spheres" of the granular component. Similarly, Murray *et al.* (1965) showed that treatment of intact *Escherichia coli* cells with egg-white lysozyme in the presence of ethylenediaminetetra-acetate (EDTA) reduces the thickness of the inner layer of the cell wall seen in thin sections. Weidel and Katz (1961) found by chemical analysis that treatment of isolated *E. coli* cell walls with extracted phage lysozyme causes the loss of mucopolymer from the rigid layer of the wall.

Further insight into the effects of phage on the cell wall has come from the observation that when a large number of intact T-even particles are mixed with a bacterial suspension the bacteria quickly lyse with a concomitant decrease in virus titer (Delbrück, 1940). This phenomenon, called "lysis from without," apparently results from the destructive action of phage lysozyme on the cell wall. Recently, direct electron microscopic observation has shown that the adsorption of high multiplicities of T2 and T4 bacteriophages to *E. coli* cells causes the dissolution of material from the innermost granular layer of the cell wall (Simon, 1966). The granular layer corresponds to the protein units associated with the mucopolymer which gives rigidity to the cell wall (Bayer and Anderson, 1965); loss of material from this layer as a result of phage attachment would weaken the wall. These observations support the conclusion that the lysozyme of infecting T2 and T4 phages causes the dissolution of the rigid, inner, mucopolymer layer of the cell wall.

The repair of such damage seems to be somewhat dependent on the phage's DNA. For example, osmotically shocked T2 phages (Barlow and Herriott, 1954) as well as ultraviolet light (UV)-inactivated T2 phages (Watson, 1950) cause lysis from without at lower multiplicities of adsorbing phages than are required for normal phages. Similarly, if the host bacteria are treated with metabolic inhibitors, such as 2,4-dinitrophenol (Heagy, 1950), either prior to or at the time of infection, fewer than the usual number of adsorbing particles are required to produce lysis from without. It seems, therefore, that although the capsids of infecting T-even particles damage the *E. coli* cell wall, the DNA of these phages, which normally is injected into the bacteria, initiates a metabolic mechanism for repairing the cell wall; if this repair mechanism does not function, because of metabolic inhibitors or due to the inactivation or absence of phage DNA, fewer phage than usual will cause lysis from without.

The initial destructive effect of the T-even phages on the bacterial surface causes the cells to release UV absorbing material (Prater, 1951). By using radioactive tracers, Puck and Lee (1954, 1955) were able to show that within 20 seconds following T2 infection at 37°C, low molecular weight sulfur and phosphorus compounds begin leaking from the host *E. coli* cells. This leakage can be distinguished from lysis from without by comparing the compounds released under both conditions. Within 2 to 3 minutes after its onset, the leakage produced by the phage under conditions of infection stops, the initial damage inflicted by the T2 particles having been repaired.

Some permeability changes in infected cells seem to be controlled, at least in part, by the phages themselves. For example, the uptake of the acridine dye, acriflavine, by *E. coli* cells is greatly increased after infection by wild-type T-even phages but remains unchanged after infection by acriflavine-resistant mutants of these phages (Silver, 1965). The following observations indicate that the uptake of acriflavine is not a simple function of the lytic activity of the adsorbed viruses: (*1*) the rate at which acriflavine is taken up (at 28°C) is independent of when the dye is added to the infected bacteria, for at least 20 minutes following T2 attachment; since phage-induced damage to the cell wall is repaired within a few minutes after low multiplicity T2 infection, it seems that acriflavine uptake is not directly dependent on the phage's lytic activity on the bacterial wall; (*2*) *E. coli* cells infected with UV-killed T2 phages take up substantially less acriflavine than similar cells infected with viable T2 particles. This fact and the observation that UV-killed phages do not induce the repair of cell wall damage (Adams, 1959) lead to the conclusion that the infecting phage's DNA, not its lytic property, is responsible for initiating most of the acriflavine uptake. This idea is supported by the fact that cells infected with acriflavine-resistant mutants of the T-even phages do not take up the dye.

Silver (1965) attempted to show that the uptake of acriflavine was dependent on phage-specific protein synthesis which occurred shortly after T2 infection. Chloramphenicol, an antibiotic which can inhibit protein synthesis, was added (at a final concentration of 25 mg/ml) to an *E. coli* culture at the time of infection with T2 particles. The chloramphenicol reduced the amount of acriflavine subsequently taken up by the bacteria. Therefore, it seems that phage-directed protein synthesis is, in fact, responsible for the increase in acriflavine uptake following T-even infection. Perhaps T-even infection causes the synthesis of a protein(s) that completely represses a bacterial mechanism which actively excludes acriflavine from the cell; or, alternatively, the T-even DNA might initiate the synthesis of proteins that are directly responsible for the active uptake of acriflavine. Although sufficient data are not presently available to distinguish between these and other alternatives, it is clear that the DNA of infecting phages may control the permeability of infected bacteria.

Alterations of the bacterial permeability following viral infection may also be responsible for the breakdown of the DNA of superinfecting phages referred to as *superinfection breakdown*. Superinfection breakdown is inhibited by treating the bacteria with potassium cyanide just before adding the first phages; however, if cyanide is added to the bacteria 5 minutes after infection, superinfection breakdown occurs. Therefore, superinfection breakdown is dependent on active metabolic functions which take place during the first few minutes after phage infection (French *et al.*, 1951, 1952; Adams, 1959). Deoxyribonuclease inhibitors prevent the breakdown of the DNA of superinfecting phages; however, even under these conditions, bacteria are refractory to superinfecting DNA, *excluding* it from genetic activity in the production of progeny phages (Dulbecco, 1949, 1952) and preventing most of it from penetrating the interior of the cell (Graham, 1953). Also significant is the demonstration that shortly after an initial phage infection, bacterial cells become much less sensitive to lysis from without (Visconti, 1953). It seems that the failure of superinfecting phages to cause lysis from without and to penetrate the bacterial cells is due to metabolic-dependent changes, initiated by the phage DNA, which repair and strengthen the cell wall following phage infection.

VIII. Discussion

It is instructive to compare the molecular mechanisms used by bacteriophages to infect host cells with what is known at the cellular level and, occasionally, at the molecular level of the fertilization of eggs by sperm. Some of the analogies between the two systems are listed in Table I.

First we may consider how sperm and phages reach their target cells. Most functional sperm are actively motile and derive the energy to op-

TABLE I

A COMPARISON OF SPERM WITH BACTERIOPHAGES

Property	Sperm	Bacteriophage
Size	ca. 1000 × 100,000 Å	60 × 8500 Å (fd) 200 × 200 Å (f2, R-17) 600 × 2000 Å (T4)
Motion	Motile	Diffusion and convection
Energy source	Oxidative phosphorylation in mitochondria	None
Receptor on target cell	Fertilizin	A. Mucopolysaccharides, etc. B. Flagella C. F pili (fd, f2, R-17)
Enzymes	Egg membrane lysin	Lysozyme
Kind of nucleic acid	Double-stranded DNA in haploid set of chromosomes from male	Double-stranded DNA (all T phages) Single-stranded DNA (ΦX174, fd) Single-stranded RNA (f2, R-17)
Activity of nucleic acid in sperm or virion	None	None
Activity of sperm and bacteriophage after adsorption on target cell	Fusion with target cell to form a zygote	Nucleic acid injected into target cell, viral protein then dispensable
Activity of nucleic acid after entry into target cell	Chromosomes of sperm and egg form a diploid set	Converts cell to phage production; or Combines with chromosome of bacterium to make it lysogenic; and/or Combines with chromosome of bacterium to bring in markers from maternal bacterium (transduction)

erate their flagella by oxidative phosphorylation in their mitochondria. In animals there seems to be no tropism reported that leads them to the egg (certain hydroids excepted; Miller, 1966). Millions of sperm simply wander and only a few reach an egg. Virions, on the other hand, have no respiration and have no sources of energy for motility. They have to depend on diffusion and convection to bring them to the surfaces of susceptible target cells.

The problem of recognition of target cells is interesting. In both cases, the surfaces of target cells contain specific receptor substances to which areas on the virion or sperm are thought to be complementary and to which they adhere as antigens adhere to their specific antibodies or as enzymes adhere to their substrates. In fact, some virions such as those of influenza contain many, slowly acting, enzyme molecules on their surfaces that adhere to the mucopolysaccharide receptor spots on the surfaces of red cells, and in forming bridges between the cells, cause them to clump. Eventually, however, the enzymes destroy the receptors so the bridging virions are eluted in an active state and the clumps of cells, now stripped of their receptors, disintegrate.

In the same way the surfaces of eggs contain substances called "fertilizins" with which substances called "antifertilizins" on the surface of the sperm react specifically. Furthermore, the treatment of sperm with a solution containing fertilizin impairs the fertilizing ability of the sperm presumably by covering up its antifertilizin so it cannot combine with the fertilizin on the surface of the egg and/or prematurely initiating the acrosomal reaction. As we have seen, bacteria also have receptor sites that specifically interact with the binding sites on bacteriophages. If some phages are treated with solutions of extracted receptor sites from the walls of sensitive bacteria, they likewise become incapable of initiating subsequent bacterial infection (Weidel, 1958). Both sperm and some phages, therefore, adsorb to specific extractable substances on the surfaces of their target cells.

Following fusion of the sperm and egg membranes, a fertilization cone forms and the sperm nucleus penetrates the egg. Many types of sperm contain enzymes associated with the acrosome called "egg-membrane lysins" which can lyse the vitelline membrane of the egg (see Dan in Chapter 6, Vol. I, of this treatise). This enzyme is probably released from the acrosome upon interaction with the egg's fertilizin; it then dissolves or softens the vitelline membrane, permitting the sperm to pass into the interior of the egg. Similarly, certain phages appear to contain a lysozyme which probably is important in producing a localized breakdown of the bacterial wall, thereby facilitating the penetration of the wall by the phage. This hypothesis is in agreement with the observation that the adsorption of a large number of infecting phage particles on a bacterium will cause it to lyse immediately (Delbrück, 1948). This lytic reaction, called "lysis from without," presumably results from rather extensive destruction of the bacterial cell wall by the phages' lysozyme.

Following penetration in both egg and bacterium, processes are initiated that reduce polyspermy, on the one hand, and polyphagy, on the other. In animals, a number of mechanisms have evolved to prevent polyspermy (see Monroy, 1965, for review). Phage-infected bacteria also

the genetics of the host; for, occasionally, during the maturation of certain temperate bacteriophages a fragment of the DNA of the host may replace some or all of the phage DNA. The particle may then carry both viral and bacterial DNA. For example, λ phage particles often carry part of the *gal* region adjacent to the site the prophage occupied on the host's chromosome, whereas P22 particles can apparently carry any piece of host DNA at random. "Infection" by this type of virus results in the introduction of both some phage and some bacterial genes into the recipient cell. The latter genes may then become incorporated into the genome of the host or its descendents. In this process, called "transduction," the bacteriophage particle, thus (rather untidily), acts as a carrier of genetic messages between bacteria in a manner that is very similar to the way in which sperm cells convey genetic information from one generation to the next. The message carried by phage is only a fragment of the bacterial genome, however, whereas that carried by sperm is normally complete.

Both sperm and transducing phages, thus, permit their parental species to transmit the blueprints for new enzymes or for novel structural components so that related organisms can test the suitability of various combinations of genes under different environmental conditions. In fact, Campbell (1961) has gone so far as to write that although "phages probably have evolved *as phages* rather than representing transient offshoots of some other evolutionary line, their role in bacterial recombination may be the most important factor in assuring their survival."

IX. Summary

Bacteriophages attach to sensitive cells by a variety of mechanisms, depending on the structure of both the virus and the receptor site on the bacterium. Receptor sites for various phages are located on the bacterial wall, on certain episomally controlled pili, and on active flagella. The chemical compositions of some of these receptor materials have been determined. Bacteriophage adsorption, at least for the relatively complex T-even viruses, involves an ordered series of events; some mechanisms controlling these events have been elucidated at the morphological level. Infecting virus particles may produce changes in the walls of sensitive bacteria; these alterations are due to both the lytic activity produced by the phage capsid and to processes initiated by phage DNA within the bacteria. Some of the changes produced by the phage DNA involve resistance to superinfection by additional viruses (polyphagy) and alterations in cellular permeability.

REFERENCES

Adams, M. (1959). "Bacteriophages." Wiley (Interscience), New York.

Anderson, T. F. (1945). The role of tryptophan in the adsorption of two bacterial viruses on their host, E. coli. J. Cellular Comp. Physiol. 25, 17.

Anderson, T. F. (1948). The activation of bacterial virus T4 by L-tryptophan. J. Bacteriol. 55, 637.

Anderson, T. F. (1949). On the mechanism of adsorption of bacteriophages on host cells. In "The Nature of the Bacterial Surface" (A. A. Miles and N. W. Pirie, eds.), pp. 76–88. Blackwell, Oxford, England.

Anderson, T. F. (1951). Techniques for the preservation of three dimensional structure in preparing specimens for the electron microscope. Trans. N. Y. Acad. Sci. 13, 130.

Anderson, T. F. (1952). Stereoscopic studies of cells and viruses in the electron microscope. Am. Naturalist 86, 91.

Anderson, T. F. (1953). The morphology and osmotic properties of bacteriophage systems. Cold Spring Harbor Symp. Quant. Biol. 18, 197.

Anderson, T. F. (1963). Structure and genetic properties of bacterial viruses. "Viruses, Nucleic Acids and Cancer," pp. 129–140. Williams & Wilkins, Baltimore, Maryland.

Anderson, T. F., and Krimm, S. (1966). Diffraction of light from electron micrographs of helical structures in bacteriophage tail sheaths. 6th Intern. Congr. Electron Microscopy, Kyoto, p. 145. (Maruzen, Tokyo.)

Anderson, T. F., and Stephens, R. (1964). Decomposition of T6 bacteriophage in alkaline solutions. Virology 23, 113.

Barksdale, L. (1959). Lysogenic conversions in bacteria. Bacteriol. Rev. 23, 202.

Barlow, J. L., and Herriott, R. M. (1954). The inhibitory action of divalent cations on Escherichia coli B multiply infected with T2 phage or ghosts. Bacteriol. Proc. Soc. Am. Bacteriologists 54, 46.

Barrington, L. F., and Kozloff, L. M. (1956). Action of bacteriophage on isolated host cell walls. J. Biol. Chem. 223, 615.

Bayer, M. E. (1967). Sites of adsorption of the bacteriophages T3 and T5 on the wall of Escherichia coli B. Proc. Electron Microscopy Soc. Am., pp. 96–97. Claitor's Baton Rouge, Louisiana.

Bayer, M. E. (1968). Adsorption of bacteriophages to adhesions between wall and membrane of Escherichia coli. J. Virol. 2, 346.

Bayer, M. E., and Anderson, T. F. (1965). The surface structure of Escherichia coli. Proc. Natl. Acad. Sci. U. S. 54, 1592.

Bradley, D. E. (1963). The structure of coliphages. J. Gen. Microbiol. 31, 435.

Brenner, S., Streisinger, G., Horne, R. W., Champe, S. P., Barnett, L., Benzer, S., and Rees, M. W. (1959). Structural components of bacteriophage. J. Mol. Biol. 1, 281.

Brenner, S., Champe, S. P., Streisinger, G., and Barnett, L. (1962). On the interaction of adsorption cofactors with bacteriophages T2 and T4. Virology 17, 30.

Brinton, C. C., Jr. (1965). The structure, function, synthesis and genetic control of bacterial pili and a molecular model for DNA and RNA transport in gram negative bacteria. Trans. N. Y. Acad. Sci. 27, 1003.

Brinton, C. C., Jr., Gemski, P., Jr., and Carnahan, J. (1964). A new type of bacterial pilus genetically controlled by the fertility factor of E. coli K12 and its role in chromosome transfer. Proc. Natl. Acad. Sci. U. S. 52, 776.

Brown, D. D., and Kozloff, L. M. (1957). Localization of the bacteriophage tail enzyme. *J. Biol. Chem.* **225**, 1.

Burnet, F. M. (1934). The bacteriophages. *Biol. Rev. Biol. Proc. Cambridge Phil. Soc.* **9**, 332–350.

Campbell, A. (1961). Conditions for the existence of bacteriophage. *Evolution* **15**, 153.

Caro, L. G., and Schnös, M. (1966). The attachment of the male specific bacteriophage fl to sensitive strains of *Escherichia coli*. *Proc. Natl. Acad. Sci. U. S.* **56**, 126.

Christensen, J. R. (1965). The kinetics of reversible and irreversible attachment of bacteriophage T1. *Virology* **26**, 727.

Christensen, J. R., and Tolmach, L. J. (1955). On the early stages of infection of *Escherichia coli* B by bacteriophage T1. *Arch. Biochem. Biophys.* **57**, 195.

Crawford, E. M., and Gesteland, R. F. (1964). The adsorption of bacteriophage R-17. *Virology* **22**, 165.

Cummings, D. J. (1964). Sedimentation and biological properties of T-phages of *Escherichia coli*. *Virology* **23**, 408.

Datta, N., Lawn, A. M., and Meynell, E. (1966). The relationship of F type piliation and F phage sensitivity to drug resistance transfer in R^+ F^- *Escherichia coli* K12. *J. Gen. Microbiol.* **45**, 365.

de Klerk, H. C., Coetzee, J. N., and Fourie, J. T. (1965). The fine structure of *Lactobacillus* bacteriophages. *J. Gen. Microbiol.* **38**, 35.

Delbrück, M. (1940). The growth of bacteriophage and lysis of the host. *J. Gen. Microbiol.* **23**, 643.

Delbrück, M. (1942). Bacterial viruses (bacteriophages). *Adv. Enzymol.* **2**, 1.

Delbrück, M. (1948). Biochemical mutants of bacterial viruses. *J. Bacteriol.* **56**, 1.

Doetsch, R. N. (1966). Some speculations accounting for the movement of bacterial flagella. *J. Theoret. Biol.* **11**, 411.

Driskell-Zamenhof, P. J., and Adelberg, E. A. (1963). Studies on the chemical nature and size of sex factors of *Escherichia coli* K12. *J. Mol. Biol.* **6**, 483.

Dulbecco, R. (1949). The number of particles of bacteriophage T2 that can participate in intracellular growth. *Genetics* **34**, 126.

Dulbecco, R. (1952). Mutual exclusion between related phages. *J. Bacteriol.* **63**, 209.

Edgar, R. S., and Lielausis, I. (1965). Serological studies with mutants of phage T4D defective in genes determining tail fiber structure. *Genetics* **52**, 1187.

Eiserling, F. A. (1967). The structure of *Bacillus subtilis* bacteriophage PBS1. *J. Ultrastruct. Res.* **17**, 342.

Eiserling, F. A., and Romig, W. R. (1962). Studies of *Bacillus subtilis* bacteriophages. Structural characterization by electron microscopy. *J. Ultrastruct. Res.* **6**, 540.

Epstein, R. H., Bolle, A., Steinberg, C. M., Kellenberger, E., Boy de la Tour, E., Chevalley, R., Edgar, R. S., Susman, M., Denhardt, G. H., and Lielausis, A. (1963). Physiological studies of conditional lethal mutants of bacteriophage T4D. *Cold Spring Harbor Symp. Quant. Biol.* **28**, 375.

Fernández-Morán, H. (1962). New approaches in the study of biological ultrastructure by high-resolution electron microscopy. *In* "The Interpretation of Ultrastructure" (R. J. C. Harris, ed.), Vol. 1, pp. 411–427. Academic Press, New York and London.

Frankel, R. W., and Joys, T. M. (1966). Adsorption specificity of bacteriophage PBS1. *J. Bacteriol.* **92**, 388.

Franklin, N. C. (1961). Serological study of tail structure and function in coliphages T2 and T4. *Virology* **14**, 417.

Freeman, V. J. (1951). Studies on the virulence of bacteriophage infected strains of *Corynebacterium diphtheriae*. *J. Bacteriol.* **61**, 675.

French, R. C., Lesley, S. M., Graham, A. F., and Van Rooyen, C. E. (1951). Studies on the relationship between virus and host cell. III. *Can. J. Med. Sci.* **29**, 144.

French, R. C., Graham, A. F., Lesley, S. M., and Van Rooyen, C. E. (1952). The contribution of phosphorus from T2r⁺ bacteriophage to progeny. *J. Bacteriol.* **64**, 597.

Garen, A., and Puck, T. T. (1951). The first two steps of the invasion of host cells by bacterial viruses. II. *J. Exptl. Med.* **94**, 177.

Goebel, W. F., and Jesaitis, M. A. (1952). The somatic antigen of a phage-resistant variant of Phase II *Shigella sonnei*. *J. Exptl. Med.* **96**, 425.

Graham, A. F. (1953). The fate of the infecting phage particle. *Ann. Inst. Pasteur* **84**, 90.

Guthrie, G. D., and Sinsheimer, R. L. (1960). Infection of protoplasts of *Escherichia coli* by subviral particles of bacteriophage ΦX 174. *J. Mol. Biol.* **2**, 297.

Heagy, F. C. (1950). The effect of 2,4-dinitrophenol and phage T2 on *Escherichia coli* B. *J. Bacteriol.* **59**, 367.

Hershey, A. D. (1957). Bacteriophages as genetic and biochemical systems. *Advan. Virus Res.* **4**, 25.

Hershey, A. D., and Chase, M. (1952). Independent functions of viral protein and nucleic acid in growth of bacteriophage. *J. Gen. Physiol.* **36**, 39.

Hershey, A. D., Kalmanson, G., and Bronfenbrenner, J. (1944). Coordinate effects of electrolyte and antibody on infectivity of bacteriophage. *J. Immunol.* **48**, 221.

Hirota, Y. (1960). The effect of acridine dyes on mating type factors in *Escherichia coli*. *Proc. Natl. Acad. Sci. U. S.* **46**, 57.

Hirota, Y., Nishimura, Y., Ørskov, F., and Ørskov, I. (1964). Effect of drug resistance factor R on the F properties of *Escherichia coli*. *J. Bacteriol.* **87**, 341.

Hummeler, K., Anderson, T. F., and Brown, R. A. (1962). Identification of poliovirus particles of different antigenicity by specific agglutination as seen in the electron microscope. *Virology* **16**, 84.

Jerne, N. K. (1956). The presence in normal serum of specific antibody against bacteriophage T4 and its increase during the earliest stages of immunization. *J. Immunol.* **76**, 209.

Jesaitis, M. A., and Goebel, W. F. (1952). The chemical and anti-viral properties of the somatic antigen of Phase II *Shigella sonnei*. *J. Exptl. Med.* **96**, 425.

Jesaitis, M. A., and Goebel, W. F. (1953). The interaction between T4 phage and the specific lipocarbohydrate of Phase II *Shigella sonnei*. *Cold Spring Harbor Symp. Quant. Biol.* **18**, 205.

Joys, T. M. (1965). Correlation between susceptibility to phage PBS1 and motility in *Bacillus subtilis*. *J. Bacteriol.* **90**, 1575.

Kalckar, H. M., Laursen, P., and Rapin, A. M. C. (1966). Inactivation of phage C21 by various preparations from lipopolysaccharide of *E. coli* K-12. *Proc. Natl. Acad. Sci. U. S.* **56**, 1852.

Kanner, L. C., and Kozloff, L. M. (1964). The reaction of indole and T2 bacteriophage. *Biochemistry* **3**, 215.

Kellenberger, E., and Arber, W. (1955). Die Struktur des Schwanzes der Phagen T2 und T4 und der Mechanismus der irreversiblen Adsorption. *Z. Naturforsch.* **10B**, 698.

Kellenberger, E., and Boy de la Tour, E. (1964). On the fine structure of normal and "polymerized" tail sheaths of phage T4. *J. Ultrastruct. Res.* **11**, 545.

Kellenberger, E., and Séchaud, J. (1957). Electron microscopical studies of phage multiplication. II. Production of phage-related structures during multiplication of phages T2 and T4. *Virology* **3**, 256.

Kellenberger, E., Bolle, A., Boy de la Tour, E., Epstein, R. H., Franklin, N. C., Jerne, N. K., Reale-Scafati, A., Séchaud, J., Bendet, I., Goldstein, D., and Lauffer, M. A. (1965). Functions and properties related to the tail fibers of bacteriophage T4. *Virology* **26**, 419.

Knolle, P., and Ørskov, I. (1967). The identity of the F^+ antigen and the cellular receptor for the RNA phage fr. *Mol. Gen. Genet.* **99**, 109.

Koch, G., and Dreyer, W. J. (1958). Characterization of an enzyme of phage T2 as a lysozyme. *Virology* **6**, 291.

Koch, G., and Jordan, E. M. (1957). Killing of *E. coli* by phage-free T2 lysates. *Biochim. Biophys. Acta* **25**, 437.

Koch, G., and Weidel, W. (1956). Abspaltung chemischer Komponenten der Coli-Membrane durch daran adsorbiente Phagen I. Mit.: Allgemeine Charakterisierung des Effekts und Particulanalyse einer der abgespaltenen Komponenten. *Z. Naturforsch.* **11B**, 345.

Kozloff, L. M., and Henderson, K. (1955). Action of complexes of the zinc group metals on the tail proteins of bacteriophage $T2^{++}$. *Nature* **176**, 1169.

Kozloff, L. M., Lute, M., and Henderson, K. (1957). Viral invasion. I. Rupture of thiol ester bonds in the bacteriophage tail. *J. Biol. Chem.* **228**, 511.

Krimm, S., and Anderson, T. F. (1967). The structure of normal and contracted tail sheaths of T4 bacteriophage. *J. Mol. Biol.* **27**, 197.

Lawn, A. M. (1966). Morphological features of the pili associated with *Escherichia coli* K12 carrying R factors or the F factor. *J. Gen. Microbiol.* **45**, 377.

Luria, S. E., and Anderson, T. F. (1942). Identification and characterization of bacteriophages with the electron microscope. *Proc. Natl. Acad. Sci. U. S.* **28**, 127.

Maass, D., and Weidel, W. (1963). Final proof for the identity of enzymatic specificities of egg-white lysozyme and phage T2 enzyme. *Biochim. Biophys. Acta* **78**, 369.

Marvin, D. A., and Schaller, H. (1966). The topology of DNA from the small filamentous bacteriophage fd. *J. Mol. Biol.* **15**, 1.

Meyer, F., Mackal, R. P., Tao, M., and Evans, E. A., Jr. (1961). Infectious deoxyribonucleic acid from λ bacteriophage. *J. Biol. Chem.* **236**, 1141.

Meynell, E. W. (1961). A phage, ΦX, which attacks motile bacteria. *J. Gen. Microbiol.* **25**, 253.

Miller, R. L. (1966). Chemotaxis during fertilization in the hydroid campanularia. *J. Exptl. Zool.* **162**, 23.

Monroy, A. (1965). "Chemistry and Physiology of Fertilization." Holt, New York.

Moody, M. F. (1967a). Structure of the sheath of bacteriophage T4. I. Structure of contracted sheath and polysheath. *J. Mol. Biol.* **25**, 167.

Moody, M. F. (1967b). Structure of the sheath of bacteriophage T4. II. Rearrangement of the sheath subunits during contraction. *J. Mol. Biol.* **25**, 201.

Murray, R. G. E., Steed, P., and Elson, H. E. (1965). The location of the mucopeptide in sections of the cell wall of *Escherichia coli* and other gram-negative bacteria. *Can. J. Microbiol.* **11**, 547.

Ørskov, I., and Ørskov, F. (1960). An antigen termed f^+ occurring in F^+ E. coli strains. Acta Pathol. Microbiol. Scand. 48, 37.

Paranchych, W., and Graham, A. F. (1962). Isolation and properties of an RNA-containing bacteriophage. J. Cellular Comp. Physiol. 60, 199.

Prater, C. D. (1951). A study of an ultraviolet light absorbing material released by Escherichia coli on infection with T2 or T4 bacteriophage. Ph.D. Thesis, University of Pennsylvania, Philadelphia, Pennsylvania.

Ptashne, M. (1967). Isolation of the λ phage repressor. Proc. Natl. Acad. Sci. U. S. 57, 306.

Puck, T. T. (1953). The first steps in virus invasion. Cold Spring Harbor Symp. Quant. Biol. 18, 149.

Puck, T. T., and Lee, H. H. (1954). Mechanism of cell wall penetration by viruses: I. An increase in host cell permeability induced by bacteriophage infection. J. Exptl. Med. 99, 481.

Puck, T. T., and Lee, H. H. (1955). Mechanism of cell wall penetration by viruses: II. Demonstration of cyclic permeability change accompanying virus infection of Escherichia coli B cells. J. Exptl. Med. 101, 151.

Puck, T. T., Garen, A., and Cline, J. (1951). Mechanism of virus attachment to host cells. J. Exptl. Med. 93, 65.

Putnam, F. W. (1954). Ultracentrifugation of bacterial viruses. J. Polymer Sci. 12, 391.

Ruska, H. (1941). Über ein neues bei der bakteriophagen Lyse auftretendes Formelement. Naturwissenschaften 29, 367.

Salivar, W. O., Tzagoloff, H., and Pratt, D. (1964). Some physical-chemical and biological properties of the rod-shaped coliphage M13. Virology 24, 359.

Sarkar, N., Sarkar, S., and Kozloff, L. M. (1964). Tail components of T2 bacteriophage. I. Properties of the isolated contractile tail sheath. Biochemistry 3, 511.

Schade, S., Adler, J., and Ris, H. (1967). How bacteriophage χ attacks motile bacteria. J. Virol. 1, 599.

Setlow, R. B., and Pollard, E. C. (1962). "Molecular Biophysics," p. 501. Addison-Wesley, Reading, Massachusetts.

Silver, S. (1965). Acriflavine resistance: a bacteriophage mutation affecting the uptake of dye by the infected bacterial cells. Proc. Natl. Acad. Sci. U. S. 53, 24.

Simon, L. D. (1966). The infection of Escherichia coli by T2 and T4 bacteriophages as seen in the electron microscope. Ph.D. Thesis, The University of Rochester, Rochester, New York.

Simon, L. D., and Anderson, T. F. (1967a). The infection of Escherichia coli by T2 and T4 bacteriophages as seen in the electron microscope. I. Attachment and penetration. Virology 32, 279.

Simon, L. D., and Anderson, T. F. (1967b). The infection of Escherichia coli by T2 and T4 bacteriophages as seen in the electron microscope. II. Structure and function of the baseplate. Virology 32, 298.

Sneath, P. H. A., and Lederberg, J. (1961). Inhibition by periodate of mating in Escherichia coli K-12. Proc. Natl. Acad. Sci. U. S. 47, 86.

Stent, G. S., and Wollman, E. L. (1952). On the two-step nature of bacteriophage adsorption. Biochim. Biophys. Acta 8, 260.

Stirm, S., Ørskov, F., Ørskov, I., and Birch-Andersen, A. (1967a). Episome-carried surface antigen K88 of Escherichia coli. III. Morphology. J. Bacteriol. 93, 740.

Stirm, S., Ørskov, F., Ørskov, I., and Mansa, B. (1967b). Episome-carried surface antigen K88 of *Escherichia coli*. II. Isolation and chemical analysis. *J. Bacteriol.* **93**, 731.

Tzagoloff, H., and Pratt, D. (1964). The initial steps in infection with coliphage M13. *Virology* **24**, 372.

Uetake, H., Luria, S. E., and Burrous, J. W. (1958). Conversion of somatic antigens in *Salmonella* by phage infection leading to lysis or lysogeny. *Virology* **5**, 68.

Valentine, A. F., and Chapman, G. B. (1966). Fine structure and host–virus relationship of a marine bacterium and its bacteriophage. *J. Bacteriol.* **92**, 1535.

Valentine, R. C., and Strand, M. (1965). Complexes of F-pili and RNA bacteriophage. *Science* **148**, 511.

Vieu, J.-F., Croissant, O., and Dauguet, C. (1964). Ultrastructure du bacteriophage Staphylococcique Twort. *J. Microscop.* **3**, 403.

Visconti, N. (1953). Resistance to lysis from without in bacteria infected with T2 bacteriophage. *J. Bacteriol.* **66**, 247.

Watanabe, T., Nishida, H., Ogata, C., Arai, T., and Sato, S. (1964). Episome-mediated transfer of drug resistance in interobacteriaceae. VII. Two types of naturally occurring R factors. *J. Bacteriol.* **88**, 716.

Watson, J. D. (1950). The properties of X-ray inactivated bacteriophage: I. Inactivation by direct effect. *J. Bacteriol.* **60**, 697.

Watson, J. D. (1965). "Molecular Biology of the Gene." Benjamin, New York.

Weidel, W. (1958). Bacterial viruses. *Ann. Rev. Microbiol.* **12**, 27.

Weidel, W., and Katz, W. (1961). Reindarstellung und Charakterisierung des für die lyse T2-infizierter Zellen verantwortlichen Enzyms. *Z. Naturforsch.* **16B**, 156.

Weidel, W. F., and Kellenberger, E. (1955). The *E. coli* B receptor for the phage T5. *Biochim. Biophys. Acta* **17**, 1.

Weidel, W., and Primosigh, J. (1958). Biochemical parallels between lysis by virulent phage and lysis by penicillin. *J. Gen. Microbiol.* **18**, 513.

Weidel, W., Koch, G., and Bobosch, K. (1954). Über die Rezeptorsubstanz für den Phagen T5. I. Extraktion und Reindarstellung aus *E. coli* B. Physikalische, chemische und funktionelle Charakterisierung. *Z. Naturforsch.* **9B**, 573.

Weidel, W., Frank, H., and Martin, H. H. (1960). The rigid layer of the cell wall of *Escherichia coli* strain B. *J. Gen. Microbiol.* **22**, 158.

Wildy, P., and Anderson, T. F. (1964). Clumping of susceptible bacteria by bacteriophage tail fibers. *J. Gen. Microbiol.* **34**, 273.

Williams, R. C., and Fraser, D. (1956). Structural and functional differentiation in T2 bacteriophage. *Virology* **2**, 289.

Wollman, E. L., and Stent, G. S. (1950). Studies on activation of T4 bacteriophage by cofactor: I. The degree of activity. *Biochim. Biophys. Acta* **6**, 292.

Wollman, E. L., and Stent, G. S. (1952). Studies on activation of T4 bacteriophage by cofactor. IV. Nascent activity. *Biochim. Biophys. Acta* **9**, 538.

CHAPTER 2

Bacteria

Giuseppe Sermonti

INSTITUTE OF GENETICS, THE UNIVERSITY, PALERMO, ITALY

I. Characteristics of Bacterial Fertilization

Processes of gene recombination have been demonstrated in bacteria since 1945, although cases of genetic transfer had already been reported as early as 1909 (see Lederberg, 1948). Genetic transfer is accomplished in bacteria by a variety of different mechanisms, including conjugation, plasmid transfer, sexduction, lysogenization, phage conversion, transformation, transfection, and transduction. All such processes have some characteristics in common. The transfer is always unidirectional and apparently progressive: one cell acts as donor and another as recipient The contribution by the donor cell is as a rule fractionary—ranging from an almost entire genome, in special cases of conjugation, to a small cluster of genes, in transduction or transformation. The zygotic nucleus is, consequently, incomplete, comprising a whole recipient genome and a more-or-less extended segment of the donor genome.

The term "fertilization," when applied to all the known processes eventually leading to bacterial gene recombination, must be used in its broadest sense, including any event through which nuclear [deoxyribonucleic acid (DNA)] materials from different cells come into effective contact within one cell. Contact of the involved cells is not compulsory for bacterial fertilization. In some processes, as in transduction or transformation, a segment of the bacterial genome is transported from cell to cell by a viral carrier or as free DNA in the medium of the recipient cell. The establishment of a symbiotic virus (lysogenization) in the bacterial cell could also be classed as one of the fertilization processes, since the virus DNA (prophage) shows a restricted genetic homology to the host DNA, and appears to be integrated in the bacterial "chromosome" in quite the same manner as any truly bacterial genomic segment (Section IV,A). A clear-cut boundary between fertilization and infection cannot be traced in the genetic bacterial systems, and this is perhaps the most striking peculiarity of the transfer of genetic information in bacteria.

Some exhaustive treatises on bacterial genetics have been published covering the matter discussed in this review up to early 1964 (Jacob and Wollman, 1961; Hayes, 1964; Gunsalus and Stanier, 1964).

II. Fertilization by Cell Contact: Conjugation

A. DISCOVERY OF BACTERIAL CONJUGATION IN *Escherichia coli* K12

The first attempt to reveal sexuality in bacteria dates back to 1937 (Sherman and Wing, 1937) and was soon followed by similar experiments by Gowen and Lincoln (1942). Both trials were unsuccessful, and in retrospect, we can understand the main reason for such failures. It was the unsuitability of the markers that were utilized for selection of very infrequent new phenotypes, and possibly the unfertility of the strain combinations, which prevented their recovery. It was only after the discovery of nutritionally deficient mutants in molds (Beadle and Tatum, 1941) and in bacteria (Tatum, 1945) that suitable markers became available, and genetic recombination was unquestionably demonstrated in the coli bacterium *Escherichia coli* K12 (Lederberg and Tatum, 1946a). Nutritional mutants can easily be selected after mutagenic treatments in *E. coli* because of the ability of this bacterium to grow on a simple synthetic medium, and many mutational sites are fairly stable. Multiple mutants can be obtained by successive treatments, and they virtually never revert to wild types, the reversion of each character being independent (Tatum, 1945).

1. *The Lederberg–Tatum Experiment*

In one of their first hybridization experiments with *Escherichia coli* K12, Lederberg and Tatum (1946a,b) employed two triple mutant strains. The first required biotin (bio⁻), phenylalanine (phe⁻), and cystine (cys⁻); the other required leucine (leu⁻), threonine (thr⁻), and thiamine (thi⁻).* Neither strain gave rise to growth when plated on minimal agar media at a density of 10^9 per dish or more. When the two strains (bio⁻ phe⁻ cys⁻ thr⁺ leu⁺ thi⁺ and bio⁺ phe⁺ cys⁺ thr⁻ leu⁻ thi⁻) were mixed and plated together, prototrophic recombinant colonies (bio⁺ phe⁺ cys⁺ thr⁺ leu⁺ thi⁺) arose on the minimal medium, with a frequency of about 1 per 10^6–10^7 parental bacteria plated. The recombinants, bearing the wild-type alleles of all the involved genes, turned out to be stable, and recessive mutant alleles were not segregated even after exposure to ultraviolet light irradiation. They were, thus, considered as the haploid products of the meiotic reduction of transient zygotes.

In successive crosses (Tatum and Lederberg, 1947; Lederberg, 1947), *unselected markers* were introduced with parents, i.e., markers the presence of which among recombinants was not made compulsory by the plating medium. These were virus-resistant mutations, as well as the nutritional mutations themselves, when partially supplemented media were adopted for recombinant selection. If, for instance, in the above-mentioned cross, biotin had been present in the plating medium, then the bio⁻/bio⁺ pair of alleles would have been utilized as unselected markers, both bio⁻ and bio⁺ recombinants being able to grow. Many unselected markers were, in fact, shown to segregate among recombinants, and a large variety of recombinant phenotypes were obtained. The segregation of the unselected markers was not at random, and evidence of linkage between some of the genes was unequivocally shown (Lederberg, 1947). With the extension of genetic analysis and the introduction of numerous unselected markers, a tentative linkage map of a single bacterial chromosome was elaborated (Lederberg *et al.*, 1951; Rothfels, 1952), but at the same time many anomalies were revealed showing that the sexual system was not as orthodox as had been thought at the beginning.

---

* The now obsolete symbols of B⁻, Ph⁻, C⁻, T⁻, L⁻, and B₁⁻ were originally employed. Other symbols of loci mentioned in this paper include: azi [resistance (r) vs sensitivity (s) to azide], gal (galactose fermentation), his (histidine synthesis), ile (isoleucine synthesis), lac (lactose fermentation), pro (proline synthesis), str (resistance vs sensitivity to streptomycin), T1 (resistance vs sensitivity to T1 virus). The plus (⁺) marks the wild-type allele, and the minus (⁻) the mutant allele.

2. Donor and Recipient Cells

Tho formation of recombinants requires cell contact. This was shown in experiments (Davis, 1950) in which tho cells of one strain were added to the cell-free filtrate of the other—no recombinants were obtained. No extracellular vector of genetic information was, thus, involved in the *Escherichia coli* fertilization. That the two parent strains performed different tasks in the fertilization process was soon shown (Hayes, 1952). In a cross A $\times$ B the parent strain A could be treated by streptomycin so as to abolish its colony-forming ability, without affecting the normal recombination frequency; treatment of strain B virtually prevented recombination. It was thus assumed that the role of A cells acting as the *genetic donors,* was to transfer their genetic material to B cells, whereas the role of B cells (*genetic recipients*) was to receive the genetic material and to become the seat of the zygote and of the subsequent recombination process (Hayes, 1953). The one-way transfer of the genetic material in bacterial conjugation has been repeatedly confirmed since then.

A striking characteristic of bacterial conjugation was shown to be the unequal contribution of the parent cells to the progeny (Hayes, 1953). It soon appeared evident that the donor cell contributed fewer markers to the progeny than did the recipient cell; the selected recombinants carried the recipient alleles and only a very few donor alleles (besides the selected one). It was concluded that the temporary zygote was probably incomplete with respect to the donor's markers (Hayes, 1953).

The transient nature of the zygotes found an early and remarkable exception in crosses involving a particular mutant strain (Lederberg, 1949). An appreciable proportion of recombinants behaved as relatively stable heterozygotes, giving rise on complete medium to occasional segregant phenotypes. The diploids were always incomplete, lacking an extended segment of the genome of one of the parents. This was direct proof of the incomplete contribution by one parent to the zygote, but it was thought at the time that the segment was missing possibly as a result of postzygotic exclusion.

B. SEXUAL DIFFERENTIATION

Sexual differentiation in *Escherichia coli* was clearly established when certain mutants, isolated by chance from the A strain (bio⁻ met⁻), were shown to be no longer able to form recombinants with the B strain (thr⁻ leu⁻ thi⁻). These mutants, however, retained their fertility with the wild-type strain (Lederberg *et al.,* 1952; Cavalli-Sforza *et al.,* 1953; Hayes, 1953). Two types of strains could, thus, be distinguished: F⁻ strains (to which the thr⁻ leu⁻ thi⁻ strain belonged, as well as occasional sterile mutants), which could not be crossed with one another, and F⁺ strains (to which the wild type and the A strains belonged), which were fertile in crosses with F⁻ strains and showed a limited fertility with other F⁺ strains. The F⁺ strains corresponded to donor bacteria, and F⁻ to recipient bacteria, according to the terminology of Hayes (1952). The fertility of the crosses among different mating types is as follows:

$$F^- \times F^- \quad \text{sterile}$$
$$F^- \times F^+ \quad \text{fertile}$$
$$F^+ \times F^+ \quad \text{less fertile}$$

1. *Inheritance of Fertility*

The fertility determinant, or *F factor,* behaves genetically in a way quite distinct from previously known markers. If F⁻ cells are incubated for 1 hour together with an excess of F⁺ cells, a large fraction of these F⁻ cells, sometimes approaching 100%, is converted to F⁺, whereas donor markers are transferred under the same condition to less than 1/1 million recipient cells. Transfer of F factor, like marker transfer, requires cell contact (Lederberg *et al.,* 1951; Cavalli-Sforza *et al.,* 1953; Hayes, 1953). This fact clearly means that cell contact is widespread in heterosexual cultures, with virtually all the minority F⁻ cells being involved in conjugation. Transfer of standard markers occurs only in a very small fraction of conjugal pairs. All recombinants are of the F⁺ type, i.e., they have received the F factor.

The F factor can be lost by F⁺ cells, spontaneously or following irradiation, and the F⁺ cells then become irreversibly converted to F⁻ cells. Acridine dyes were later shown to "cure" the F⁺ strains very efficiently of the F particle (Hirota, 1960). This loss and the above-mentioned acquisition of the F factor by F⁻ cells, make it easy to change an F⁻ cell into F⁺ and vice versa. "Reciprocal crosses" can thus be performed, i.e., $(X)F^- \times (Y)F^+$ and $(X)F^+ \times (Y)F^-$. In these crosses it may be clearly seen that the F⁺ (donor) strain, whatever its markers, makes a limited contribution to the progeny genotypes (Hayes, 1953; Lederberg *et al.,* 1952; Cavalli-Sforza *et al.,* 1953). From these observations it can be concluded that donor F⁺ bacteria transfer to recipient F⁻ bacteria only part of their genome, which results in incomplete zygotes. This conclusion was later substantiated.

2. *The Superfertile (Hfr) Strains*

Besides being able to give rise to sterile derivatives, the F⁺ strains have also been shown to yield occasionally superfertile strains (Cavalli-Sforza, 1950), giving a frequency of recombinants about 1000 times greater than normal strains; these are called "high frequency of recombi-

Male markers	thr	lac	gal	trp	his	str	met
frequency among recombinants (% Hfr input)	46	36	30	18	6	1	0.1
time of transfer (minutes)	8	18	24	33	59	90	115

FIG. 1. The correlation between the frequency of transmission of male markers to genetic recombinants (undisturbed matings) and their time of transfer (as deduced from interrupted matings) in a Hfr × F⁻ cross of *Escherichia coli* K12. Key: thr, threonine; lac, lactose; gal, galactose; trp, tryptophan; his, histidine; str, streptomycin; met, methionine. (From Wollman *et al.*, 1956.)

nation" (Hfr) strains. The Hfr strains work as donors, their fertility not being affected by streptomycin (Hayes, 1953). A high frequency of recombination was observed in a particular Hfr strain only when specific types of recombinants were selected. Unlike normal F⁺ donors, the Hfr strains do not transmit the fertility factor to the recipient bacteria, nor can the fertility be eliminated by acridine treatment (Hirota, 1960). The F factor, however, is still present in the Hfr strains, as proved by its exceptional transfer to recipient bacteria (in such cases the recipients are converted to Hfr) (Hayes, 1953) and by the occasional reversal of Hfr cells to the F⁺ state (Cavalli-Sforza, 1950). The F factor thus appears to be stored in Hfr bacteria in a masked, uninfectious form (Cavalli-Sforza *et al.*, 1953).

Different Hfr clones can be isolated from the same F⁺ culture (Jacob and Wollman, 1956a), each being able to transfer its peculiar group of markers with high frequency. The donor markers are, indeed, transferred to recombinants according to a gradient of transfer frequency, peculiar to each different Hfr strain (see Section II,D and Fig. 1). When transmitted, the Hfr determinant is linked to the marker which is transferred with the lowest frequency in that particular Hfr strain (Cavalli-Sforza and Jinks, 1956; Jacob and Wollman, 1957).

3. *Recombination in F⁺ × F⁻ Crosses*

Since rare Hfr cells are regularly present in F⁺ strains, it has been assumed that the F⁺ cells are unable to perform chromosome transfer, and the limited fertility of an F⁺ strain is due to the occurrence in it of Hfr mutants (Wollman *et al.*, 1956; Jacob and Wollman, 1957). The mutational origin of Hfr strains is proved by two lines of evidence (Jacob and Wollman, 1956a). The fertility of independent F⁺ cultures derived from very small inocula shows a large clonal variation, as expected after the chance occurrence of mutants at various times in the course of growth of the cultures (Luria and Delbrück, 1943). Various Hfr clones can be isolated, among F⁺ microcolonies, by replica plating of the culture onto a selective medium densely inoculated with F⁻ cells. The Hfr clones can be localized by correspondence to the points of the selective medium where recombinants have arisen. The assumption that the fertility of F⁺ strains is due to Hfr mutants is, however, contradicted by the already mentioned observation that Hfr cells transmit only very rarely the F factor to recombinants, whereas F⁺ strains do. Moreover, it has been demonstrated that the F factor is transmitted by the same cells that contribute the nuclear markers, which are not thus typical Hfr cells (Reeves, 1960). Therefore it seems reasonable to assume that fertility of F⁺ strains is due either to candidate Hfr cells having a chromosome-attached F factor but still retaining cytoplasmic F factors, pending their elimination as a result of cell multiplication, or to transient attachment of the F factor to the chromosome of F⁺ cells (Clark and Adelberg, 1962), a stable attachment being only a rare event.

Standard F⁺ cells would also be unable to act as recipients. The observed reduced fertility of F⁺ × F⁺ as well as of Hfr × F⁺ crosses is very likely due to the occurrence of F⁻ mutants as well as of F⁻ *phenocopies*, i.e., of cells that maintain their F factor but have temporarily lost their ability to work as donors in the F⁺ cultures (Lederberg *et al.*, 1952).

4. *Heterothallism and Heterosexuality*

In conclusion, *Escherichia coli* K12 appears to be a heterothallic species, with two distinct sexes: donor or *male* cells, corresponding to the Hfr strains, and recipient or *female* cells, corresponding to the F⁻ strains.

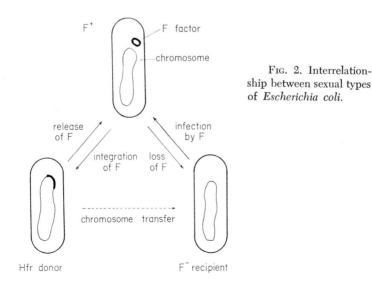

Fig. 2. Interrelationship between sexual types of *Escherichia coli*.

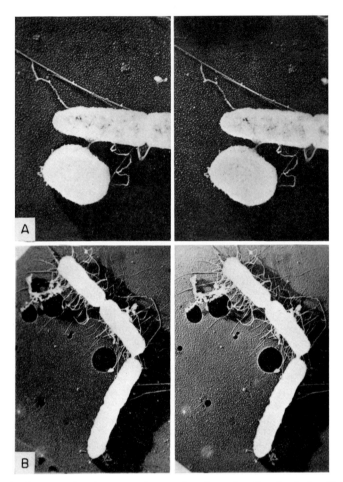

FIG. 3. Stereoscopic electron micrographs of conjugating bacteria. Notice the connecting bridges. (A) The elongated bacterium is the male (*Escherichia coli* K12 Hfr H); the plump bacterium is the female (*E. coli* C F⁻). (B) The male bacterium (*E. coli* K12, Hfr H) is "labeled" with phage λ; the female is recognizable because of its many fimbriae. Pili protruding from the male cells are evident, especially in (A). (Reproduced from Anderson *et al.*, 1957, by permission of the authors.)

Both sexes derive from the wild-type strain (F⁺) which harbors in its cells the free F factor (Fig. 2). The female cells result from the loss of the F factor from the temporary repression of its activity (F⁻ pheno-copies). The male cells result from the stable fixation or from the tran-sient attachment of the F factor to a chromosomal site. Chromosomal transfer occurs unidirectionally from male to female cells. Although the actual donors of genetic markers are the Hfr cells, the F⁺ strains are also currently named "male," because of their mating pattern and their occasional conversion into true males.

C. KINETICS AND PHYSIOLOGY OF CELLULAR UNIONS

1. Cell Pairing

The first event in heterosexual bacterial culture leading to fertilization is the collision between cells. It occurs at random, as a simple function of cell density (Nelson, 1951). Specific pairs result from random collision of donor and recipient cells and may be observed under the microscope (Lederberg, 1956; Anderson, 1958). They occur with the same frequency in $F^+ \times F^-$ crosses as in Hfr $\times F^-$ crosses. Multiple clumps of cells are also commonly observed, with the progress of time, occasionally leading to triparental recombinants (Fischer-Fantuzzi and Di Girolamo, 1961). Specific pairs are converted into "effective pairs" (Clark and Adelberg, 1962) by the establishment of intercellular connections. A thin bridge has been repeatedly observed, under the electron microscope, to join heterosexual pairs (Anderson et al., 1957) (Fig. 3), but its function as an actual channel for chromosome transfer has been questioned recently. This function has been attributed to special pili regularly observed on male cells (Brinton et al., 1964). The relevance of pili in conjugal transfer is supported by the observation that male-specific viruses (Maccacaro and Comolli, 1956) are only absorbed on male sexual pili (Crawford and Gesteland, 1964; Datta et al., 1966; Simon and Anderson in Chapter 1 of this volume) (Fig. 4) and that sexual pili are eliminated from F^+ cells treated with acridine orange and acquired by F^- cells infected with the F factor. Mechanical removal of pili from Hfr cells prior to mating virtually prevents chromosome transfer (Brinton et al., 1964).

2. Zygotic Induction

The number of effective pairs in a mixed culture has been estimated by the detection of zygotic inductions, occurring when the male parent is lysogenic (see Section V,B) for a given phage (λ^+) and the female is

---

FIG. 4. Sexual (F-type) pili in bacteria. (A) A bacterial cell exhibiting numerous short common pili and two long F-type pili clearly distinguishable by the presence of absorbed male-specific RNA phages (M 12). The cell belongs to a strain of *Proteus mirabilis*, originally F^- and having only common pili—upon infection by the *Escherichia coli* fertility factor (F lac$^+$) the cell acquired F-pili and fertility. (From Brinton et al., 1964, by permission of the authors.) (B) Common and F-type pili on the surface of a cell of *Escherichia coli* K12 carrying one R factor (R 237). F-type pili can be identified by the absorbed male-specific RNA phage (MS2). (From Lawn, 1966, by permission of the authors.) (Figure on following pages.)

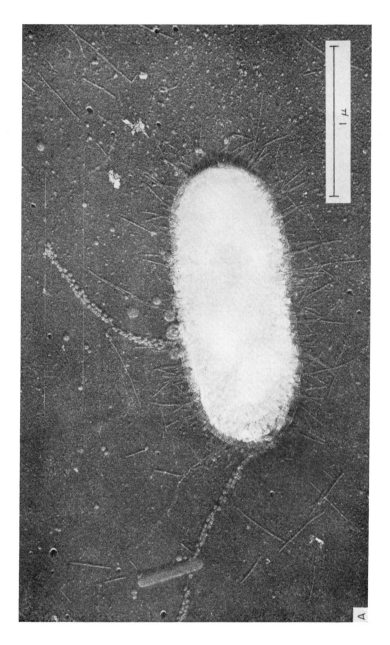

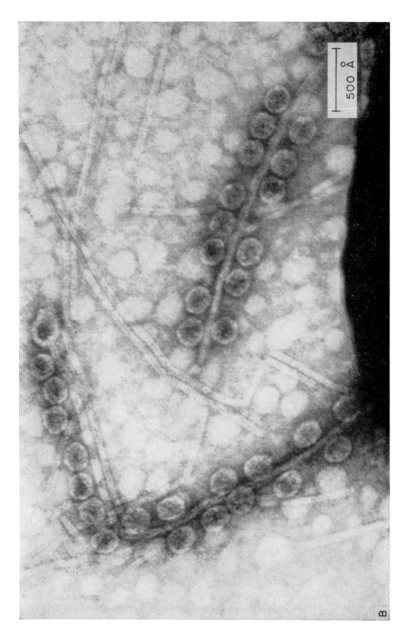

not (Jacob and Wollman, 1956b). The formation of the intercellular connection between a specific pair, followed by the unidirectional chromosome transfer, regularly involves the transmission of the prophage from the male to the female cell. As soon as it enters the phage-free cell, the prophage is "induced" and starts explosive multiplication eventually leading to the host cell burst. All markers distal to the phage site will, thus, never be contributed to the progeny. To detect zygotic induction, a sample of the mixed culture (Hfr $\lambda^+ \times$ F$^-$ λ^-) is gently diluted in order to leave the conjugal pairs undisturbed and to prevent further pairing. The diluted cell suspension is thus mixed to a dense phage-sensitive culture and plated. The recipient cell of each effective pair will eventually lyse giving rise to a lytic center detectable as a plaque on the indicator culture (Wollman and Jacob, 1958). Effective pairing, as detected by zygotic induction, starts at zero time and is practically complete after about half an hour (Figs. 5 and 6) (Wollman and Jacob, 1958). The frequency of zygotic induction is thus greater than 50% of the male partners. This does not mean that fertilization is complete by that time, but all the events that could possibly be affected by gentle dilution have

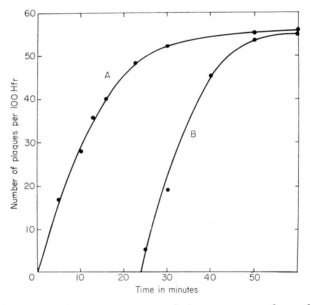

FIG. 5. Kinetics of effective pairing and chromosome transfer as deduced from zygotic induction in an Hfr H $\lambda^+ \times$ F$^-$ λ^- cross of *Escherichia coli* K12. Plaques formed by λ-infected recipient bacteria are plotted against time of sampling. (A) Untreated samples—new matings were prevented by dilution. (B) Blended samples—chromosome transfer was interrupted at the time of sampling. Effective pairing begins at zero time; transfer of λ phage takes 25 minutes to initiate. (From Wollman and Jacob, 1958.)

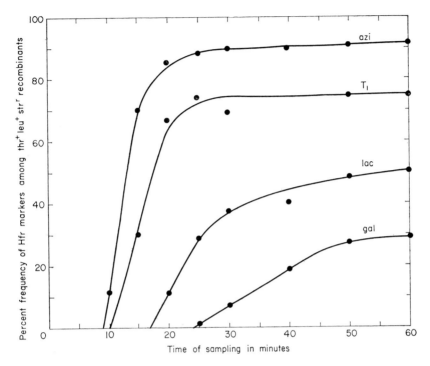

Fig. 6. Order of penetration of proximal markers of Hfr H bacteria during conjugation. The parents were marked as follows: Hfr H (origin) thr⁺ leu⁺ azi-s T1-s lac⁺ gal⁺ str-s and F⁻ thr⁻ leu⁻ azi-r Tl-r lac⁻ gal⁻ str-r. Samples taken at different intervals (abscissa) were blended and plated on a medium selecting thr⁺ leu⁺ str-r, and the frequency of donor markers was recorded (ordinate) as percent of total selected recombinants. Key: thr, threonine; leu, leucine; str, streptomycin; azi, azide; lac, lactose; gal, galactose; r, resistance; s, sensitivity.

already taken place. Fertilization is, in fact, still susceptible to interference by energetic mechanical stirring for about 2 hours (see below).

3. *Surface Differences Between Sexes*

Effective pairing is regulated by surface differences between male and female cells. A characteristic surface antigen, named f⁺ has been detected on male cells (Ørskov and Ørskov, 1960), probably located on the above-mentioned sexual pili (Ishibashi, 1967) and on the site of absorption of male-specific phages. Several such phages, all containing ribonucleic acid (RNA), have been isolated (Loeb, 1960; Dettori *et al.*, 1961). Mutant strains made resistant to male-specific phages are less fertile (see Clark and Adelberg, 1962) whereas F⁻ phenocopies absorb male-specific phages less efficiently (Dettori *et al.*, 1961).

Evidence of a surface component of the male cell, probably a polysaccharide, has been provided by treating male cells with periodate. Their pairing ability is temporarily reduced (Sneath and Lederberg, 1961) as well as their ability to absorb specific male phages (Dettori *et al.*, 1961). The chemical nature of the sexual pili has, however, not yet been determined. Phages that multiply in female cells only have also been isolated (Dettori *et al.*, 1961).

Other surface differences between sexes have been observed: male cells have greater affinity for acidic dyes; they precipitate out of suspension in less acid media, they have a greater tendency to agglutinate, and they are less motile (Maccacaro and Comolli, 1956). The F$^-$ cells are negatively charged to a greater extent than F$^+$ cells, as proved by microelectrophoresis (Turri and Maccacaro, 1960).

Formation of effective pairs is an energy-dependent process, requiring a carbon source and the high-energy bonds generated by oxidative phosphorylation. The process is temperature-sensitive, pH-dependent, and severely inhibited by anaerobic conditions and dinitrophenol (Fisher, 1961).

D. Chromosome Transfer as an Oriented Process

Soon after the establishment of effective pairing, transfer of genetic material from male (Hfr) to female (F$^-$) cells begins. In some pairs, however, there may be a delay of up to 15 minutes before transfer initiates (de Haan and Gross, 1962). Chromosome transfer then proceeds gradually, taking over 100 minutes before the chromosome is transferred in its entirety.

1. *The Interrupted Mating Experiment*

Evidence of a gradual and oriented transfer of chromosomes during conjugation has been provided by experiments involving the physical interruption of the mating, either by violent mechanical agitation (Wollman and Jacob, 1955) or by killing, selectively, the donor cells by virulent phages (Hayes, 1957). The sooner the interruption of the mating takes place, the smaller the number of donor markers detectable among the progeny. When a given Hfr strain is used as donor, a gradient of transmission of the donor markers is observed, each marker making its first appearance among the progeny of the exconjugants after a fixed time of undisturbed mating (Figs. 1 and 6) (Wollman and Jacob, 1955, 1958; Hayes, 1957). This fact has been interpreted as evidence that the male chromosome is transferred to the female cell progressively, starting from a fixed point (the origin) and proceeding in a fixed direction. When

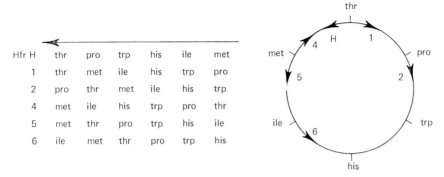

Hfr H	thr	pro	trp	his	ile	met
1	thr	met	ile	his	trp	pro
2	pro	thr	met	ile	his	trp
4	met	ile	his	trp	pro	thr
5	met	thr	pro	trp	his	ile
6	ile	met	thr	pro	trp	his

FIG. 7. The order of markers transferred in different Hfr strains of *Escherichia coli*. Only loci controlling amino acid synthesis are reported. On the right, diagram showing arrangement of markers on a circular structure. The arrow heads indicate the origin and the sequence of chromosome transfer of different Hfr strains. Key: thr, threonine; pro, proline; trp, tryptophan; his, histidine; ile, isoleucine; met, methione. (Simplified, after Jacob and Wollman, 1961.)

the mating is interrupted, the male markers which have already passed into the recipient cell will have a probability of appearing among the recombinant progeny, but those located further from the origin will be excluded. After one donor marker has made its first appearance in a blended sample, it will appear with higher frequency in later samples, the number of pairs having spent the required interval from the beginning of transfer undisturbed gradually increasing with time (Fig. 6). For any given Hfr strain, a time of transfer for each marker can be established, which corresponds strictly to the order of markers obtained on the basis of recombination experiments (Wollman *et al.*, 1956). The origin varies from male to male strain (Fig. 7), but it is constant among cells of the same male (Hfr) strain. The different order of markers in various Hfr strains represent different permutations of a circular arrangement of markers. The circular structure of the F+ chromosome was first inferred on this basis (Jacob and Wollman, 1956a). The rate of chromosome transfer is fixed in the individual cells of some Hfr strains, although it may vary in others (de Haan and Gross, 1962).

Breakage of the migrating chromosomes also occurs spontaneously, so that the probability of a male marker appearing among the recombinant progeny, in uninterrupted matings, is inversely proportional to the distance of that marker from the origin (Wollman and Jacob, 1958). The gradient of transmission of the male markers in uninterrupted matings is simply correlated to their time of transfer as detected by interrupted mating experiments (Fig. 1). When a conjugating pair is artificially separated, the male chromosome may not immediately be broken and a certain extent of chromosome withdrawal can be observed under special experimental conditions (de Haan and Gross, 1962).

2. Physical Parameters in Escherichia coli Conjugation

Much physical evidence of the gradual and oriented transfer of male DNA to recipient cells in conjugating pairs has been provided.

When ^{32}P is incorporated into the genetic material of male cells and allowed to decay before mating, the transfer of donor markers is reduced to a greater extent the farther the marker is located from the origin (Fuerst et al., 1956). One can assume that this effect is due to chromosome breakage, known to occur as a consequence of ^{32}P decay in bacterial DNA. When mating of ^{32}P-containing males with unlabeled females occurs prior to decay, radioactive material corresponding to 10% male DNA is transferred to recipient cells in 1 hour (Garen and Skaar, 1958). Labeled thymidine is also transferred to a similar extent (Silver, 1963). Assuming an average of the three nuclei per cell, this would amount to an average of one-fourth of a chromosome transferred per male cell.

The rate of chromosomal transfer can be expressed in physical parameters (Jacob and Wollman, 1961). The transfer of 20 map units takes 1 minute, and, since the entire chromosome takes 100 minutes to be transferred, the chromosome can be estimated as 2000 map units long. The DNA of a chromosome, as judged by chemical estimations, is of the order of 10^7 base-pairs; 10^5 base-pairs are, therefore, transferred in 1 minute, corresponding to 34μ of a DNA duplex. Every second, a length of donor DNA exceeding the smaller diameter of the cell penetrates the recipient bacterium.

E. CONJUGATION IN BACTERIA OTHER THAN Escherichia coli K12

Chromosomal transfer by cellular conjugation has been described in a few genera of bacteria besides Escherichia coli, i.e., Salmonella, Pseudomonas, Streptomyces, Serratia, and Vibrio. In no case, however, has a picture identical to that observed in Escherichia coli K12 been recorded. Only in Salmonella typhimurium has a conjugation system of the Escherichia coli K12 types been established, after conjugal transfer of the F factor from E. coli K12 to Salmonella cells and after its integration in the Salmonella chromosome to give Hfr donors (Zinder, 1960; Miyake, 1962). An arrangement of markers in the same order as in E. coli has been shown (Sanderson and Demerec, 1965).

In the other bacterial conjugation systems described, no Hfr strains have ever been found, and only low-frequency recombination has emerged. A group of strains infertile with each other but fertile with strains of a second group has been described in all systems. The two groups, comparable to F⁻ and F⁺ strains of *E. coli* K12, have been de-signed as FP⁻ and FP⁺ in *Pseudomonas* (Holloway, 1956), P⁻ and P⁺ in *Vibrio* (Bhaskaran, 1960), and R⁻ and R⁺ in *Streptomyces* (Sermonti and Hopwood, 1964). Similar groupings have been reported in *Serratia* (Belser and Bunting, 1956) and in a system of conjugation in *Salmonella,* where chromosome transfer is mediated by colicinogeny factors (Ozeki *et al.,* 1961) (see Section III,C). Strains of the interfertile group appear to act as "donor" of chromosomal material to cells of the insterile group, but some evidence of inverted transfer has been reported in *Serratia* (Belser and Bunting, 1956).

The best evidence of unidirectional transfer comes from *Pseudomonas* crosses. Destruction with virulent phages of the FP⁺ conjugant cells after ½ hour of contact does not reduce the number of recombinants, whereas destruction of the FP⁻ cells completely abolishes formation of recombi-nants (Holloway and Fargie, 1960).

Chromosomal transfer is largely incomplete in all the bacterial con-jugation systems examined. Although large chromosomal segments may be transferred in *Salmonella* (colicinogeny-factor-mediated conjugation), *Pseudomonas,* and *Streptomyces,* only very small segments are trans-mitted in *Vibrio* and *Serratia.* In the first three systems the occurrence of a single chromosome has been shown on the basis of recombination data. Evidence of a circular structure of the chromosome has been provided for *Salmonella* (Smith and Stocker, 1962) and for *Streptomyces* (Hopwood, 1965). Preliminary data suggesting a progressive chromosome transfer have been reported in *Streptomyces* (Sermonti *et al.,* 1966). The ends of the transferred chromosomal segment vary in *Streptomyces* in the dif-ferent zygotes of the same cross (Sermonti *et al.,* 1966; Hopwood, 1966), a situation comparable to that occurring in F⁺ × F⁻ crosses of *Escherichia coli* K12.

Recent reviews on conjugation in *eubacteria* have been prepared by Clark and Adelberg (1962) and by Gross (1964) (see, also, Hayes, 1964). *Streptomyces* conjugation has been reviewed by Sermonti and Hopwood (1964).

III. Conjugal Transfer of Extrachromosomal Elements

A. PLASMIDS

The bacterial cell may harbor autonomous extrachromosomal ele-ments, besides its chromosomal material. These consist of small segments of DNA (roughly 1–2% of the bacterial chromosome) that self-reproduce and self-regulate independently of the host chromosome and can be transferred by cell contact from cell to cell. The F factor (Section II,B) is the first and the best known among these elements, called "plasmids[*]" (Lederberg, 1952), and a series of them have since been reported. Various sex factors have been described (Section II,A) in different bacterial systems. Other plasmids are the colicinogeny factors (Section III,C), the resistance transfer factors (Section III,D), as well as factors controlling mutability (Gundersen et al., 1962), sporogenesis (Jacob and Wollman, 1961), penicillinase synthesis (Novick, 1963), and hemolysin production (William Smith and Halls, 1967). All the plasmid-controlled properties are expendable. Extrachromosomal elements can be acquired by conjugal "infection," and usually are lost by "disinfection" with acridine orange treatment or other procedures. Intergeneric transmission of plasmids is very common. A plasmid can be transferred to a much wider range of bacterial species than can the bacterial chromosome, even to species which have a DNA base ratio quite different from that of the plasmid.

1. Integration of Some Plasmids with Chromosomes: Episomes

Some of the plasmids can promote the synthesis by the host cell of specific surface components, which induce cellular pairing and plasmid transfer. Competition or mutual exclusion can occur between different plasmids. Those plasmids that have some homology with the host chromosome can exchange segments with it, so that gene markers can be carried and transferred by an extrachromosomal element or, alternatively, a plasmid region may be inserted into the chromosome. Some plasmids can be integrated more or less in a stable manner, into the host chromosome, mobilizing it (or a fragment of it) and affording its transfer to a recipient cell lacking the involved plasmid. Indeed, the bacterial chromosome by itself can never promote cell pairing and its own mobilization, and its transmissibility is strictly connected to its previous association with a carrier plasmid.

Those plasmids that can occur either in the autonomous or in the integrated state, i.e., either at an extrachromosomal or at a chromosomal site, have been named "episomes" (Jacob and Wollman, 1958). They are of great evolutionary interest, due to their tendency to pass from cell to

---

[*] The term plasmid originally was used to designate all extranuclear structures that are able to reproduce in an autonomous fashion (Lederberg, 1952). The *episomes,* due to their possible nuclear integration, were ascribed to a different class of elements (Jacob and Wollman, 1958). Since, however, the evidence of the possible integrated state of some elements is often unclear, it is recommended that the term plasmid be used to denote all elements capable of autonomous reproduction until the alternative attached state has clearly been demonstrated (Clark and Adelberg, 1962). We prefer to designate all self-reproducing extranuclear elements as plasmids and to consider episomes as a subclass of plasmids.

cell, even those belonging to different genera, and their ability to be integrated eventually into the cellular genome, increasing the net genetic information of the species. Episome fertilization may represent a flow of genetic information between loosely related bacterial species (for a review on episomes, see Driskell-Zamenhof, 1964).

Different intracellular episomes have been designated as fertilizing factors or as carriers of specific characteristics, according to which property was first discovered or preeminently manifested. Most of them are fertilization promoters, as well as determinants of bacterial characters, encoded in their DNA or in an associated fragment of chromosomal DNA.

B. Sex Factors and the Process of Sexduction

1. *Nature and Transfer of Sex Factor*

As we have already seen (Section II,B) the sex factor, or F particle, can be transmitted at high frequency from F^+ to F^- cells of *Escherichia coli* K12. By interrupted mating experiments, it has been established that the minimum time for complete transfer of the sex factor is 4–5 minutes (Jacob and Wollman, 1955). In Hfr $\times$ F^- crosses the sex factor is not transferred except after about 2 hours, linked to the very last-transferred chromosomal marker (Section II,B). Experiments with ^{32}P-labeled donors indicate that the *E. coli* K12 sex factor is a short DNA segment, about 2% of the total length of the bacterial chromosome (Driskell-Zamenhof and Adelberg, 1963). The exponential nature of the inactivation curves of the sex factor provides evidence that only one particle is transferred to any one recipient. It has also been calculated that the passage of this "F chromosomelet" from the donor to the recipient cell would take about 1 minute (Driskell-Zamenhof and Adelberg, 1963). The physical transfer of what was very probably the F factor from F^+ cells labeled with tritiated thymine to unlabeled F^- cells has been shown by autoradiography (Hermann and Forro, 1964). Transfer of ^{14}C-thymidine-labeled DNA from F^+ to F^- cells was also detected (Silver, 1963).

The autonomous F factor can "infect" a wide range of bacterial species. In various strains of the species *E. coli* and *Salmonella typhimurium* the transfer of the F factor from *E. coli* K12 can be followed by its integration with the host genome and by chromosomal mobilization (see Section II,E). In many cases, however, the F factor cannot be integrated in the host genome and promote fertility. Its occurrence in the infected cells is detected by the acquired ability of such cells to transfer it back to *E. coli* K12. This transfer, without integration of the F factor, has been observed, from *E. coli* K12 F^+ to *Shigella* (Luria and Burrous, 1957) and to many strains of *E. coli* (Lederberg *et al.*, 1952). Special experimental conditions later permitted the detection of this type of transfer to a much wider range of bacterial genera (see below). Sex factors comparable to the F factor of *E. coli* K12 have been found in some *E. coli* strains, which were shown to be able to transfer fertility to F⁻ strains of *E. coli* K12. These factors were not, however, identical with the F⁻ factor, giving rise in *E. coli* K12 only to unstable donors (Lederberg *et al.*, 1952).

2. Sexduction

We have seen (Section II,B) that the F factor can be integrated into the bacterial chromosome, losing its autonomy and giving rise to an Hfr cell. It can, occasionally, regain its free state by spontaneous detachment from its chromosomal site. It has been shown that a released sex particle can bear only rarely a minute fragment of the bacterial chromosome, from the distal region to which it was linked (Jacob and Adelberg, 1959). One such particle, bearing the lac⁺ marker (F lac⁺), was detected in a cross Hfr lac⁺ × F⁻ lac⁻, where the donor strain normally transferred lac⁺ as its final marker only about 2 hours after the beginning of mating. The mating was interrupted after 30 to 60 minutes, and rare lac⁺ recombinants were detected which could not be accounted for by the normal process of transfer. They bore the F lac⁺ autonomous factor, superimposed on the lac⁻ allele of the recipient strain. Occasional loss of the F lac⁺ particle produced the rare but regular segregaton of lac⁻ bacteria ($\sim 10^{-3}$). The transfer process of a bacterial marker integrated into the sex factor is called "F-duction" or- "sexduction" (Jacob and Wollman, 1961). The complex particles (F-genotes) are transferred at high frequency to F⁻ recipients by the carrier strains. They are also able to mobilize the host chromosome and to promote its transfer to F⁻ cells with the same orientation as in the original Hfr, although with one-tenth of the frequency.

3. Intermediate Males

The recurrent integration of the F-genotes into the bacterial chromosome at the constant site indicates striking affinity for a host chromosome region of the chromosome fragment attached to the sex factor (see Section IV,A). Strains bearing a substituted sex factor (F-genote, also called F′) have been designated as *intermediate donors* because of their ability to transfer either their autonomous sex factor or their chromosome, the latter having fixed orientation (Adelberg and Burns, 1960). The segment of the bacterial chromosome integrated into the F′ factors may vary in size and occasionally include several markers from the distal end of the original Hfr strain (Hirota and Sneath, 1961). The larger the size of the chromosomal segment integrated into the F′ factor, the more un-stable is the carrier strain. An F lac$^+$ factor can be "cured" after standard acridine orange treatment, but an F′ factor carrying a number of chromo-somal genes proved resistant to cure by this method (Clowes *et al.*, 1965).

Intermediate donor strains have also been isolated bearing an F par-ticle without associated chromosomal markers (Adelberg and Burns, 1960). It has been assumed that such a particle carries, in fact, a minute chromosomal segment, but no known marker is located in it. This seg-ment would enable the F′ factor to recognize a region of homology on the host chromosome. Another type of intermediate donor has been de-scribed, in which the homology between the F factor and a specific re-gion of the chromosome was secured by the occurrence in the host chromosome of a small inserted segment of the sex factor, establishing a region of "sex factor affinity" (sfa) (Adelberg and Burns, 1960) (see Secton IV,A).

4. *Interspecific Transfer of Sex Factor*

The integrated chromosomal marker works as a genetic label for the F′ particle. When a lac$^-$ strain is used as recipient, cells infected by an F lac$^+$ particle give rise to phenotypically lac$^+$ colonies which contrast sharply with the unaffected lac$^-$ colonies, when grown on an indicator medium in the presence of lactose. This fact has permitted the detection of the transfer of the sex factor even in systems where it occurs as an exceptional event. By this means the transfer of an F factor (F′ lac$^+$) from *Salmonella typhosa* to various Enterobacteriaceae has been detected (Falkow and Baron, 1962) as well as the transfer of the *Escherichia coli* K12 sex factor to the genera *Serratia* (Marmur *et al.*, 1961) and *Pas-teurella* (Martin and Jacob, 1962). These findings are particularly inter-esting because the F′ factor can be transferred to and from and can multiply and exert its function (lactose fermentation) in "foreign" en-vironments such as *Pasteurella* and *Serratia* cells. The *Serratia* DNA has a guanine and cytosine content (58%) quite distinct from that of *Esche-richia coli* (50%). The DNA extracted from *Serratia* clones bearing the F lac$^+$ factor from *E. coli* may be separated, by ultracentrifugation in a caesium chloride density gradient, into two peaks: a large one corre-sponding to *Serratia* DNA and a small peak corresponding to *E. coli* DNA. This provides additional evidence that the F′ factor is composed of DNA (Marmur *et al.*, 1961). In *Serratia* and *Pasteurella* (as well as in many enterobacteria) the F′ factor fails to mediate chromosomal trans-fer, very probably because it is prevented, owing to lack of affinity, from finding a site of attachment on the host chromosome.

C. Colicinogeny

Various members of the Enterobacteriaceae (e.g., *Salmonella, Shigella, Escherichia*) produce bacteriocidal substances of a protein, peptide, or lipocarbohydrate-protein nature, which kill susceptible cells of other members of the family. These antibiotics, or *colicins*, are produced by dying cells, and can be related to the proteins of the bacteriophage coat or tail. About twenty groups of colicins are known, designated as col E1, col E2, col I, col K, etc. For reviews on colicins, see Frédéricq (1957) and Ivanovics (1962). The ability to produce colicins (*colicinogeny*) is a stable characteristic and can be transmitted from col$^+$ to col$^-$ cells. The transfer of col E1 occurs from F$^+$ to F$^-$ cells, starting 5 minutes after the beginning of mating and shows no linkage either with the chromosome or with the sex factor (see Frédéricq, 1957). Another colicinogenic factor, col I, can be transferred from one F$^-$ cell to another, being able by itself to promote conjugation and its own transfer as well as that of any other col factor harbored by its host (Ozeki and Howarth, 1961). A cell freshly infected by the col I factor effects conjugation and col transfer very efficiently, but the cell loses this ability almost completely after seven generations. As already mentioned (Section II,E) the col I factor can mediate chromosomal transfer, although only at low frequency. This indicates a close analogy to the F factor. The possibility of a chromosomal location of some col factors has been suggested (see Smith *et al.*, 1963). More recent experiments do not, however, support this conclusion (Clowes, 1963). The failure of stock cultures to transmit the col factors appears not to be due to the state of integration of the agent but, rather, to a col-determined cytoplasmic repressor that prevents the synthesis of a surface product required for the promotion of mating (see Clark and Adelberg, 1962). From experiments comparable to those that have revealed the physical nature of the F factor (see Section III,B), it has been concluded that the colicinogenic factors are composed of a DNA molecule roughly of the same size as the F factor and that only one copy of each factor is transferred during conjugation (Silver and Ozeki, 1962). Colicinogeny cannot be cured by acridine orange treatment, but treatment with cobalt salts or thymine starvation of thymineless col$^+$ mutants (Clowes *et al.*, 1965) efficiently eliminates the col factors. The colicinogenic factors thus represent a clear case of extrachromosomal self-regulating elements, sharing many properties with the F factor, but very likely unable to assume a stable chromosomal site. They also exhibit considerable similarity with prophages (see Section V,A), the production of colicins being inducible by the same agents that promote the induction of vegetative phages (Jacob *et al.*, 1952).

D. Resistance Transfer Factors

Some strains of the family Enterobacteriaceae are simultaneously resistant to one or more of the four therapeutic agents streptomycin, chloramphenicol, tetracycline, and sulfonamide, in various combinations. The multiple resistance factor (R) can be transferred as a unit, through cell contact, from resistant to sensitive cells, regardless of their sex, as an autonomous extrachromosomal agent (for a review, see Watanabe, 1963). The R factors are carried by a determinant, termed "resistance transfer factor" (RTF), in a relationship comparable to that of the sex factor with chromosomal genes in F-genotes (Section III,B). There are striking analogies between RTF and the sex factor of *Escherichia coli* K12. The RTF consists of DNA of a size comparable to F′ particles, it is able to promote conjugation and its own transfer independently of the host chromosome, can be transferred to a wide range of Enterobacteriaceae and to *Vibrio cholerae*, and is capable of bringing about transfer of chromosomal determinants at low frequency (Sugino and Hirota, 1962). Like the col I factor, R factors are transmitted with greater efficiency from cells that have recently acquired resistance. The R factors can be cured by acridine orange treatment, especially in cells previously treated with ultraviolet light irradiation, which also stimulates transfer (Watanabe and Fusakawa, 1961).

1. *Homology of RTF with F Factor*

The RTF is able efficiently to promote chromosome transfer starting from a fixed origin (Sugino and Hirota, 1962), in a way comparable to that of an F′ factor harbored in an intermediate donor (Section III,B). A homology between the F factor and the RTF is thus shown. Most RTF inhibit by a mechanism of repression the transfer of F or of the bacterial chromosome from F⁺ or Hfr strains of *Escherichia coli* K12 by interfering with the synthesis of the surface component involved in F-mediated transfer (Watanabe, 1963). RTF also interferes with the autonomous replication of col factors.

There is good evidence that R factors may be inserted or attached at a particular site of the host chromosome, but this integration often appears to be very unstable. The occurrence of genetic exchanges between R factors and the host chromosome has also been reported (Watanabe, 1963) as well as recombination between different R factors (Hashimoto and Hirota, 1966) and between R and F factor (Watanabe and Ogata, 1966).

The same surface component is synthesized by F and R factors, as proved by the ability of F⁻ bacteria freshly infected by the R factors to support growth of F⁺-specific bacteriophages (Fig. 4A). However, R factors always repress the synthesis of the receptor produced by themselves and in many instances extend their repression to F (Meynell and Datta, 1966a). The unrepressed behavior of F itself appears to be exceptional among conjugation factors (Meynell and Datta, 1966b).

IV. General Models for Chromosome Transfer

Fertilization in bacterial conjugation is characterized by the process of longitudinal chromosome transfer, which is peculiar to the bacterial genetic systems. A mechanical model of this process has been sketched, accounting for practically all known facts, and amenable to particular experimental verifications. This is substantially agreed upon by all authors, although alternative hypotheses for some steps are still under consideration.

Chromosome transfer involves the following events: (1) insertion of the sex factor into the bacterial chromosome; (2) opening of the integrated chromosome in correspondence to the inserted sex factor; (3) recognition, by the front end of the open chromosome, of the mouth of the channel established between mated cells; (4) longitudinal displacement of the extended chromosome, through the conjugation bridge, toward the recipient cell.

The outlined model is generally applicable to all processes of bacterial DNA transfer by conjugation, including sexduction and any kind of episome-mediated transfer (see Hayes, 1966).

A. INTEGRATION OF SEX FACTOR AND CHROMOSOME

Chromosomal transfer in *Escherichia coli* K12 must always be preceded by the integration of the chromosome with the F factor. Similar, although less stable, associations have been described in other conjugal transfer system (see Section II,E). The ability to open and initiate transfer is peculiar to the sex factor and the whole bacterial chromosome may open and start transfer when the two structures are physically joined.

1. *Integration of Sex Factor and Chromosome by Crossing-over*

The bacterial chromosome has been shown to be a circular continuous structure (Cairns, 1963), and the sex factor has been assumed, by analogy, to be a small DNA loop. With a very low frequency in the F⁺ cells the two circles could come into contact, in some region of reciprocal affinity, and could, then, be integrated by a crossover between the paired regions, to give the Hfr chromosome, which is still a circular structure

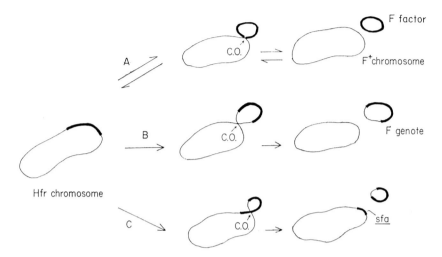

FIG. 8. Release of F factor (heavy tract) from an Hfr chromosome (thin line) by a mechanism of equal (A) or unequal (B and C) crossing-over (c.o.). In (A), reversible detachment of a normal F particle; in (B), release of an F particle bearing a piece of chromosome (F-genote); in (C), release of an incomplete F particle leaving a fragment (sfa) on the chromosome.

(see Adelberg and Pittard, 1965). The reverse event would produce the release of the F particle and the reappearance of an F^+ cell in an Hfr population (Figs. 2 and 8A). These events are very rare (10^{-4}–10^{-5} per cell generation), because the regions of homology between the F factor and the chromosome amenable for pairing are mostly either extremely short or irregular.

The hypothesis that the F' factor (and, by inference, the normal sex factor) is inserted into the bacterial chromosome by recombination has recently been supported by the observation that a recombination-deficient mutant female may be infected by an F' factor but this cannot promote chromosome transfer (Clark and Margulies, 1965).

2. *Unequal Crossover at the Origin of Intermediate Males*

Occasionally the Hfr chromosome might "reloop" imperfectly, by a crossing-over not coinciding with the points of insertion of the sex factor within the chromosome (Fig. 8B). The released particle would not, in these cases, correspond to the original F, and could include a segment of the bacterial chromosome, forming an F-genote (see Section III,B). It is also possible that, because of a shortened relooping, a sex factor fragment remains inserted in the bacterial chromosome (Fig. 8C), there establishing a sex factor affinity (sfa) region (see Adelberg and Pittard,

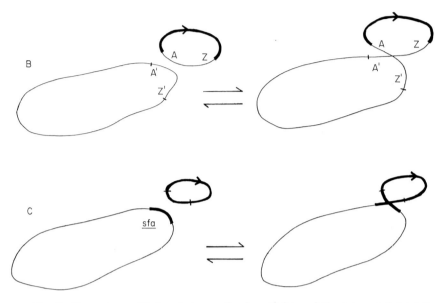

FIG. 9. Dynamic equilibrium between the free (left) and the integrated (right) state of the sex factor in two types of intermediate males. (The equilibrium in the common male is shown in Fig. 8A.) In (B), a chromosomal segment inserted in the F particle (bearing the alleles A and Z) finds a homologous region on the normal chromosome (A′ Z′); in (C), the normal F particle contains a region homologous to an F fragment (sfa) inserted in the chromosome. Integration and release take place through one crossing-over in the homology regions. The mechanisms of insertion of the AZ segment and of the sfa region are depicted in Fig. 8B and C. The arrow heads indicate the origin and sequence of transfer of the F factor which is free or attached to the chromosome.

1965). The result of imperfect relooping would, thus, be either a substituted sex factor bearing a short chromosomal fragment or a peculiar chromosome bearing a sex factor segment. The substituted sex factor, confronted with a standard chromosome, exhibits a net region of homology (Fig. 9B), and should thus readily pair with the host chromosome and become integrated after a crossing-over. With the same readiness it should reloop and be released (Cuzin and Jacob, 1963). This is what happens in the so-called *intermediate males* (Section III,B). The same reasoning can be applied to a standard sex factor (Fig. 9C) confronted with a chromosome bearing a sex factor segment (sfa) (Section III,B).

According to the mentioned model, which was first postulated for phage integration by Campbell (1962), intermediate males differ from the system Hfr $\rightleftharpoons$ F$^+$ in the rate of integration–release (approximately 0.1 and 0.9, respectively, per cell per generation) of the sex factor to and from the chromosome. The higher rates of the two processes, conferring to the intermediate male strains their peculiar instability, are easily explained on the basis of an increased pairing rate determined by the occurrence of extended homology regions in the two DNA structures (see Adelberg and Pittard, 1965).

B. REPLICATION AND TRANSFER OF BACTERIAL CHROMOSOME

1. *The Replicon*

Chromosome transfer by Hfr strains is associated with chromosome replication. Replication has clearly been shown, by autoradiographic observation (Cairns, 1963), to proceed from only one point of the circular chromosome (Fig. 10A) and to continue throughout most of the cell cycle until two new circular structures are formed. The point at which replication begins has been called replicator (Jacob and Brenner, 1963), and it has been assumed that it represents a specific element of recognition of a replication stimulus. Any DNA structure (a chromosome or a plasmid) is capable of independent replication if endowed with a replicator. Such a unit is called the *replicon*. A Hfr chromosome would bear several replicators, one on the F particle and some on the original

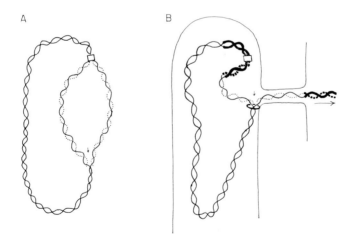

A   B

FIG. 10. Replication and transfer of the bacterial chromosome. (A) The Cairns model of replication. (B) Jacob–Brenner model of chromosome transfer during replication elicited by conjugation. The broken lines represent the newly synthesized DNA strands. The heavy tracts represent the inserted F particle. The small arrows indicate the point at which DNA synthesis is taking place. This point would be the location of the DNA polymerase which is pictured in (B) as a ring fixed to a point in the bacterial membrane. The entire chromosome moves past that point.

chromosome. It is not clear which replicator operates in normally dividing Hfr cells (for a discussion, see Adelberg and Pittard, 1965).

2. *Chromosome Transfer during Replication (Jacob–Brenner Model)*

According to Jacob and Brenner (1963) the event of conjugation would trigger the start of a round of replication beginning at the F replicator (Fig. 10B). One daughter replica is detached from the point of initiation of the replication and penetrates the recipient cell, taking its energy from the process of polymerization of nucleoside triphosphates. This model is supported by the observation that acridine orange, known to block replication of the F element, also prevents chromosome transfer in Hfr $\times$ F$^-$ crosses. The model is also consistent with the observation (Fulton, 1965) that after completion of one round of genetic transfer, donors can initiate a second round. It has also been verified (Jacob *et al.*, 1963) that the DNA transferred to zygotes by males grown on a heavy medium ($^{13}C^{15}N$) and mated in a light medium ($^{12}C^{14}N$) actually contains one light and one heavy strand. This has been recently confirmed by autoradiographic studies (Gross and Caro, 1965).

3. *Chromosome Transfer after Replication (Bouck–Adelberg Model)*

An alternative model of chromosome transfer has been presented by Bouck and Adelberg (1963). They propose that, as a consequence of mating, each Hfr chromosome completes its replication (started at the F replicator), but one of the replicas fails to undergo ring closure and its free end becomes the origin from which transfer starts. This model predicts that transfer proceeds in a sequence opposite to the sequence of replication, in accordance with some observations of Nagata (1963) on synchronously growing Hfr cell cultures.

4. *Association of Sex Factor with Membrane*

How is it that the mobilized chromosome finds its way through the conjugal bridge or pilus? The most reasonable assumption is that the sex factor, whether free or inserted into the chromosome, is constantly associated to the cell membrane (Jacob *et al.*, 1966), producing its specific surface antigen (or pilus) at its site of attachment to the membrane. Thus the conjugal cell union would take place exactly at the point where the sex factor is connected to the membrane (presumably by its replicator) (Jacob *et al.*, 1963).

5. *Episome Transfer*

The Jacob-Brenner model of chromosome transfer (as well as the Bouck–Adelberg model) account very well for the transmission of free sex factors and of F-genotes in the sexduction process (Section III,B). The models concern, indeed, the sex factor transfer, chromosome transfer being simply a consequence of the physical association of the chromosome with the sex factor, to form a single replicon. In sexduction the association would involve only a limited segment of the chromosome.

The leading function of the F factor apparently contradicts the reported observation (Section III,B) that the sex factor is transferred after the very last marker in an Hfr $\times$ F$^-$ cross. This fact can easily be accounted for if one assumes that the sex factor is actually transferred in two pieces: a piece going from the F replicator to the chromosome origin initiates and leads the transfer and the residual piece terminates the transfer. Completion of F factor transfer from Hfr cells is, thus, only achieved after total transfer of chromosomal markers (Fig. 10B). The transfer of colicinogeny factors (Section III,C) and of resistance factors (Section IV,B) is also explained by the reported models, assuming that they are also replicons associated to the cell membrane in correspondence with their specific surface antigens.

V. Fertilization without Cell Contact

Transfer of genetic information from cell to cell may take place in bacteria even in the absence of cell contact. Such a process requires that some vector of genetic information is released from a cell (which invariably dies), floats in the medium, reaches another cell, penetrates its wall, and is eventually incorporated into the recipient genome. This kind of fertilization always concerns minute fragments of the donor cell genome and has many traits in common with a process of "infection." Indeed, it may range from a true viral infection (*lysogenization* and *phage conversion*) to infection by a viral particle carrying a defective phage genome associated with bacterial genes (*transduction*) to the intercellular transmission of a free DNA molecular fragment (*transformation*). These forms of infective inheritance are operationally characterized by the fact that a cell-free filtrate of a donor strain contains some filtrable "principle" which can carry out specific genetic modification of cells of a donor strain after addition to their culture. The principle may be specifically inactivated by DNase in transformations or by the corresponding phage antibodies in phage-mediated conversions or transductions.

A. TRANSFORMATION

1. *Deoxyribonucleic Acid as Transforming Principle*

Bacterial transformation is the process by which a free DNA particle released from a donor cell is introduced into a recipient cell and is integrated within its genome. It was the first phenomenon of transfer of genetic information to be described in bacteria (Griffith, 1928). The identification of the transforming principle as DNA (Avery *et al.*, 1944) marked a milestone not only in bacterial genetics, but in the whole field of general biology. Recent and comprehensive reviews on bacterial transformation have been presented by Ravin (1961), Schaeffer (1964), and by Spizizen *et al.* (1966). We shall only refer to transformation as a peculiar fertilization process, involving the most elementary fertilizing procedure, i.e., the transfer to a recipient cell of genetic information in its pure molecular form. Transformation has been described in species of the bacterial genera *Pneumococcus* (*Diplococcus*), *Hemophilus*, *Streptococcus*, *Neisseria*, *Bacillus*, *Rhizobium*, *Azotobacter*, etc. Interspecific and intergeneric transformations have also been reported even between unrelated genera such as *Neisseria* (a coccus) and *Moraxella* (a rod) (Catlin, 1964). Transformation is routinely obtained by extracting DNA from the cells of a culture of a donor strain and adding the purified DNA to the culture of a differently marked strain. The recipient culture is then diluted and plated to detect the possible occurrence of a donor marker in some cellular clones (see Marmur, 1961). Only one unlinked marker per cell is transferred, as a rule, to a fraction of the recipient cells. Selective procedures are usually adopted to detect transformant clones. No sexual differentiation of the parental strains is required, and the two strains can, in principle, be freely interchanged.

2. Competence

The uptake and penetration of DNA is only possible when cells of the strain acting as recipient attain a stage of "competence." In a synchronously growing culture of *Diplococcus*, "waves" of competence have been observed, thus showing that competence arises at a specific phase of cell growth (Hotchkiss, 1954). In *Bacillus subtilis* the competence state appears to be associated with certain phases of the process leading to sporulation, certain asporogenous strains being unable to take up DNA (Reilly and Spizizen, 1965). Competence formation involves synthesis of a competence receptor (Hotchkiss, 1954). A specific antigen essential for DNA uptake has been suggested in *Diplococcus*, whereas extracellular competence activators have been detected in *Bacillus subtilis*, *Bacillus cereus*, and *Diplococcus*. Competence development is inhibited by periodate, possibly affecting the DNA-binding site (Ranhand and Lichstein, 1966). It has been suggested that one aspect of competence may involve the attachment of DNA to a specific protein (Spizizen *et al.*, 1966). Inhibition of protein synthesis, as obtained by chloramphenicol treatment (Fox and Hotchkiss, 1957) or by starvation of any of the amino acids, prevents the rise of competence. Protein synthesis may also be essential for the formation of an autolytic enzyme, producing certain wall changes (Young and Spizizen, 1963) required for DNA penetration. Cell wall changes are certainly involved in competence formation (reduction in a wall peptide, decrease of net negative charge, permeability changes, etc.). Local wall lysis leading to partial protoplasts has also been suggested (Thomas, 1955).

3. Deoxyribonucleic Acid Penetration

Energy is required for DNA attachment and penetration. Metabolic inhibitors, such as dinitrophenol and cyanide, prevent DNA uptake (Young and Spizizen, 1961) and an energy source such as glucose must be present. A minimal molecular weight of DNA is necessary for its uptake. It has been estimated as $1-2 \times 10^6$, but more recent experiments indicate minimal values higher than 10^7 (see Spizizen *et al.*, 1966). Some recent data (Strauss, 1965, 1966) suggest that the molecule of transforming DNA penetrates linearly and progressively in the recipient cell. Only double-stranded DNA molecules appear to be able to penetrate the cell wall. Heat denaturation causes a rapid loss of uptake ability, the critical temperature varying for different markers (Roger and Hotchkiss, 1961).

Soon after irreversible uptake, DNA is converted to a single-stranded stage (*eclipse period*) which couples with the double-stranded resident DNA. It seems that only a unique strand of donor DNA may be incorporated in the recipient genome (Guild and Robison, 1963). Complete integration of donor DNA fragments into resident DNA appears to be preceded by formation of a transient, perhaps triple-stranded complex. A model for strand interactions and recombination in transformation has been presented by Lacks (1966).

Two or more closely linked markers can occasionally be borne by a single molecular fragment of DNA and be jointly transferred (*cotransfer*). Up to thirteen genetic loci have been shown to occur on a single transforming molecule (Nester *et al.*, 1963). When DNA is isolated under specially mild conditions, markers normally found to be unlinked may be cotransferred (Kelly and Pritchard, 1965). At any rate, transformation appears to be the fertilization process with the lowest upper limit of transferable molecular size.

4. Transfection

Bacterial cells can be infected by DNA isolated from a virus (see Section V,B), and the production of complete viruses can result (Tagaki and Ikeda, 1962). This process, known as *transfection*, requires penetration but not integration of the foreign DNA into the recipient cell. Trans-

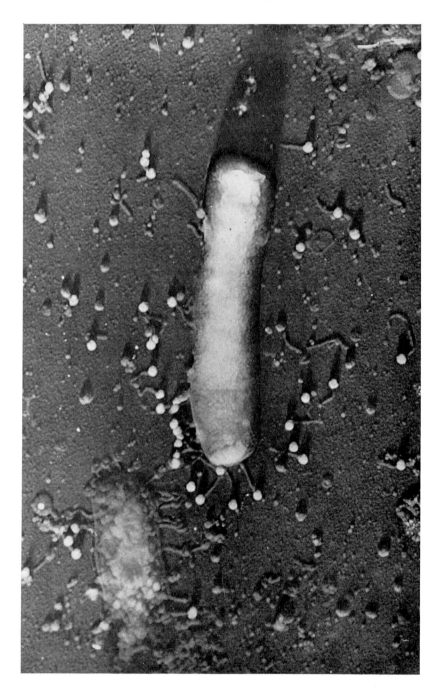

fection is characterized by a low absolute efficiency of infection, possibly because of the relatively large size of the molecule of the phage DNA that must enter the cell. A recent review comprising bacterial transfection has been prepared by Spizizen and collaborators (1966).

B. LYSOGENIZATION AND PHAGE CONVERSION

As we have already argued (Section I), virus infection deserves consideration in the context of bacterial fertilization, because it involves transfer of genetic information, encoded in DNA molecules, from one bacterial cell to another (see, also, Simon and Anderson in Chapter 1 of this volume). Infection by temperate viruses appears to be especially pertinent, the transferred information being integrated within the bacterial chromosome.

1. *Life Cycles of Bacteriophage*

A bacterial virus or *bacteriophage* (also called a *phage*), is a submicroscopic particle made up by a DNA core (the F^+-specific phages contain RNA; see Section II,B) surrounded by a protein sheath endowed with a hollow tail, by which the virus makes contact with specific receptors located on the surface of a sensitive bacterium (for a treatise on bacterial viruses, see Hayes, 1964; also see Simon and Anderson in Chapter 1 of this volume). After its attachment to the bacterial wall and the local lysis of the latter, the virus injects its DNA (consisting of a single circular "chromosome" about one-hundredth the size of the bacterial genome) into the cell cytoplasm. The injected DNA may thus follow two alternative pathways: it either undergoes rapid replication at the expense of the bacterial cell, eventually leading to the formation of mature particles which are released by the bursting cell (*lytic cycle*) (Fig. 11), or it can be attached or inserted into the bacterial chromosome (*lysogenic cycle*). A mechanism strictly comparable to sex factor integration (Section IV,A) has been postulated for phage integration (Campbell, 1962). As to which pathway is undertaken depends on the genetic properties of the phage, as well as on the host cell and environmental conditions. Some phages (*virulent phages*) may only follow the lytic cycles, others (*temperate phages*) can enter either the lysogenic cycle or the lytic, according to the physiological conditions of the host. Virulent mutants of temperate phages can easily be selected.

FIG. 11. An electron micrograph showing an intact bacterium (center) assailed by phage particles attached by their tails to the cell wall and, in the lower left corner, a lysed bacterium with ghost phages attached to its empty wall. (After Penso, 1953, by permission of the author.)

The dormant state of the phage in the lysogenic cell is due to specific repressor molecules genetically coded by a gene of the phage itself (c_1). The repressor prevents the expression of the phage genome except for the c_1 gene and confers immunity from outside infection by related phages. A mutation in the c_1 gene converts the temperate phage into virulent. The repressor protein of the λ phage has been recently isolated by Ptashne (1967) who also showed that it can bind to λ DNA.

2. Lysogenization

The phage DNA inserted within the bacterial chromosome is called *prophage* and the carrier bacterium is called *lysogenic* (see Jacob and Wollman, 1959). *Lysogenization* is the process by which a bacterium becomes a prophage carrier following infection. A prophage can occasionally be released from the bacterial chromosome—very likely by a process comparable to sex factor release (see Section IV,A)—and start a lytic cycle ending with the bacterial burst and the liberation of a crop of mature phage particles. Temperate phages have been classified as episomic elements (Jacob and Wollman, 1958).

The requirement of crossing-over for prophage release is supported by the observation that lysogenic recombination-deficient mutants of *Escherichia coli* fail to release phages either spontaneously or after ultraviolet irradiation (Hertman and Luria, quoted by Hertman, 1967).

Thus a lysogenic culture always generates free phages (hence the name), each bacterium being a potential phage releaser. Prophage release and consequent bacterial lysis and phage production can be greatly stimulated by exposing lysogenic cells to small doses of ultraviolet light (or to many other mutagens), a process known as phage (or prophage) *induction* (see Jacob and Wollman, 1959). There is good evidence that prophage induction results from the destruction of a cytoplasmic immunity substance, or *repressor*, produced by the prophage itself. In Section II,C, it was shown that conjugal transfer of a prophage to an uninfected cell leads to phage induction (*zygotic induction*), very likely because a prophage is abruptly brought into a repressor-free cytoplasm (Fig. 5). Prophage transfer by Hfr cells follows the pattern of that of any chromosomal marker, each prophage having a time of entry peculiar to it on any particular Hfr strain.

3. Phage Conversion

A lysogenized bacterial clone acquires a number of inheritable characters, including the ability to release a phage and immunity from lytic infection by phages of the type it carries. In some cases, lysogenization may produce the appearance in the lysogenic bacterium of structures or functions which are typically bacterial. Thus toxinogenicity of *Coryne-*

bacterium diphtheriae is the direct result of lysogenization by a particular phage (Freeman, 1951), and the production of certain somatic antigens by strains of *Salmonella* is elicited by a group of temperate phages (Uetake *et al.*, 1955). These processes were originally called *lysogenic conversion* (for a review, see Barksdale, 1959). When it was later discovered that even virulent mutants of converting phages can induce the same syntheses, although only for a short time in cells destined to be lysed, the new, more general term *phage conversion* has been coined. Transfer of a toxinogenic factor in *Corynebacterium* has been recently attributed to the phage-mediated transduction (see next section) of a plasmid (Rajaahyaksha and Srinivase Rao, 1965).

4. *Phage Infection and Fertilization*

Phage conversion emphasizes the substantial affinity between the prophage and a segment of bacterial chromosome. It has been speculated that phages may have evolved from normal chromosomal DNA (Jacob *et al.*, 1960). As a matter of fact, several types of defective prophages are known which can no longer give rise to mature phages and may only be detected under very special conditions. They can hardly be distinguished from a bacterial DNA segment. Phage infection can thus be compared to a process of fertilization, in which genetic information is transferred from the donor (lysed) cell to the recipient one, through the medium, within a protein coat, resembling in some respects a sperm cell that carries only a minute fraction of the donor's hereditary material. This similarity becomes still more cogent if we consider the process of transduction.

C. TRANSDUCTION

By *transduction* is meant the transfer of a small fragment of bacterial chromosome from cell to cell, without cell contact, by means of a phage vector (for reviews on transduction, see Clowes, 1960; Hartman, 1963; Campbell, 1964). The transduced fragment is a "passenger" carried by the virus vector, in place of part of the phage genome. The segment of bacterial DNA that a transducing phage can carry may be smaller or larger than the missing phage DNA. The carried segment is, in any case, continuous—markers not closely linked never being cotransduced. Larger pieces of chromosome are transduced by phage-mediated transduction than by transformation (Takahashi, 1966).

1. *Generalized and Localized Transduction*

Two distinct types of transduction are known, the so-called *unrestricted* or *generalized* (or *common*) transduction, discovered by Zinder and Lederberg (1952) in the phage P22–*Salmonella* system, and the *restricted* or *localized transduction* discovered by Morse (1954) in the phage λ–*Escherichia coli* K12 system. In generalized transduction, any marker can be transmitted singly by vector phage, whereas in localized transduction, only the limited region of bacterial chromosome adjacent to the point of insertion of the prophage can be transduced by phages obtained by induction of a lysogenic culture. Generalized transduction can also be carried out by phages emerging from lytic infection. In both types of transduction, about one transductant for any particular marker is found for every million infectious particles.

The two types of transduction are also differentiated by a striking feature. Generalized transduction gives rise to stable recombinants carrying an exogenous marker (as in transformation; see Section V,A); localized transduction gives rise to unstable clones (*heterogenotes*), the exogenous marker not having replaced the resident allele, but having been added near it, at a homologous site (as in sexduction; see Section III,B). When cultures of heterogenotes are induced by ultraviolet light, it has been determined that the phage lysates are able to transduce the mobilized marker at a rate of about one transductant per two particles. Such phage lysates are termed "high frequency of transduction" (Hft).

2. The Transducing Molecule

In restricted transduction the linkage between phage DNA and the coupled marker, the occurrence of Hft, as well as other evidence which cannot be discussed here (see the mentioned reviews) indicate that the transducing particles consist of a single continuous DNA molecule (improperly called "hybrid") comprising the phage genome and the bacterial genes, covered by a protein coat. It has been assumed that the transducing molecule is released from the chromosome–prophage system by an imperfect relooping (unequal crossing-over) liberating a DNA segment which includes, together with the prophage genome, some connected bacterial genes (Campbell, 1962). The same model has been extended to the formation of F-genotes (see Fig. 8C and Section IV,A). A constant feature of the transducing molecule is that the phage genome is regularly *defective* for one-quarter to one-third of its length. In the λ-mediated transduction of the gal⁺ marker, the transducing particle is termed λdg.

Quite a different model has been proposed for generalized transduction. It has been assumed (Clowes, 1958) that the formation of the transducing particle was the result of the accidental incorporation into the phage coat of a small fragment of bacterial chromosome broken during the lytic cycle. However, the prevailing view is now, that the transducing particles giving restricted and unrestricted transduction, both consisting of a continuous segment of DNA comprising a defective phage genome and some bacterial genes, stem from a similar process. In generalized transduction the phage genome is supposed to have affinities for many regions of the bacterial chromosome with which it can mate during vegetative growth (generalized transduction is mediated by phages not having a mappable chromosomal location). It has, indeed, been shown that the transducing particles in unrestricted transduction regularly carry a defective phage genome together with bacterial genes. Under special conditions (Luria *et al.*, 1958), Hft preparations have been also obtained with phages usually amenable to generalized transduction.

The different fates of the transducing particles in the two types of transduction (integration or heterogenote formation) would depend on the relative contribution of the defective phage genome and of bacterial genes to the particle. In localized transduction the phage genome is more complete retaining many of its functions which may interfere with integration. In generalized transduction, most functions of the phage are lost and a relatively more extended bacterial genome fragment would provide for a region of homology favoring integration with the recipient chromosome (Luria *et al.*, 1960).

Transduction (mainly generalized) has been reported in *Salmonella, Escherichia coli, Shigella, Pseudomonas, Staphylococcus, Proteus,* and *Bacillus subtilis.* Interspecific and intergeneric transductions have also been reported.

3. *Evolutionary Origin of Phages and Plasmids*

The process of transduction not only illustrates the possible function of a phage as vector of genetic information from cell to cell, by a peculiar fertilization procedure, but it also shows that bacterial genes can occasionally be picked up by the viral genome and become intimately and stably associated with it. The reverse can also occur—genes from viruses can become integrated into the bacterial chromosome as a genuine addition to the bacterial genome. These assumptions are of great evolutionary significance providing models for the development and the mutual contribution of bacterial and viral genomes. We have already mentioned the hypothesis of the possible chromosomal origin of the bacterial viruses which may be generalized assuming that viruses (including RNA viruses) originate from bacterial elements. The hypothesis of the possible viral origin of bacterial plasmids has also been suggested. A defective prophage particle devoid of viral properties except those involving autonomous replication would not be distinguishable from a plasmid (Luria *et al.*, 1960). The F particle, colicinogeny determinants, and resistant transfer factors have been compared to highly defective phages.

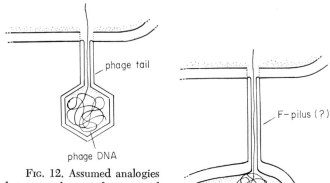

Fig. 12. Assumed analogies between phage infection and F transfer. Both structures can carry an associated bacterial deoxyribonucleic acid.

They do not elaborate a viral coat for a free intercellular transfer, and may have replaced this structure by a surface antigen synthesized on the host cell wall (F-pilus?) which promotes cell conjugation and permits their intercellular transfer (Hayes, 1966; Fig. 12).

VI. Genetic Recombination in Merozygotes

The formation of an incomplete zygote (*merozygote*) is the final result of any type of fertilization in bacteria. Following the transfer of donor DNA to a recipient cell, a merozygote is formed, made up by the complete resident genome and by a more-or-less large continuous segment of the exogenous genetic complement. The size of the donor fragment may be less that one-hundredth of the bacterial chromosome (as in transformation; Section V,A), may reach 2 or 3 times this amount in transduction (Section V,C), may rise to up to one-tenth of the host genome in sexduction (Section III,B), and vary to around one-third of it after conjugation (Section II), in extreme cases amounting to an entire complement. Rare instances of incomplete recipient contribution to the zygote have been reported in *Salmonella* (Smith and Stocker, 1962), *Serratia* (Belser and Bunting, 1956), and *Streptomyces* (Sermonti *et al.*, 1966). After transformation the biparental association is limited to a three-stranded hybrid DNA structure confined to a minute chromosome segment (Section V,A).

A. DIFFERENT FATES OF EXOGENOUS DEOXYRIBONUCLEIC ACID

The exogenous DNA, whether of chromosomic or extrachromosomic origin, can be faced with different destinies, after fertilization, depending on the degree of its homology with the resident genome and on its ability to self-reproduce autonomously (i.e., as a complete replicon; see Section IV,B).

1. Autonomous Replication

A self-reproducing particle, bearing no homology with the host genome, goes on replicating autonomously after infection (in cases such as found after virulent phage infection and after nonepisomic plasmid transfer, the term "fertilization" may not be appropriate).

2. Insertion and Release

When the exogenous DNA bears but little homology with the resident genome and has the nature of a replicon, it may replicate for a number of generations after fertilization and be occasionally inserted within the host chromosome, from which it may later be released. This is the case with the sex factor of *Escherichia coli* K12 (Section III,B), and similar behavior is found in temperate phages, the autonomous replication of which ends, however, in the death of the host cell (Section V,A).

Replicons endowed with greater homology with the host genome are those plasmids (episomes) that have incorporated a segment of the donor genome, e.g., transducing phages in restricted transduction (Section V,C) and F-genotes in intermediate males (Section III,B). These may multiply autonomously for a period of time and be incorporated and released at a much higher rate than the unsubstituted particles. The chromosomic regions of this substituted episome can also undergo a true reciprocal recombination process with the homologous regions of the host genome, not by being inserted but by replacing some endogenous genes with corresponding exogenous genes (Hermann, 1965).

3. Linear Inheritance and Reciprocal Recombination

In the merozygotes found after conjugation in Hfr $\times$ F⁻ crosses of *Escherichia coli*, the contributed donor DNA presents complete homology with the recipient genome but no self-reproducing ability (Jacob and Brenner, 1963). The contributed donor segment fails to multiply in the recipient cell, so that the merozygote is not self-reproduced (Anderson, 1958; Lederberg, 1957). By an even number of crossing-overs, one or more tracts of the donor contribution can replace homologous regions of the circular recipient chromosome (see Jacob and Wollman, 1961). The result would be a recombinant circular structure and a recombinant incomplete segment which is eventually lost. The recombination process is under the control of an enzyme, lacking or defective in some recombination-deficient mutants (Clark and Margulies, 1965). Absence of replication and recombination of the transferred fragment are also assumed to occur after transformation (where the minute donor fragment is reduced to a single DNA strand; see Section V,A) and after unrestricted transduction (Section V,C). Unrestricted transduction may not end in recombination, the exogenous fragment being inherited unilinearly by only one daughter cell at each division (*abortive transduction;* Stocker *et al.,* 1953). Linear inheritance has also been observed after transformation (Iyer, 1965).

B. SEMISTABLE MEROZYGOTES

Cases have been described of relatively stable merozygotes resulting from the conjugation process. In such cases the donor fragment does certainly reproduce for a number of generations. Relatively stable heterozygotes have been observed in crosses of *Escherichia coli* K12, when a particular mutant strain was adopted (Lederberg, 1949; Section II,A). It has been assumed that the locus controlling diploid persistence might correspond to a replicator (Jacob and Brenner, 1963). Relatively stable merozygotes are the rule in *Streptomyces* crosses (Sermonti *et al.,* 1966), where the donor fragment is able to self-reproduce for several generations giving rise to heterozygous clones (*heteroclones*) (Hopwood *et al.,* 1963).

Rather stable partial diploids of *Escherichia coli* have recently been demonstrated in matings involving strains with some aberration in the pro-1 pro-2 region of the chromosome (Curtiss, 1964). It was proposed that the "exogenote" was an independent chromosomal segment, associated noncovalently with the complete chromosome, in such a way as to be always replicated with it. A definite structural model for these self-replicating merozygotes has still to be provided (for a discussion, see Hopwood, 1967).

REFERENCES

Adelberg, E. A., and Burns, S. N. (1960). Genetic variation in the sex factor of *Escherichia coli*. *J. Bacteriol.* **79,** 32.

Adelberg, E. A., and Pittard, J. (1965). Chromosomal transfer in bacterial conjugation. *Bacteriol. Rev.* **29,** 161.

Anderson, T. F. (1958). Recombination and segregation in *Escherichia coli*. *Cold Spring Harbor Symp. Quant. Biol.* **23,** 97.

Anderson, T. F., Wollman, E. L., and Jacob, F. (1957). Sur les processus de conjugaison et de recombinaison chez *E. coli*. III. Aspects morphologiques en microscopie électronique. *Ann. Inst. Pasteur* **93,** 950.

Avery, O. T., Macleod, C. M., and McCarty, M. (1944). Studies on the chemical nature of the substance inducing transformation of pneumococcal types. I. Induction of transformation by a desoxyribonucleic acid fraction isolated from pneumococcus. III. *J. Exptl. Med.* **79,** 137.

Barksdale, L. (1959). Lysogenic conversions in bacteria. *Bacteriol. Rev.* **23**, 202.

Beadle, G. W., and Tatum, E. L. (1941). Genetic control of biochemical reactions in *Neurospora. Proc. Natl. Acad. Sci. U. S.* **27**, 499.

Belser, W. L., and Bunting, M. I. (1956). Studies on a mechanism providing for genetic transfer in *Serratia marcescens. J. Bacteriol.* **72**, 582.

Bhaskaran, K. (1960). Recombination of characters between mutant stocks of *Vibrio chloerae,* strain 162. *J. Gen Microbiol.* **23**, 47.

Bouck, N., and Adelberg, E. A. (1963). The relationship between DNA synthesis and conjugation in *Escherichia coli. Biochem. Biophys. Res. Commun.* **11**, 24.

Brinton, C. C., Jr., Gemski, P., Jr., and Carnahan, J. (1964). A new type of bacterial pilus genetically controlled by the fertility factor F of *E. coli* K12 and its role in chromosome transfer. *Proc. Natl. Acad. Sci. U. S.* **52**, 776.

Cairns, J. (1963). The chromosome of *Escherichia coli. Cold Spring Harbor Symp. Quant. Biol.* **28**, 43.

Campbell, A. (1962). Episomes. *Advan. Genet.* **11**, 101.

Campbell, A. (1964). Transduction. *In* "The Bacteria" (I. C. Gunsalus and R. Y. Stanier, eds.), Vol. V: Heredity, pp. 155–222. Academic Press, New York.

Catlin, B. W. (1964). Reciprocal genetic transformation between *Neisseria catarrhalis* and *Moraxella nonliquefaciens. J. Gen. Microbiol.* **37**, 369.

Cavalli-Sforza, L. L. (1950). Le sessualità nei batteri. *Boll. Ist. Sieroterap. Milan.* **29**, 281.

Cavalli-Sforza, L. L., and Jinks, J. L. (1956). Studies on the genetic system of *E. coli* K12. *J. Genet.* **54**, 87.

Cavalli-Sforza, L. L., Lederberg, J., and Lederberg, E. M. (1953). An infective factor controlling sex compatibility in *Bacterium coli. J. Gen. Microbiol.* **8**, 89.

Clark, A. J., and Adelberg, E. A. (1962). Bacterial conjugation. *Ann. Rev. Microbiol.* **16**, 283.

Clark, A. J., and Margulies, A. D. (1965). Isolation and characterization of recombination deficient mutants of *E. coli* K12. *Proc. Natl. Acad. Sci. U. S.* **53**, 451.

Clowes, R. C. (1958). The nature of the vector involved in bacterial transduction. *7th Intern. Congr. Microbiol., Stockholm,* Abstracts, p. 53.

Clowes, R. C. (1960). Fine genetic structure as revealed by transduction. *Symp. Soc. Gen. Microbiol.* **10**, 92.

Clowes, R. C. (1963). Colicin factors and episomes. *Genet. Res. (Cambridge)* **4**, 162.

Clowes, R. C., Moody, E. E. M., and Pritchard, R. H. (1965). The elimination of extrachromosomal elements in thymineless strains of *Escherichia coli* K12. *Genet. Res. (Cambridge)* **6**, 147.

Crawford, E. M., and Gesteland, R. F. (1964). The adsorption of bacteriophage R17. *Virology* **22**, 165.

Curtiss, R. (1964). A stable partial diploid strain of *Escherichia coli. Genetics* **50**, 679.

Cuzin, F., and Jacob, F. (1963). Intégration réversible de l'épisome sexuel F′ chez *Escherichia coli* K12. *Compt. Rend. Acad. Sci.* **257**, 795.

Datta, N., Lawn, A. M., and Meynell, E. (1966). The relationship of F-type piliation phage sensitivity to drug resistance transfer in R⁺F⁻ *Escherichia coli* K12. *J. Gen. Microbiol.* **45**, 365.

Davis, B. D. (1950). Non-filterability of the agents of genetic recombination in *E. coli. J. Bacteriol.* **60**, 507.

de Haan, P. G., and Gross, J. D. (1962). Transfer delay and chromosome withdrawal during conjugation in *Escherichia coli. Genet. Res.* (*Cambridge*) 3, 251.

Dettori, R., Maccacaro, G. A., and Piccinin, G. L. (1961). Sex-specific bacteriophages of *Escherichia coli* K12. *Giorn. Microbiol.* 9, 141.

Driskell-Zamenhof, P. J. (1964). Bacterial episomes. In "The Bacteria" (I. C. Gunsalus and R. Y. Stanier, eds.), Vol. V: Heredity, pp. 155–222. Academic Press, New York.

Driskell-Zamenof, P. J., and Adelberg, E. A. (1963). Studies on the chemical nature and size of sex factors of *Escherichia coli* K12. *J. Mol. Biol.* 6, 483.

Falkow, S., and Baron, L. S. (1962). Episomic element in a strain of *Salmonella typhosa. J. Bacteriol.* 84, 581.

Fischer-Fantuzzi, L., and Di Girolamo, M. (1961). Triparental matings in *Escherichia coli. Genetics* 46, 1305.

Fisher, K. W. (1961). Environmental influence on genetic recombination in bacteria and their viruses. *Symp. Soc. Gen. Microbiol.* 11, 272.

Fox, M. S., and Hotchkiss, R. D. (1957). Initiation of bacterial transformation. *Nature* 179, 1322.

Fredéricq, P. (1957). Colicins. *Ann. Rev. Microbiol.* 11, 7.

Freeman, V. J. (1951). Studies on the virulence of bacteriophage-infected strains of *Corynebacterium diphtheriae. J. Bacteriol.* 61, 675.

Fuerst, C. R., Jacob, F., and Wollman, E. L. (1956). Déterminations de liaisons génétique chez *E. coli* K12, à l'aide du radio phosphore. *Compt. Rend. Acad. Sci.* 243, 2162.

Fulton, C. (1965). Continuous chromosome transfer in *Escherichia coli. Genetics* 52, 55.

Garen, A., and Skaar, P. (1958). Transfer of phosphorus-containing material associated with mating in *Escherichia coli. Biochim. Biophys. Acta* 27, 457.

Gowen, J. W., and Lincoln, R. E. (1942). A test for sexual fusion in bacteria. *J. Bacteriol.* 44, 551.

Griffith, F. (1928). Significance of pneumococcal types. *J. Hyg.* 27, 113.

Gross, J. D. (1964). Conjugation in bacteria. In "The Bacteria" (I. C. Gunsalus and R. Y. Stanier, eds.), Vol. V: Heredity, pp. 1–48. Academic Press, New York.

Gross, J. D., and Caro, L. (1965). Genetic transfer in bacterial mating. *Science* 150, 1679.

Guild, W. R., and Robison, M. (1963). Evidence for message reading from a unique strand of pneumococcal DNA. *Proc. Natl. Acad. Sci. U. S.* 50, 106.

Gunderson, W. B., Jyssum, K., and Lie, S. (1962). Genetic instability with episome-mediated transfer in *Escherichia coli. J. Bacteriol.* 83, 616.

Gunsalus, I. C., and Stanier, R. Y. (eds.) (1964). Heredity. "The Bacteria," Vol. V. Academic Press, New York.

Hartman, P. E. (1963). Methodology in transduction. In "Methodology in Mammalian Genetics" (W. J. Burdette, ed.), p. 103. Holden-Day, San Francisco, California.

Hashimoto, H., and Hirota, Y. (1966). Gene recombination and segregation of resistance factor R in *Escherichia coli. J. Bacteriol.* 91, 51.

Hayes, W. (1952). Recombination in *Bacterium coli* K12: unidirectional transfer of genetic material. *Nature* 169, 118.

Hayes, W. (1953). The mechanism of genetic recombination in *E. coli. Cold Spring Harbor Symp. Quant. Biol.* 18, 75.

Hayes, W. (1957). The kinetics of the mating process in *E. coli. J. Gen. Microbiol.* 16, 97.

Hayes, W. (1964). "The Genetics of Bacteria and their Viruses. Studies on Basic Genetics and Molecular Biology." Blackwell, Oxford, England.

Hayes, W. (1966). Sex factors and viruses. *Proc. Roy. Soc.* **B164**, 230.

Hermann, R. K. (1965). Reciprocal recombination of chromosome and F-merogenote in *Escherichia coli*. *J. Bacteriol.* **90**, 1664.

Hermann, R. K., and Forro, F., Jr. (1964). Autoradiographic study of transfer of DNA during bacterial conjugation. *Biophys. J.* **4**, 335.

Hertman, I. M. (1967). Isolation and characterization of a recombination-deficient Hfr strain. *J. Bacteriol.* **93**, 580.

Hirota, Y. (1960). The effect of acridine dyes on mating type factors in *Escherichia coli*. *Proc. Natl. Acad. Sci. U. S.* **46**, 57.

Hirota, Y., and Sneath, P. H. A. (1961). F' and F mediated transduction in *Escherichia coli* K12. *Nippon Idengaku Zasshi* **36**, 307.

Holloway, B. W. (1956). Self-fertility in *Pseudomonas aeruginosa*. *J. Gen. Microbiol.* **25**, 221.

Holloway, B. W., and Fargie, B. (1960). Fertility factors and genetic linkage in *Pseudomonas aeruginosa*. *J. Bacteriol.* **80**, 362.

Hopwood, D. A. (1965). A circular linkage map in the Actinomycete *Streptomyces coelicolor*. *J. Mol. Biol.* **12**, 514.

Hopwood, D. A. (1966). Lack of constant genome ends in *Streptomyces coelicolor*. *Genetics* **54**, 1177.

Hopwood, D. A. (1967). Genetic analysis and genome structure in *Streptomyces coelicolor*. *Bacteriol. Rev.* **31**, 373.

Hopwood, D. A., Sermonti, G., and Spada-Sermonti, I. (1963). Heterozygous clones in *Streptomyces coelicolor*. *J. Gen. Microbiol.* **30**, 249.

Hotchkiss, R. D. (1954). Cyclical behaviour in pneumococcal growth and transformability occasioned by environmental changes. *Proc. Natl. Acad. Sci. U. S.* **40**, 49.

Ishibashi, M. (1967). F pilus as f⁺ antigen. *J. Bacteriol.* **93**, 379.

Ivanovics, G. (1962). Bacteriocins and bacteriocin-like substances. *Bacteriol. Rev.* **26**, 108.

Iyer, V. N. (1965). Unstable genetic transformation in *Bacillus subtilis* and the mode of inheritance in unstable clones. *J. Bacteriol.* **90**, 495.

Jacob, F., and Adelberg, E. A. (1959). Transfert de caractères génétiques par incorporation au facteur sexuel d'*Escherichia coli*. *Compt. Rend. Acad. Sci.* **249**, 189.

Jacob, F., and Brenner, S. (1963). Sur la régulation de la synthèse du DNA chez les bactéries: hypothèse du replicon. *Compt. Rend. Acad. Sci.* **248**, 3219.

Jacob, F., and Wollman, E. L. (1955). Etapes de la recombinaison génétique chez *E. coli* K12. *Compt. Rend. Acad. Sci.* **240**, 2566.

Jacob, F., and Wollman, E. L. (1956a). Recombinaison génétique et mutants de fertilité chez *E. coli* K12. *Compt. Rend. Acad. Sci.* **242**, 303.

Jacob, F., and Wollman, E. L. (1956b). Sur les processus de conjugaison et de recombinaison génétique chez *E. coli* I. L'induction par conjugaison au induction zygotique. *Ann. Inst. Pasteur* **91**, 486.

Jacob, F., and Wollman, E. L. (1957). Analyse des groupes de liaison génétique de différentes souches donatrices. *Compt. Rend. Acad. Sci.* **245**, 1840.

Jacob, F., and Wollman, E. L. (1958). Les épisomes, éléments génétiques ajoutés. *Compt. Rend. Acad. Sci.* **247**, 154.

Jacob, F., and Wollman, E. L. (1959). Lysogeny. *In* "The Viruses" (F. M. Burnet and W. M. Stanley, eds.), Vol. 2, p. 319. Academic Press, New York.

Jacob, F., and Wollman, E. L. (1961). "Sexuality and the Genetics of Bacteria." Academic Press, New York.

Jacob, F., Siminovitch, L., and Wollman, E. L. (1952). Sur la biosynthèse d'une colicine et sur son mode d'action. *Ann Inst. Pasteur* **83**, 295.

Jacob, F., Schaeffer, P., and Wollman, E. L. (1960). Episomic elements in bacteria, *Symp. Soc. Gen. Microbiol.* **10**, 67.

Jacob, F., Brenner, S., and Cuzin, F. (1963). On the regulation of DNA replication in bacteria. *Cold Spring Harbor Symp. Quant. Biol.* **28**, 329.

Jacob, F., Ryter, A., and Cuzin, F. (1966). On the association between DNA and membrane in bacteria. *Proc. Roy. Soc.* **B164**, 267.

Kelly, M. S., and Pritchard, R. H. (1965). Unstable linkage between genetic markers in transformation. *J. Bacteriol.* **83**, 288.

Lacks, S. (1966). Integration efficiency and genetic recombination in pneumococcal transformation. *Genetics* **53**, 207.

Lawn, A. M. (1966). Morphological features of the pili associated with R^+F^- and R^-F^+ bacteria. *J. Gen. Microbiol.* **45**, 377.

Lederberg, J. (1947). Gene recombination and linked segregations in *Escherichia coli. Genetics* **32**, 505.

Lederberg, J. (1948). Problems in microbial genetics. *Heredity* **2**, 195.

Lederberg, J. (1949). Aberrant heterozygotes in *Escherichia coli. Proc. Natl. Acad. Sci. U. S.* **35**, 178.

Lederberg, J. (1952). Cell genetics and hereditary symbiosis. *Physiol. Rev.* **32**, 403.

Lederberg, J. (1956). Conjugal pairing in *E. coli. J. Bacteriol.* **71**, 497.

Lederberg, J. (1957). Sibling recombinants in zygote pedigrees of *Escherichia coli. Proc. Natl. Acad. Sci. U. S.* **43**, 1060.

Lederberg, J., and Tatum, E. L. (1946a). Novel genotypes in mixed cultures of biochemical mutants of bacteria. *Cold Spring Harbor Symp. Quant. Biol.* **21**, 150.

Lederberg, J., and Tatum, E. L. (1946b). Gene recombination in *E. coli. Nature* **158**, 558.

Lederberg, J., Lederberg, E. M., Zinder, N. D., and Lively, E. R. (1951). Recombination analysis of bacterial heredity. *Cold Spring Harbor Symp. Quant. Biol.* **16**, 413.

Lederberg, J., Cavalli-Sforza, L. L., and Lederberg, E. M. (1952). Sex compatibility in *E. coli. Genetics* **37**, 720.

Loeb, T. (1960). Isolation of a bacteriophage specific for the F^+ and Hfr mating types of *Escherichia coli* K12. *Science* **131**, 932.

Luria, S. E., and Burrous, J. W. (1957). Hybridisation between *Escherichia coli* and *Shigella. J. Bacteriol.* **74**, 461.

Luria, S. E., and Delbrück, M. (1943). Mutations of bacteria from virus sensitivity to virus resistance. *Genetics* **28**, 491.

Luria, S. E., Fraser, D. K., Adams, J. N., and Burrous, J. W. (1958). Lysogenisation, transduction, and genetic recombination in bacteria. *Cold Spring Harbor Symp. Quant. Biol.* **23**, 71.

Luria, S. E., Adams, J. N., and Ting, R. C. (1960). Transduction of lactose-utilising ability among strains of *E. coli* and *S. dysenteriae* and the properties of the transducing phage particles. *Virology* **12**, 348.

Maccacaro, G. A., and Comolli, R. (1956). Surface properties correlated with sex compatibility in *E. coli. J. Gen. Microbiol.* **15**, 121.

Marmur, J. (1961). A procedure for the isolation of deoxyribonucleic acid from micro-organisms. *J. Mol. Biol.* **3**, 208.

Marmur, J., Rownd, R., Falkow, S., Baron, L. S., Schildkraut, C., and Doty, P.

(1961). The nature of intergeneric episomic infection. *Proc. Natl. Acad. Sci. U. S.* 47, 372.

Martin, G., and Jacob, F. (1962). Transfer de l'épisome sexual d'*Escherichia coli*. à *Pasteurella pestis*. *Comp. Rend. Acad. Sci.* 259, 3589.

Meynell, E., and Datta, N. (1966a). The relation of resistance transfer factors to the F factor (sex-factor) of *Escherichia coli* K12. *Genet. Res.* (*Cambridge*) 7, 139.

Meynell, E., and Datta, N. (1966b). The nature and incidence of conjugation factors in *Escherichia coli*. *Genet. Res.* (*Cambridge*) 7, 141.

Miyake, T. (1962). Exchange of genetic material between *Salmonella typhimurium* and *Escherichia coli* K12. *Genetics* 47, 1043.

Morse, M. L. (1954). Transduction of certain loci in *Escherichia coli* K12. *Genetics* 39, 984.

Nagata, T. (1963). The molecular synchrony and sequential replication of DNA in *Escherichia coli*. *Proc. Natl. Acad. Sci. U. S.* 49, 551.

Nelson, T. C. (1951). Kinetics of genetic recombination in *E. coli*. *Genetics* 36, 162.

Nester, E. W., Schafer, M., and Lederberg, J. (1963). Gene linkage in DNA transfer: a cluster of genes concerned with aromatic biosynthesis in *Bacillus subtilis*. *Genetics* 48, 529.

Novick, R. P. (1963). Analysis by transduction of mutations affecting penicillinase formation in *Staphylococcus aureus*. *J. Gen. Microbiol.* 33, 121.

Ørskov, I., and Ørskov, F. (1960). An antigen termed f^+ occurring in F⁺ *E. coli* strains. *Acta Pathol. Microbiol. Scand.* 48, 37.

Ozeki, H., and Howarth, S. (1961). Colicine factors as fertility factors in bacteria: *Salmonella typhimurium*, strain LT2. *Nature* 190, 986.

Penso, G. (1953). Le cycle de dévelopement d'un mycobacteriophage dans sa cellule hôte. *Rend. Ist. Super. Sanità.* Suppl. 1, p. 58.

Ptashne, M. (1967). Isolation of the λ phage repressor. *Proc. Natl. Acad. Sci. U. S.* 57, 306.

Rajaahyaksha, A. B., and Srinivase Rao, S. (1965). The role of phage in the transduction of the toxinogenic factor in *Corynebacterium diphtheriae*. *J. Gen. Microbiol.* 40, 421.

Ranhand, J. M., and Lichstein, H. C. (1966). Periodate inhibition of transformation and competence development in *Haemophilus influenzae*. *J. Bacteriol.* 92, 956.

Ravin, A. W. (1961). The genetics of transformation. *Advan. Genet.* 10, 61.

Reeves, P. (1960). Role of Hfr mutants in F⁺ × F⁻ crosses in *E. coli* K12. *Nature* 185, 265.

Reilly, B. E., and Spizizen, J. (1965). Bacteriophage deoxyribonucleate infection of competent *Bacillus subtilis*. *J. Bacteriol.* 89, 782.

Roger, M., and Hotchkiss, R. D. (1961). Selective heat inactivation of pneumococcal transforming deoxyribonucleate. *Proc. Natl. Acad. Sci. U. S.* 47, 653.

Rothfels, K. H. (1952). Gene linearity and negative interference in crosses of *Escherichia coli*. *Genetics* 37, 297.

Sanderson, K. E., and Demerec, M. (1965). The linkage map of *Salmonella typhimurium*. *Genetics* 51, 897.

Schaeffer, P. (1964). Transformation. *In* "The Bacteria" (I. C. Gunsalus and R. Y. Stanier, eds.), Vol. V: Heredity, pp. 87–153. Academic Press, New York.

Sermonti, G., and Hopwood, D. A. (1964). Genetic recombination in *Streptomyces*. *In* "The Bacteria" (I. C. Gunsalus and R. Y. Stanier, eds.), Vol. V: Heredity, pp. 223–251. Academic Press, New York.

Sermonti, G., Bandiera, B., and Spada-Sermonti, I. (1966). New approach to the genetics of *Streptomyces coelicolor*. *J. Bacteriol.* 91, 384.

Sherman, J. M., and Wing, H. U. (1937). Attempts to reveal sex in bacteria; with some light on fermentative variability in the coli-aerogenes group. J. Bacteriol. 33, 315.

Silver, S. D. (1963). Transfer of material during mating in Escherichia coli. Transfer of DNA and upper limits on the transfer of RNA and protein. J. Mol. Biol. 6, 349.

Silver, S. D., and Ozeki, H. (1962). Transfer of deoxyribonucleic acid accompanying the transmission of colicinogenic properties by cell mating. Nature 195, 873.

Smith, S. M., and Stocker, B. A. D. (1962). Colicinogeny and recombination. Brit. Med. Bull. 18, 45.

Smith, S. M., Ozeki, H., and Stocker, B. A. D. (1963). Transfer of colE1 and colE2 during high frequency transmission of coli in Salmonella typhimurium. J. Gen. Microbiol. 33, 231.

Sneath, P. H. A., and Lederberg, J. (1961). Inhibition by periodate of mating in Escherichia coli K12. Proc. Natl. Acad. Sci. U. S. 47, 86.

Spizizen, J., Reilly, B. E., and Evans, A. H. (1966). Microbial transformation and transfection. Ann. Rev. Microbiol. 20, 371.

Stocker, B. A. D., Zinder, N. D., and Lederberg, J. (1953). Transduction of flagellar characters. J. Gen. Microbiol. 9, 410.

Strauss, N. (1965). Configuration of transforming deoxyribonucleic acid during entry into Bacillus subtilis. J. Bacteriol. 89, 288.

Strauss, N. (1966). Further evidence concerning the configuration of transforming deoxyribonucleic acid during entry into Bacillus subtilis. J. Bacteriol. 91, 702.

Sugino, Y., and Hirota, Y. (1962). Conjugal fertility associated with resistance-factor R in Escherichia coli. J. Bacteriol. 89, 902.

Tagaki, J., and Ikeda, Y. (1962). Transformation of genetic traits by the DNA prepared from a lysogenic phage S-1. Biochem. Biophys. Res. Commun. 7, 482.

Takahashi, I. (1966). Joint transfer of genetic markers in Bacillus subtilis. J. Bacteriol. 91, 101.

Tatum, E. L. (1945). X-ray induced mutant strains of E. coli. Proc. Natl. Acad. Sci. U. S. 31, 215.

Tatum, E. L., and Lederberg, J. (1947). Gene recombination in the bacterium Escherichia coli. J. Bacteriol. 53, 673.

Thomas, R. (1955). Recherches sur la cinétique des transformations bactériennes. Biochim. Biophys. Acta. 18, 467.

Turri, M., and Maccacaro, G. A. (1960). Osservazioni microelettroforetiche su cellule di E. coli. K12 di diversa compatibilita sessuale. Giorn. Microbiol. 8, 1.

Uetake, H., Nakagawa, T., and Akiba, T. (1955). The relationship of bacteriophage to antigenic change in group E Salmonella. J. Bacteriol. 69, 571.

Watanabe, T. (1963). Infectious heredity of multiple drug resistance in bacteria. Bacteriol. Rev. 27, 87.

Watanabe, T., and Fusakawa, T. (1961). Episome-mediated transfer of drug resistance in Enterobacteriaceae. II. Elimination of resistance factors with acridines. J. Bacteriol. 81, 679.

Watanabe, T., and Ogata, C. (1966). Episome-mediated transfer of drug-resistance in Enterobacteriaceae. IX. Recombination of an R factor with F. J. Bacteriol 91, 43.

William Smith, H., and Halls, S. (1967). The transmissible nature of the genetic factor in Escherichia coli that controls haemolysin production. J. Gen. Microbiol. 47, 153.

Wollman, E. L., and Jacob, F. (1955). Sur le mécanisme du transfert de matériel génétique au cours de la recombinaison chez *E. coli*. K12. *Compt. Rend. Acad. Sci.* **240**, 2443.

Wollman, E. L., and Jacob, F. (1958). Sur les processus de conjugaison et de recombinaison chez *E. coli*. V. Le mécanisme du transfert de matériel génétique. *Ann. Inst. Pasteur* **95**, 641.

Wollman, E. L., Jacob, F., and Hayes, W. (1956). Conjugation and genetic recombination in *Escherichia coli*. *Cold Spring Harbor Symp. Quant. Biol.* **21**, 141.

Young, F. E., and Spizizen, J. (1961). Physiological and genetic factors affecting transformation of *Bacillus subtilis*. *J. Bacteriol.* **81**, 823.

Young, F. E., and Spizizen, J. (1963). Incorporation of deoxyribonucleic acid in *Bacillus subtilis* transformation system. *J. Bacteriol.* **86**, 392.

Zinder, N. D. (1960). Sexuality and mating in *Salmonella*. *Science* **131**, 924.

Zinder, N. D., and Lederberg, J. (1952). Genetic exchange in *Salmonella*. *J. Bacteriol.* **64**, 679.

Addendum

An active role in fertilization of the recipient DNA has been shown in different processes. The Jacob-Brenner model of chromosome transfer in Hfr $\times$ F⁻ conjugation has been substantiated by various lines of evidence (Barbour, 1967; Cuzin and Jacob, 1966, 1967). However, an active role of the F⁻ parent, which was not considered in the mentioned model, is pointed out by some recent data (Curtiss and Charamella, 1966; Curtiss *et al.*, 1968). It appears (Curtiss *et al.*, 1968) that extensive transfer of the donor genome first requires homologous pairing of the lead region of the immigrant chromosome with the comparable region of the recipient chromosome. The F⁻ parent would then wind in the donor chromosome with a resultant expenditure of energy. Only the transfer of the lead region of the donor chromosome requires DNA synthesis in the Hfr parent. Chromosome transfer could then be independent on continual DNA synthesis in the donor (see also Pritchard, 1965).

The F episome is transferred even to "minicells" which do not contain DNA (Cohen *et al.*, 1968). The immigrant DNA appears to consist of single-stranded and double-stranded fractions, suggesting that DNA may be transferred during conjugation as a single strand, as it occurs in transformation.

An active role of the recipient DNA in the entrance of foreign nucleic acids has also been suggested in transformation (Erikson and Braun, 1968). Working with synchronized cells of *Bacillus subtilis*, it has been shown that "the temporal sequence in attainment of the peak of transformation frequency for a given marker corresponded to its known relative location on the linkage map." These results have been interpreted by assuming that integration of DNA will occur at the point of DNA replication (Bodmer, 1966), which takes place at the membrane level. At this point the replicating recipient DNA might drag the homologous transforming DNA into the cell. It has been proposed that all processes of bacterial fertilization may involve entrance of donor DNA through replication sites in the membrane (Erikson and Braun, 1968).

Some recent contributions point out that the transient nature of the partial diploids (merodiploids) formed in bacteria after conjugation does not depend on their intrinsic inability to self-reproduce, but rather to a very efficient recombination mechanism which immediately originates haploid recombinants. When some step of the recombination process is defective, as in Rec⁻ mutants, persistant merodiploids are formed (Low, 1968). Merodiploids also result from intergeneric bacterial matings (i.e. *Escherichia coli* $\times$ *Salmonella* or *E. coli* $\times$ *Proteus mirabilis*) as a consequence of a restricted homology between the immigrant and the resident DNA's, or conceivably because of other factors as the specificity of recombination enzymes (Baron *et al.,* 1968). It has been assumed that the exogenous DNA may be tandemly inserted into the resident chromosome and replicated as part of a single DNA duplex (Baron *et al.,* 1968), as also postulated by Hopwood (1967) for *Streptomyces* merodiploids. Such mechanism of insertion is comparable to the Campbell (1962) model of prophage integration (see Fig. 8A).

REFERENCES TO ADDENDUM

Barbour, S. D. (1967). Effect of nalidixic acid on conjugational transfer and expression of episomal *lac* genes in *Escherichia coli* K12. *J. Mol. Biol.* **28**, 373.

Baron, L. S., Gemski, P., Jr., Johnson, E. M., and Wohlhieter, J. A. (1968). Intergeneric bacterial mating. *Bacteriol. Rev.* **32**, 362.

Bodmer, W F. (1966). Integration of deoxyribonuclease-treated DNA in *Bacillus subtilis* transformation: Involvement of DNA synthesis. *J. Mol. Biol.* **14**, 534.

Cohen, A., Fisher, W. D., Curtiss, R., and Adler, H. I. (1968). DNA isolated from *Escherichia coli* minicells after mating with F⁺ cells. *Proc. Natl. Acad. Sci. U. S.* **61**, 61.

Curtiss, R., and Charamella, L. J. (1966). Role of the F⁻ parent during bacterial conjugation in *Escherichia coli*. *Genetics* **54**, 329.

Curtiss, R., Charamella, L. J., Stallions, D. R., and Mays, J. A. (1968). Parental functions during conjugation in *Escherichia coli* K12. *Bacteriol. Rev.* **32**, 320.

Cuzin, F., and Jacob, F. (1966). Inhibition par les acridines du transfert génétique par les souches donatrices d'*Escherichia coli* K12. *Ann. Inst. Pasteur* **111**, 427.

Cuzin, F., and Jacob, F. (1967). Mutations de l'episome F d'*Escherichia coli* K12. II. Mutants a réplication thermosensible. *Ann. Inst. Pasteur* **112**, 397.

Erikson, R. J., and Braun, W. (1968). Apparent dependence of transformation on the stage of deoxyribonucleic acid replication of recipient cells. *Bacteriol. Rev.* **32**, 291.

Low, B. (1968). Formation of merodiploids in matings with a class of rec⁻ recipient strains of *Escherichia coli* K12. *Proc. Natl. Acad. Sci. U. S.* **60**, 160.

Pritchard, R. H. (1965). The relationship between conjugation, recombination, and DNA synthesis in *Escherichia coli*. *Proc. 11th Intern. Congr. Genet. The Hague, 1963,* p. 55. Macmillan (Pergamon), London.

CHAPTER 3

Fungi

E. A. Horenstein and E. C. Cantino

DIVISION OF RESEARCH GRANTS, NATIONAL INSTITUTES OF HEALTH, BETHESDA,
MARYLAND, AND DEPARTMENT OF BOTANY AND PLANT PATHOLOGY, MICHIGAN
STATE UNIVERSITY, EAST LANSING, MICHIGAN

I. Introduction

In recent years, remarkable progress has been made in establishing the existence of, and then in analyzing, the complex relationships between genetics and sexual potency in a handful of well-known fungal types. In sharp contrast—and in spite of an abundance of well-documented sexual life histories among all major groups of fungi—scores of years have passed without providing a significant amount of discriminating insight into the nongenetic aspects of their fertilization mechanisms and the physiology of their gametes. More than a decade ago, Raper (1952) remarked that, whereas the general pattern of relationship between nutritional availability and sexual expression had been established by Klebs many years previously, little of real significance had been added since his day. Yet now, with an additional 15 years gone by, the situation is little changed. This is not to say that mycologists and some

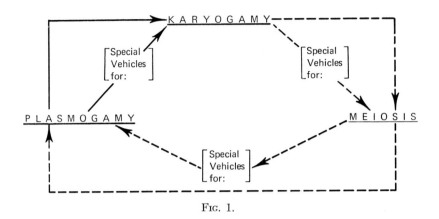

FIG. 1.

other interested experimentalists have not approached nor set foot upon the issue; rather, it is more as if the latter had been either stepped around or somewhat trampled on, thus, in no small measure, obscuring our perception of the points in question.

The flow sheet (Fig. 1) encompasses all the sexually reproducing fungi when their various life histories are arbitrarily reduced to simple common denominators. Fusion of cellular elements (plasmogamy) leads (a) either directly or by way of specialized vehicles to karyogamy; (b) from this, directly or by way of specialized vehicles, to meiosis; and, finally (c), again either directly or by way of specialized vehicles, back to plasmogamy. In the following pages, we shall limit our discussion to the events that transpire between plasmogamy and karyogamy. The many studies which have been made on effects of the environment upon morphogenesis of fungal fruit bodies—and to a lesser extent, genesis of the sex cells themselves—have shed little light upon the subject we have been asked to cover and they will not be dealt with here (with one exception; Section II,B,2). This review will reveal how very little comprehensible and unambiguous information about the physiology of fertilization we really have at our disposal; indeed, were we to paraphrase Grobstein (1966), the title of this chapter would read, quite appropriately, "What We Do Not Know about Sex in Fungi."

A. METHODS OF PLASMOGAMY

What are the basic methods by which fungi initiate the fertilization process via plasmogamy? This subject, too, has been treated extensively in reviews, notably those of Raper (1954, 1960, 1966); therefore, it is only necessary, here, to provide enough description to render intelligible what we may have to say in the following pages. In essence, plasmogamy in fungi consists of the union of two protoplasts via the fusion of two cells which may be visibly similar or dissimilar. The kinds of cells that can exhibit fusions have been given various labels by mycologists (i.e., myxamoebae, planogametes, antheridia, oogonia, ascogonia, spermatia, micronidia, conidia, oidia, and receptive hyphae). Nonetheless, sexual plasmogamy can be visualized as involving paired copulations among three cell types—gametes (sex cells), gametangia (sex organs), and hyphal cells—in all possible combinations (Fig. 2).

B. A Measure of the Task before Us

For any one of the six categories in Fig. 2, the morphology of the cell types involved in copulation will not always be identical. For example, the superficial structure of the sex organs characteristic of the Saprolegniales differs obviously from that of the copulating gametangia in the order Mucorales; and even within this single latter group, easily noticeable morphological differences exist among the cellular elements which fuse in various species. Thus, the pattern (Fig. 2) does not, in its simplicity, reflect the wide range of detailed differences associated with these sexual phenomena. The depth and breadth of these differences become further magnified when a functional point of view is taken; sequential and integrated events of unknown complexity must be involved every time a planogamete or a spermatium or an antheridium initiates and/or participates in the subsequent activities of the cellular element with which it fuses. Presumably, most any couple of fungal sex cells will be

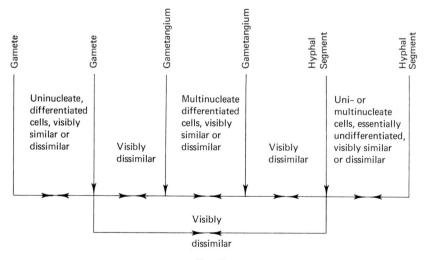

Fig. 2.

carrying an assemblage of discrete and visible entities—nuclei (with their nucleoli, chromosomes, and porous nuclear membranes), cytoplasmic plasma membranes and organelles within them such as mitochondria and cytoplasmic inclusions (droplets, granules, vesicles, and crystals, and so on), etc. Yet we know practically nothing of what happens to these organelles as two such cells prepare to enter into, and then consumate, their sexual union. Thus, the task of dealing with the comparative physiology of fertilization mechanisms in the fungi is potentially enormous. But, unfortunately, many of the initial descriptions of fertilization methods employed by fungi were not studied in depth—and often not at all— and, thus, our insight into the sex life of numerous molds and mushrooms remains superficial to this day. Furthermore, among the relatively few exceptions to this generalization, additional study has seldom led to greater understanding of the mechanisms by which fertilization is accomplished. Add to this the fact that what little *has* been done on the physiology of sex has centered on the yeasts and on hormonal aspects of the problem (Chapter 4, Vol. I, of this treatise), and we are left with little more than a collection of descriptive life histories. In every one of these, the mechanism of fertilization must involve some very complicated networks of temporal relationships among an unknown number of intracellular events which fluctuate as sexual ontogeny progresses. Unfortunately, the only glimpse we have of this stems almost exclusively from partial images created mainly by genetic studies with a handful of fungi. In the sections which follow, we shall try, nonetheless, to look for substance in this partial vacuum.

II. Literature Review

A. Myxomycetes

1. *Plasmogamy*

Whether or not sexual reproduction can occur in *all* the Myxomycetes remains, to this day, a mildly controversial issue. Nonetheless, although the possibility that apogamy may occur therein cannot be positively excluded, "every species that has been critically investigated initiates its plasmodial stage through syngamy and the formation of zygotes" (Alexopoulos, 1966). Furthermore, the recent discovery of heterothallism (self-sterility) in several slime molds reveals that syngamy is, in fact, obligatory for some of them (Dee, 1960; Collins, 1961). The sexual process in Myxomycetes is circumscribed by the following events: (a) spores release from one to four uninucleate $1n$ myxamoebae or flagellated swarm cells; (b) these cells may copulate in pairs, either immediately or after one or more divisions; and (*c*) karyogamy occurs very soon after cytoplasmic fusion to yield a uninucleate 2*n* zygote; the latter, either immediately or after coalescence with other zygotes, develops into a plasmodium.

Whether it is myxamoebae or flagellated swarm cells that copulate probably depends (Alexopoulos, 1966) on the particular organism and prevailing environmental conditions. But in either case, the gametes are morphologically indistinguishable. Presumably, some sort of physiological difference characterizes the gametes produced by genetically distinct mating types in heterothallic species. Indeed, in two out of three such species which have been analyzed spectrophotometrically by Therrien (cited in Alexopoulos, 1966), opposite mating types differed quantitatively from one another in their complement of deoxyribonucleic acid (DNA). But there is no other direct evidence that chemical or metabolic differences exist. As for gametes from homothallic (self-fertile) species, aside from claims (Alexopoulos, 1966) that multiple fusions may lead to formation of zygotes, there is even less to go on.

And yet, the slime mold should be a "natural"—perhaps even the choicest of fungi—for studies of gamete physiology and fertilization mechanisms. Of all the fungi, how many produce cells that under normal circumstances function *only* as gametes (i.e., cannot produce a vegetative thallus), yet can propagate *themselves* by fission as independent entities apart from parental influence and attachments? In spite of this, experimental studies designed to take advantage of these attributes have been rare, perhaps in part because very few species have been made to complete their life cycles in culture. Fortunately, this obstacle has been partially overcome by recent pioneering efforts of Rusch and his colleagues (see Baldwin and Rusch, 1965, for review) which have led to the formulation of synthetic media (Daniel and Rusch, 1961; Daniel *et al.*, 1962, 1963) for growing pure cultures of plasmodia of *Physarum polycephalum* (not to mention other numerous and important experimental studies of this organism). However, the myxamoebal stage does remain a major bottleneck; *no* Myxomycete has, as yet, been cultured for one complete cycle, i.e., from spore to spore, in the absence of other microorganisms (Alexopoulos, 1963). The myxamoebae of *Didymium nigripes* grow and reproduce in a broth supplemented with either formalin-killed bacteria or the supernatant derived by sonic fragmentation of bacteria (Kerr, 1963); but, they do not form plasmodial clones on this axenic medium in the absence of living organisms. Regrettably, it is not clear exactly what process is being repressed—zygote formation itself, the mitotic divisions which result in a multinucleate plasmodium, or the coalescence of zygotes which could also generate plasmodia. Earlier, Kerr (1960) had reported that in the presence of brucine, populations of myxamoebae did not form plasmodia. However, when brucine was removed, the myxamoebae washed, incubated for 12 hours in phosphate buffer, and then spread onto plates, plasmodia were formed (Kerr, 1961). Since divalent cations (e.g., Sr^{++}, Ca^{++}) stimulated formation of zygotes, Kerr suggested that the inhibitory effect of brucine may have been due to its function as a chelating agent. The entire fusion process is consumated in 150 minutes or more, but it is during the first 90 minutes of incubation in buffer that myxamoebae become flagellated and that pairs of flagellated swarm cells in the process of fusing make their appearance. If cells from these populations are transferred to new plates before fusions are complete, the cells break apart and no plasmodia are formed. Kerr (1965) also reported that formation of flagella is inhibited by streptomycin and other drugs. Since it appears, in this species at any rate, as if it is the flagellated cells that function as gametes, it now becomes important to know if such streptomycin-inhibited flagellaless myxamoebae can, nonetheless, still function as gametes.

It seems likely that, in the future, heterothallic Myxomycetes are going to provide answers to certain questions not approachable with homothallic strains. *Physarum polycephalum* is one likely candidate; myxamoebae derived from a single spore do not yield plasmodia in culture, whereas mixed populations of amebas from several spores derived from a single plasmodium do (Dee, 1960). Since these myxamoebal gametes reproduce themselves for fairly long periods under suitable environmental conditions, it should be possible to prepare large populations of each mating type for intensive studies of their sexual processes.

2. *Karyogamy*

The fusion of gametic protoplasts does not complete the fertilization process; the two parental 1n nuclei, now in a common cytoplasm, must fuse into one zygotic 2n nucleus. In Myxomycetes, this nuclear fusion quickly follows plasmogamy, and it results in a 2n unicellular zygote. Ross (1957) detected this in stained preparations of many species. More recently, by photographing the entire process in *Physarum gyrosum*, Koevenig (cited in Alexopoulos, 1963) demonstrated conclusively that karyogamy takes place right after plasmogamy. And with evidence based upon spectrophotometric data (cited in Alexopoulos, 1966), Therrien concluded that the nuclei of plasmodia contain twice the amount of DNA found in the nuclei of myxamoebae.

In summary, the myxomycete provides an experimentally exploitable system with great potential; once myxamoebae have been grown in defined media, they should provide the means for physiological studies in

depth of gametes, plasmogamy, and karyogamy in the zygote. It is a vir-
tual certainty that much more will be forthcoming from these creatures
in the future.

B. PHYCOMYCETES

1. *Allomyces*

The biology of this water mold is well understood and has been documented in great detail, largely through the efforts of Ralph Emerson, his
colleagues, and his students (see Emerson, 1954; some aspects of comparative sexual phenomena in the Blastocladiaceae have been reviewed
more recently by Cantino, 1966). A great deal has been accomplished
experimentally with two species, in particular, *Allomyces arbuscula* and
Allomyces macrogynus. But, in spite of all the data acquired, the potential value of this aquatic Phycomycete as a tool for studies of gamete
physiology and fertilization mechanisms has barely been exploited.

The life cycle of *Allomyces* displays an alternating 2n sporophytic
and 1n gametophytic generation, either of which can be grown separately
under controllable and defined conditions (see Lovett, 1967, for review
of cultural procedures and references). Gametangia are formed in pairs
at the ends of gametophytic hyphae whereon the male sex organ may be
either terminal or subterminal to the female, depending on the species.
Since the organism is homothallic, male gametes can fuse with female
gametes derived from a common parent. Female gametes, on the other
hand, can also develop parthenogenetically and thus display their genetic
and physiological totipotency. Self-sterile, essentially unisexual, interspecific hybrids are also available. Copulation involves dissimilar uniflagellate gametes; the male is orange, small, and swims actively, whereas
the female is colorless, larger, and more sluggish. Physiologically, they
also differ in that the female produces a chemotactic hormone (to which
it is, itself, insensitive; Carlile and Machlis, 1965) which attracts the male
gamete and, thereby, initiates fertilization (see Machlis and Rawitscher-
Kunkel in Chapter 4, Vol. I, of this treatise). These and other differences
between them must stem from something other than a heterogeneity in
nuclear genotype because compatible male and female sex cells arise on
the same 1n parent.

2. *Gametogenesis in Allomyces*

Strictly speaking, gamete formation is irrelevant to the subject we
delimited for coverage by our review; however, in this particular case its
potential importance for an eventual understanding of the sexual fusions
which follow in ontogeny serves to justify its brief inclusion here. At

maturity, the entire multinucleate protoplasts in a pair of differentiated male and female gametangia are cleaved into uninucleate male and female gametes, respectively; following their release, only empty gametangial sacs remain. Studies have been made of their fine structure (Blondel and Turian, 1960). The intricate details of organization displayed therein reflect the extensive cytoplasmic and nuclear differentiation which must occur prior to or/and during gametogenesis; if these male and female gametes differ in physiological and biochemical properties, this divergence must represent the end product of two dissimilar biochemical machines which underlie gametangial morphogenesis. Therefore, what we learn about their formation may provide indirect insight into their function.

Carotenoids are responsible for the orange color of male gametes in *Allomyces* (Emerson and Fox, 1940), but there is no conclusive evidence for a direct physiological involvement of these pigments in sexual reproduction. Inhibition (with diphenylamine) of carotene synthesis in male gametangia did not prevent copulation by the decolorized gametes subsequently released (Turian, 1952); either very small amounts of carotenoids were required for some vital role in sexual reproduction, or carotenogenesis merely accompanied some more fundamental metabolic alteration responsible for differentiation of male gametes (Turian and Haxo, 1954). However, more severe inhibition of carotenogenesis with higher concentrations of diphenylamine was accompanied by a selective disturbance of morphogenesis in the male gametangia of *Allomyces macrogynus* and *Allomyces arbuscula* (Turian, 1957). By using albino mutants of *A. arbuscula*, Foley (1958) showed that in certain strains, male gametangia reappeared whenever the physiological capacity for production of pigment was regained; thus, the metabolic pathways responsible for these two events are obviously connected. Further work along these lines has not been done, however, and to what extent if any carotenoids play a causal role in sexual reproduction remains unsettled.

In a comparative study of mature sex organs in male and female hybrid strains of *Allomyces*, Turian concluded that DNA/ribonucleic acid (RNA) ratios for the male were almost twice as high as for the female (for discussion and references, see Cantino and Turian, 1959; Cantino, 1966). This difference appears to stem from two microscopically detectable events which occur during gametangial differentiation. First, more nuclear divisions occur in the male than the female once the sex organs have been delimited from each other by a septum (Hatch, 1935); this results in the formation of roughly 4 times as many male gametes as female gametes during gametogenesis (Turian, 1962). The second point of difference concerns the nuclear cap, a large and unique organelle found only in the motile cells—zoospores (Cantino *et al.*, 1963) as well as gametes (Turian and Kellenberger, 1956; Blondel and Turian, 1960)—in the Blastocladiales and some of the Chytridiales. It is in intimate contact with and partially encloses the nucleus, and, at least in *Blastocladiella emersonii*, it contains almost all of the cell's ribosomes (Lovett, 1963). The nuclear cap in a mature female gamete is several times larger than in the male (Turian, 1962) and apparently reflects differentially increased net synthesis of RNA during the last stages of gametogenesis in the female sex organ as compared to the male. Finally, after syngamy has occurred, male and female nuclei fuse, as do their nuclear caps; the resulting zygote does not germinate, however, until the combined caps disintegrate and release their ribosomal contents (Hatch, 1938; Turian, 1964).

3. Gametogenesis in Blastocladiella

Blastocladiella emersonii also produces motile orange cells which are smaller than the colorless ones and which are "phenotypically" determined as in *Allomyces*. The pigment appears to be predominantly carotene, and its synthesis is also depressed by diphenylamine. Mutant strains thereof, much like *Allomyces* male hybrids, produce orange cells exclusively. Cytoplasmic gamma particles, possibly lysosome-like, are detectable in the motile cells of both wild type and mutant; such particles are less numerous in orange cells than in colorless ones, and their number also responds to treatment with diphenylamine as does synthesis of carotene itself. And finally, although the orange and colorless motile cells of *B. emersonii* do not participate in conventional gametic copulation, they *do* indulge in a transient kind of plasmogamy involving cytoplasmic bridges through which materials seem to be exchanged (Cantino and Horenstein, 1954).

4. Gamete Physiology in Allomyces and Blastocladiella

A good deal is known about metabolic differences and similarities between the orange and colorless thalli of *Blastocladiella* and *Allomyces*. In so far as these thalli are the progenitors of the orange and colorless sex cells with which we are concerned in this review, what has been learned about them is important. But they provide only indirect clues about the physiology of the gametes themselves. Therefore, for such details the reader is referred to the review by Cantino (1966) wherein these organisms are discussed at length.

Finally, although the colorless motile cells of *Blastocladiella emersonii* have been analyzed in biochemical, physiological, microscopical, and other terms, the same has not been done for the orange swarmers. And, as for *Allomyces*, virtually no first-hand knowledge is available about the physiology of its motile gametes nor (aside from Machlis' work with sex hormones, see Machlis and Rawitscher-Kunkel, in Chapter 4, Vol. I, of this treatise) of their fertilization mechanisms. Although it was discovered long ago (Turian, 1954) that the addition of boric acid to suspensions of male and female gametes prevents copulation, the mechanism by which it exerts its effect has not been studied nor has this technique been exploited for investigations of plasmogamy. Yet clearly, *Allomyces* and *Blastocladiella* have very great potential, for along with members of the Myxomycetes they provide experimentalists with a ready and controllable source of motile sex cells for investigations of gamete physiology and the sexual mechanisms in which they become involved. There is every prospect that these organisms will make major contributions in the future.

C. Ascomycetes

The Ascomycetes introduce us to some innovations involved in sexual reproduction which do not have to be contended with among the lower fungi. Disregarding many minor details which vary from one species to another, there remain certain features more or less characteristic of this class which bear directly on this discussion. Ascomycetes produce no motile sex cells; in the majority of them, in fact, gametes are reduced to little more than nuclei, and they become coupled with one another by several mechanisms. And, again in contrast with the Myxomycetes and Phycomycetes, karyogamy is delayed via the process of dikaryosis. Parental 1n nuclei, brought together in a common cytoplasm by plasmogamy, become associated in pairs; conjugate nuclear divisions follow and result in the successive formation of dikaryotic cells. This condition, which persists for varying periods of time, occurs in specialized ascogenous hyphae wherein the paired nuclei migrate. Finally, karyogamy takes place in one of these dikaryotic cells—the ascus mother cell—and meiosis follows immediately. Formation of ascospores ensues thereafter.

Among hermaphroditic, heterothallic (self-sterile) Ascomycetes, obligatory cross-mating is assured at the genetic level by a two-allele mating system at a single locus rather than by phenotypic sexual differentiation at a morphological or cytoplasmic level. Such fungi can function as either donors or recipients of fertilizing nuclei, but plasmogamy occurs only between compatible strains, i.e., those of opposite mating type. Although attempts are being made (Turian, 1966) to establish metabolic bases for the genesis of sex organs (in *Neurospora*), physiological explanations of fertilization itself are not as yet available; however, attacks upon the problem have begun. Bistis (1956, 1957; Bistis and Raper,

1963) has focused attention on the early stages of this process in *Ascobolus stercorarius;* i.e., upon sexual activation and plasmogamy. In a compatible mating "it appears that any viable, nucleated unit of protoplasm may function in plasmogamy with a trichogyne of opposite mating type" (Bistis, 1956). Thus, an oidium—a cell derived by fragmentation of a somatic hypha—displays a dual capacity: it can either germinate to form a new mycelium or serve as a spermatizing agent. When an oidium is placed on, or sufficiently near, a compatible mycelium it is inhibited from germinating to produce new vegetative hyphae. Concomitantly with this growth inhibition, the oidium undergoes sexual activation. Apparently, this change of state results from physiological alteration within the oidium itself, and it is probably induced by the mycelium of opposite compatibility. Formation of the ascogonium and elongation of its trichogyne (fertilization tube) follow oidial activation. When the trichogyne has grown sufficiently to make contact with the oidium, plasmogamy occurs. Let us examine these stages more closely.

1. *Inhibition of Germination*

This seems to consist of at least two different processes. Germination of an oidium to yield new vegetative growth is inhibited if the cell is placed in close proximity to any mycelium—*even one of the same mating type!* (Bistis and Raper, 1963). This repression of growth is reversible, all the way up to the time of plasmogamy. Thus, if the inhibited oidium is removed from the mycelium's sphere of influence, it will germinate after a lag period. A particularly critical point has not as yet been established, however; namely, whether or not inhibition of germination by a compatible mycelium is of the same degree and kind as that brought on by an incompatible mycelium. If the latter were to turn out to be the case, then this first crucial step in the sexual mechanism would not be mating-type specific.

2. *Sexual Activation*

This phenomenon *is* controlled by the genetic incompatibility factor; activation of an ungerminated oidium occurs only in the presence of a mycelium of opposite mating type. The physiological changes that underlie and drive oidial activation remain to be identified.

3. *Ascogonial Induction*

A sexually activated oidium can, at the present time, only be recognized by the response which it elicits in a compatible mycelium. The latter responds—presumably to a diffusible substance—first by developing an ascogonium and, then, by directing the growth of its trichogyne toward the oidium. But it is not yet clear just how much of this developmental process, if any, is under mating-type control. Perhaps it should be emphasized at this point that, notwithstanding the intriguing nature of Bistis' observations (Bistis and Raper, 1963) with *Ascobolus,* no generalizations are possible at the moment. In *Neurospora,* for example, the female apparatus (the protoperithecium) developed *independently* of influence or stimulation by its opposite mating type, within a limited range of environmental conditions, including temperature (Hirsch, 1954) and nutrition (McNelly-Ingle and Frost, 1965). Whether or not sexual reproduction occurs is simply contingent upon direct contact between the trichogyne of a protoperithecium and a spermatizing agent of suitable genotype.

4. *Plasmogamy*

In *Ascobolus,* the actual fusion of an oidium with a trichogyne apparently does not require that their mating types be opposite. When an oidium, previously activated by a compatible mycelium, is transferred to a trichogyne of an incompatible mycelium, the trichogyne continues growth and plasmogamy eventually occurs. However, development of a fruit body is halted, which indicates that one or more stages in ontogeny subsequent to plasmogamy must be under mating-type control. But here again, generalizations are impossible. For example, in another Ascomycete, *Podospora anserina,* the fusion of a spermatizing agent with a trichogyne occurs *only* between compatible strains (Esser, 1959a). Also, while failure of plasmogamy between two Ascomycetous strains ordinarily precludes a successful cross, Carr and Olive (1959) managed to induce formation of fruit bodies with two self-fertile strains of *Sordaria fimicola* that were unable to anastomose with each other by introducing a third strain with which both anastomosed—a novel approach, indeed, and one that could possibly be exploited in studies of other fungi wherein factors responsible for incompatibility mechanisms are expressed at the primary stage of plasmogamy.

But to return to *Ascobolus,* another observation has emerged from these studies which, if and when it is pursued further, could bring new pertinent information to bear upon this problem. Although it might seem, *a priori,* that the same genetic cross could be accomplished in either of two ways (*A* oidium vs *a* mycelium and *a* oidium vs *A* mycelium), the former, an *A/a* mating, is demonstrably more successful than the latter. There is, in fact, a basic difference between these two crosses which may be responsible for influences exerted at later stages in development. Essentially, the oidial parent donates only a nucleus; but the mycelial or receptor parent provides, in addition to a nucleus, the cytoplasmic milieu in which the future course of physiological activities will take place. Bistis (1965) suggests that only the *a* strain can synthesize an antheridial inducer. Whether or not this particular speculation has validity, it is not unreasonable to assume that cytoplasmic factors, in addition to nuclear factors, do play some decisive roles in the mating responses of these organisms. There are, after all, several known instances of the induction of abnormalities in Ascomycetes by introduction of "defective" cytoplasm. Garnjobst *et al.* (1965) transmitted abnormal characteristics to normal strains of *Neurospora* by injecting them with nuclei-free cytoplasm from abnormal strains, and Diacumakos *et al.* (1965) brought about abnormalities in the same organism by injecting mitochondria from an abnormal strain into a normal one.

5. Events after Plasmogamy

During this "interphase" between plasmogamy and karyogamy in Ascomycetes, dikaryosis becomes established; the two parental nuclei first coexist, then divide, and finally their respective sister nuclei migrate side-by-side into ascogenous hyphae (in the developing fruit body) wherefrom ascus mother cells are formed, and wherein karyogamy occurs. There is no significant experimental evidence concerning activities during this intervening period; nothing is known of the events that trigger the termination of this phase nor the onset of the final nuclear fusion in the fertilization process. Not only *Ascobolus*, but even *Sordaria* and *Neurospora*, have failed, so far, to provide a truly solid grasp of this particular problem at a physiological level. They have, however, yielded a few hints.

Following ultraviolet irradiation of a self-fertile strain of *Sordaria macrospora*, Esser and Straub (1958) isolated eighteen mutant strains which were sterile when grown alone. According to the stage of sexual development that each could achieve individually, these variants fell into fourteen distinct classes. All the interclass pairings resulted in the production of normal perithecia, indicating that at least fourteen mutable stages were involved in the normal developmental sequence. Carr and Olive (1959) found what seems to be a comparable situation in the normally self-fertile *Sordaria fimicola*. Certain pairs of self-sterile mutant strains were compatible with each other through complementation, when the genes responsible for sterility were at different loci. Their results suggested that specific blocks at any one of several stages during the sexual process may have been responsible for incompatibility. The sexual progression in these two species has been dissected into several, separable, functional, genetic units. The immediate job, now, will be to relate these various genetic loci to specific physiological functions in the nuclear and cytoplasmic regions of the cell.

Finally, there is one last feature, one that is associated with Basidiomycetes as well as Ascomycetes, which deserves brief attention here: it is heterokaryosis. This is characterized by the simultaneous existence and replication of nuclei of different genotype in a common cytoplasm. True, dikaryosis brings about this same condition, but it does so in a temporally limited fashion and it functions primarily as an essential prerequisite for sexual reproduction. By contrast, heterokaryosis occurs in long-lived vegetative mycelia and it is initiated, usually, by anastomosis of genetically different hyphae. For our purpose, the salient point here is this: in *Neurospora crassa,* dikaryosis via sexual plasmogamy is *initiated* only when the mating-type factors in the two participants are different (i.e., compatible), whereas heterokaryosis via hyphal fusion is *prevented* when the two participants are compatible in their mating types. It is other genetic loci that control this latter process (Garnjobst, 1953, 1955). Not only are two such strains prevented from forming a heterokaryon, but, in fact, when their two protoplasms are intermixed the reaction is generally lethal; a cytoplasmic, soluble, proteinlike substance appears to be responsible (Wilson *et al.,* 1961). Obviously, cytoplasmic differentiation does occur in these strains. And in particular, the conclusion seems inescapable that the cytoplasm in the female gametangium of an ascomycete undergoes a physiological change which renders it competent to accept certain "foreign" materials while, in other regions of its thallus, it cannot do so.

D. Basidiomycetes

Most of the recent experimental studies of sexuality in these fungi have been directed toward, and have led to an increasingly sophisticated comprehension of, incompatibility phenomena. But even though the genetics of the situation has been so intensively investigated and although biochemical investigations of development are underway (Wessels, 1965; Niederpruem *et al.,* 1965; and references therein), there remains a paucity of definitive information about the physiological mechanisms whereby genetic factors exert their effects, and at what stage or stages in the sexual sequence they are operative. Although aberrant morphological responses have been described for various stages in ontogeny, these effects usually have been merely symptomatic of incompatibility rather than explanatory.

1. *Plasmogamy and Dikaryosis*

Sexual reproduction in Basidiomycetes—said by some to be the most advanced—is at once both simple and complex. Its simplicity lies in the fact that this class of fungi produces no specialized sex organs; ordinary vegetative hyphae have assumed the function of plasmogamy. Or, to quote Raper (1959): they ". . . have evolved the most outlandish, free wheeling, sexual promiscuity . . . every vegetative cell . . . acts as both donor and as recipient of fertilizing nuclei." As far as we are aware, there is no direct experimental evidence that a hypha which anastomoses with another one, and thus initiates the sexual process, is any different morphologically or physiologically from the rest of its mycelium which grows vegetatively. Although some searches have been made, no diffusible substance has ever been detected which could be implicated in either activation (*sensu* Bistis; Bistis, 1956, 1957, Section II,C,2) or initiation of the sexual fusion in the Basidiomycete life history. Even at a purely descriptive level, few if any details are available about the primary fusion process and all that it must entail. Judging from the literature and word-of-mouth reports, an *in vivo* anastomosis between a pair of hyphae has actually been followed visually in only a very few instances (A. Ellingboe, personal communication); the conclusion that fusions have occurred has almost always been inferred from observations made at later stages in the ontogeny of the parents or their progeny.

In a compatible mating, there occurs between plasmogamy and karyogamy an orderly progression of morphological events which culminate in the establishment of "the hallmark of compatibility" (Dick, 1965); i.e., dikaryosis. In most of the heterothallic Basidiomycetes, such a mating is genetically controlled by two sets of incompatibility factors: at one locus, the A series, and at a second locus, the B series, each consisting of numerous alternate alleles. It has been estimated that in natural populations of *Schizophyllum commune*, there are 339 of the A alleles and 64 of the B alleles (Raper *et al.*, 1958). This contrasts sharply with the mating system of most heterothallic Ascomycetes, with its single factor and its two alternate alleles. In Basidiomycetes, a fully compatible reaction is achieved only when both A and B factors in the paired mycelia differ from one another. Results obtained from semicompatible matings (i.e., crosses involving either common-A or common-B factors) have led to the identification of some of the morphological and nuclear events, occurring in the interval between plasmogamy and karyogamy, which are controlled by these two factors. This is illustrated in Fig. 3 (modified from Dick, 1965; see also, Snider and Raper, 1965, and references therein).

Functional control of clamp formation by the A factor and of nuclear migration by the B factor has been reported for eleven different species of Basidiomycetes (Day, 1965). Under normal circumstances, the pairing of incompatible strains does not result in the production of fruit bodies; therefore, all of the events depicted in Fig. 3, being regulated somehow by these genetic factors, must be involved directly or indirectly in this

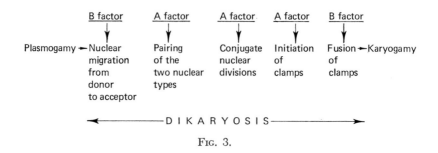

FIG. 3.

final expression of sexuality. For this reason, further intensive investigations of dikaryosis in Basidiomycetes are inevitable, and they are bound to expand our understanding of the sexual mechanism. Recent and current studies include work on nuclear migration, the transfer of nuclei from the donor strain into and throughout the recipient strain, and the early stages of dikaryosis. Let us examine briefly the more important conclusions reached thus far.

2. Nuclear Migration

According to Snider (1965), this occurs at a rate several times greater than that for hyphal growth, and it is quite temperature dependent; for example, the Q_{10} for nuclear migration is 6.0 for *Schizophyllum* and 3.1 for *Coprinus*. In Ascomycetes, the mechanism responsible for nuclear migration may involve cytoplasmic streaming (Snider, 1965); but in Basidiomycetes, there is no evidence that these two phenomena are related. Thus, returning to our incompatibility system, it remains to be established by what means this "overall, master controlling device" (Raper, 1965)—and, specifically, the *B* factor—exerts its profound control on nuclear migration. Why is it that a mating between two participants that share the same *B* factor does not result in nuclear migration? Heterokaryosis, although transient, can occur even in incompatible matings, and, thus, it is not simply lack of plasmogamy which constitutes a bottleneck to nuclear mobility. One possible mechanism currently being considered by some invokes the thought that the *B* factor exerts control over the structural alteration of hyphal cross walls; this, in turn, may impose or release physical barriers to migration. For example, electron photomicrographs reveal that the septal pore in certain Basidiomycetes sometimes displays an elaborate architecture, but that, at other times, it does not (Giesy and Day, 1965; Moore and McAlear, 1962). In *Coprinus lagopus*, to cite Giesy and Day (1965) directly: "The septal pores of homokaryotic and dikaryotic mycelia of *Coprinus lagopus* have the complex structure characteristic of certain Basidiomycetes. The edge of the pore is thickened and covered on either side by a dome-shaped, perforated, membranous cap. However, in heterokaryotic mycelia in which nuclear migration was believed to be taking place, in addition to the complex pores, simple pores were also observed as well as various intermediate stages interpreted as steps in the breakdown of complex to simple pores. The simple pores may facilitate nuclear migration."

Along another tack, Sicari and Ellingboe (1967) examined hyphal tips of both *Schizophyllum* and *Coprinus* at sites adjacent to anastomoses, and in certain instances they obtained a positive correlation between appearance of hyaline hyphal tips and occurrence of incompatibility. They suggested that the difference between a compatible and incompatible reaction, as conferred by the *A* and *B* factors, is not constitutive but, instead, is induced at the time of mating.

3. *Physiological Differences*

Finally, at a more biochemical level of organization, Dick and Raper found (unpublished data; Dick, 1965) that no immunochemical differences were detectable between the polysaccharides in two homokaryotic strains of *Schizophyllum* and the polysaccharides in a dikaryon resulting from their interaction. However, reasoning that the chemical control of incompatibility would probably be mediated by protein, and applying immunological methods for assays, Raper and Esser (1961) demonstrated that protein differences did exist between homokaryotic strains (isogenic except for incompatibility factors) and a dikaryotic mycelium resulting from their anastomoses. At this preliminary stage of these investigations, it is, of course, difficult to assess the broad implications of these results or, in particular, their relation to the biochemical specificities of the incompatibility factors. But this first demonstration of possible protein fingerprints for mating-type factors in *Schizophyllum* holds promise for continued breakthroughs.

While writing of the kinds of studies that had been made with various unrelated organisms, and at different stages in their sexual processes, Raper (1952) emphasized ". . . whether it is indeed the case that there exist many different regulatory mechanisms, perhaps as many as there are morphologically different sexual apparatuses?, or that there is a basic mechanism common to all plants, and that the variations from group to group are only differences in detail, is a moot question, answerable only at some time in the future when sufficient data have been amassed." It seems to us—discouraging as it may be—that this outlook is still almost as fully applicable as it was when Raper wrote it, a decade and a half ago!

III. Discussion: An Outlook

Developmental processes always end in differentiation of cells and tissues and the genesis of form, yet they manifest themselves in endless fashion. The sexual phenomenon in fungi, as defined in this review, encompasses a variable number of these developmental steps. Thus, the mycological sexologist who seeks to determine how a primary contactual stimulus is translated via intrahyphal impulses into a sexual response is confronted with a manifold problem—but one with dimensions no less numerous nor tangled than those that confront an ornithological neurologist who seeks to determine how a primary visual stimulus is translated via nerve impulses into a muscular response.

The essential elements of copulation can best be visualized by reconstructing the process, although oversimplified, from the features which, superficially, seem to be its basic components. Once two cells are in juxtaposition with one another—whether or not one or both have been directed to do so by chemotaxis (Chapter 4, Vol. I, of this treatise) or have met in random fashion, is not really our concern here—the first step in the sexual mechanism consists of three discrete and sequential events: contact, dissolution of delimiting boundaries, and intermixing of protoplasmic ingredients. Unfortunately, virtually nothing is known as yet about the chemistry, physics, and general dynamics of these first critical stages in the sexual activities of the fungi covered in this review. Clearly, these three phenomena provoke a great many questions, but judging from the lack of literature on the subject, either the questions have not been asked, or suitable experiments have not been designed to answer them.

A. CONTACT

This is at once the potential beginning for the sexual act or, alternatively, its major block; as such, it deserves proportionate attention. Surely, not every meeting between compatible gametes, gametangia, or hyphae results in copulation! Why not? Over forty years ago, Burgeff (1924) reported that when plus and minus strains of *Mucor mucedo* ceased to grow at the periphery of the zone of restraint that formed between them, a *few* hyphae from each colony continued to grow, penetrated the zone, and then developed the swollen tips destined to become copulating gametangia. Under these environmental conditions, at least, not all hyphal tips of this bread mold displayed sexual competence. But what of the many fungi which do *not* display zones of restraint? The cells in a mycelial mat are hardly all identical; each hypha is a cylindrical tube which continuously lays down new wall material at its tip as it grows and in which protoplasm continuously moves forward (Zalokar, 1959; cf.

Hickman, 1965, for discussion) leaving other parts in the rear empty. Hyphal tips differ substantially from older portions of mycelium in their content of protein, enzymic activities, etc. They differ, too, in their subcellular organelles; for example, the mitochondria in hyphal tips are visibly different—as seen both through the light microscope (Zalokar, 1959) and the electron microscope (Tanaka and Yanagita, 1963)—than those further back along similar hyphae. Undoubtedly there are also differences in the surface properties of hyphal tips, just as there must be special properties on any copulating element which permit contact to be made and maintained. For many Phycomycetes and all Ascomycetes and Basidiomycetes, this adhesive force between copulating pairs must be maintained at least long enough to permit dissolution of their delimiting cell walls or membranes. But among aquatic fungi which produce motile sex cells, this adhesive force must also be strong enough to withstand the planogamete's motive power to tug and pull away from its partner and swim off again. Thus, the motility of a gamete introduces an additional dimension. The attractive force, which permits the amoeboid gliding of one cell upon its mate and yet ensures maintenance of intimate contact between the two, may differ from that which is involved in the more static relationship which ensues when two hyphal cells, or a conidium and a trichogynal element, make and then maintain contact. Unfortunately, meaningful data on this subject are practically nonexistent in the literature on fungi (exclusive of yeasts). For neither situation is there direct evidence regarding the chemical or physical nature of these binding forces. Even temporal relationships—how long it takes two fungal cells to fuse, for example—have seldom if ever been documented quantitatively. Perhaps the best we can do at the moment is to put it subjectively: pairs of flagellated gametes seem to fuse more quickly than nonmotile elements.

1. *Nature of Binding Forces*

The lack of information about cell contacts among the fungi contrasts sharply with an extensive and growing literature on interactions—although most of them are nonsexual—among animal cells. According to Humphreys (1963), in spite of the fact that the nature of selective cell adhesions is still unclear, the views of workers in this field suggest some five possible models: (a) stabilization of intercellular cementing substances by divalent cations; (b) bonding of sterically complementary surface groups on adjacent cells; (c) bonding between cell surfaces due to calcium bridges; (d) bonding of cells by long-range attractions between their membranes; and (e) binding at the cell surfaces by special metabolic products. These models could serve as a convenient starting point for studies of sexual interactions in fungi. Some of the techniques employed by workers in other fields, for example those used by Moscona (1962) in his studies with dissociated animal cells, should be adaptable for investigations of plasmogamy, especially in fungi with differentiated gametes. Also, the mycelia of various fungi—even some nonseptate phycomycetes (e.g., see Davies, 1959)—often can be broken up into populations of cells or chains of cells of high viability. Would certain cells in such mycelial suspensions from organisms which normally copulate by hyphal fusions exhibit plasmogamy, particularly the "contact" stage, when mixed with similarly prepared suspensions from another mycelium of compatible genotype? If so, would they then dissociate again like Moscona's (1962) mammalian cells if they were treated with tryptic enzymes before plasmogamy had progressed beyond its "point of no return," thus suggesting involvement of peptide linkages?

In particular, ion bridges may play a role between adjacent cell surfaces, perhaps involving carboxyl end groups (Curtis, 1962), calcium (refer to Steinberg and others in DeHaan and Ebert, 1964), and/or other ions and functioning to stabilize these surfaces by cross-linkages with cell membranes (e.g., L. Weiss, 1962). Possibly copulating fungal elements are also held together in this way. Would manipulation of the concentration of exogenous cations such as Ca^{2+} affect their aggregation? Moscona's work (1962) suggests that some animal cells do not adhere to one another in the absence of Ca. Although it would be premature to conclude that the data available bear *directly* upon the applicability of this notion to the fungi covered in this review, it is of interest to examine briefly recent work with several of them. We have already referred (Section II,A,1) to Kerr's (1960) report that divalent cations such as Ca^{2+} and Mg^{2+} stimulate formation of zygotes in the slime mold *Didymium*. Although their sites of action have not been established, these results could mean that Ca^{2+} and/or Mg^{2+} play some role in the binding of gametic myxamoebae.

Also, using replacement culture techniques, Yang and Mitchell (1965) conclude that Ca, at certain concentrations, is necessary for formation of oospores (i.e., zygotes) in the oogonia (i.e., female sex organs) of *Pythium*. Thus, Ca appears to be involved somewhere along the sequence: formation of the antheridium (male sex organ), penetration of the female sex organ by fertilization tubes from the male sex organ, fertilization of the egg in the oogonium, and/or development of the fertilized egg into a differentiated zygote, the oospore. Unfortunately, it is unclear whether or not male organs are produced, and if so, what they are doing, if anything, when a Ca deficiency prevents formation of the oospore. Thus, the precise place at which Ca plays its part in *Pythium's* sexual ontogeny has yet to be determined. We would like to make a guess: that it serves to help bind the male gametangium to the female prior to penetration by fertilization tubes. If this were so, *Pythium* cells would not be unlike those studied by Moscona (1962), Humphreys' (1963) sponge cells, etc., wherein intercellular adhesions depend upon the presence of Ca and, in some instances, other divalent cations.

Finally, relative to the model involving metabolic products, one last thought; would aqueous extracts of cells from one fungal strain contain some glycoproteinaceous equivalent of the "fertilizin" of metazoa (see Austin, 1965, and Metz in Chapter 5, Vol. I, of this treatise, for discussion) which, when added to cells of the opposite mating type, could react with some acidic protein on these cells and thus agglutinate them? Could it be that the "sirenin" produced by *Allomyces* and other chemotactic agents formed by molds (cf. Chapter 4, Vol. I of this treatise) play such a role? Of such populations of fungal cells, many questions could be asked and experiments done to answer them. But to the best of our knowledge, no serious effort has been made in this direction.

2. *Temporal Aspects of Binding-Site Formation*

When are binding sites produced? Is their fabrication one of the criteria of sexual maturity or competence? If so, this would be an automatic consequence of ontogeny and, as such, would not depend on special external triggers or stimuli, but would be dependent upon the functional or developmental stage of the cells involved. With animal cells, for example (Moscona, 1962), the question was asked: "Do cells which are progressively differentiating also exhibit a gradual change in their contact reactions?" The answer seems to be "yes." Cells from the same tissue but differing in developmental time exhibit a progressive age-dependent decline in their capacity for aggregation—their cohesive competence decreases as they differentiate. With suitably uniform mycelial homogenates of sexually competent strains of fungi or with homogenous suspensions of conidia or motile gametes (sufficiently synchronized, such as those prepared with spores of *Blastocladiella emersonii*; Cantino and Lovett, 1964), similar studies could certainly be made. This approach would permit mycologists to determine if, in the process whereby a fungal cell is becoming competent for plasmogamy, its metabolic machinery also becomes progressively committed to its specialized sexual task—perhaps even to the exclusion of others more mundane! The oidia of *Ascobolus* (with which work has already begun; Section II,C,1) should be ideally suited for this kind of investigation, once methods are devised for producing large homogeneous populations of them.

Or, are binding sites generated only at that specific time when they will be "needed" by cells destined to copulate? If this is the case, where does the stimulus come from? Is the trigger unilateral, i.e., does the stimulus come from only one of the two cellular elements? Clearly, at *gross morphological* levels, unilateral effects are observable just before copulation. In some fungi, such as *Anthrocobia melaloma* (Rosinski, 1956), the tip of the ascogonial trichogyne coils around an antheridial hypha; after contact, one of the antheridial cells swells into a spherical shape, following which wall dissolution and plasmogamy occur. And in *Ascobolus* (Bistis, 1957; see Section II,C,1), the "activation" of an ungerminated oidium occurs only in the presence of mycelium of opposite mating type; thus, at a *gross physiological* level, unilateral effects are also detectable. By the same token, then, genesis of binding sites could follow a similar directional pattern. Is it possible that the altered "physiological state" (Bistis and Raper, 1963) which stems from "activation" of the male cell in *Ascobolus* is, in fact, the result of genesis of binding sites? There is no experimental evidence for or against this point of view at the moment.

On the other hand, the mechanism could also be bilateral, each copulating element simultaneously activating in its partner the cellular machinery for genesis of these sites—perhaps even via a sequential system in which unilateral stimulation of site formation in one cell triggers site formation in its mate.

Furthermore, whichever of these it may be, is the phenomenon set in motion only after physical contact between the cells has occurred or may it be set in motion before that time by an altered composition of the extracellular environment as the copulating elements approach one another? Although mycologists have not as yet provided answers to these questions, some clues from other quarters are worth brief consideration here. Certain animal cells, when mixed together, may at first display nonspecific adhesions with one another to yield randomly mixed aggregates. This initial aggregation is followed by a sorting out process wherein cells of like histotype gradually regroup together (e.g., Moscona, 1960), possibly because they liberate material with cell-binding activity. It is not yet clear if this material acts directly or indirectly to produce cohesiveness or, for that matter, whether or not it functions in specific fashion (see references in DeHaan and Ebert, 1964). But it seems to us quite possible that among fungi, too—whether it be hyphal cell with hyphal cell, motile gamete with motile gamete, or some other combination—a process of random "hits" between them could be involved, but that (like the regrouping of histotypes above) only "like" types would unite on contact because they alone could make the cementing substances whereas the rest would not be able to do so. Only tests and time will tell.

3. Spacial and Physiological Aspects of Binding-Site Formation

Are binding sites uniformly distributed or localized on the surfaces of copulating fungal elements? And, if localized, then where are these places on the cell? In plants and animals, contact stimuli play many vital roles; in nerve endings, for example, excitation at receptor membranes is confined to mechanically distorted regions (Loewenstein, 1960). Perhaps the binding sites in fungal elements also occur in particular regions which exhibit mechanical distortion when cells make contact. *In vivo* observations (Robertson, 1965) suggest that apical extension of fungal hyphae occurs only at their tips, that hyphal plasticity progressively decreases in basipetal fashion, and that new branches on such hyphae are generated where plasticity reappears in already hardened walls. It seems reasonable to expect, therefore, that plasmogamy may also be associated with localized changes in plasticity. With appropriate procedures, perhaps such as those described by Rosenberg (1963) for studying the reactions of cells to mechanical constraint and distortion, it should be possible to examine the relation between extensibility of a fungal cell, or part of it, and its potential for plasmogamy. All of this, in turn, implies that such regions in fungal walls should exhibit an altered structural and/or chemical makeup. In *Neurospora crassa*, it is already known that morphology and wall composition show relationships with one another; the facts (Mahadevan and Tatum, 1965) suggest that the kind of "growth" exhibited by its cells may be due to changes in the relative levels of structural components in their walls. Bartnicki-Garcia (1966) recently began analyses to gain some insight into the morphogenetic role of cell wall chemistry in the life cycle of *Phytophthora*. Similar approaches, but aimed specifically at copulating fungal elements, could possibly reveal that localized changes in chemical composition do, indeed, occur at the sexual stage.

From the foregoing, it follows that a change in plasticity might provoke or permit many other things to happen which would confer surface specificities upon a fungal sex cell—and it provokes us to speculate even further. For instance, changes in permeability on specific regions of the cell surface might follow. This calls to mind a suggestion, made by Danielli many years ago (see Danielli, 1958 for review), that some small portion of the membranes of certain cells might be more permeable than the rest, possibly because their bimolecular lipid layers were exposed in these special regions. Many others have postulated the presence of regions of special activity in cell membranes, such as Rothstein's (1954) demonstration that some enzyme activities may be limited to certain regions on cell surfaces. With new fluorescent reagents (e.g., see Maddy,

1964) and techniques available for labeling cell membranes, it may now be possible to find and identify sites of fungal cell wall expansion and/or physiological differences in surface regions specifically involved in copulation.

Finally, if indeed there are binding sites in fungi which are localized at specific regions, how would these regions be recognized by the copulating elements involved? What, so to speak, would the code depend upon? The phenomenon of mutual recognition by cells has received much attention from zoologists (e.g., see P. Weiss, 1961) but practically none from mycologists. Would it, for example, depend upon the presence of specific enzymes associated with a need for some specific cation? This suggestion recalls an earlier discussion (Section III,A,1) which dealt with cationic effects. Specifically, we are reminded of Moscona's work (1962) with animal cells which require Ca for mutual adherence; he also found that periodate reversibly blocks such cells from aggregating. Since bound sugars can be selectively degraded with periodate without detectably affecting active sites on enzymically active proteins (Pazur et al., 1963; however, see Brock, 1965, and Taylor, 1964, for a discussion of periodate effects in yeasts in vivo and in vitro), it should not be difficult to determine if periodate-reactive groupings—possibly on protein-bound sugars—participate in binding the sex cells of fungi. In this light, it seems highly significant that the motile gametes of Allomyces are prevented from fusing with one another in the presence of borate (Turian, 1954; Foley, 1958), a substance known to complex polyhydroxy components such as sugars (Zittle, 1951), whereas syngamy is not inhibited by various other treatments, e.g., severe depletion of the male gamete's carotenoid pool (Turian, 1952) and treatments with sugars, auxins, etc. (Machlis and Crasemann, 1956).

Humphreys' work (1963) prompts one last comparison, that is, with the observations of Hawker et al. (1957) and Hepden and Hawker (1961) on the bread mold Rhizopus. In the latter (as well as in species of Mucor and Zygorrhyncus), one or more of the early developmental stages in zygospore (zygote) formation is prevented by exposure to reduced temperatures (10°C or less); it is not known—at least, it is not clear—if this block in the initiation of conjugation occurs before or after the gametangia (sex organs) have made contact. In any case, the adhesions which occur among Humphreys' sponge cells are also blocked by exposures to reduced temperatures (5°C), even when the necessary Ca is present. For both organisms, substances that counteracted the inhibitory effects of low temperature were detected. For Rhizopus, extracts of the fungus itself were inactive, but an extract of carrots was effective. For sponge cells, homogenates were at first also inactive, but when Ca was removed from these cells, active extracts were obtained which, with Ca added back, permitted adhesion at 5°C. It was suggested that sponge cells were held together by an intercellular material bound to each surface by bonds involving divalent cations; perhaps the gametangia of *Rhizopus* depend upon a similar prerequisite for copulation!

It should not be too surprising if we learn, as work progresses, that the initial stages of plasmogamy in some fungi depend upon cation-linked physiological phenomena localized in specific regions on the surface of copulating elements.

4. Sequential Triggers in Plasmogamy

Irrespective of the nature of the binding forces (Section III,A,1), the functional aspects of binding sites (Section III,A,3), or the final trigger that turns these forces on, the question is unavoidable: "What controls the quality, quantity, and duration of these forces?" For one thing, the answer to this question will depend upon identification of the location and nature of the internal mechanism which manufactures the binding sites on the surface of two cells destined to copulate. Assuming that answers will be provided to these questions, further progress will not be made until it is established *how* a contact between two sex cells then triggers subsequent events: the dissolution of cell boundaries, the intermixing of protoplasts, and so on.

In some of the bread molds, contact between two progametangial cells *precedes* their swelling, their eventual differentiation into gametangia, and the final dissolution of the walls between them. In other fungi, morphological change may be less pronounced or even undetectable. What are the events which lead from the initial physical contact to the final, presumably enzymic digestion of delimiting boundaries? If, in fact, a copulating cell does secrete enzyme protein that digests its neighbor's wall, is the onset of this event reflected in a detectably altered structure within the cell itself or some particular portion thereof? In other living systems, the interesting correlation has been drawn (Petermann, 1964) that in some cells which are actively secreting protein, ribosomes are often associated with their membranes whereas in cells which are not liberating extracellular proteins ribosomes are not membrane-bound. In their discussion of the ways in which cellular gateways handle molecular traffic, Hokin and Hokin (1965; see, also, 1963) refer to the phenomenon of "reverse phagocytosis" whereby internal zymogen sacs have been seen to travel to a cell's outer membrane, whereupon they then presumably spill their digestive enzymes. Then again, many male gametes (cf. Austin, 1965, for discussion) possess an organelle called the acrosome—a baglike structure with a unit membrane which seems to contain lysins—which enables them to traverse barriers surrounding the female gametes. Do fungi contain any of these structures or their functional equivalents? Almost certainly, studies of the fine structure of copulating fungal elements would shed much light upon our problem, and they are badly needed.

Furthermore, along a different vein, does the transition from contact to dissolution of boundaries involve conversion of one form of energy into another, i.e., a transduction mechanism in which mechanoreceptors play a vital role? Does the signal system operate with all-or-none pulses? That is, does it operate on a digital principle as do some other biological sensory systems (see Loewenstein, 1960)? And, how *fast* is the trigger? Do things happen as quickly as in the sea urchin where, within a minute after a sperm attaches itself to an egg, spectacular changes are detectable in the cortex of the egg (Allen, 1959)? Do fungi fit the pattern set by animals wherein, after making contact, the fertilizing sperm triggers reactions in the egg which exclude supernumerary sperm and activate the cytoplasm for the events to follow? In the fungi, can plasmogamy occur between a zygote and a third sex cell, and, if so, is further development deranged, as it almost invariably is when more than one sperm succeeds in entering the egg of an animal? There are a few very old reports (Olive, 1953; see Alexopoulos, 1966, for review) that such phenomena may occur in slime molds, but by and large our view of this today is most unclear. We shall have to set our sights upon questions such as these if progress is to come.

B. "Interphase"; Events between Plasmogamy and Karyogamy

Any developing system involves quantitative and qualitative shifts which culminate eventually in replication of likeness. The "interphase" between plasmogamy and karyogamy, a temporal and spatial vehicle in which genotypic capacity for nuclear fusions is converted to its phenotypic expression, is part of such a developmental system. Its duration varies over wide limits among the fungi. Even within a single species— for example, Greis' 1941 monoecious strain of *Chaetomium* wherein an ascogonium, once fertilized by an antheridium or a hypha, may give rise to a perithecium either directly or, alternatively, only after dikaryotic ascogenous hyphae have been formed—plasmogamy may be widely separated from karyogamy or hardly at all. The events that transpire during "interphase," in effect, determine whether or not parental components, hereditary and other, are eventually brought together and then put in position to be passed on to the next generation.

In animals, it is known (Austin, 1965) that cytoplasmic fusion is followed by changes in structure (e.g., protoplasmic streaming, rearrangement of organelles, cell shape and volume, and membrane capacitance) and metabolic function (e.g., oxygen consumption, DNA synthesis, and other physiological activities). Granted, the manner in which male gametes provoke these alterations is still conjectural (Austin, 1965); but, in fungi, the physiological complexity of this interphase between plasmogamy and karyogamy is presently almost beyond our comprehension. In fact, it is only in rare examples such as *Schizophyllum* (Dick, 1965; see Section II,D,1) that some of the morphological events controlled by different genetic compatibility factors have been tentatively identified and to some extent localized in time and place during this interval in ontogeny. But even here, the modes of action of the genes which exert this control—the physiological mechanisms by which the information they carry is translated into the events themselves—remain to be established (see Parag, 1965, for discussion). Nonetheless, as obscure as things are, let us try to see what may be barely visible at the moment.

1. Nuclear Differentiation

To what degree, if any, does a population of nuclei "differentiate" after plasmogamy? And if, indeed, they do, then to what extent and in what way would such a division of labor among them direct, select for, or otherwise affect the ultimate event of karyogamy? For the fungi, there are no certain answers, unless disintegration of extraneous nuclei can be considered a differentiating process. In some species, for example in *Pyronema omphalodes* (Bessey, 1950, p. 207), entry of male nuclei into the female does trigger—but exactly how quickly is not known—microscopically detectable events. By the time most of the antheridial nuclei have passed into the receptive trichogyne of the ascogonium, trichogynal nuclei have already begun to degenerate visibly. It is also known that in other plants, nuclei may vary in size but not in weight; they do not display uniform degrees of hydration. Groups of functionally specialized cells may also become polyploid as compared with neighboring cells, suggesting that endopolyploidization may be involved in the developmental processes of a plant cell after it has ceased to divide (see Stange, 1965, for review). And in animals, Viola-Magni's (1966) radioautographic studies with ^{3}H-thymidine lend support to the belief that the quantity of DNA/nucleus may be related to the functional activities of some cell types. Other studies (Barth, 1964) reveal that the gross quality of nuclei (size, shape, etc.) may also differ markedly among various kinds of differentiated cells. The "puffing" of chromosomes is also illustrative of this point. For example, the structure of a given band in a salivary gland chromosome seems to depend upon the state of secretory activity of the cell. These and many other facts about plants and animals make one wonder if, in fungi too, differential changes in nuclear morphology are associated with those cells, or those regions in a cell, where functionally specialized nuclei are actively preparing for karyogamy.

2. Differentiation via Nucleocytoplasmic Interactions

Whether or not changes in nuclear chemistry or morphology precede the eventual nuclear coalition, the fusion product of plasmogamy is always—at the organizational level of the cell—a "hybrid" entity. At least some of the factors that regulate the subsequent activities of genes therein must be extragenic, and many examples can be cited which support this point of view. In the semiincompatible reaction displayed by Podospora anserina (Esser, 1959b)—wherein the barrage phenomenon is not induced, apparently, by diffusible substances but, in fact, occurs only after hyphal contact—synthesis of melanins is inhibited in the cells at the zone of contact between mycelia even though phenoloxidases are present. Thus, when the alleles responsible for the barrage reaction are brought into coexistence with one another's cytoplasm, a visible metabolic change occurs. In Schizophyllum and Coprinus, too (Sicari and Ellingboe, 1967), hyphal tips become hyaline at sites adjacent to anastomoses between incompatible mycelia. In the strains of Podospora anserina studied by Beisson-Schecroun (1962), who presents another case for the role of the cytoplasm in the determination of phenotype, the metabolic change which follows mixing of protoplasm is not only visible but lethal. In the dikaryotic mycelium resulting from a cross between two fully compatible but otherwise highly isogenic homokaryons of Schizophyllum, serologically detectable, protein fractions are produced which are not detectable in the parental strains (Raper and Esser, 1961). And, in Ascobolus (Section II,C,4), wherein quantitatively unequal protoplasts— whether or not they are qualitatively different is not known—are brought together by a pair of copulating elements, the course of subsequent development seems to depend upon which of the two mating types contributes the greater quantity of cytoplasm in plasmogamy. Following plasmogamy in Schizophyllum (Papazian, 1956), certain morphological mutants thereof tend to donate but not accept nuclei from compatible mating types; since, therefore, such nuclei can exist together in the cytoplasm of an acceptor strain but not in a donor strain, the cytoplasm probably plays a major role in establishing the phenotype. In a similar vein, Ellingboe's (1964) work also shows that some internuclear selection must occur in dikaryotic–homokaryotic crosses involving certain mating-type strains of Schizophyllum. Specifically, the data suggest that nuclei of one genotype begin to migrate before their genetically different counterparts, that this must occur early—i.e., during initiation of nuclear migration—

and that this effect may be due to either a preferential entrance of nuclei following the anastomosis or a differential rate of initiation of migration. Again, the extranuclear environment is implicated. The behavior of *Neurospora crassa* also points directly to the importance of cytoplasmic factors. In this species, hyphae containing nuclei of like mating type (i.e., sexually incompatible) can anastomose to form vegetative heterokaryons, whereas those of opposite mating type (i.e., sexually compatible) cannot. Yet, plasmogamy between conidia and protoperithecia does occur in the latter. Thus, "that vegetative associations of sexually compatible nuclei do not occur with equal facility in this species . . . points to a special property of the female reproductive structure, where nuclei of opposite mating type necessarily coexist" (Davis, 1966). Obviously, the intercellular but extranuclear environment plays a decisive role in regulating the disposition of nuclei in this organism. Evidence for this has come from work by Wilson *et al.* (1961, and references therein) with *N. crassa* (wherein the heterokaryotic association referred to above occurs only if two pairs of alleles are identical at both loci); the fusion itself is not prevented when two strains are incompatible with one another, but *after* the anastomosis is completed one or both cells usually die. A soluble cytoplasmic protein fraction seems to be responsible for this lethal reaction. The foregoing observations, combined with the induction of unusual responses in normal strains of *Neurospora* by injections of cytoplasm (Garnjobst *et al.*, 1965) and mitochondria (Diacumakos *et al.*, 1965) derived from abnormal strains, point strongly to the conclusion that changes can be brought about in fungi by introduction of "defective" cytoplasmic components.

In the future, many questions will have to be posed and answered about the nucleocytoplasmic interactions that occur upon creation of a hybrid "embryo" of a fungus. What actually *happens* when, as a result of plasmogamy, some nuclei and cytoplasm are introduced from one hypha into another—from one motile gamete into another or from one gametangium into another—after which chromosomes are forced to replicate in a "foreign" cytoplasm? Does the recipient cell possess any capacity to protect itself from "contamination" by the donor cell; for example, from proteins or nucleic acids different than its own? And, if so, does it translate this capacity into a detectable reaction which has survival value and which can be measured? Does competition for paternal and maternal gene products control the time and place of karyogamy? Is it quantitative and/or qualitative modification of gene action, resulting from nuclear reproduction in a partially foreign cytoplasm, which ultimately controls the time and place of karyogamy? With comparative biology as our source book, an endless variety of questions could, of course, be asked, but it would be a pointless gesture now. Simple descriptions of what goes on during the interphase between plasmogamy and karyogamy in fungi are badly needed to provide perspective; until this bottleneck is cleared away, it will be difficult to discriminate between questions and hypotheses of high quality and potentially worthy of experimental attack and those of dubious value which would best be left untested for the moment.

But judging from what has been done with other organisms, there is promise that, eventually, we will hear from the fungi. Mutant strains blocked at different stages of the sexual mechanism (Hirsch, 1954; Westergaard and Hirsch, 1954; and more recent examples cited in this review) will undoubtedly provide important tools. Advances in radioautographic techniques (see Baserga and Kisieleski, 1963, for example) may facilitate the tracing of ontogeny in fungal cells in transit between plasmogamy and karyogamy. It should be possible to track, therein, fluctuating levels of RNA, DNA, and protein synthesis and from these results, perhaps, differential losses in nuclear "function"; i.e., which nuclei remain potentially capable of karyogamy and which do not. Radioautography, combined with reciprocal nuclear transplantations (possibly with microinjection techniques as used by Wilson, 1961, with *Neurospora*), could yield real insight into the relative roles of nucleus and cytoplasm in the sexual mechanisms of the fungi. Perhaps equally applicable to fungi is the recent work of Byers *et al.* (1963a,b) with amebas, in which evidence is provided for the existence of "cytonucleoproteins," which are much more concentrated in the nucleus than the cytoplasm and which shuttle back and forth between nucleus and cytoplasm in interphase amebas, and for another class of proteins which appears to be nonmigratory and restricted to the nucleus. Byers *et al.* suggest that the mobile cytonucleoproteins may be involved in important nucleocytoplasmic interactions, for example, in communications among various compartments of the cell, and that the nonmigrating proteins may be involved in differentiation. Somewhere along the line, observations and experiments of this type may have to be done with fungi, because a description of the phenomena that intervene between the interrelated activities of structural genes, operator genes, regulator genes, mRNA, and extrachromosomal repressors, on the one hand, and the ultimate synthesis of end products, on the other, will have to precede a full understanding of the overall genetic control of karyogamy in fungal systems. In one important respect, the mycologist can take heart and have hope. The essential problem that confronts him is the same as—and probably is no less "solved" than—the one that confronts all biologists: "What is the regulatory mechanism whereby specific substances are synthesized at the right time and place, and at increasing levels of organization, such as to culminate in the gross structures of the cell?"

C. KARYOGAMY

This event, the final and crucial act in the sexual mechanism, mimics superficially the three successive steps in plasmogamy: two nuclei make contact, their delimiting boundaries dissolve, and their contents intermix. It would seem that karyogamy recapitulates plasmogamy! This phenomenon, like the interphase discussed above, elicits countless questions, yet surprisingly little is known about it in the fungi. It provokes in us a final plea, which we take from a recent essay by Professor P. Weiss (1963) and which, we think, will have to be kept in mind as mycologists spy upon their captive copulating fungi: "the disciplined study of the *systemic* properties of the cell—of 'cell biology'—that is, the manner in which its molecular components, which are the prime obects of 'molecular biology,' are subordinated to ordered group coexistence in a system of 'molecular ecology'—is one of the major challenges and tasks of modern science. To meet it, we must face it. To face it, we must see it. To see it, we may even at times have to put on blinders so as to reduce the dimming effect of contrast engendered by all the brilliant light that emanates from 'molecular biology.' "

REFERENCES

Alexopoulos, C. J. (1963). The Myxomycetes II. *Botan. Rev.* **29**, 1.

Alexopoulos, C. J. (1966). Morphogenesis in the Myxomycetes. *In* "The Fungi" (G. C. Ainsworth and A. S. Sussman, eds.), Vol. II, pp. 211–233. Academic Press, New York.

Allen, R. D. (1959). The moment of fertilization. *Sci. Am.* **201**, 124.

Austin, C. R. (1965). "Fertilization." Prentice Hall, Englewood Cliffs, New Jersey.

Baldwin, H. H., and Rusch, H. P. (1965). The chemistry of differentiation in lower organisms. *Ann. Rev. Biochem.* **34**, 565.

Barth, L. J. (1964). "Development." Addison-Wesley, Reading, Massachusetts.

Bartnicki-Garcia, S. (1966). Chemistry of hyphal walls of *Phytophthora. J. Gen. Microbiol.* **42**, 57.

Baserga, R., and Kisieleski, W. E. (1963). Autobiographies of cells. *Sci. Am.* **209**, 103.

Beisson-Schecroun, J. (1962). Incompatibilité cellulaire et interactions nucléo-cytoplasmiques dans les phénomènes de "barrage" chez le *Podospora anserina. Ann. Genet.* **4**, 4.

Bessey, E. A. (1950). "Morphology and Taxonomy of Fungi." Blakiston, Philadelphia, Pennsylvania.

Bistis, G. (1956). Sexuality in *Ascobolus stercorarius.* I. Morphology of the ascogonium; plasmogamy; evidence for a sexual hormonal mechanism. *Am. J. Botany* **43**, 389.

Bistis, G. (1957). Sexuality in *Ascobolus stercorarius.* II. Preliminary experiments on various aspects of the sexual process. *Am. J. Botany* **44**, 436.

Bistis, G. N. (1965). The function of the mating-type locus in filamentous Ascomycetes. In "Incompatibility in Fungi" (K. Esser and J. R. Raper, eds.), pp. 23–31. Springer, Berlin.

Bistis, G. N., and Raper, J. R. (1963). Heterothallism and sexuality in *Ascobolus stercorarius*. Am. J. Botany **50**, 880.

Blondel, B., and Turian, G. (1960). Relation between basophilia and fine structure of cytoplasm in the fungus *Allomyces macrogynus* Em. J. Biophys. Biochem. Cytol. **7**, 127.

Brock, T. D. (1965). The purification and characterization of an intracellular sex-specific mannan protein from yeast. Proc. Natl. Acad. Sci. U. S. **54**, 1104.

Burgeff, H. (1924). Untersuchungen über Sexualität und Parasitismus bei Mucorineen. I. Botan. Abhandl. **4**, 5.

Byers, T. J., Platt, D. B., and Goldstein, L. (1963a). The cytonucleoproteins of amebae. I. Some chemical properties and intracellular distribution. J. Cell Biol. **19**, 453.

Byers, T. J., Platt, D. B., and Goldstein, L. (1963b). The cytonucleoproteins of amebae. II. Some aspects of cytonucleoprotein behavior and synthesis. J. Cell Biol. **19**, 467.

Cantino, E. C. (1966). Morphogenesis in aquatic fungi. In "The Fungi" (G. C. Ainsworth and A. S. Sussman, eds.), Vol. II, pp. 283–337. Academic Press, New York.

Cantino, E. C., and Horenstein, E. A. (1954). Cytoplasmic exchange without gametic copulation in the water mold *Blastocladiella emersonii*. Am. Naturalist **88**, 143.

Cantino, E. C., and Lovett, J. S. (1964). Nonfilamentous aquatic fungi: model systems for biochemical studies of morphological differentiation. Advan. Morphogenesis **3**, 33.

Cantino, E. C., and Turian, G. F. (1959). Physiology and development of lower fungi (Phycomycetes). Ann. Rev. Microbiol. **13**, 97.

Cantino, E. C., Lovett, J. S., Leak, L. V., and Lythgoe, J. (1963). The single mitochondrion, fine structure, and germination of the spore of *Blastocladiella emersonii*. J. Gen. Microbiol. **31**, 393.

Carlile, M. J., and Machlis, L. (1965). A comparative study of the chemotaxis of the motile phases of *Allomyces*. Am. J. Botany **52**, 484.

Carr, A. J. H., and Olive, L. S. (1959). Genetics of *Sordaria fimicola*. III. Cross-compatibility among self-fertile and self-sterile cultures. Am. J. Botany **46**, 81.

Collins, O. R. (1961). Heterothallism and homothallism in two myxomycetes. Am. J. Botany **48**, 674.

Curtis, A. S. G. (1962). Cell contact and adhesion. Biol. Rev. **37**, 82.

Daniel, J. W., and Rusch, H. P. (1961). The pure culture of *Physarum polycephalum* on a partially defined soluble medium. J. Gen. Microbiol. **25**, 47.

Daniel, J. W., Kelley, J., and Rusch, H. P. (1962). Hematin-requiring plasmodial myxomycete. J. Bacteriol. **84**, 1104.

Daniel, J. W., Babcock, K. L., Sievert, A. H., and Rusch, H. P. (1963). Organic requirements and synthetic media for growth of the Myxomycete *Physarum polycephalum*. J. Bacteriol. **86**, 324.

Danielli, J. F. (1958). Surface chemistry of cell membranes. In "Surface Phenomena in Chemistry and Biology" (J. F. Danielli, K. G. A. Pankhurst, and A. C. Riddleford, eds.), pp. 246–265. Macmillan (Pergamon), New York.

Davies, M. E. (1959). The nutrition of *Phytophthora fragariae*. Brit. Mycol. Soc. Trans. **42**, 193.

Davis, R. H. (1966). Mechanisms of inheritance. 2. Heterokaryosis. *In* "The Fungi" (G. C. Ainsworth and A. S. Sussman, eds.), Vol. II, pp. 567–588. Academic Press, New York.

Day, P. R. (1965). The genetics of tetrapolar incompatibility. *In* "Incompatibility in Fungi" (K. Esser and J. R. Raper, eds.), pp. 31–36. Springer, Berlin.

Dee, J. (1960). A mating-type system in an acellular slime-mould. *Nature* **185**, 780.

DeHaan, R. L., and Ebert, J. D. (1964). Morphogenesis. *Ann. Rev. Physiol.* **26**, 15.

Diacumakos, E. G., Garnjobst, L., and Tatum, E. L. (1965). A cytoplasmic character in *Neurospora crassa.* The role of nuclei and mitochondria. *J. Cell Biol.* **26**, 427.

Dick, S. (1965). Physiological aspects of tetrapolar incompatibility. *In* "Incompatibility in Fungi" (K. Esser and J. R. Raper, eds.), pp. 72–80. Springer, Berlin.

Ellingboe, A. H. (1964). Nuclear migration in dikaryotic-homokaryotic matings in *Schizophyllum commune. Am. J. Botany* **51**, 133.

Emerson, R. (1954). The biology of water molds. *In* "Aspects of Synthesis and Order in Growth" (D. Rudnick, ed.), pp. 171–208. Princeton Univ. Press, Princeton, New Jersey.

Emerson, R., and Fox, D. L. (1940). γ-Carotene in the sexual phase of the aquatic fungus *Allomyces. Proc. Roy. Soc.* **B128**, 275.

Esser, K. (1959a). Die Incompatibilitätsbeziehungen zwischchen geographischen Rassen von *Podospora anserina* (Ces.) REHM. II. Die Wirkungsweise der semi-incompatibilitäts-Gene. *Z. Vererbungslehre* **90**, 29.

Esser, K. (1959b). Die Incompatibilitätsbeziehungen zwischen geographischen Rassen von *Podospora anserina.* III. Untersuchungen zur Genphysiologie der Barragebildung und der Semi-incompatibilität. *Z. Vererbungslehre* **90**, 445.

Esser, K., and Straub, J. (1958). Genetische Untersuchungen an *Sordaria macrospora* Auersw., Kompensation und Induktion bei genbedingten Entwicklungsdefekten. *Z. Vererbungslehre* **89**, 729.

Foley, J. M. (1958). The occurrence, characteristics and genetic behavior of albino gametophytes in *Allomyces. Am. J. Botany* **45**, 639.

Garnjobst, L. (1953). Genetic control of heterocaryosis in *Neurospora crassa. Am. J. Botany* **40**, 607.

Garnjobst, L. (1955). Further analysis of genetic control of heterocaryosis in *Neurospora crassa. Am. J. Botany* **42**, 444.

Garnjobst, L., Wilson, J. F., and Tatum, E. L. (1965). Studies on a cytoplasmic character in *Neurospora crassa. J. Cell Biol.* **26**, 413.

Giesy, R. M., and Day, P. R. (1965). The septal pores of *Coprinus lagopus* in relation to nuclear migration. *Am. J. Botany* **52**, 287.

Greis, H. (1941). Befruchtungsvorgänge in der Gattung *Chaetomium. Jahrb. Wiss. Botany* **90**, 233.

Grobstein, C. (1966). What we do not know about differentiation. *Am. Zoologist* **6**, 89.

Hatch, W. R. (1935). Gametogenesis in *Allomyces arbuscula. Ann. Botany* (*London*) **49**, 623.

Hatch, W. R. (1938). Conjugation and zygote germination in *Allomyces arbuscula. Ann. Botany* (*London*) **2**[N.S], 583.

Hawker, L. E., Hepden, P. M., and Perkins, S. M. (1957). The inhibitory effect of low temperature on early stages of zygospore production in *Rhizopus sexualis. J. Gen. Microbiol.* **17**, 758.

Hepden, P. M., and Hawker, L. E. (1961). A volatile substance controlling early stages of zygospore formation in *Rhizopus sexualis. J. Gen. Microbiol.* **24**, 155.

Hickman, C. J. (1965). Fungal structure and organization. In "The Fungi" (G. C. Ainsworth and A. S. Sussman, eds.), Vol. I, pp. 21–45. Academic Press, New York.

Hirsch, H. M. (1954). Environmental factors influencing the differentiation of protoperithecia and their relation to tyrosinase and melanin formation in *Neurospora crassa*. *Physiol. Plantarum* 7, 72.

Hokin, L. E., and Hokin, M. R. (1963). Biological transport. *Ann. Rev. Biochem.* 32, 553.

Hokin, L. E., and Hokin, M. R. (1965). The chemistry of cell membranes. *Sci. Am.* 213, 78.

Humphreys, T. (1963). Chemical dissolution and *in vitro* reconstruction of sponge cell adhesions: isolation and functional demonstration of the components involved. *Develop. Biol.* 8, 27.

Kerr, N. S. (1960). Effect of chelating agents on plasmodium formation by the true slime mould, *Didymium nigripes*. *Nature* 188, 1206.

Kerr, N. S. (1961). A study of plasmodium formation by the true slime mould, *Didymium nigripes*. *Exptl. Cell Res.* 23, 603.

Kerr, N. S. (1963). The growth of myxamoebae of the true slime mould, *Didymium nigripes*, in axenic culture. *J. Gen. Microbiol.* 32, 409.

Kerr, N. S. (1965). Inhibition by streptomycin of flagella formation in a true slime mold. *J. Protozool.* 12, 276.

Loewenstein, W. R. (1960). Biological transducers. *Sci. Am.* 203, 99.

Lovett, J. S. (1963). Chemical and physical characterization of "nuclear caps" isolated from *Blastocladiella* zoospores. *J. Bacteriol.* 85, 1235.

Lovett, J. S. (1967). Aquatic Fungi. In "Experimental Techniques of Development" (F. Wilt and N. Wessels, ed.), pp. 341–358. Crowell, New York.

Machlis, L., and Crasemann, J. M. (1956). Physiological variations between the generations and among the strains of watermolds in the subgenus Euallomyces. *Am. J. Botany* 43, 601.

McNelly-Ingle, C. A., and Frost, L. C. (1965). The effect of temperature on the production of perithecia by *Neurospora crassa*. *J. Gen. Microbiol.* 39, 33.

Maddy, A. H. (1964). A fluorescent label for the outer components of the plasma membrane. *Biochim. Biophys. Acta* 88, 390.

Mahadevan, P. R., and Tatum, E. L. (1965). Relationship of the major constituents of the *Neurospora crassa* cell wall to wild-type and colonial morphology. *J. Bacteriol.* 90, 1073.

Moore, R. T., and McAlear, J. H. (1962). Fine structure of mycota. 7. Observations on septa of Ascomycetes and Basidiomycetes. *Am. J. Botany* 49, 86.

Moscona, A. (1960). Patterns and mechanisms of tissue reconstruction from dissociated cells. In "Developing Cell Systems and their Control" (D. Rudnick, ed.), pp. 45–70. Ronald Press, New York.

Moscona, A. (1962). Analysis of cell recombinations in experimental synthesis of tissues *in vitro*. *J. Cellular Comp. Physiol.* 60 (Suppl. 1), 65.

Niederpruem, D. J., Hafiz, A., and Henry, L. (1965). Polyol metabolism in the basidiomycete *Schizophyllum commune*. *J. Bacteriol.* 89, 954.

Olive, L. S. (1953). The structure and behavior of fungus nuclei. *Botan. Rev.* 19, 439.

Papazian, H. P. (1956). Sex and cytoplasm in the fungi. *Trans. N. Y. Acad. Sci.* 18, 388.

Parag, Y. (1965). Genetic investigation into the mode of action of the genes controlling self-incompatibility and heterothallism in Basiodiomycetes. *In* "Incompatibility in Fungi" (K. Esser and J. R. Raper, eds.), pp. 80–98. Springer, New York.

Pazur, J., Kleppe, K., and Ball, E. (1963). The glycoprotein nature of some fungal carbohydrases. *Arch. Biochem. Biophys.* **103,** 515.

Petermann, M. L. (1964). "The Physical and Chemical Properties of Ribosomes." Elsevier, Amsterdam.

Raper, J. R. (1952). Chemical regulation of sexual processes in the thallophytes. *Botan. Rev.* **18,** 447.

Raper, J. R. (1954). Life cycles, sexuality, and sexual mechanisms in fungi. "Sex in Microorganisms," pp. 42–81. Am. Assoc. Advan. Sci., Washington, D. C.

Raper, J. R. (1959). Sexual versatility and evolutionary processes in fungi. *Mycologia* **51,** 107.

Raper, J. R. (1960). The control of sex in fungi. *Am. J. Botany* **47,** 794.

Raper, J. R. (1965). Introduction. *In* "Incompatibility in Fungi" (K. Esser and J. R. Raper, eds.), pp. 1–6. Springer, New York.

Raper, J. R. (1966). "Genetics of Sexuality in Higher Fungi." Ronald Press, New York.

Raper, J. R., and Esser, K. (1961). Antigenic differences due to the incompatibility factors in *Schizophyllum commune*. *Z. Vererbungslehre* **92,** 439.

Raper, J. R., Krongelb, G. S., and Baxter, M. G. (1958). The number and distribution of incompatibility factors in *Schizophyllum*. *Am. Naturalist* **92,** 221.

Robertson, N. F. (1965). The mechanism of cellular extension and branching. *In* "The Fungi" (G. C. Ainsworth and A. S. Sussman, eds.), Vol. I, pp. 613–623. Academic Press, New York.

Rosenberg, M. D. (1963). The relative extensibility of cell surfaces. *J. Cell Biol.* **17,** 289.

Rosinski, M. A. (1956). Development of the ascocarp of *Anthrocobia melaloma*. *Mycologia* **48,** 506.

Ross, I. K. (1957). Syngamy and plasmodium formation in the Myxogastres. *Am. J. Botany* **44,** 843.

Rothstein, A. (1954). The enzymology of the cell surface. *Protoplasmatologia* **2,** 4.

Sicari, L. M., and Ellingboe, A. H. (1967). Microscopical observations of initial interactions in various matings of *Schizophyllum commune* and of *Coprinus lagopus*. *Am. J. Botany* **54,** 437.

Snider, P. J. (1965). Incompatibility and nuclear migration. *In* "Incompatibility in Fungi" (K. Esser and J. R. Raper, eds.), pp. 52–70. Springer, Berlin.

Snider, P. J., and Raper, J. R. (1965). Nuclear ratios and complementation in common-A heterokaryons of *Schizophyllum commune*. *Am. J. Botany* **52,** 547.

Stange, L. (1965). Plant cell differentiation. *Ann. Rev. Plant Physiol.* **16,** 119.

Tanaka, K., and Yanagita, T. (1963). Electron microscopy on ultrathin sections of *Aspergillus niger*. I. Fine structure of hyphal cells. *J. Gen. Appl. Microbiol.* (*Tokyo*) **9,** 101.

Taylor, N. W. (1964). Inactivation of sexual agglutination in *Hansenula wingei* and *Saccharomyces kluyveri* by disulfide-cleaving agents. *J. Bacteriol.* **88,** 929.

Turian, G. (1952). Carotenoides et differenciation sexuelle chez *Allomyces*. *Experientia* **8,** 302.

Turian, G. (1954). L'acide borique, inhibiteur de la copulation gamétique chez *Allomyces*. *Experientia* **10,** 498.

Turian, G. (1957). Recherches sur l'action anticaroténogène de la diphenylamine et ses conséquences sur la morphogenèse reproductive chez *Allomyces* et *Neurospora. Physiol. Plantarum* 10, 667.

Turian, G. (1962). Differential synthesis of nucleic acids in sexual differentiation of *Allomyces. Nature* 196, 493.

Turian, G. (1964). Compléments sur la morphogenèse normale et anormale de l'*Allomyces. Bull. Soc. Botan. Suisse* 74, 242.

Turian, G. (1966). Morphogenesis in Ascomycetes. In "The Fungi" (G. C. Ainsworth and A. S. Sussman, eds.), Vol. II, pp. 339–385. Academic Press, New York.

Turian, G., and Haxo, F. T. (1954). Minor polyene components in the sexual phase of *Allomyces javanicus. Botan. Gaz.* 115, 254.

Turian, G., and Kellenberger, E. (1956). Ultrastructure du corps paranucléaire, des mitochondries et de la membrane nucléaire des gamètes d'*Allomyces macrogynus. Exptl. Cell Res.* 11, 417.

Viola-Magni, M. P. (1966). A radioautographic study with H^3-thymidine on adrenal medulla nuclei of rats intermittently exposed to cold. *J. Cell Biol.* 28, 9.

Weiss, L. (1962). Cell movement and cell surfaces: a working hypothesis. *J. Theoret. Biol.* 2, 236.

Weiss, P. (1961). Guiding principles in cell locomotion and cell aggregation. *Exptl. Cell Res. Suppl.* 8, 260.

Weiss, P. (1963). The cell as unit. *J. Theoret. Biol.* 5, 389.

Wessels, J. G. H. (1965). Morphogenesis and biochemical processes in *Schizophyllum commune* fr. *Wentia* 13, 1.

Westergaard, M., and Hirsch, H. (1954). Environmental and genetic control of differentiation in *Neurospora. Colston Papers* 7, 171.

Wilson, J. F. (1961). Microsurgical techniques for *Neurospora. Am. J. Botany* 48, 46.

Wilson, J. F., Garnjobst, L., and Tatum, E. L. (1961). Heterokaryon incompatibility in *Neurospora crassa*—micro-injection studies. *Am. J. Botany* 48, 299.

Yang, C. Y., and Mitchell, J. E. (1965). Cation effect on reproduction of *Pythium* spp. *Phytopathology* 55, 1127.

Zalokar, M. (1959). Growth and differentiation of *Neurospora* hyphae. *Am. J. Botany* 46, 602.

Zittle, C. A. (1951). Reaction of borate with substances of biological interest. *Advan. Enzymol.* 12, 493.

Addendum

Since the completion of this review, a few additional reports on this subject and some related reviews have appeared; they are outlined below.

Although we excluded research on the *induction* of sex organs—the formation of pythiaceous gametangia in response to sterols, for example—the interested reader will want to note the appearance of some recent articles on the influence of environment upon reproduction (Hawker, 1966), and on various aspects of sexual activity among the fungi (Raper, 1966), and the book by Esser and Kuenen (1967) and that by Burnett (1968), especially pp. 401–435. Also, because the yeasts were not included in our article, attention is called to the timely review by Crandall and Brock (1968) covering the past studies of Brock and his colleagues, those of N. W. Taylor and his associates, and others on the molecular basis of mating in *Hansenula wingei*.

With respect to other areas of research, the Myxomycetes have continued to receive attention. Therrien's microspectrophotometric measurements of nuclear deoxyribonucleic acid (cited in Alexopoulos, 1966) have now been published (Therrien, 1966)—and, in turn, have already provoked some additional comment (Kerr, 1968).

Along other lines, and taking advantage of the achievement of greater control over the timing of syngamy that the heterothallic condition makes possible in *Didymium iridis,* Ross (1967a) studied sexual fusions and the formation of plasmodia in this organism. He described the events seen with phase microscopy *in vivo* during plasmogamy, nuclear fusion generally being detectable from 4 to 20 minutes thereafter. The work also prompted thoughts about possible mechanisms for explaining alternate developmental pathways in *D. iridis;* for example, the absence of immediate syngamy among myxamoebae and the fact that not all myxamoebae participated in it suggested that inducing and/or inhibitory substances may have helped determine the incidence of fusion in a population of such cells. In another paper (Ross, 1967b), it is noted that the breaking and fusing of cell membranes during plasmogamy might also play a part in triggering events after fusion. Unfortunately, there is little direct evidence yet to support these hypotheses.

Finally, although it does not relate *directly* to sexual fusions between myxamoebae, the work of Carlile and Dee (1967) and Collins and Clark (1968) could come to bear upon questions about syngamy. The first of these papers is a preliminary account of the lethal reaction which follows the fusion between two plasmodia of *Physarum polycephalum,* the first indication of the response being the cessation of streaming near the point of fusion, and this being followed by at least partial destruction of the protoplast. The second paper contains, among other things, thoughts about plasmodial compatibility in terms of possible changes in their slime sheaths and plasma membranes.

Coming next to the sex organs of the Pythiaceae, and especially the fact that numerous investigators in the past have given much attention to the physiology of their *formation* while few have succeeded in even grappling with the events that *followed* plasmogamy, several reports suggest that a start in this direction may yet be made. To begin with, the need to simply track some of the initial sexual events is being met; Marchant (1968), in an ultrastructural study of *Pythium ultimum,* followed penetration of the oogonium by the antheridium, changes in storage material and wall thickness during oospore development, and shifts in quantity and disposition of some organelles during maturation of the oospore.

At a more physiological level, Elliott *et al.* (1966) examined the capacity of various sterols and related substances to promote sexual reproduction in *Phytophthora cactorum* and found that in the presence of certain compounds, genesis of oogonia and antheridia was permitted while the normal consummation of the sexual process (i.e., subsequent formation of oospores) was prevented. As the authors pointed out, a knowledge of why the course of development broke down under certain environmental conditions could go far toward helping to explain the mechanism by which sterols regulate sexual processes in these fungi. There is some evidence, too (Leal *et al.,* 1967) that other nutritional factors (e.g., C/N ratios, the amount of available NH_3) may also play a role in this phenomenon.

As for the Mucoraceae, a long history of interest in the possible functional relationships between carotenoids and sex has heretofore failed to provide significant insight into the mechanism of the sexual act itself. Even the most recent reports—such as the suggestion by Sutter and Rafelson (1968) that a *physical contact* between (+) and (−) strains of *Blakeslea tri pora* is what initiates synthesis of carotenoid in mated cultures, and the counter arguments by Van den Ende (1968) that such results can also be explained in terms of the inducing effects of soluble substances—continue to illustrate some of the difficult aspects of the problem.

With respect to the Basidiomycete *Schizophyllum commune*, Wessels and Niederpruem (1967) studied a cell-wall glucan-degrading enzyme which displayed more activity in dikaryons than in homokaryons, and postulated that this enzyme might play a role in the mating reaction in the sense that unlike B factors could be necessary for derepression of this enzyme.

Finally, attention must be called to the appearance of a Ph.D thesis by John V. Leary (1969) which encompasses one of the most penetrating probes made so far into the "molecular biology" of the early mating reactions in any filamentous fungus. Techniques were developed for mating mycelial fragments in liquid culture, and with them the kinetics of the initial nuclear exchange between homokaryons in compatible and noncompatible mating combinations was determined. Evidence for nuclear exchange was based on analyses of fragments derived from developing heterokaryons which possessed nuclei from all 4 types of matings. In all 4 (compatible, common-A, common-B, and common-AB), the percentage of fragments possessing both types of nuclei was nearly equal initially, but after 12–24 hours significant differences became evident in the kinetics of the compatible vs the 3 noncompatible matings. Furthermore, the common-A pattern was distinguishable from the common-B and common-AB. The kinetics of nuclear exchange was independent of the mating-type alleles and nutritional markers used. In the majority of fragments analyzed in compatible matings at zero time and 48 hours after mating, the number of cells/fragment was fewer than 5, indicating that the interactions observed were closer to a cell-to-cell basis than had been possible before.

These studies (a) indicated that the earliest mating interactions detected, i.e., hyphal anastomosis and nuclear exchange, were independent of the mating type factors but that subsequent events were determined by them; (b) supported the hypothesis that the incompatibility reaction was not constitutive but induced; and (c) provided a basis for the design of experiments to investigate the incompatibility reaction at the molecular level. Methods were developed for isolating large amounts of functional 80 S ribosomes and polysomes from lyophilized, dry-ground mycelia; the polysomes consisted of 2, 3, and 4 ribosomes, were stable at low Mg concentrations, and were sensitive to ribonucleases. The ribosomes were functional in a cell-free system; they incorporated amino acids in the absence of exogenous messenger RNA, reflecting the stability of the ribosome-mRNA complex to lyophilization and grinding.

The molecular basis for the incompatibility reaction was investigated by testing the possibility that cytoplasmic exchange occurred at the same time as nuclear exchange. Homokaryotic mating partners were differentially labeled with deuterium and radioactive amino acids; thus, the appearance of radioactivity in the heavy ribosomes would constitute evidence for cytoplasmic exchange. Ribosomes isolated from such mycelia were separated by CsCl equilibrium density-gradient centrifugation; both the heavy and light ribosomes were radioactive. Thus, exchange of cytoplasm had occurred. The heavy ribosomes were greatly reduced in number after 24 hours of mating in H_2O media, indicating considerable synthesis and rapid turnover of ribosomes. This was confirmed by analysis; after 12 hours of mating in H_2O media, the concentration of heavy ribosomes decreased and ribosomes of hybrid density were present, while after an additional 12 hours, hybrid ribosomes were almost gone and the "light" ribosomes had increased. Additional evidence for cytoplasmic exchange came from the observation that the "light" ribosomes increased in density progressively after 12 and 24 hours of mating. This was interpreted to mean that cytoplasmic D_2O in the D_2O-labeled homokaryon was made available to the mating partner *via* cytoplasmic exchange, and that ribosome synthesis in the developing dikaryon occurred in a common cytoplasm in which D_2O increased progressively as cytoplasmic exchange increased.

REFERENCES TO ADDENDUM

Burnett, J. H. (1968). "Fundamentals of Mycology." St. Martin's Press, New York.

Carlile, M. J., and Dee, J. (1967). Plasmodial fusion and lethal interaction between strains in a Myxomycete. *Nature* **215**, 832.

Collins, O. R., and Clark, J. (1968). Genetics of plasmodial compatibility and heterokaryosis in *Didymium iridis. Mycologia* **60**, 90.

Crandall, M. A., and Brock, T. D. (1968). Molecular basis of mating in the yeast *Hansenula wingei. Bacteriol. Rev.* **32**, 139.

Dennen, D. W., and Niederpruem, D. J. (1967). Regulation of glutamate dehydrogenase during morphogenesis of *Schizophyllum commune. J. Bacteriol.* **93**, 904.

Elliott, C. G., Hendrie, M. R., and Knights, B. A. (1966). The sterol requirement of *Phytophthora cactorum. J. Gen. Microbiol.* **42**, 425.

Esser, K., and Kuenen, R. (1967). "Genetics of Fungi." Springer Verlag, New York.

Hawker, L. E. (1966). Environmental influences on reproduction. *In* "The Fungi" (G. C. Ainsworth and A. S. Sussman, eds.), Vol. II, pp. 435–469. Academic Press, New York.

Kerr, S. (1968). Ploidy level in the true slime mould *Didymium nigripes. J. Gen. Microbiol.* **53**, 9.

Leal, J. A., Gallegly, M. E., and Lilly, V. G. (1967). The relation of the carbon-nitrogen ratio in the basal medium to sexual reproduction in species of *Phytophthora. Mycologia* **59**, 953.

Leary, J. V. (1969). Early mating interactions in *Schizophyllum commune.* Ph.D Thesis, Michigan State University, East Lansing, Michigan.

Marchant, R. (1968). An ultrastructural study of sexual reproduction in *Pythium ultimum. New Phytol.* **67**, 167.

Raper, J. R. (1966). Life cycles, basic patterns of sexuality, and sexual mechanisms. *In* "The Fungi" (G. C. Ainsworth and A. S. Sussman, eds.), Vol. II, pp. 473–511. Academic Press, New York.

Ross, I. K. (1967a). Syngamy and plasmodium formation in the myxomycete *Didymium iridis. Protoplasma* **64**, 104.

Ross, I. K. (1967b). Abnormal cell behavior in the heterothallic myxomycete *Didymium iridis. Mycologia* **59**, 235.

Sutter, R. P., and Rafelson, M. E., Jr. (1968). Separation of β-factor synthesis from simulated β-carotene synthesis in mated cultures of *Blakeslea trispora. J. Bacteriol.* **95**, 426.

Therrien, C. D. (1966). Microspectrophotometric measurement of nuclear deoxyribonucleic acid content in two myxomycetes. *Can. J. Botany* **44**, 1667.

Van den Ende, H. (1968). Relationship between sexuality and carotene synthesis in *Blakeslea trispora. J. Bacteriol.* **96**, 1298.

Wessels, J. G. H., and Niederpruem, D. J. (1967). Role of a cell-wall glucan-degrading enzyme in mating of *Schizophyllum commune. J. Bacteriol.* **94**, 1594.

CHAPTER 4

Algae

L. Wiese

DEPARTMENT OF BIOLOGICAL SCIENCES, FLORIDA STATE UNIVERSITY,
TALLAHASSEE, FLORIDA

Fertilization in algae includes a wide variety of phenomena. In all cases—apart from derivable exceptions—fertilization means the union of two sexually different cells and their nuclei and, by this, the combination of two haplophases into a diplophase. The latter effect is reversed by the segregation of the diplophase into haplophases and in algae always involves a two-step meiosis. The relative location of fertilization and meiosis determines the life cycle as haplont, haplodiplont, or diplont. Fertilization is invariably connected with sexuality. In its simplest manifestation, sexuality is the bipolar differentiation which causes the affinity between the interacting stages at fertilization. The combination of fertilization with propagation in sexual reproduction, however, is a secondary phenomenon.

I. Manifestation Forms of Fertilization in Algae

Fertilization may occur between sex cells (gametes) differentiated as egg and sperm (*oogamy*) (Fig. 1), between two gametes differing from each other only in degree (*anisogamy*) (Fig. 2), or between two gametes which are morphologically alike (*isogamy*) (Fig. 3). Even with perfect morphological isogamy, the copulating gametes are of different sex (cf. Section IIB,C). To avoid the terms sex and male or female to describe

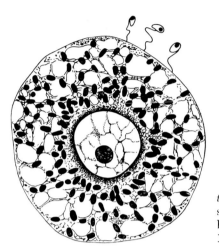

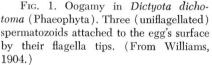

Fig. 1. Oogamy in *Dictyota dichotoma* (Phaeophyta). Three (uniflagellated) spermatozoids attached to the egg's surface by their flagella tips. (From Williams, 1904.)

morphological indifferences, some authors prefer to speak of mating types and to designate the involved bipolarity with ($+$) and ($-$). In unicellular forms the entire organism may act as a gamete and fuse with another one to form the zygote (*hologamy*). In certain such cases, the potential to react as a gamete seems to be a permanent feature. Usually, however, the gametic state is temporary, i.e., the cell normally, in its vegetative state, grows and divides by mitosis and only under special conditions differentiates intracellularly into a gamete (cf. investigations on *Chlamydomonas* species, Sager and Granick, 1954; Förster and Wiese, 1954b; Förster, 1957, 1959; Stifter, 1959; Trainor, 1959; Hartmann, 1962; Kates and Jones, 1964).

Parthenogenesis in its various modes is wide-spread in algae and is more common in some systematic units than in others (cf. Ettl *et al.,* 1967). Parthenogenetic development of the sex cell, normally of the gynogamete, may be facultative occurring in the absence of a sex partner or connected with a disturbance of the fertilization process. In cases of isogamy or of slight anisogamy, the gametes of both sexes may possess the potency to develop parthenogenetically (*Ectocarpus,* cf. Hartmann, 1956; *Cladostephus,* Schreiber, 1931; *Halopteris,* Ernst-Schwarzenbach, 1957; *Monostroma grevillei,* Kornmann, 1962; *Percursaria percursa,* Kornmann, 1956; *Mougeotia heterogama,* Geitler, 1958a; *Chlamydomonas* species as *C. chlamydogama,* Bold, 1949; *Spirogyra* species, cf. Czurda, 1937; Transeau, 1951, etc.). In other algae parthenogenesis is constitutional (certain diatoms, *Chara crinata,* Ernst, 1918; *Spirogyra mirabilis,* Czurda, 1937; *Spirogyra dimorpha,* Geitler, 1949b, etc.).

Included in the algae is one of the last groups of organisms, the

Cyanophyta, in which sexuality and fertilization either do not exist or have not yet been demonstrated. In some groups, for example in Eugleno-phyta and in the genus *Chlorella*, the existence of sex and fertilization is still dubious; in other taxonomic units fertilization was known only sporadically, however, closer inspection has revealed more and more cases as in the centric diatoms, Dinophyta and Chrysophyta (cf. Fott, 1959, 1964; Kristiansen, 1963). The only recent discovery of fertilization in such an intensively studied object as *Scenedesmus* (Trainor and Burg, 1965) should caution investigators that many forms which are now considered "sex-less" may simply never have been exposed to appropriate conditions for the manifestation of sex and fertilization.

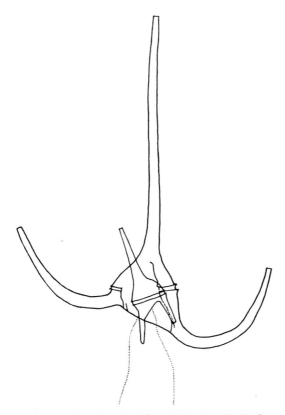

Fig. 2. Anisogamy in *Ceratium horridum* (Dinophyta). Early copulation stage between the large gynogamete and the small androgamete. The gynogamete can be distinguished from a vegetative cell only by its different nuclear structure, whereas the androgametes arise by repeated mitotic division from one vegetative cell. (From von Stosch, 1964.)

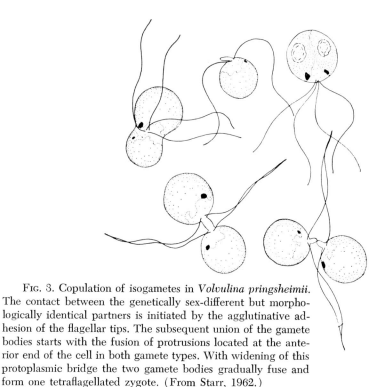

Fig. 3. Copulation of isogametes in *Volvulina pringsheimii*. The contact between the genetically sex-different but morphologically identical partners is initiated by the agglutinative adhesion of the flagellar tips. The subsequent union of the gamete bodies starts with the fusion of protrusions located at the anterior end of the cell in both gamete types. With widening of this protoplasmic bridge the two gamete bodies gradually fuse and form one tetraflagellated zygote. (From Starr, 1962.)

II. Morphology, Physiology, and Biochemistry of Fertilization in Algae

Depending on the particular mode of gamete union in algae, rather different mechanisms are involved in effecting the approach of the gametes, their contact, and their final fusion into a zygote. The mode may vary greatly within a restricted taxonomic unit, for instance in the genus *Chlamydomonas* it varies both morphologically and physiologically (isogamy, anisogamy, oogamy; with or without chemotaxis; with or without flagella agglutination). Larger taxonomic units may show a uniform type (egg–sperm fertilization in the Charophyta; egg–spermatium fertilization in the Rhodophyta). The mode may even characterize a particular taxonomic unit (Zygnematales).

In each case, fertilization proceeds as an ordered, harmoniously correlated sequence of individual steps. Many features and mechanisms can be resolved by direct observation of the copulation process. Some steps are under the control of special substances specific for species, sex, and phase. The algae are unusually favorable objects for the exploration of the physical and chemical interactions between the gametes. Those forms that can be cultivated axenically under defined conditions are of special value. With such genetically and physiologically controlled material the effects of endogenous and exogenous factors on gametogenesis and gamete copulation can be identified in clear-cut experiments.

A. Induction Effects and Induction Substances

Induction effects between the sex partners and between the gametes are responsible for the harmonious course of fertilization by conditioning the sequence of the individual steps and their coordination. Induction effects can be initiated by direct contact of the interacting stages (for instance, completion of the development of the oogonium mother cell in *Oedogonium* after the attachment of the androspore; cf. Section II,B) or in the aqueous milieu by diffusing substances. In accordance with their different functions such substances are specific in character. Different kinds of induction effects will be mentioned in Sections II,B,D,E, and Section III.

Substances inducing sexualization exist in several algae. They are secreted by the vegetative cells of one sex and initiate sexual differentiation in competent cells of the opposite sex. A pair of such substances was demonstrated by specific filtrate effects in *Glenodinium lubiensiforme* (Diwald, 1938). The substances are pH-sensitive and thermolabile. In *Volvulina pringsheimii*, gametogenesis occurs only after mixing of the two sexually different strains (Starr, 1962), the same holds true for *Gonium* (Stein, 1958b) and *Pandorina* (Coleman, 1959). The diatom *Rhabdonema adriaticum* forms sexual stages only if cells of both sexes are cultivated together under crowded conditions (von Stosch, 1958). Biebel (1964) reports a similar effect in the desmid, *Netrium digitus*. These phenomena can best be explained by material induction between cells of different sex.

A substance which controls the type of sexual differentiation has been evidenced in *Volvox aureus* (Darden, 1966). In a homothallic, dioecious strain (colonies of the identical genotype are either male or female), this component causes male differentiation of young colonies. Male-inducing substance (MIS) is nondialyzable (mol. wt. over 200,000), inactivated by proteases, and heat-stabile (100°C, 30 minutes). Even high-titered preparations never induce more than 50% males indicating some alternative sexual predetermination or some other regulatory mechanism. In another strain of *V. aureus*, also homothallic and in principle dioecious, males are almost completely absent. The eggs develop by parthenogenesis. The parthenogenetic strain, however, also produces MIS as demonstrable with test material of the first strain. A nonrespondence of the parthenogenetic strain's cells must account for the absence of males (Darden, 1968).

Recent investigations on other species and strains in *Volvox* detected a number of similar substances all of which are secreted by sexually differentiated male colonies and cause sexualization of female colonies (FIS, female inducing substances) (Kochert, 1968; Starr, 1968). In the heterothallic *V. carteri*, for instance, no sexualization of female colonies occurs in absence of this substance. The sexual differentiation of male colonies occurs in response to environmental conditions independent of the presence of female colonies. These substances affect specific stages in development which may be different from species to species. All these substances which are effective in low concentrations, are species- or even strain-specific. Several of these components reveal their protein character by temperature lability and by sensitivity to proteases.

Addition of androgametes to swarming gynogametes of *Ectocarpus siliculosus* causes the gynogametes to settle down (cf. Section II,B). As yet, no special "settling-down substance" excreted by the male gametes has been demonstrated by an appropriate filtrate test. Settling down of one gamete type upon addition of the opposite one precedes copulation in the isogamous heterothallic *Chlamydomonas chlamydogama* (Bold, 1949).

B. Chemotaxis, Chemotropism, and Chemotactica

Chemotactic reactions between sexually different gametes are widespread in the various groups of algae and occur especially in Chlorophyta and Phaeophyta.

Chemotaxis at fertilization is always of the kind in which one gamete type emits a particular chemotactic substance and the other responds positively. The report of mutual attraction between $(+)$ and $(-)$ gametes in *Chlamydomonas eugametos* (Moewus, 1938, 1939) was not confirmed by others (Förster and Wiese, 1954a; cf. Kuhn and Löw, 1960). Reciprocal chemotactic attraction could, nevertheless, exist on certain levels of organization and might be responsible for the mutual approach of gamonts in certain pennate diatoms such as *Eunotia arcus*. In *Cosmarium botrytis*, sex cells of one clone act as emanators of a chemotactic substance in combination with certain clones of the opposite sex, but as recipients in the combination with others (Fig. 4; Brandham, 1967).

Chemotaxis, based upon a concentration gradient of the chemotactic substance, occurs most frequently in cases where the emanating gametes are sessile and the attracted ones are motile, as seen in most pronounced

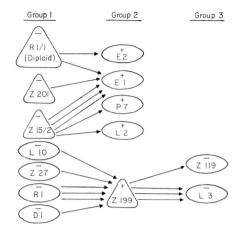

FIG. 4. Diagram showing anisogamous mating behavior in fourteen clones of *Cosmarium botrytis*. Biradiate clones are represented by ellipses each containing the clone number and the mating type. Morphologically abnormal clones, both haploid and diploid, are represented by triangles. Arrows extend from clones which are moving at conjugation to those which are stationary. Multiple arrows indicate replicate crosses. More details in the text. (From Brandham, 1967.)

form in egg–sperm fertilization* (*Sphaeroplea, Chlorogonium, Oedogonium, Fucus, Ascophyllum, Chlamydomonas suboogama,* etc.). In cases in which gametes of both sexes are motile, such a situation is often created secondarily.

For example, in *Ectocarpus siliculosus*, male and female gametes are morphologically alike and initially both are motile. At mixing, a physiological sex difference is manifested: one gamete type settles down attaching itself by its anterior flagellum to the substrate, and only then chemotactically attracts the opposite one to form the clusters discussed below (see Fig. 7) (Berthold, 1881; M. Hartmann, 1925, 1934, 1937). The chemotactic substance has recently been isolated (Müller, 1967, 1968).

In *Cutleria multifida*, immobilization of gametes of one sex, preceding chemotaxis, occurs autonomously as part of a physiological maturation process (M. Hartmann, 1950). The egglike gynogametes are initially motile (two flagella) and unfertilizable even when in direct contact with the spermlike androgametes. After a certain period of time, the gyno-

---

* Extrapolating from oogamy and anisogamy to cases of isogamy, the attracting gametes are designated ♀, the attracted ones ♂. However, the dual function of one gamete type in *Cosmarium* greatly disqualifies this extrapolation.

gametes settle down. Only then do they release the agent that attracts the androgametes and only then can they be fertilized.

Another relation between chemotaxis and the immobilization of cells of one sex was detected in the isogamous heterothallic *Cosmarium botrytis* (Brandham, 1967). In the vegetative stage, both sexes of these unicellular forms are motile and move independently of each other in mixtures. At sexualization, the cells of one sex cease locomotion and release the chemotactic material as revealed by the immediate, directed approach of their motile prospective copulation partners. Although chemotaxis is unilateral in each combination of two sexually different clones, the reaction pattern of all the clones examined (Fig. 4) reveals the unique situation mentioned above. Cells of one particular (+) clone (Z 199+), recognizable by a morphological marker (*triradiate*), attract the cells of several (−) clones (group 1). However, gametes of this same (+) clone are attracted by two other (−) clones (group 3). Additional combination tests should establish if this potential dual behavior is shared by all (+) clones.

The chemotactic cluster formation reported in *Chlamydomonas paupera* (Pascher, 1931b) and *Hydrodictyon* (Mainx, 1931; Pockock, 1960) needs reinvestigation to correlate the reported phase specificity with either the *Ectocarpus*, *Cutleria*, or *Cosmarium* type. *Colpomenia sinuosa* behaves as *Ectocarpus* (Kunieda and Suto, 1938).

In other cases, the emission of chemotactic agents is not restricted to a certain phase. In *Fucus* (A. H. Cook *et al.*, 1948) and *Chlamydomonas suboogama* (Tschermak-Woess, 1959), the chemotactic agents are already diffusing from the oogonia before the eggs are released.

Chemotaxis is also possible when gametes of both sexes are and remain motile but exhibit enough difference in their locomotive activity that a gradient may become effective as in the anisogamous heterothallic *Derbesia tenuissima* (Ziegler and Kingsbury, 1964) and in the reported "pursuing" of sexually mature female colonies by the organized sperm packets in *Eudorina conradii* (Goldstein, 1964). Also in other species of *Eudorina* and in *Volvox* (Darden, 1966), the course of fertilization suggests chemotactic attraction of the sperm packets to the female colonies.

Chemotactic attraction between equally motile isogametes exists in *Chlamydomonas moewusii* var. *rotunda* (Tsubo, 1961b). The (−) gametes or their cell-free supernatant attract (+) gametes but there is no reciprocal attraction. The reactivity of the (+) sex and the emanation of the chemotactic material by the (−) sex are restricted to the gametic stage. Nevertheless, it is an open question whether chemotaxis really plays an essential part at fertilization of this form since the proper sex cell contact and adhesion is initiated by flagella agglutination as in

Chlamydomonas spp. without chemotaxis. Tsubo distinguished chemotaxis from the agglutination mechanism by demonstrating that osmium-killed (−) gametes agglutinate the (+) gametes in the absence of any chemotactic agent. Four other isogamous and heterothallic species (*Chlamydomonas eugametos, Chlamydomonas moewusii, Chlamydomonas moewusii* var. *tenuichloris,* and *Chlamydomonas reinhardti*) do not possess any chemotactic activity between their sex cells and are inert against their supernatants in all possible intra- and interspecific combinations. When checked for reactivity against the supernatants of *C. moewusii* var. *rotunda, C. reinhardti* gametes of both sexes failed to respond, whereas the gametes of both sexes from the other three species responded positively to the (−) supernatant (cf. Table I). This responsiveness is not accompanied by another compatibility step; namely, flagella agglutination does not occur. In *C. moewusii* var. *tenuichloris,* vegetative cells do not react to the chemotactic material. The reactivity in this species, too, is restricted to the gamete stage yet not specified for one sex as in *C. moewusii* var. *rotunda.* The reaction pattern in the five species suggests how chemotaxis may have evolved to assume a specific function at fertilization.

How efficiently selection can place the chemotactic reaction is displayed by the various types of fertilization in *Oedogonium.* In direct egg–sperm fertilization as in *Oedogonium cardiacum,* the spermatozoids are chemotactically attracted to the egg, enter through a pore in the oogonium wall, and are trapped in the space around the globular egg (Hoffman, 1960, 1961) or are guided by a tubelike device to the egg (cf. Hirn, 1900). In a heterothallic, nannandrous species two chemotactic reactions in dependent sequence constitute single steps in the series of fertilization events (Rawitscher-Kunkel and Machlis, 1962) (Fig. 5). In nannandrous species, the male-determined cells release motile reproduction cells, the androspores, which are not destined to fertilize the egg but develop into a few-celled filament, the dwarf male. The initiation of sexual differentiation is environment-induced in both sexes and independent of the presence of the complementary sex. The male filaments produce the androspores which develop autonomously up to the production of sperm. Female differentiation proceeds autonomously only until the oogonium mother cell (OMC) is formed. The OMC's chemotactically attract the androspores which attach by their apical poles. A chemotactic substance was confirmed by the capillary method. The attachment initiates a closely interrelated sequence of steps. First, the OMC is induced to divide into an upper cell, the oogonium, and a lower suffultory cell. From this time onward, the chemotactic material is no longer emitted. If the attachment of androspores is prevented by coating the OMC with a

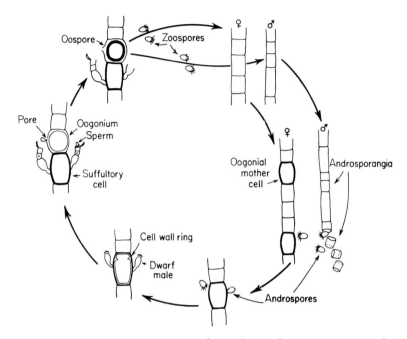

Fig. 5. Diagrammatic representation of sexual reproduction in a nannandrous, heterothallic species of *Oedogonium*. For explanation see text. (From Rawitscher-Kunkel and Machlis, 1962.)

thin agar layer, the OMC continues to attract androspores (experimentally checked for 21 days) and the division into the two cells does not occur. Female differentiation beyond the OMC stage must be triggered by the direct contact of the attached androspore. Thus the autonomous differentiation of the male sex is a prerequisite for the induction-requiring female differentiation resulting in a rather plastic system and in a time coordination of the steps. The attached androspore next elongates and develops into a two-celled dwarf male, the apical cell of which, the antheridium, forms two sperms. For a reason not yet known this growth of the dwarf male occurs along the female filament. This influence is not exerted by agar-coated OMC's; androspores attached to them grow perpendicularly to the filament. During the maturation of the dwarf males, the OMC undergoes the mentioned cell division and the oogonium secretes a thick gelatinous sheath which includes the tip of the bent dwarf males. The released sperm are trapped in this gel. After about 1 hour of movement at random in the gel, the sperm begin to aggregate at a certain spot in the upper third of the oogonium. "Suddenly, too fast for observation so far, a protoplasmic papilla appears through the wall of the oogonium and all sperm in the vicinity tend to attach to this papilla by their anterior ends. In a second, or even less, the papilla is withdrawn through the pore in the wall taking with it one single sperm. The remaining sperm hover about the pore for perhaps 5 minutes before they disperse in the gel." (Rawitscher-Kunkel and Machlis, 1962.) Four hormone-controlled steps precede this unique engulfment mechanism: (*1*) the chemotaxis of the androspores; (*2*) the directed growth of the dwarf males; (*3*) the triggering of the division of the OMC; and (*4*) the chemotaxis of the sperm to the prospective opening of the oogonium.

Tschermak-Woess (1959, 1962) reports chemotaxis in the oogamous homothallic *Chlamydomonas suboogama*. In a peculiar mode of sex determination, the gametangia mother cell normally divides into four cells, three of which represent oogonia. The fourth cell, as spermatogonium, divides twice to yield four biflagellated spermatozoids. Each oogonium releases one gynogamete with two nonfunctional flagella. The gynogametes secrete a thick jelly coat of their own within the larger gel surrounding the gametangia mother cell (Fig. 6). The spermatozoids are chemotactically attracted by the oogonia and by the gynogametes which they fertilize; however, they never appear to be attracted by their three aunt gametes. The author interprets this feature which favors outbreeding as depending on some obligatory maturing process that enables the spermatozoid to react with the female cells after the former are released from the gametangia mother cell. Attracted gametes are mechanically trapped when they enter the jelly coat of another gametangia mother cell and that of the gynogamete. The spermatozoid penetrates with its apical "perforatorium." However, it should be noted that a perforatorium comparable to

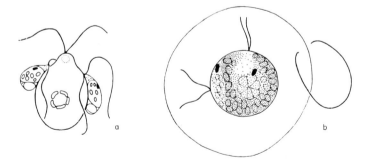

Fɪɢ. 6. Fertilization in *Chlamydomonas suboogama*. (a) Two chemotactically attracted spermlike androgametes gliding on the gynogamete's surface. The gelatinous coat of the gynogamete is not shown. (b) Young zygote within the jelly coat secreted by the gynogamete before copulation. The empty cell wall of the oogonium is attached. Within the zygote the part originating from the spermatozoid can be distinguished from that of the gynogamete. (From Tschermak-Woess, 1959.)

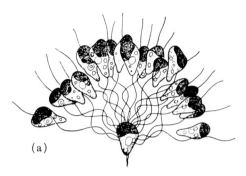

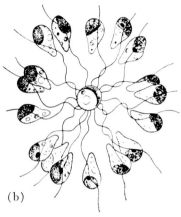

Fig. 7. Clusters of gametes in *Ectocarpus siliculosus*. (a) Side view; (b) from above. Many androgametes are attached by means of the tips of their front flagella to the gynogamete's body. The gynogamete is attached to substrate by its (shortened) front flagellum. For further explanation see text. (From Berthold, 1881.)

the acrosome of Metazoa, still needs to be demonstrated in algae (cf. Geitler, 1954).

In cases of iso- and anisogamy between motile gametes, the chemotactic approach of one gamete type toward the opposite one results in the formation of gamete clusters in which the attracting gamete is surrounded by a number of the attracted ones. This was shown in the earliest report on this type of cluster formation in *Ectocarpus siliculosus* (Berthold, 1881) (Fig. 7) and experimentally evidenced by differential vital staining of the two sexes (Bauer and Hämmerling cited in M. Hartmann, 1934). In such anisogamous species as *Giffordia secunda,* the proportion of the two gamete types in the clusters is morphologically evident. Only one zygote emerges from each cluster unit in contrast to the many zygotes that arise from clusters based upon flagella agglutination between many gametes of both sexes (cf. Section II,C).

In the above clusters in *Ectocarpus,* as generally in the Phaeophyceae, each androgamete attaches itself to the gynogamete's body by the tip of its anterior flagellum. When one androgamete starts to fuse with the gynogamete, the flagella of the other androgametes become detached from the surface of the gynogamete; this suggests that the attachment is rather loose. The same attachment, however, leads to the peculiar rotation of the *Fucus* egg (cf. Farmer and Williams, 1898) by the combined action of all sperm's posterior flagella indicating a true adhesion stabile enough to permit the transmission of the impulses. It needs to be emphasized, however, that chemotaxis explains only the directed approach of the male to the female gamete; this often leads to some trapping device (cf. Machlis and Rawitscher-Kunkel, 1967). The mechanism of interaction of the male flagella with the gynogamete's surface represents an additional and unsolved problem.

Our present restricted knowledge leaves undecided at what point chemotaxis ends. It may simply produce an approximation of the motile gametes to the sessile one, leaving the final contact and adhesion to another mechanism, or, as indicated by some data, it may be more directly involved with the surface contact. A functional surface structure may be envisaged as responsible for the last step of chemotaxis and for the initial agglutinative fusion step. Just as the chemotactic substance is supposed to interact with some adsorbing receptor on the androgamete, one may visualize a similar interaction of these receptors with the emitting surface of the gynogamete. No such case is sufficiently investigated. Whatever the material basis for the contact in the Phaeophyceae may be, it is characterized by its temporariness. Certain data suggest that in the gamete contact in the brown algae two distinct processes may be operating, namely, the chemotactic attraction and the final agglutinative contact. Kotte (1923) was not able to demonstrate chemotaxis with a viscous supernatant of *Fucus* eggs in capillaries, but he observed a specific sticking of the spermazoid flagella to a slimy component in experiments which need reinvestigation especially in respect to the role of the jelly coat *in situ.* Köhler (1967) reports agglutination of the (uniflagellated) spermatozoids by isolated egg jelly in *Dictyota.* The clustering in *Ectocarpus,* then, could be the result of a chemotactic attraction and an agglutinative adhesion. The reversal of contact of the supernumerary androgametes at fertilization may be attributed to a mechanism comparable to the flagella separation after pairing in *Chlamydomonas moewusii* (cf. Section II,C).

Also in *Chlamydomonas suboogama* (Tschermak-Woess, 1959), *Cosmarium* (Brandham, 1967), *Eudorina,* and *Volvox,* chemotaxis may—in addition to the directed approximation of the partners—effect their close association enabling them to perform the next step. In *Eudorina,* the attached sperm packet ruptures the outer colonial envelope by an active "burrowing" process (Goldstein, 1964). Excellent photographs of the fertilization in *Volvox* point to an enzymatic lysis of the colonial matrix (Darden, 1966). In both genera, the sperm packets later dissociate and the individual spermatozoids approach and fertilize the eggs. Gamete fusion has not been observed but is initiated by the spermatozoid's epibolical gliding along the egg's surface.

Another unexplained phenomenon is the sudden withdrawal of the attracted gametes as soon as the attracting one has been fertilized (*Fucus, Ascophyllum, Derbesia, Ectocarpus, Cladostephus, Chlamydomonas sub-*

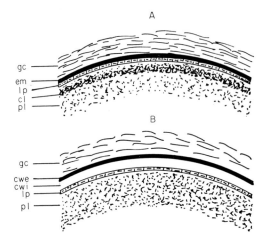

FIG. 8. Diagram of the egg surface in *Fucus*. (A) Mature, unfertilized egg. (B) Fertilized egg. gc—Gelatinous coat; em—egg membrane; lp—lipoprotein membrane; cl—cortical layer; pl—cytoplasm; cwe—cell wall, outermost part; cwi—cell wall, innermost part. (From Levring, 1952.)

oogama, Chlamydomonas paradoxa, etc.). The cause of this phenomenon which aids in preventing polyspermy is unknown. Suggested mechanisms are (*a*) a sudden stop in the emission of the chemotactic agent, (*b*) the emission of a substance causing negative chemotaxis (Levring, 1952), and/or (*c*) an alteration of the egg's surface after fertilization (Levring, 1949, 1952) which in the Phaeophyceae strikingly resembles the situation in the sea urchin and other Metazoa (Fig. 8).

1. Chemotropism

In algae, nothing is known about substances that direct the chemotropic reaction. Their existence, however, is suggested by certain phenomena such as the directed growth and bending of the antheridia toward the oogonia in *Vaucheria,* and in the directed growth of the copulation tubes in diatoms (see Figs. 19 and 20), desmids, and in *Spirogyra.*

2. Sex Chemotactica

Additional questions regarding the chemotactically active substances concern their chemical nature, their specificity, and their mode of action.

The data now available (cf. Machlis and Rawitscher-Kunkel, 1967) fail to point to any class of substances specifically involved in this function. Experimental data on *Fucus* (A. H. Cook et al., 1948; A. H. Cook and Elvidge, 1951), *Ectocarpus* (Müller, 1968), and *Chlamydomonas moewusii* var. *rotunda* (Tsubo, 1961b) demonstrate that simple and unrelated compounds can act as sex chemotactica. In all probability, these compounds merely mimic the effect of the natural one(s), and their very different chemical nature (cf. Table I) indicates that the reaction can be nonspecifically induced. The demonstrated lack of specificity of these agents, however, is in sharp contrast to other cases in which the reacting gametes distinguish between species-specific chemotactic substances and between stereoisomers (cf. Machlis and Rawitscher-Kunkel, 1967). Unequivocal information can only be obtained by the analysis of the effective natural principle in the gametes' supernatant. To complete current research in *Oedogonium* and *Allomyces* (cf. Machlis *et al.*, 1966), the objects of choice in algae should be dioecious oogamous *Chlorogonium* species and especially *Sphaeroplea annulina* where Pascher (1931a) demonstrated the existence of a sperm attractant in the oogonial fluid.

Species specificity of the chemotactic system has been evidenced between two dioecious species of *Oedogonium* (Hoffman, 1960, 1961). The heat-stabile, nonvolatile agents are emitted only during the sexual phase of the female filaments; they act exclusively on spermatozoids but not on zoospores and remain in the water phase after extraction with ether and chloroform. Species specificity of gamete chemotaxis has also been shown between *Oedogonium cardiacum* and *Oedogonium geniculatum* (cf. Machlis and Rawitscher-Kunkel, 1967).

In *Fucus*, as in *Chlamydomonas moewusii* var. *rotunda*, the chemotactic agent is characterized by its hydrophobia. In spite of intense research, the analytical data in *Fucus* (A. H. Cook *et al.*, 1948; A. H. Cook and Elvidge, 1951) gave no conclusive results as to the nature of the volatile, egg-secreted agent(s). The substances isolated from *Fucus vesiculosus* and from *Fucus serratus* attract sperm of both species and additionally that of *Fucus spiralis*. The active principle in the supernatant of *Fucus* eggs is highly stable chemically and extractable by chloroform and ether. Its sperm-attracting effect is simulated by several hydrocarbons, such as n-hexane (in a dilution of 10^{-6} to $10^{-7}\,M$) and also by certain ethers and esters. However, none of these substances corresponds in all its characteristics to those of the biological compound. The authors conclude that the majority of substances that attract *Fucus* spermatozoids in a manner indistinguishable from that of the natural principle do so by a mechanism which is physical in essence.

The sex chemotacticum in *Ectocarpus siliculosus* is a small molecular volatile substance with a characteristic odor. In the capillary test the isolated substance attracts the androgametes and affects their movement as do settled gynogametes. Very similar to the situation in *Fucus*, the action of this component can be simulated by a series of different substances as n-hexane, cyclohexane, etc. (Müller, 1967, 1968).

The chemotactic agent(s) in *Chlamydomonas moewusii* var. *rotunda* is volatile and heat-resistant (130°C, 30 minutes) and can be replaced by a variety of compounds (Table I). n-Hexane gives a positive reaction in the characteristic pattern with $(+)$ but not with $(-)$ gametes or with vegetative cells.

TABLE I

CHEMOTACTIC RESPONSE OF FIVE SPECIES OF *Chlamydomonas* TO GASEOUS SUBSTANCES AND TO THE (−) AND OF THE (+) GAMETES OF *Chlamydooronas moevusii* VAR. *rotunda*[a]
SUPERNATANT OF THE

Chlamydomonas	Sex	CO	CO₂	C₂H₄	C₂H₂	H₂S	N₂	(Coal gas)	(−) Supernatant	(+) Supernatant
C. moevusii var. rotunda	plus	+	0	+	+	0	0	+	+	0
	minus	0	0	0	0	0	0	0	0	0
C. moevusii var. tenuichloris	plus	±	±	+	+	/	/	+	+	+
	minus	±	±	+	+	/	/	+	+	+
C. moevusii	plus	±	±	+	+	/	/	+	+	+
	minus	±	±	+	+	/	/	+	+	+
C. eugametos	male	±	±	+	+	0	/	+	+	+
	female	±	±	0	0	0	/	0	0	0
C. reinhardti	plus	/	/	0	0	/	/	0	0	0
	minus	/	/	0	0	/	/	0	0	0

N_2

[a] +, Positive reaction; ±, positive, but weak reaction; 0, negative reaction; /, not examined. (After Tsubo, 1961b.)

C. Agglutinative Sex Cell Adhesion; Gamete Agglutinins; Mating-Type Reaction and Mating-Type Substances

In many cases, gamete contact apparently ensues without preceding chemotactic attraction. Sexually different gametes meet at random and immediately adhere by means of an extremely efficient agglutination mechanism. This type of gamete contact is most conspicuous in forms in which both gamete types are motile and especially where the initial contact takes place at the flagella tips. Analogous to the comparable reaction between the sex partners in ciliates, the agglutinative flagella adhesion between gametes of different sex is designated as mating type reaction.

The responsible mechanism must work on the principle of complementarity. Since only gametes of different sex agglutinate, the interacting components must be sex-specific. In addition the mechanism possesses features that cause an extreme selectivity of gamete adhesion as displayed in the species specificity of fertilization.

In dense suspensions of sexually reactive, complementary gametes, many cells may adhere in the form of clusters, clumps, or agglutinates. This can occur when gametes are released simultaneously from sexually different gametangia or when an "epidemic" sexualization of a population results from a suitable constellation of environmental factors, or under certain experimental conditions with dense gamete suspensions. Many gametes of both sexes agglutinate together by means of their flagella tips. The size of the cluster depends on the density of the suspensions and may involve from three to a dozen or more gametes. The individual clusters can join to form larger aggregates of several hundred cells. Ordinarily, as many zygotes are subsequently formed as there are pairs of complementary gametes. This type of cluster, based on random collision between sexually different gametes, is different in principle from a cluster caused by chemotaxis.

All copulations of this kind occur in two steps. The flagellar agglutination constitutes the decisive initial contact and is succeeded by pairing; within the clusters, two sexually different gametes come together and form a second and irreversible contact by lateral or head-on fusion of the two gamete bodies (cf. Section II,D). As soon as the second contact is established, the flagellar contact is disjoined and the pair detaches from the cluster. The joined cells behave as one physiological unit and fuse to form the zygote. Very often, two agglutinated gametes free themselves from the cluster before pairing has ensued and fuse subsequently.

The specificity of the agglutination mechanism guarantees the adhesion of sexually different gametes and, in addition, the species specificity of this contact, i.e., the recognition pattern of one species' gametes is dependent on the peculiarities of the agglutination mechanism. Since flagellar agglutination is temporary and is the first in the sequence of steps leading to zygote formation, flagellar agglutination may trigger the chain of processes involved.

The separated flagella of paired gametes exhibit altered surface properties—they lose entirely their capacity to agglutinate with unmated gametes. This change does not result from flagellar contact with opposite sex cells; rather it occurs only after this contact leads to final pair formation. The ultimate fate of the flagella varies among species; they may be shed or retracted.

As demonstrated in *Ulva* (Levring, 1949), *Chaetomorpha* (Köhler, 1956), and *Chlamydomonas eugametos* and *Chlamydomonas moewusii* (Wiese and Jones, 1963; Wiese and Metz, 1969), flagella agglutination and gamete pairing result from different mechanisms as evidenced by their different sensitivity to dodecylsulfate, SH reagents, and proteases.

Gamete union by a separate initial contact step (spatially isolated at the flagella tips) and a subsequent fusion of the gamete bodies is common in Chlorophyta (*Dasycladus, Ulva, Chaetomorpha, Cladophora, Tetraspora,* and *Acetabularia, Chlamydomonas* species, *Dunaliella,* etc). In colonial Volvocales, fertilization initiated by flagella contact is restricted to the isogamous genera (*Gonium, Pandorina, Volvulina, Astrephomene*), whereas gamete contact at oogamy (*Eudorina, Volvox,* and unicellular oogamous species of *Chlamydomonas* and *Chlorogonium*) proceeds differently even in forms with flagellated eggs. In the former group, flagellar contact may occur between the gametes while still in their colony [*Pandorina morum* (Coleman, 1959), *Astrephomene* (Pockock, 1953; Stein, 1958a), *Volvulina steinii* (Pockock, 1953; Stein, 1958a; Carefoot, 1966)] or after their release [*Gonium pectorale* (Stein, 1958b) and *Volvulina pringsheimii* (Starr, 1962)]. The contact between sexually mature colonies of different sex in *Pandorina morum* triggers the swelling of the colony, the softening of the intercellular matrix, and the gametes' release. In addition it initiates copulation between gametes after their liberation (Coleman, 1959).

Details of this type of fertilization were investigated in dioecious, iso-, and hologamous chlamydomonads. Upon mixing the two gamete types, sexually different gametes agglutinate at once forming the characteristic clusters in less than a second. No reaction occurs in a suspension of gametes of only one sex or in any possible combination of gametes with vegetative cells or upon mixing vegetative cells of both sexes. Thus the capacity to agglutinate characterizes, and is restricted to, the gametic stage of the *Chlamydomonas* cell.

During differentiation of the vegetative cell into a gamete, the sexually specific agglutination mechanism develops in the form of special mating-type substances. Since agglutination occurs exclusively at the flagella tips, this must be the site of mating-type substance localization.

Finally, the agglutination is species-specific. In summary, the agglutination reaction occurs only between sex-specific mating-type substances of gametes of complementary mating type; as such the reaction is species-, sex-, phase-, and site-specific.

The mating-type reaction is further characterized by a pronounced dependence on the electrolyte composition of the medium and by its pH-dependency indicating that electrostatic interactions are involved. The reaction needs Ca^{++} ions which can be replaced by Mg^{++} ions (Lewin, 1954a; Tsubo, 1961a; cf. Wiese, 1965). Monovalent cations inhibit the reaction in concentrations higher than $5 \times 10^{-3} M$; NH_4^+ ions even at $5 \times 10^{-5} M$. The Ca^{++} is also required for the flagella agglutination in *Volvulina* (Carefoot, 1966).

Sex cell agglutination does not depend on the living condition. Gametes killed under appropriate conditions will agglutinate with living complementary ones [*Chlamydomonas moewusii* (Hutner and Provasoli, 1951), *C. moewusii* var. *rotunda* (Tsubo, 1961b), *Chlamydomonas eugametos* and *C. moewusii* (Wiese and Jones, 1963)]. Also, isolated gamete flagella agglutinate complementary living gametes (Wiese and Jones, 1963).

Within the clusters of agglutinated gametes, each cell becomes joined with its final mate. Mate change between the involved flagella can easily be observed in the smaller clusters with phase contrast optics. Each cell engages in clumping only until it joins the complementary cell with which it will finally fuse. Supernumerary unmated cells will still agglutinate with fresh complementary gametes. In paired gametes of *C. eugametos* and *C. moewusii,* flagella agglutinability terminates with the establishment of papilla or body contact, whereas the flagella of unpaired cells continue to agglutinate. Thus the agglutination mechanism continues to operate until interrupted by pairing. If pairing is prevented by reagents which do not interfere with the agglutinability (dodecyl sulfate, β-mercaptoethanol, 0.0125% trypsin), the agglutination of gametes persists indefinitely (Wiese, unpublished observations).

The components responsible for flagellar agglutination, namely the mating-type substances, are localized and function at the flagella tips. When experimentally isolated or spontaneously detached the substances still possess the capacity to interact. When added to the opposite gamete type, they cause *isoagglutination*—a component in the supernatant of female gametes makes male gametes agglutinate, and a component in the supernatant of males causes isoagglutination between female gametes

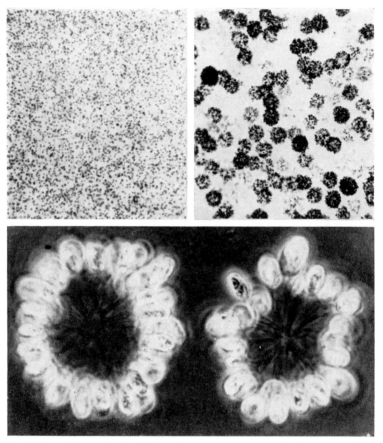

FIG. 9. Isoagglutination in *Chlamydomonas eugametos*. Androgametes (top left) in cluster formation (top right) after addition of the isoagglutinin prepared from the supernatant of gynogametes (Magnification: × 80). Typical clusters of isoagglutinated androgametes (bottom) (Magnification: × 900). Because of overlapping adhesion of their flagella tips, the single gametes stick together and are arranged in a hollow sphere with the diameter of about twice the flagella length.

(Fig. 9). No reaction occurs with gametes of the same sex or with vegetative cells of either sex or with gametes of incompatible species. Thus, these isoagglutinins share the specificity pattern of the mating-type substances *in situ* with regard to species, sex, phase, and site in all details and are, thereby, identified as the components of the flagellar surface secondarily released into the medium (Wiese, 1965). Isoagglutination is explained by the assumption that the isoagglutinin is bi- or multivalent with respect to combining sites specific for interaction with complementary combining sites of cells that are agglutinated (cf. Metz, 1967).

In both isoagglutinations, then, the specific initial sex cell interaction is isolated, and the mechanism and the chemistry of the interaction can be investigated by analyzing the isoagglutinins and their mode of action on the gametes of the opposite sex.

The isoagglutination of gametes of either sex reverses spontaneously and the isoagglutinating activity of the supernatants disappears. Furthermore, the gametes released from isoagglutination can isoagglutinate again on addition of more isoagglutinin and can mate with complementary gametes with undiminished capacity. Accordingly, the isoagglutinin must be inactivated into a form which no longer causes and maintains isoagglutination. Complete reversal of the isoagglutination indicates the loss of the isoagglutinin.

The duration of an isoagglutination, e.g., the rate of the isoagglutinin's loss, depends on the ratio between the number of reacting gametes and the amount of the isoagglutinin, and additionally on the temperature. This loss of the agglutinin is due to two processes—an adsorption of the isoagglutinin to the gametes of the opposite sex (reflecting the flagella contact at normal fertilization) and to subsequent temperature-dependent inactivation (believed to reflect the process at the disagglutination of the flagella). The temperature dependency suggests enzymic breakdown of the agglutinin after the adsorption has ensued (Wiese, unpublished observations).

The adsorption is species- and sex-specific. The (−) isoagglutinin of *Chlamydomonas moewusii* (and the female isoagglutinin of *Chlamydomonas eugametos*) is exclusively adsorbed by (+) gametes of *C. moewusii* and by the corresponding male gametes of the sexually compatible *C. eugametos,* but not by the (−) gametes of *C. moewusii* and the female gametes of *C. eugametos* nor by both gamete types of the sexually incompatible syngen II of *C. moewusii* (Indiana Collection, Nos. 792 and 793) or of *Chlamydomonas reinhardti.* Correspondingly, the (+) isoagglutinin of *C. moewusii* is exclusively adsorbed by the (−) gametes of *C. moewusii* and the female gametes of *C. eugametos.* It is not adsorbed in any other of the above combinations (Wiese, unpublished observations).

The two components in *Chlamydomonas,* which, *in situ,* cause flagellar adhesion between sexually different gametes and in solution cause isoagglutination, correspond to the components in metazoan fertilization known as fertilizin and antifertilizin. In addition, one mating-type substance parallels fertilizin in being a sulfated glycoprotein. If the two isoagglutinins produce instantaneous flagellar agglutination reactions on the basis of a direct complementarity, they should neutralize each other *in vitro* by forming the complex which is indicated in the sex- and species-specific adsorption. Quantitative tests in *Chlamydomonas* reveal that there is no such neutralization.

According to the analysis of the isoagglutinins in *C. eugametos, C. moewusii,* and *C. reinhardti,* the mating-type reaction involves glycoproteins. No species-specific differences between the corresponding mating-type substances of *C. eugametos* and *C. moewusii* could be detected by chemical or physical means. The isoagglutinins, in electrophoretically pure state, have a particle weight of about 10^8 and cause isoagglutination in a concentration as low as 0.01 ng/ml. The male and female components differ in their N value and in their physicochemical properties (pH and temperature sensitivity, and chromatographic behavior). The protein content is about 36% in the female and 21% in the male isoagglutinin. The sugars (rhamnose, arabinose, xylose, galactose, and an unidentified one) are identical in both substances. The female component has 3.9% S, and the male substance includes some lipid and a volatile compound in bound form (Förster and Wiese, 1954b; Förster *et al.,* 1956; Wiese, 1961 and unpublished observations).

Rabbit antiserum against electrophoretically pure female isoagglutinin contains no sex-specific antibodies. Such antisera do agglutinate flagella of gynogametes, androgametes, and vegetative cells of both sexes. This sexually nonspecific flagellar agglutination occurs at the flagella tip in a manner similar to the sex reaction (Wiese, 1965).

Isoagglutinins were first detected in algae by Geitler (1931) in *Tetraspora lubrica* and related to the fertilizin–antifertilizin system of Metazoa fertilization. Moewus (1933) demonstrated the presence of two sex-specific isoagglutinins in *Chlamydomonas eugametos.* Their role was later misinterpreted by confusing the isoagglutination-causing effect with a proposed chemotactic one (Moewus, 1938, 1939). In addition, isoagglutinins have been demonstrated in *Dunaliella* (Lerche, 1937), *C. eugametos* and *Chlamydomonas reinhardti* (Förster and Wiese, 1954b, 1955), in two syngens of *Chlamydomonas moewusii,* in *Chlamydomonas mexicana* (Wiese, unpublished observations), and in *Pandorina morum* (Coleman, 1959). In *Pandorina* the sex-specific mating-type substances in the supernatant of sexually mature colonies not only cause isoagglutination between free complementary gametes or colonies but also effect the gametes' release following swelling and softening of the intercellular jelly. This latter effect is probably indirect. In *Chaetomorpha area,* no isoagglutinating activity could be demonstrated (Köhler, 1956).

The absence of demonstrable isoagglutinins may simply reveal that in the particular species the flagellar agglutinin may not be detached. Agglutination in such forms might also depend on functionally univalent components which, even if detached, do not cause isoagglutination and need to be detected by a test for their specific receptor-blocking power on the opposite gamete's combining sites. No such case has been investigated.

The demonstration of sex-specific gamete agglutinins in isogamous dioecious species raises the question as to whether a similar bipolar mechanism also initiates copulation in isogamous monoecious forms. This question is of great theoretical interest in the most crucial situation, the intraclonal copulation between isogametes of an unicellular hologamous haploid monoecist. Two possibilities are feasible: At sexualization of such a strain two types of gametes, differentiated as (+) and (−) as in dioecists, are formed and fuse on the basis of their sexual complementarity; or, there arises only one kind of gametic cells possessing the general capacity to copulate so that each gamete will fuse with every other one.

There exists no chemical information on the material basis of gamete contact in monoecious species. However, a sexual difference between isogametes of a monoecist (as postulated by Hartmann, 1956) was evidenced physiologically in *Acetabularia* the gametes of which are released in batches by a pulsewise maturation of a greater number of gametangia (cysts) (Hämmerling, 1934). In individual batches of gametes from several plants, an intensive clustering and pairing ensues. After exhaustive copulation in each batch, a smaller number of cells always remains unmated and can easily be separated by their positive phototaxis from the negatively phototactic pairs. When these unpaired cells of the various batches are combined systematically they agglutinate and fuse in certain combinations but not in others. The capacity to copulate reveals the gametic character of these leftover cells, and the pattern of positive and negative reactions indicates that they are sexually differentiated as (+) and (−). So, two gamete types are produced at gametogenesis and both types are released in approximately but never exactly the same number. Thus, there is always a slight predominance of one gamete type in each given sample. After each gamete of the sex in minority has found its copulation partner a certain number of gametes of the other sex remains unmated. Combination of these "rest gametes" from different suspensions either provides sexual partners or joins them with "wall flowers" of their own kind. (The sexual difference between the isogametes in *Acetabularia* is also indicated by the fact that gametes of one cyst do not copulate with each other. Copulation occurs only at combination of gametes from different cysts. The cysts are apparently sexually determined as (+) and (−) gametangia.)

With this method Lerche (1937) detected the existence of two gamete types in the copulation between isogametes of the unicellular and haploid *Haematococcus pluvialis*. In addition, she analyzed an interesting compli-

cation (cf. Hartmann, 1955, 1956). *Haematococcus* develops red pig-
mented cells in N- and P-depleted cultures. Under conditions which elicit
gametogenesis these cultures produce large red cells which do not copulate with cells of the same or of similarly treated cultures. The pigment-
labeled cells, however, do copulate when mixed with rest gametes of
normal green suspensions and do so only with one of the two types.
Obviously, the N- and P-depletion interfers with the normal gametogenesis, and gametes of only one sex were formed. The conclusiveness of these
data has been questioned but with unconvincing argumentation (cf.
Pringsheim, 1963).

D. Pairing, Plasmogamy, and Zygote Formation

In fertilization initiated by flagella agglutination, the subsequent adhesion and fusion of the gametes' bodies need not necessarily be as
specific as the flagellar interaction. As mentioned above for *Ulva* (Levring,
1955), *Chaetomorpha* (Köhler, 1956), and *Chlamydomonas* (Wiese and
Jones, 1963), this second interaction involves another chemical system.
In many species of *Chlamydomonas* and other unicellular and higher
algae, the second contact starts at the apical end of the gamete body,
near the flagellar insertion. Meeting of the apices is often facilitated by a
spiralization of the flagella starting from their agglutinated tips [*Chaetomorpha* and *Dasycladus* (Köhler, 1956)]. In other cases as in *Astrephomene* (Brooks, 1966) and in *Chloromonas saprophila* (Tschermak-Woess,
1963), the flagella agglutinate over their entire length and bring the
papillae of both partners in direct contact. Contact may also be made
after repeated collisions of the agglutinated gametes.

The fusion of the gamete bodies assumes several different forms. The
two gametes may unite head on or laterally. At head-on fusion, a thin
plasmatic thread is often formed which interconnects the two apices
irreversibly. The thread enlarges later to a plasmatic bridge which
gradually widens to effect the complete fusion of both gametes. In certain
species, for example *Chlamydomonas moewusii*, such a bridge connects
the two gamete bodies for a longer-lasting prezygotic stage, the so-called
vis-à-vis pair (Moewus, 1933; Gerloff, 1940; Lewin, 1952; Lewin and
Meinhart, 1953; Gibbs *et al.*, 1958). The two united cells behave as a
physiological unit, with ordered locomotion due to the flagellar action of
the $(+)$ partner. The $(-)$ partner discontinues its flagella beat after the
bridge is formed (Lewin, 1954a,b).

The union of sexually different gametes may involve the fusion of
morphologically recognizable protrusions as in *Volvulina* (Fig. 3) and
Astrephomene. In *Volvulina*, these copulation papillae may be present as
soon as the vegetative cell differentiates into a gamete [*Volvulina pring-*

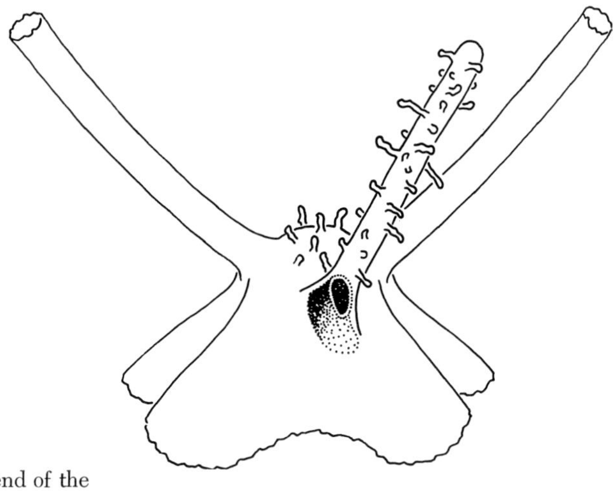

Fig. 10. Anterior end of the (+) gamete in *Chlamydomonas reinhardti*. Explanations see text. (From Friedmann *et al.*, 1968.)

sheimii (Starr, 1962)] or they may be formed only after the release of the gametes from their colonies, i.e., after flagella contact with the opposite sex [*Volvulina steinii* (Carefoot, 1966)]. In the forms just mentioned, the copulation papillae are present on gametes of both sexes.

The pairing in *Chlamydomonas reinhardti* was considered as a prototype of a simple lateral fusion of naked gametes. Electron-microscopic studies, however, reveal the initiation of the fusion by a special copulation tube near the apical end of the (+) gametes (Fig. 10) (Friedmann *et al.*, 1968). This tube (2μ long and 0.2μ in diameter) possesses a varying number of slender tubular projections up to 0.5μ long and 0.025μ in diameter. Such projections also exist at the apical end of the gamete body. The tube of the (+) gamete somehow fuses with an electron-dense area of the (−) gamete establishing continuity between both partner's plasma membranes. The connection widens at shortening and proceeds to final fusion of both gametes. In the conical base of the fertilization tube, a small ringlike organelle, the choanoid body, has been detected which disappears during a later stage of fertilization. Its function is still unknown.

The physiology of pairing is only incompletely understood. The naked gametes of Metazoa unite by membrane fusion (cf. L. H. Colwin and A. L. Colwin, Chapter 8 in Vol. I of this treatise). This also is true for

C. reinhardti according to the recent studies of Friedmann *et al.* (1968). In gametes with cell walls, a gelatinization, tumefaction, or a partial or total dissolution of the wall material precedes gamete fusion, indicating the action of lytic enzymes. In other cases the entire cell wall is shed and only then do the secondarily naked gametes fuse, cf. *Chlamydomonas gymnogama* (Deason, 1967).

In *Chloromonas saprophila* the flagella of copulating gametes agglutinate over their entire lengths. A special initial agglutination of the flagellar tips has not been described (Tschermak-Woess, 1963). The two copulation partners form a type of vis-à-vis pair connected merely by their agglutinated flagella but not by the typical plasmatic bridge. The coordinated swimming of the paired gametes reveals, as in true vis-à-vis pairs, the passivity of the frontal partner's flagella. The frontal gamete sheds its cell wall. Subsequently, after loosening of the flagellar contact, the naked gamete attaches laterally to the membrane of its partner. At the point of attachment the latter's wall dissolves locally and its protoplast passes over into that of the naked gamete to be enclosed in a common membrane. Similar phenomena exist in certain other species of *Chlamydomonas* (cf. Pascher, 1943).

Mechanisms of gamete contact are largely unknown in algae with non-flagellated immotile gametes or with flagellated, motile gametes the flagella of which do not participate in making gamete contact. In *Golenkinia*, for instance, biflagellated spermatozoids contact the eggs by means of their apical end, the flagella being folded to the side of the spermatozoid's body (Starr, 1963).

If the motile gametes exhibit chemotaxis, such action can, at least in part, be responsible for the species specificity of fertilization. The exact contact mechanism and the processes at gamete fusion after a preceding chemotaxis are unexplored as to their nature and may involve different mechanisms in various algae. The basic problem of distinguishing between chemotaxis and the specific attachment (or fusion) has already been discussed with respect to the cluster formation of the Phaeophyceae (Section II,B). At least in certain cases, lytic enzymes are involved which dissolve the gynogamete's envelope [*Chlamydomonas coccifera* (Skuja, 1949)] or liquefy the jelly [*Eudorina* (Goldstein, 1964) and *Volvox* (Darden, 1966)]. Many more examples could be mentioned. Lytic enzymes also must be active during the sperm's penetration of the silicified egg membranes in certain diatoms (see below). The lytic agent–substrate relationship may introduce a specificity factor in these forms.

A perforatorium has been described at the tip of spermlike androgametes in only a few algae [*Chlamydomonas pseudogigantea* (Geitler, 1954) and *Chlamydomonas suboogama* (Tschermak-Woess, 1959)].

Electron-microscopical investigation is required to determine if these organelles resemble acrosomes.

The union of gametes may ensue at any place on the gamete surface or it may be restricted to predetermined sites. Certain features of the underlying mechanisms have been elaborated at oogamy in *Oedogonium,* in centric diatoms, and a few other algae. The engulfment of the sperm by a pulsatory papilla in *Oedogonium* spec. (cf. Section II,B) is unique in algae. Preformed reception sites for gamete fusion have been described in other species of *Oedogonium* and many other algae. Such reception sites are generally characterized by a subsurface concentration of plasma connected with a local retraction of the chromatophore. In *Sphaeroplea,* the spermatozoids are specifically attracted to the reception spot (Pascher, 1939). Lytic enzymes are often responsible for the access of the sperm to the egg. In *Vaucheria* sp., the oogonium opens apically and a hyaline reception site protrudes through the opening. The circular opening arises by a dissolving of the membrane by an enzymic influence of the spermatozoids or of the neighboring antheridium. In other *Vaucheria* species the oogonium protoplast is first transformed into an egg and the opening of the oogonium may be independent from the presence of the male sex (cf. Rieth, 1954).

Insight into other events of sex cell fusion have been obtained at oogamy of centric diatoms (in which chemotaxis at fertilization needs experimental proof). Fertilization in centric diatoms (and in the few oogamous Pennales) varies with the type of the egg and the sperm and with the latter's access to the egg. The sperm of the Centrales is uninucleate and uniflagellate (von Stosch, 1951; Geitler, 1952a; von Stosch and Drebes, 1964; Holmes, 1967; Schultz and Trainor, 1968). The flagellum of *Lithodesmium undulatum* possesses lateral hairs (Flimmergeissel) and lacks the two central fibrils (Manton and von Stosch, 1966). In the pennate *Rhabdonema adriaticum,* two flagellaless binucleated androgametes arise per spermatogonium. They are capable of restricted locomotion by thin pseudopodia (see Fig. 15) (von Stosch, 1958). By complete separation of the oogonial thecae, naked eggs may be shed as in *Cyclotella, Bellerochea, Lithodesmium,* and *Streptotheca,* or the eggs may offer a great part of their surfaces to the sperms as in *Biddulphia mobiliensis* and *Biddulphia regia* (von Stosch, 1954, 1956). In most cases the sperm gains access to the egg inside the oogonium during a temporary or permanent opening of the thecae which is caused by alterations of the turgor during the meiotic divisions. In *Melosira varians,* the frustules dissociate permanently at anaphase I, exposing the naked plasmatic egg surface along an annular narrow slit through which the sperm enters. Once incorporated, the sperm nucleus "waits" for the maturation of its female

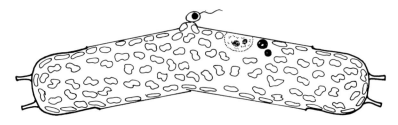

Fig. 11. Oogamy in *Stephanopyxis turris*. Mature oogonium with the egg nucleus and two pycnotic nuclei. Frustules of the oogonium moved apart; one sperm in contact with the exposed surface of the egg. (From von Stosch and Drebes, 1964.)

partner, an indication that the affinity between the two nuclei is not developed until the egg pronucleus is formed (von Stosch, 1951). A temporary opening of the thecae permits the entrance of the sperm in *Stephanopyxis turris*. By swelling of its protoplast at telophase II, the ripe oogonium stretches and, by bending, exposes a small plasma area (Fig. 11). Here, the (chemotactically attracted?) sperm makes contact and glides along for some time until it enters. Sometimes, the first sperm fails to enter in which case a subsequent sperm succeeds quickly possibly because of the preparing action of its predecessor. The type of surface interaction is unknown. After the sperm's entry, the thecae close by a plasma contraction (von Stosch and Drebes, 1964; Drebes, 1964).

An enzymic action is assumed in those cases in which the oogonium does not open and the sperm has to penetrate closed thecae (*Cyclotella*) or silicified membranes between the thecae as in *Biddulphia rhombus* (Geitler, 1952a; von Stosch, 1956).

All investigated centric diatoms are monoecious with the exception of *Coscinodiscus granii* in which Drebes (1968) describes male and female clones. In male clones, occasionally a few oogonia are differentiated (subdioecy).

Auxospores may develop without fertilization in *Melosira nummuloides* (Erben, 1959), *Melosira octagona,* and *Actinoptychus undulatus* (von Stosch, 1956). The type of parthenogenesis needs clarification especially in *M. nummuloides* and *A. undulatus* in which meiosis has been observed. Autogamy is reported in *Cyclotella meneghiana.* Two of the four meiotic nuclei fuse, the two others become pycnotic (Iyengar and Subrahmanyan, 1944). In an American strain of *C. meneghiana* and *C. cryptica,* Schultz and Trainor (1968) detected auxospores and flagellated small cells considered to be male gametes.

In all other cases (without chemotaxis or flagella agglutination and with random meeting), the gametes probably fuse by means of mechanisms which initiate contact and adhesion and involve species specificity in a manner similar to or identical with the agglutination mechanism in *Chlamydomonas*. In algae, such a system can only be deduced, whereas in the yeast *Hansenula* its chemical basis has been analyzed (cf. Chapter 3). Certain forms with flagellated gametes which exhibit neither chemotaxis nor flagellar agglutination but agglutinate directly with their bodies also belong in this category. Reports on such cases often relate to erroneous interpretations of abnormalities or division stages, but there is no doubt that this type is regularly found in other forms as for instance *Gloeococcus bavaricus* (von Witsch *et al.*, 1966).

The physiology and the chemical basis of the attachment of the spermatium to the trichogyne in the Rhodophyta is entirely unknown.

It is hard to classify the peculiar type of fertilization in the green alga, *Prasiola stipitata* (Friedmann, 1959, 1960; Manton and Friedmann, 1960). By a special type of sex determination, many male and female gamete-producing areas arise in a mosaiclike arrangement within the same thallus. Gametes of both sexes are released pulsewise and intermingle so closely that gamete contact is inevitable. Zygote formation is completed within seconds after the gametes' liberation and neither clumping nor chemotaxis can be observed. The gynogametes are egglike and lack flagella; the androgametes possess two equal flagella each of which may make contact with the egg in the following manner: the androgamete attaches by only one flagellum and, during a short interruptable stage, glides along the egg's surface. Then, the flagellum tip coalesces with the egg and the androgamete is quickly engulfed (Fig. 12). The young zygote first swims by the action of the unincorporated flagellum of the androgamete. Later the zygote rounds up and becomes immotile. Since interactions between two sperms or between two eggs have never been observed, the coalescence of the membranes obviously seems to require some sex difference between the surfaces involved.

Fertilization in *Prasiola* resembles contact and fusion of sex cells in the brown algae. The mode of fertilization in the Phaeophyta varies morphologically from isogamy and various degrees of anisogamy to true oogamy. Regardless of the morphology, gamete contact ensues between the tip of the anterior flagellum of the (normally) biflagellated androgamete and the body of the gynogamete (cf. Section II,B). The frontal flagellum coalesces with the gynogametes surface and fusion occurs. [In some isogamous brown algae, as *Cylindrocarpus* and *Striaria*, gamete contact is initiated by an entanglement of the flagella of both partners (Caram, 1957, 1965).]

Details on contact and fusion are largely unknown. The front flagellum of most investigated brown algae carries laterally two garnitures of hairs (Flimmergeissel) (cf. Manton *et al.*, 1953). In *Ectocarpus siliculosus* and

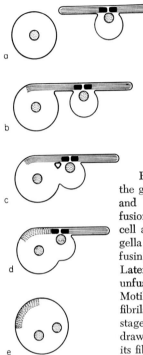

FIG. 12. Plasmatic fusion in the fertilization of the green alga *Prasiola stipitata* Suhr. (a) Egg cell and biflagellated spermatozoid. (b) Beginning of fusion—coalescence of the membranes of the egg cell and of the tip of one of the spermatozoid flagella and engulfment of the fibrillar core of the fusing flagellum by the egg cell protoplast. (c) Later stage in fusion with a "bridge" and an unfused depression between the gametes. (d) Motile stage of the zygote—disintegration of the fibrils of the engulfed flagellum. (e) Immobile stage after the "rounding up" of the zygote—withdrawal of the motile flagellum and disintegration of its fibrils. (From Friedmann, 1962.)

Sorocarpus uvaeformis the anterior flagellum ends with a long thin process which can be seen in living gametes and which establishes the contact (Müller, 1965, 1967).

Also unknown is the function of some special organelles encountered in the spermatozoids of oogamous brown algae. In *Fucus, Ascophyllum,* and *Pelvetia,* a mobile funnel-shaped organelle, the proboscis, surrounds the base of the front flagellum and is characterized by a peculiar fine structure with a species-typical number of concentric thickenings (Manton and Clarke, 1950, 1951, 1956; Manton *et al.*, 1953). Considered to be comparable with the proboscis of the mentioned species is the "root" at the basis of the front flagellum in *Dictyota* (cf. Manton, 1959). Also the role of the spines in *Dictyota* and of the one large hook near the end of the front flagellum in *Himanthalia* is unknown. Cinematographic analysis of the fertilization in *Ascophyllum nodosum* revealed some of the physiological interactions between egg and spermatozoid. The free-swimming spermatozoid is first drop-shaped. The front flagellum exerts rapid undulating movement whereas the erect hind flagellum appears to be trailing behind the cell. On the egg's surface the spermatozoid alters its shape and crawls slowly over the surface, the proboscis facing the egg. The actual fusion has not been observed (Friedmann, 1961).

E. GAMETANGIOGAMY

Gametangiogamy is the peculiar situation in which fertilization is initiated by sexual contact between the gamete mother cells (gametangia, gamonts) and can be manifested in either the haplophase (Zygnematales) or in the diplophase (Pennales). In the typical cases, gamete formation is started only if some interaction between the mother cells has ensued. Physiologically, gametangiogamy involves the same principles as are active at the copulation of gametes, namely induction effects and induction substances, chemotaxis, chemotropism, adhesion substances, and lytic enzymes. Excretion of copulation jelly is a common phenomenon and often the first indication of the sexual phase.

1. *Desmidiaceae and Mesotaeniaceae*

Fertilization in desmids (unicellular haplonts) involves—as in all Zygnematales—conjugation between isomorphic cells, the protoplasts of which fuse as gametes. There is considerable variation in detail, and type-specific features of the sex act and of the zygote structure shall be evaluated taxonomically (Starr, 1959; P. W. Cook, 1963, Brandham and Godward, 1965). Depending upon the particular species or form, conjugation may occur after pairing of full grown cells to produce one zygote per pair of sexualized cells (Fig. 13A) or the paired cells, as gametangia mother cells, may first divide once and their descendants conjugate subsequently producing two zygotes between nonsister cells (Fig. 13B). In other species or forms, the gametangia mother cells do not pair but divide into two gametangia, and one zygote is formed between the two sister protoplasts (Fig. 13C).

Conjugating cells may be genetically identical (homothallic forms; intraclonal conjugation, in several species regularly between sister cells) or sex may be determined genetically by monofactorial segregation at meiosis (heterothallic forms).* The detection of genetically determined (+) and (−) sexes in these strictly isogamous organisms by Starr (1954a,b) clarified the complicated situation concerning sexual differentiation in Zygnematales (Czurda, 1937). The demonstration of sex involves the cultivation of clones and the search for zygotes within the clones or in combinations of them under conditions known to induce copulation (Starr, 1955). This method demonstrated clearly both the

---

* In placoderm and some saccoderm desmids, a tetrad analysis is not possible because of abnormal meiosis.

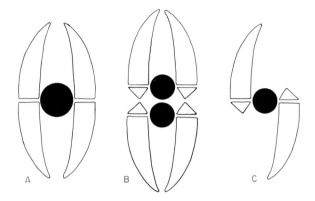

FIG. 13. Pairing, gamete copulation, and zygote formation in desmids. Diagram of basic prototypes. (A) Pairing between two full-grown conjugants, recognizable by their equal-sized semicells. Occurs in hetero- and homothallic forms, in the latter case the partners may be sister cells of the last mitosis. (B) Double copulation between nonsister cells originating by mitosis of two paired partners, usually in heterothallic forms but also in homothallic ones. Conjugation partners are immature cells recognizable by the incomplete regeneration of one semicell in each partner. (C) Conjugation between sister cells—mitotic descendants of an unpaired gametangium mother cell in homothallic forms.

existence of true heterothallism and the existence of homo- and heterothallic forms within one species [*Closterium* (Cook, 1963), *Netrium digitus* (Biebel, 1964), *Closterium moniliferum* and *Closterium ehrenbergii* (Lippert, 1967)] and, in addition, the existence within one species of pairs of complementary mating types sexually isolated from each other [cf. *Closterium* (Cook, 1963), *Netrium digitus* (Biebel, 1964), *Cosmarium turpinii* (Starr, 1954a, 1959), *Cosmarium botrytis* (Brandham and Godward, 1965)]. The sexual incompatibility between such syngens has not yet been traced to a specific block at sexualization or at fertilization. As yet, only two mating types per syngen have been demonstrated in these haploid organisms; homo- and heterothallic forms have not been crossed successfully.

Apart from the type of sex distribution, the gametes fuse either in the common mucilage secreted by the two partner cells or within a copulation tube formed between and by the two gametangia (cf. Presscott, 1948, for older literature). Fertilization is generally isogamous with respect to the size of the gametes and their locomotive activity. *Desmidium cylindricum* is an exception in which one gamete passes into the other gametangium (deBary, 1858). The fused gametes shrink by secreting water (cf. Starr, 1955; Kies, 1964) and, as in all Zygnematales, a three-layered zygote membrane is formed consisting of endo-, meso-, and exospore, the latter often being characteristically ornamented.* As seen in *Mesotaenium dodekahedron* (Geitler, 1965) and many other forms (cf. Kies, 1964; P. W. Cook, 1963), the zygote membrane develops within the thin membrane of the fused gametes.

Quite generally, the zygote is a dormancy stage (hypnozygote), and in placoderm desmids (and *Netrium digitus*) the nuclei do not fuse until zygote germination. Subsequent to karyogamy, meiosis occurs and is mostly irregular since only two gones are formed and two nonsister nuclei degenerate. The same holds true for *N. digitus* (Biebel, 1964). In *Cosmarium biureticum* (Starr, 1959), and a few other species, only one gone is produced while three nuclei degenerate as in *Spirogyra*. In the Mesotaeniaceae, a normal meiosis may yield four gones. The nuclei in *Mesotaenium dodekahedron* fuse directly after the gamete union.

Irregularities at pairing and conjugation may lead to parthenogenetic development of the released gametes or of the protoplast inside of the gametangium (cf. Krieger, 1937; Kies, 1964). Both types of spores form the three-layered membrane which is normally characteristic of the zygote. Discrepancies at early pairing stages may lead to dedifferentiation of the sexualized cells into the vegetative stage.

The physiology of pairing and conjugation is largely unknown. Phenomena and problems appear to resemble closely those in diatoms (Section II,E,3).

The old question of whether partner cells meet at random or by chemotaxis was answered in one representative case, *Cosmarium botrytis*, by cinematographical evidence of chemotaxis. Chemotactic attraction is also believed to hold the conjugants together until the jelly coat is secreted (Brandham, 1967). Chemotaxis is very likely to occur in other heterothallic forms too, and its occurrence is not excluded in certain homothallic forms. In other homothallic species in which the pairing gametangia are sister cells of the previous mitosis, or in which daughter cells of unpaired gametangia conjugate, the necessity for a directed approximation of the partners is bypassed.

The juxtaposition of the conjugants in many species is strictly fixed, and partners adjust their position before the final conjugation. Placoderm desmids form the conjugation tube at the isthmus between the two semicells. In saccoderm desmids, conjugation tubes may be initiated at any site on the partners [cf. Krieger, 1937; *Mesotaenium kramstai* (Starr and Rayburn, 1964)]. In both groups, the copulation tubes of the partners are always formed from papillae opposing each other. This arrangement

---

* In *Netrium digitus*, an autosomal recessive factor, when homozygous, causes the normally three-layered and rough zygote membrane to be two-layered and smooth (Biebel, 1961).

requires induction substances which may be of the type observed in *Achlya* (des S. Thomas and Mullins, 1967). The copulation papillae develop, in most species, synchronously in both partners; in a few species, however, the papilla appears first in one sex only, suggesting a mutual induction at this step of fertilization. Also an occasional lack of synchrony is reported (Lippert, 1967). The growth of the copulation tube may be chemotropically controlled as manifested in the copulation course and as demonstrated when copulation partners are experimentally moved apart (cf. Kies, 1962).

Gamete union within the copulation jelly or within the copulation tube ensues quickly. The process of entry of the gametes into the copulation tube, often described as ameboid movement (cf. Presscott, 1948), needs specification. In *Netrium,* the gametes' entry into the soft-walled conjugation tube is explained by turgor (Biebel, 1964; see also Lippert, 1967); in other cases secretion of mucus, pushing the gamete, is discussed.

2. *Zygnemataceae*

Fertilization in filiform Zygnematales involves the same mechanisms as in desmids (sexualization, contact between the gametangia, induction of gametogenesis, formation of copulation papillae and copulation tubes, fusion of the gametes, zygote formation, and karyogamy).

In species where the gametes are equally motile the zygote forms in the copulation tube in between the two gametangia (Mesocarpaceae). In others with one stationary and one migratory gamete, the zygote is formed within one of the two gametangia (all *Spirogyra* species).

Conjugation may ensue between neighboring cells of the same filament (lateral conjugation; Fig. 14b), or, between cells of different filaments (scalariform conjugation; Fig. 14a) leading to a ladderlike arrangement. In *Spirogyra jogensis* two neighboring cells dissolve the separating cell wall and fuse as gametes (Iyengar, 1958) (direct lateral conjugation). Whereas lateral conjugation necessarily characterizes homothallic forms, scalariform conjugation does not always indicate heterothallic ones. Many forms with lateral conjugation are also capable of conjugating scalariformly. In such a case, both filaments may each produce migrating and stationary gametes. Certain species are strictly heterothallic and never conjugate laterally (Fig. 14a); their filaments are of two types producing migrating or stationary gametes. Migratory and stationary behavior, however, is no reliable criterion for male or female differentiation and, on the other hand, sex-different gametes may exhibit equal locomotion [*Mougeotia heterogama* (Geitler, 1958a), *Zygnema circumcarinatum* (Czurda, 1937)].

Whereas in most Zygnemataceae the protoplasts of the gametangia act directly as gametes, the protoplasts of *Sirogonium melanosporum* are first transformed into male and female anisogametes (Hoshaw, 1965). In this homothallic species, two genetically identical cells, the "progametangia," adhere to each other and form copulation papillae at the contact site. Subsequently each progametangium divides into a smaller vegetative cell and the definitive gametangium. This unequal mitosis must be differential in both sexes since the male gametangium is smaller than the female one. In a pair of conjugating cells the male gametangium normally differentiates before the female one. Likewise the smaller male gamete differentiates first, often before the female gametangium is well developed,

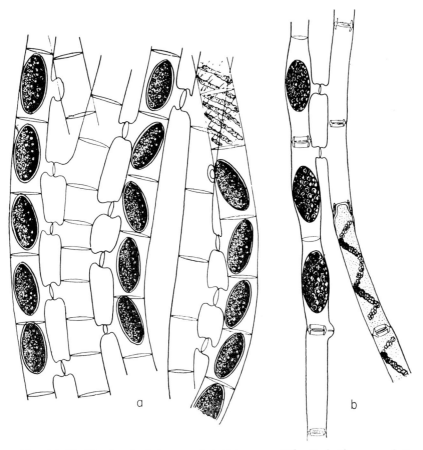

Fig. 14. Fertilization in *Spirogyra*. (a) *Spirogyra nitida*. Scalariform copulation between six filaments which are either entirely male (with migratory gametes) or female (with stationary gametes). (b) *Spirogyra grevilleana*. Scalariform and lateral copulation. (After Kniep, 1928.)

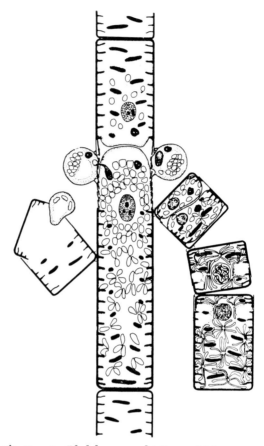

Fig. 15. Fertilization in *Rhabdonema adriaticum*. Mature oogonium with egg (central below, rich in plastids) and polar body (central above, deprived of plastids). Both cells arise by an unequal cytokinesis after meiosis I and possess one functional and one pycnotic nucleus each. Attached to the oogonium are a short chain of spermatogonia in different stages of meiosis (right) and one spermatogonium (left) which has released its two binucleated sperms and contains only a nucleus-free residual body. The flagellaless sperms are capable of locomotion by means of thin pseudopodia. The two sperms are attached to the girdle region of the oogonium, the left one injecting its functional nucleus into the egg. (From von Stosch, 1958.)

possibly indicating an induction process. At dissolution of their separating end walls, the papillae fuse forming the copulation tube. The male gamete passes through the copulation tube, regains its spherical shape, and fuses immediately with the female one. The dissolution of the separating membrane between the two copulation papillae has been studied electron microscopically in *Spirogyra* (Dawes, 1965).

3. *Bacillariophyta*

In diatoms, gametangiogamy is encountered only in the Pennales, whereas the Centrales display oogamy.* Exceptionally, oogamy occurs in the Pennales; in the special case of *Rhabdonema adriaticum* (von Stosch, 1958), an oogamy with a preceding gametangiogamy unites the characteristic features of both types (Fig. 15).

Fertilization has an unusual feature in that it restores the maximal size of the diatom cell. The cell size normally diminishes steadily during vegetative reproduction. Apparently only these smaller cells are physiologically

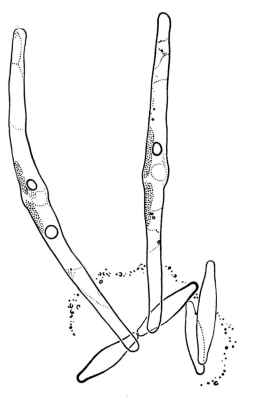

FIG. 16. Auxospore formation in *Synedra rumpens* var. *fragilarioides*. The two zygotes, formed in one of the two paired gamete mother cells, stretch to the maximal cell size typical for the species. The empty frustules of the two gametangia indicate the copulation size of the species; the mass of the copulation jelly is marked by the attached detritus. (From Geitler, 1952b.)

* In all diatoms, meiosis occurs at gamete formation and the nuclei that fuse at fertilization are the immediate products of meiosis.

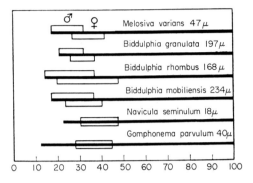

Fig. 17. Copulation size in centric and pennate diatoms; the range in centric diatoms specified for sex. The species-typical width of the gametangia is given in percent of the size of the biggest auxospore measured, the absolute value of which is stated for each species. The horizontal lines represent the entire size range of the species; the rectangles above the line indicate the range for male, those below for female differentiation. In pennate diatoms in which the absence of recognizable sex characters prevents sexing, the line halves the rectangles representing the range of potential gametangia-forming cells. (From von Stosch, 1956.)

competent for sexual differentiation since only cells in this "copulation size" respond to special environmental conditions effecting sexual differentiation and fertilization (Geitler, 1932; von Stosch, 1956). The zygote grows as an auxospore to the original cell size typical for the species and gives rise to a new, rejuvenated clone (Fig. 16). In some Pennales, the copulation size is larger than the minimum size beyond which no division occurs, and cells of clones not sexualized on time divide to death (cf. Fig. 17).

Sex determination (male vs female) in the oogamous Centrales [and in the anisogamontic pennate *Grammatophora marina* (Magne-Simon, 1962)] is also a function of cell size at sexualization. Within the copulation size range, the bigger cells differentiate into oogonia, the smaller ones into spermatogonia. At a certain intermediate size, size-dependent determination does not ensue unequivocally and the cells seem to differentiate arbitrarily into either male or female (von Stosch, 1951, 1956; Geitler, 1952a) (Fig. 17).

Internal fine-structural peculiarities inherent to certain cells may interfere with sexualization and exclude a certain cell type within a clone from fertilization (Geitler, 1952c).

The zygote membrane of the Centrales transforms directly into the auxospore membrane and silicifies by incorporation of submicroscopic siliceous scales into a pectinaceous stratum (Reimann, 1960). In the Pennales, the zygote membrane ruptures or is liquefied; the auxospore membrane (the perizonium) is a new organelle formed inside of the original zygote membrane (von Stosch, 1962).

In both groups the two new frustules (thecae) develop inside of the adult auxospore peculiarly connected with two special mitotic nuclear divisions without accompanying cytokineses (Geitler, 1953b, 1963; von Stosch, 1962; cf. Magne-Simon, 1962). Each mitosis results in the production of one theca. The functionally uninucleated stage of the auxospore is maintained at both divisions by degeneration of one of the daughter nuclei. (For details of oogamy in Centrales, see Section II,D.)

In the basic features, gametangiogamy in pennate diatoms proceeds rather uniformly as a two-step fertilization. Two diploid cells, the gametangia or gamonts, pair and form, at meiosis, gametes which fuse within a common mucilage secreted by both partners. Much species-specific variation exists in details. The lack of true sex characters prevents designation of the gamontic or gametic stages as $(+)$ and $(-)$.

Variation is based on different types of gametogenesis. In most cases, the two meiotic nuclear divisions are connected with one cytokinesis. This cytokinesis can be equational producing two binucleated gametes, or unequal producing one binucleated gamete and one polar body. Cytokinesis can be omitted, in which case the entire protoplast of the gamont develops into one gamete.* Four gametes are never formed.

Fertilization may occur amphimictically between gametes of different gamonts (*allogamy*), automictically between gametes of the same gamont (*paedogamy*), or automictically between gone nuclei of the same gamont (*autogamy*). Furthermore, *apomixis* exists as diploid-parthenogenetic development of the gamont's protoplast into an auxospore.

Allogamy is the most frequent mode. Constitutional paedogamy (*Gomphonema constrictum* var. *capitata, Achnanthes subsessilis, Cymbella aspera*) and constitutional autogamy (*Amphora normani* and *Denticula tenuis*) take place in unpaired gamonts only (cf. Geitler, 1957). Apomixis, however, may occur in unpaired gamonts (*Cocconeis placentula* var. *clinoraphis*) or, in other forms, only after pairing (*Cymatopleura solea* and *Cymatopleura elliptica*) (Geitler, 1932).

---

* Whatever the type of gametogenesis, only one nucleus ordinarily survives in each gamete; the others eventually become pycnotic. In *Navicula radiosa* and *Navicula cryptocephala* var. *veneta*, however, the gametes fuse while still in their binucleated state, and, demonstrating their equivalence, the four nuclei form two pairs one of which later degenerates (Geitler, 1952b). As in all algae, the mechanism that determines which nucleus undergoes pycnosis and which remains functional is unknown. A unique type of pycnotization occurs in the oogonium of the centric *Stephanopyxis turris*. After each meiotic division the two separated equivalent daughter nuclei approach again, adhere, and finally separate after 30 minutes as one functional and one pycnotic nucleus (von Stosch and Drebes, 1964).

Each type of fertilization is finally characterized by the behavior of the gamonts and gametes and by other specializations (type of copulation jelly, copulation tubes, etc.). Copulating gametes are always identical morphologically. Physiologically, they may differ as exemplified in the following prototypes for an allogamy with two gametes per gamont (differentially active and passive behavior exists likewise in forms with one gamete per gamont as in *Navicula seminulum* and *Navicula cryptocephala* var. *veneta;* for a more detailed list including all older literature, see Geitler, 1932, 1957):

1. Both gametes of one gamont migrate into the other gamont and fuse there with the stationary gametes of the second gamont. Two zygotes are formed, both situated in one gamont [*Synedra ulna* and *Synedra rumpens* var. *fragilarioides* (Geitler, 1939, 1952b) and *Navicula halophila* (Subrahmanyan, 1947)] (Fig. 16).

2. The gametes of both gamonts fuse crosswise two by two, but the two gametes of each gamont differ physiologically—one gamete migrates actively into the other gamont and fuses there with the other's stationary one. Two zygotes result, one in each gamont [most species of *Gomphonema, Nitzschia subtilis,* and *Amphipleura pellucida* (cf. Geitler, 1949a, 1957)].

3. Both gametes of both gamonts migrate equally, meet, and form two zygotes in between the two gamonts [*Rhoicosphenia curvata* and *Rhopalodia gibba* (cf. Geitler, 1957)] (Fig. 18).

The individual mode of fertilization is specific for the particular form. Certain morphologically defined species or subspecies subdivide into forms with different fertilization modes (cf. Geitler, 1957). For example, most subspecies of *Cocconeis placentula* are allogamous, but var. *lineata* is apomictic and var. *klinoraphis* possesses an allogamous and an apomictic form; *Denticula tenuis* has an allogamous and an autogamous form, and *Gomphonema constrictum* var. *capitata* an allogamous and a paedogamous one.

Nothing is known about the physiology of pairing and about possible sex differences between the partners. Combinations of clones are indispensable in testing for the existence of mating types. The available data indicate that the great majority of species is monoecious. In sessile species (*Gomphonema, Rhoicosphenia,* and *Anomoneis*), one gamont becomes motile and approaches a sessile one (Fig. 18). Although this difference in locomotion as well as other diversities were not regarded as true sex characters of the diplophase (Geitler, 1957), Magne-Simon (1962) supplied sufficient proof for the sex difference between pairing gamonts in *Grammatophora marina.* Smaller male gamonts attach to larger female ones. Both gamonts undergo meiosis and form, by an unequal division, one gamete and one polar body. The male gamete escapes his frustule in an unknown manner and fuses with the female one in her frustule.

No information on mating-type substances is available. Tentatively, the adherence of the spermatogonia to the oogonia in *Rhabdonema adriaticum* (von Stosch, 1958) and the attachment of the male gamont to the mucilagenous parts of its female partner in *Grammatophora marina* (Magne-Simon, 1962) may be related to a specific adhesiveness comparable to complementary mating-type substances.

The pairing creates, in most cases, a direct contact (*contact pairing*) preferably along the girdle sides where the opening of the frustules automatically brings the gametes in close proximity. The contact differs from species to species and may take place in a fixed manner or without a defined position of the partners to each other. The fixed arrangement is usually associated with less jelly and such jelly is often restricted to connecting strands. The second arrangement occurs within varying amounts of soft jelly or with the formation of copulation tubes (*Eunotia arcus;*

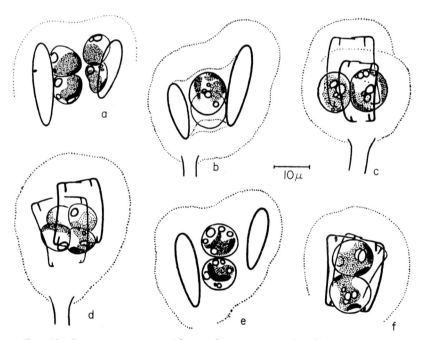

Fig. 18. Gametangiogamy in *Rhoicosphenia curvata*. (a, d) Gametes emerged from their mother cells (in d turned by 90°); (b, e, f) zygotes, just formed; (c) zygotes forming the perizonium. In a, b, e, the mother cells are seen in valve view, in c, d, f, in girdle view. Only chromatophores and oil droplets are pictured. (From Geitler, 1952d.)

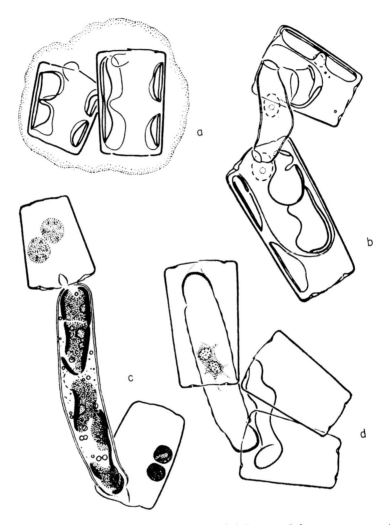

Fɪɢ. 19. Gametangiogamy in *Eunotia arcus*. (a) Pairing of the gametangia within the secreted copulation jelly—the copulation papillae being formed. (b) Later stage in copulation—the papillae have fused and form the copulation tube, and the nucleus of one gamete has already entered the copulation tube. (c) Final stage of gamete copulation—the zygote near completion, its wall still incomplete at the poles. In both mother cells the polar body is in a different degree of decomposition. (d) Copulation involving three cells. The pairing of the cell above with one of two sister cells induced the formation of a dead-ending copulation tube in the third unmated cell. (From Geitler, 1951.)

Fig. 19). Copulation tubes can be formed without additional copulation jelly (*Nitzschia subtilis* and *Nitzschia sigmoidea*) (cf. Geitler, 1957).

The contact between the gamonts seems to initiate their intracellular differentiation (meiosis, gametogenesis, formation of copulation tubes, etc.). The induction effect is manifested in a third cell if it contacts pairing gamonts and subsequently forms a blind copulation tube [*Cocconeis placentula* and *Eunotia arcus* (Geitler, 1932, 1951)] (Fig. 19).

In some apomictic forms as *Cocconeis placentula* var. *lineata* pairing of gamonts is obligatory in order to start sexual differentiation, yet each gamont then develops apomictically (a comparable situation to pseudogamy; cf. Wiese, 1966; Beatty, 1967) (cf. Geitler, 1957, 1958b). In *Denticula tenuis,* contact between two gamonts results in true gametogenesis and allogamous copulation whereas in unpaired gamonts the formation of gametes fails to occur and two of the four meiotic nuclei fuse autogamously (Geitler, 1953a).*

Mutual or unilateral induction by means of diffusing substances accounts for the coordinated differentiation of spatially separated partners in species with *distance pairing* [*Synedra rumpens* and *Cocconeis placentula* (cf. Geitler, 1957)]. An induction, initiating sexualization as such, seems to occur in *Rhabdonema adriaticum* (von Stosch, 1958).

Significantly no pairing is needed to initiate meiosis and gametogenesis in species with constitutional paedogamy or autogamy, suggesting that the acquisition of automixis may have evolved by elimination of a block which is inactivated in allogamous species by pairing.

The chemical basis of gamete copulation is likewise unknown. Although morphologically alike, the gametes will be sexually differentiated into $(+)$ and $(-)$ as in all other sufficiently investigated organisms and will fuse on the basis of a surface complementarity. However, no evidence for such a complementarity exists.

The copulation of gametes may occur free in the copulation jelly after the opening of the frustules or free in the medium as in *Grammatophora marina* or by means of copulation tubes. Copulation tubes arise by fusion of outgrowing papillae of both partners and are formed before the definitive gametes develop. Both partners participate equally [*Eunotia arcus* (Geitler, 1951)] (Fig. 19) or distinctly unequally [*Cocconeis placentula* var. *euglypta* and var. *euglyptoides* (Geitler, 1952c, 1954)] (Fig. 20). The two papillae meet and coalesce; the exact details are not completely understood (cf. Geitler, 1957). The growth is chemotropically directed as is obvious in the case of the long tubes in *E. arcus* and *C. placentula*

---

* It is undecided if the two modes characterize two genetically separate forms or if they represent merely alternatives within one form.

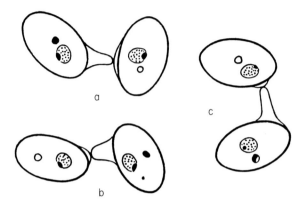

Fig. 20. Distance pairing in *Cocconeis placentula* var. *euglyptoides*. Copulation tube formed mainly by one partner. Note the directed approach of the copulation papillae between the partners. (From Geitler, 1958b.)

(Figs. 19 and 20). In species with closely neighbored partners, the papillae may be in contact from the beginning of their formation.

The mechanism of gamete locomotion is largely unknown. Ameboid movement and turgor-dependent alterations in the cell's shape seem involved in some cases; in others, swelling mucus excreted by the gametes pushes inside the preformed channel system in the copulation jelly. In addition, the polar body, swelling by hydration, may assume a pushing function.

The gametes may unite within the copulation tube (*Eunotia arcus;* Fig. 19) or the copulation tube may serve as a channel through which the migrating gamete passes to fuse with the stationary one. In forms with two gametes per gamont, the reciprocal union ensues subsequently through the same channel (*Nitzschia*) or, as in *Amphipleura pellucida* with two copulation tubes, the migrating gametes move simultaneously in opposite directions. The empty pathways are recognizable later as copulation channels.

III. Sexual Incompatibility, Syngens, and Self-Sterility

Sexual incompatibility at fertilization, as normally occurs between species, may also be manifested within one species as a sexual isolation between pairs of complementary mating types. Such a separation into syngens within one species has been detected in many unicellular and a few multicellular algae after appropriate examination [*Chlamydomonas* species (Schreiber, 1925), *Cosmarium turpinii* (Starr, 1954a), *Pandorina morum* (Coleman, 1959), *Gonium pectorale* (Stein, 1958b), *Astrepho-mene* (Stein, 1958a; Brooks, 1966), *Volvulina* (Carefoot, 1966), *Cos-marium botrytis* (Brandham and Godward, 1965), *Closterium* (Lippert, 1967)]. Concerning syngens in protozoa see Hiwatashi (this volume).

On the other hand, sexual compatibility may exist between forms which have been considered as different species [*Chlamydomonas* species (Pascher, 1916; Strehlow, 1929), *Chlamydomonas eugametos* and *Chlamydomonas moewusii* (Bernstein and Jahn, 1955; Gowans, 1963), *Chlamydomonas reinhardti* and *Chlamydomonas smithii* (Hoshaw and Ettl, 1966), and *Eudorina* species (Goldstein, 1964)]. Complete sexual compatibility between two forms requires that they cross successfully. Incompatibility may be caused by any disharmony from sexual differentiation at gametogenesis to the compatibility of the two plasms and genomes. Many studies on algae designed to detect gross incompatibility (for example the absence of the formation or germination of the zygote) regrettably fail to record details on the type and the site of blocks to fertilization. The investigation of these phenomena is still at an early stage and reliable information on higher algae is lacking.

Following an early report on incompatibility phenomena among species of *Chlamydomonas* (Schreiber, 1925), systematic investigations began in desmids (see Section II,E) and colonial Volvocales. In *Pandorina*, the combination of 60 isolated clones originating from 37 different natural populations showed two homothallic and 49 heterothallic strains. The latter could be arranged into fifteen pairs of mating types each pair sexually isolated from all others. Two such syngens may be present in the same pond and strains from far-distant loci may belong to one syngen.

The existing plasticity is best demonstrated in *Eudorina*. After reclassification of the genus, Goldstein (1964) observed compatibility between taxons regarded as separate species and incompatibility between syngens within one species. In certain combinations, sexual interaction ended at specific blocks (absence of zygote formation after sperm entry in the female colony), others were successful in one but not in the reciprocal direction. Disharmonies at meiosis in other intercrosses resulted in poly- or aneuploidy associated with reduced viability of the F_1. At crossing of the haploid, heterothallic *Eudorina elegans* and *Eudorina illinoisensis,* diploid selfing males and selfing females arise. Such subdioecious clones produce an abundance of zygotes in combination with the opposite sex but also in a certain minor percentage within one clone. Subdivision into syngens in *Astrephomene gubernaculifera* appears partially related to types with different chromosome numbers (Stein, 1958a; Brooks, 1966). One clone which mates equally well with $(-)$ and $(+)$ test clones may exhibit relative sexuality (cf. Hartmann, 1955, 1956); its behavior, however, can also indicate a self-sterile homothallic condition (Stein, 1958a).

Properties of the agglutination mechanism, which normally produces the initial interaction of the two gamete types, are in some cases responsible for the sexual isolation between related forms. For example, because of an incompatible system of mating-type substances, initial contact fails to occur on mixing of *Chlamydomonas moewusii* (Indiana Collection, Nos. 96 and 97 and Nos. 792 and 793). Such absolute isolation requires complete incongruence of the mating-type substances. Certain combinations of species or syngens permit the sex reaction to occur in one, but not in the reciprocal direction. Such pairs of systems apparently have one sex in common, suggesting partial evolutionary separation from one original system or an evolutionary convergence. Strehlow (1929) successfully crossed *Chlamydomonas paradoxa* (+) from Vladivostok with *C. botryoides* (−) from Germany, but the reciprocal cross could not be achieved. Both forms possess differently shaped zygotes, the hybrid zygote is intermediate in character. Coleman (1959) detected another case in *Pandorina*.

In species in which the cells enter the sexual phase or acquire sexual maturity only in presence of the complementary sex, some mutual or unilateral interaction has to be assumed in order for gametogenesis to be completed. In such cases the absence of the agglutination reaction may be secondary, since gametogenesis might not have been initiated or completed. This is the case in *Pandorina morum* (Coleman, 1959) and *Gonium pectorale* (Stein, 1958b).

Although experiments with clone combinations have not been reported, the existence of syngens in certain diatoms is indicated by the absence of sex reactions between different types of a morphologically well-defined species. Subspecies that differ in their fertilization behavior may be absolutely isolated from each other as exemplified in *Cocconeis placentula* with allogamous, auto-, and apomictic forms (cf. Geitler, 1958b), whereas hybridization is possible in other cases as between *C. placentula* var. *pseudolineta* with anisogamous allogamy and *C. placentula* var. *euglyptoides* with isogamous allogamy.

A very special case of self-sterility exists in a freshwater form of *Ceratium cornutum* (von Stosch, 1965). Copulation in this haploid form occurs only between two types of clones. This complementarity, however, does not represent true heterothallism with two sexes since in each clone androgametes appear. The gynogametes are morphologically indistinguishable from vegetative cells. The mating pattern thus indicates copulation between two self-sterile homothallic forms, a unique situation in haploid algae. Furthermore, one clone of one form exhibited some selfing, indicating a lowered barrier to self-sterility.

Self-sterility finally exists within the copulation partners in those allogamous, pennate diatoms which form two gametes per gamont and exhibit reciprocal exchange of their migrating gametes (*Nitzschia subtilis*). Occasionally, the self-sterility breaks down and paedogamous fusion between the two gametes of one gamont ensues. Self-sterility in these forms is intimately connected with sexual differentiation, a problem which has been repeatedly discussed by Geitler (1932, 1939, 1949a, cf. 1957).

REFERENCES

Beatty, R. A. (1967). Parthenogenesis in vertebrates. *In* "Fertilization: Comparative Morphology, Biochemistry, and Immunology" (C. B. Metz and A. Monroy, eds.), Vol. I, pp. 413–457. Academic Press, New York.

Bernstein, E., and Jahn, T. L. (1955). Certain aspects of the sexuality of two species of *Chlamydomonas*. *J. Protozool.* **2**, 81–85.

Berthold, G. (1881). Die geschlechtliche Fortpflanzung der eigentlichen Phaeosporeen. *Mitt. Zool. Stat. Neapel* **2**, 401–413.

Biebel, P. (1961). Genetic study of *Netrium digitus* var. *lamellosum*. *News Bull. Phycol. Soc. Am.* **14**, 5.

Biebel, P. (1964). The sexual cycle of *Netrium digitus*. *Am. J. Botany* **51**, 697–704.

Bold, H. C. (1949). The morphology of *Chlamydomonas chlamydogama*, sp. nov. *Bull. Torrey Botan. Club* **76**, 101–108.

Brandham, P. E. (1967). Time-lapse studies of conjugation in *Cosmarium botrytis*. II. Pseudoconjugation and an anisogamous mating behaviour involving chemotaxis. *Can. J. Botany* **45**, 483–493.

Brandham, P. E., and Godward, M. B. E. (1965). The inheritance of mating type in desmids. *New Phytol.* **64**, 428–435.

Brooks, A. E. (1966). The sexual cycle and intercrossing in the genus *Astrephomene*. *J. Protozool.* **13**, 368–375.

Caram, B. (1957). Sur la sexualité et le developement d'une Phéophycée: *Cylindrocarpus berkeleyi* (Grev) Crouan. *Compt. Rend. Acad. Sci.* **245**, 440–443.

Caram, B. (1965). Recherches sur la reproduction et le cycle sexué de quelques Phéophycées. *Vie et Milieu* **16**, 21–221.

Carefoot, R. J. (1966). Sexual reproduction and intercrossing in *Volvulina steinii*. *J. Phycol.* **2**, 150–156.

Coleman, A. W. (1959). Sexual isolation in *Pandorina morum*. *J. Protozool.* **6**, 249–264.

Cook, A. H., and Elvidge, J. A. (1951). Fertilization in the Fucaceae: Investigations on the nature of the chemotactic substance produced by eggs of *Fucus serratus* and *F. vesiculosus*. *Proc. Roy. Soc.* **B138**, 97–114.

Cook, A. H., Elvidge, J. A., and Heilborn, Sir J. (1948). Fertilization including chemotactic phenomena in the Fucaceae. *Proc. Roy. Soc.* **B135**, 293–301.

Cook, P. W. (1963). Variation in vegetative and sexual morphology among the small curved species of *Closterium*. *Phycologia* **3**, 1–18.

Czurda, V. (1937). *Conjugatae*. *In* "Handbuch der Pflanzenanatomie" (K. Linsbauer, G. Tischler, and A. Pascher, eds.), Vol. 6, 2B. Gebr. Bornträger, Berlin.

Darden, W. H. (1966). Sexual differentiation in *Volvox aureus*. *J. Protozool.* **13**, 239–255.

Darden, W. H. (1968). Production of a male-inducing hormone by a parthenosporic *Volvox aureus*. *J. Protozool.* **15**, 412–414.

Dawes, C. J. (1965). An ultrastructure study of *Spirogyra*. *J. Phycol.* **1**, 121–127.

Deason, T. D. (1967). *Chlamydomonas gymnogama*, a new homothallic species with naked gametes. *J. Phycol.* **3**, 109–112.

deBary, A. (1858). "Untersuchungen über die Familie der Conjugaten." Leipzig.

des S. Thomas, D., and Mullins, J. T. (1967). Role of enzymatic wall softening in plant morphogenesis. Hormonal induction in *Achlya. Science* **156**, 84–85.

Diwald, K. (1938). Die ungeschlechtliche und geschlechtliche Fortpflanzung von *Glenodinium lubiensiforme* sp. n. *Flora* **32**, 174–192.

Drebes, G. (1964). Über den Lebenszyklus der marinen Planktondiatomee *Stephanopyxis turris* (*Centrales*) und seine Steuerung im Experiment. *Helgolaender Wiss. Meeresuntersuch.* **10**, 152–153.

Drebes, G. (1968). Subdioecie bei der zentrischen Diatomee *Coscinodiscus granii. Naturwissenschaften* **55**, 236.

Erben, K. (1959). Untersuchungen über Auxosporenentwicklung und Meioseauslösung an *Melosira nummuloides* (Dillw.) Agardh. *Arch. Protistenk.* **104**, 165–210.

Ernst, A. (1918). Bastardierung als Ursache der Apogamie im Pflanzenreich. Gustav Fischer, Jena.

Ernst-Schwarzenbach, M. (1957). Zur Kenntnis der Fortpflanzungsmodi der Braunalge *Halopteris filicina* Kütz. *Pubbl. Staz. Zool. Napoli* **29**, 347–388.

Ettl, H., Müller, D. C., Neumann, K., von Stosch, H. A., and Weber, W. (1967). Vegetative Fortpflanzung, Parthenogenese und Apogamie bei Algen. In "Handbuch Pflanzenphysiologie" (W. Ruhland, ed.), Vol. 18, 597–776. Springer, Heidelberg.

Farmer, J. B., and Williams, J. L. (1898). Contributions to our knowledge of the *Fucaceae*. Their life-history and cytology. *Phil. Trans. Botan. Soc. London,* **B190**, 623–645.

Förster, H. (1957). Das Wirkungsspektrum der Kopulation von *Chlamydomonas eugametos. Z. Naturforsch.* **12b**, 765–770.

Förster, H. (1959). Die Wirkungsstärken einiger Wellenlängen zum Auslösen der Kopulation von *Chlamydomonas moewusii. Z. Naturforsch.* **14b**, 479–480.

Förster, H., and Wiese, L. (1954a). Untersuchungen zur Kopulationsfähigkeit von *Chlamydomonas eugametos. Z. Naturforsch.* **9b**, 470–471.

Förster, H., and Wiese, L. (1954b). Gamonwirkungen bei *Chlamydomonas eugametos. Z. Naturforsch.* **9b**, 548–550.

Förster, H., and Wiese, L. (1955). Gamonwirkung bei *Chlamydomonas reinhardti. Z. Naturforsch.* **10b**, 91–92.

Förster, H., Wiese, L., and Braunitzer, G. (1956). Über das agglutinierend wirkende Gynogamon von *Chlamydomonas eugametos. Z. Naturforsch.* **11b**, 315–317.

Fott, B. (1959). Zur Frage der Sexualität bei den Chrysomonaden. *Nova Hedwigia* **1**, 116–130.

Fott, B. (1964). Hologamic and agamic cyst formation in loricate Chrysomonads. *Phykos* **3**, 15–18.

Friedmann, I. (1959). Structure, life-history and sex determination of *Prasiola stipitata* Suhr. *Ann. Botany* (*London*) [N.S.] **23**, 571–594.

Friedmann, I. (1960). Gametes, fertilization and zygote development in *Prasiola stipitata* Suhr. I. Light microscopy. *Nova Hedwigia* **1**, 333–344.

Friedmann, I. (1961). Cinemicrography of spermatozoids and fertilization in Fucales. *Bull. Res. Council Israel,* **10D**, 73–83.

Friedmann, I. (1962). Cell membrane fusion and fertilization mechanisms in plants and animals. *Science* **136**, 711–712.

Friedmann, I., Colwin, A. L., and Colwin, L. H. (1968). Fine structural aspects of fertilization in *Chlamydomonas reinhardi. J. Cell Sci.* **3**, 115–128.

Geitler, L. (1931). Untersuchungen über das sexuelle Verhalten von *Tetraspora lubrica. Biol. Zentr.* **51**, 173–187.

Geitler, L. (1932). Der Formwechsel der pennaten Diatomeen. *Arch. Protistenk.* **78**, 1–226.

Geitler, L. (1939). Gameten- und Auxosporenbildung von *Synedra ulna* im Vergleich mit anderen Diatomeen. *Planta* **30**, 551–566.

Geitler, L. (1949a). Die Auxosporenbildung von *Nitzschia sigmoidea* und die Geschlechtsbestimmung bei den Diatomeen. *Portugaliae Acta Biol., Ser. A,* (Richard Goldschmidt volume) 79–87.

Geitler, L. (1949b). Das Kopulationsverhalten von *Mougeotia polymorpha* n. sp. und die Azygotenbildung von *Spirogyra dimorpha* n. sp. *Österr. Botan. Z.* **96**, 15–24.

Geitler, L. (1951). Kopulation und Formwechsel von *Eunotia arcus. Österr. Botan. Z.* **98**, 292–337.

Geitler, L. (1952a). Oogamie, Mitose, Meiose und metagame Teilung bei der zentrischen Diatomee *Cyclotella. Österr. Botan. Z.* **99**, 506–520.

Geitler, L. (1952b). Untersuchungen über Kopulation und Auxosporenbildung pennater Diatomeen. IV. Vierkernige Zygoten bei *Navicula cryptocephala* var. *veneta.* V. Allogamie bei *Synedra rumpens* var. *fragilarioides. Österr. Botan. Z.* **99**, 598–605.

Geitler, L. (1952c). Über differentielle Teilung und einen im Zellbau begründeten, kopulationsbegrenzenden Faktor bei der Diatomee *Cocconeis. Z. Naturforsch.* **7b**, 411–414.

Geitler, L. (1952d). Die Auxosporenbildung von *Rhoicosphenia curvata. Österr. Botan. Z.* **99**, 78–88.

Geitler, L. (1953a). Allogamie und Autogamie bei der Diatomee *Denticula tenuis* und die Geschlechtsbestimmung der Diatomeen. *Österr. Botan. Z.* **100**, 331–352.

Geitler, L. (1953b). Abhängigkeit der Membranbildung von der Zellteilung bei Diatomeen und differentielle Teilungen im Zusammenhang mit der Bildung von Innenschalen. *Planta* **43**, 75–82.

Geitler, L. (1954). Echte Oogamie bei *Chlamydomonas. Österr. Botan. Z.* **101**, 570–578.

Geitler, L. (1956). Automixis, Geschlechtsbestimmung und Pyknose von Gonenkernen bei *Cymbella aspera. Planta* **47**, 359–373.

Geitler, L. (1957). Die sexuelle Fortpflanzung der pennaten Diatomeen. *Biol. Rev.* **32**, 261–295.

Geitler, L. (1958a). Isogames Bewegungsverhalten unter morphologischer Anisogamie bei einer konjugaten Alge. *Biol. Zentr.* **77**, 202–209.

Geitler, L. (1958b). Fortpflanzungsbiologische Eigentümlichkeiten von *Cocconeis* und Vorarbeiten zu einer systematischen Gliederung von *Cocconeis placentula* nebst Beobachtungen an Bastarden. *Österr. Botan. Z.* **105**, 350–379.

Geitler, L. (1963). Alle Schalenbildungen der Diatomeen treten als Folge von Zell- oder Kernteilungen auf. *Ber. Deut. Botan. Ges.* **75**, 393–396.

Geitler, L. (1965). *Mesotaenium dodekahedron* n. sp. und die Gestalt und Entstehung seiner Zygoten. *Österr. Botan. Z.* **112**, 344–358.

Gerloff, J. (1940). Beiträge zur Kenntnis der Variabilität und Systematik der Gattung *Chlamydomonas. Arch. Protistenk.* **94**, 311–502.

Gibbs, S. P., Lewin, R. A., and Philpott, D. E. (1958). The fine structure of the flagellar apparatus of *Chlamydomonas moewusii. Exptl. Cell Res.* **15**, 619–622.

Goldstein, M. (1964). Speciation and mating behavior in *Eudorina. J. Protozool.* **11**, 317–344.

Gowans, C. S. (1963). The conspecificity of *Chlamydomonas eugametos* and *Chlamydomonas moewusii:* an experimental approach. *Phycologia* **3**, 37–44.

Hämmerling, J. (1934). Über die Geschlechtsverhältnisse von *Acetabularia mediterranea* und *Acetabularia Wettsteinii. Arch. Protistenk.* **83**, 57–97.

Hartmann, C. (1962). Die Regulation der Gametogenese von *Chlamydomonas eugametos* und *Chlamydomonas moewusii* durch exogene und endogene Faktoren. Vergleichend morphologische, physiologische und biophysikalische Untersuchungen. Dissertation, Tübingen, Germany.

Hartmann, M. (1925). Untersuchungen über relative Sexualität. I. Versuche an *Ectocarpus siliculosus. Biol. Zentr.* **45**, 449–467.

Hartmann, M. (1934). Untersuchungen über die Sexualität von *Ectocarpus siliculosus. Arch. Protistenk.* **83**, 110–153.

Hartmann, M. (1937). Ergänzende Untersuchungen über die Sexualität von *Ectocarpus siliculosus. Arch. Protistenk.* **89**, 382–392.

Hartmann, M. (1950). Beiträge zur Kenntnis der Befruchtung und Sexualität mariner Algen. I. Über die Befruchtung von *Cutleria multifida. Pubbl. Staz. Zool. Napoli* **22**, 1–9.

Hartmann, M. (1955). Sex problems in algae, fungi and protozoa. *Am. Naturalist* **89**, 321–340.

Hartmann, M. (1956). "Die Sexualität." Fischer, Stuttgart.

Hirn, K. E. (1900). Monographie der Oedogoniaceen. *Acta Soc. Sci. Fennicae* **27**, No. 1.

Hoffman, L. R. (1960). Chemotaxis of *Oedogonium* sperms. *Southwestern Naturalist* **5**, 111–116.

Hoffman, L. R. (1961). Studies of the morphology, cytology, and reproduction of *Oedogonium* and *Oedocladium.* Dissertation, Univ. of Texas, Austin, Texas.

Holmes, R. W. (1967). Auxospore formation in two marine clones of the diatom genus *Coscinodiscus. Am. J. Botany* **54**, 163–168.

Hoshaw, R. W. (1965). A cultural study of sexuality in *Sirogonium melanosporum. J. Phycol.* **1**, 134–139.

Hoshaw, R. W., and Ettl, H. (1966). *Chlamydomonas smithii* sp. nov.—a chlamydomonad interfertile with *Chlamydomonas reinhardtii. J. Phycol.* **2**, 93–96.

Hutner, S. H., and Provasoli, L. (1951). The phytoflagellates. *In* "Biochemistry and Physiology of Protozoa" (A. Lwoff, ed.), Vol. 1, pp. 27–128. Academic Press, New York.

Iyengar, M. O. P. (1958). A new type of lateral conjugation in *Spirogyra. J. Indian Botan. Soc.* **37**, 387.

Iyengar, M. O. P., and Subrahmanyan, R. (1944). On reduction division and auxospore-formation in *Cyclotella meneghiniana* Kutz. *J. Indian Botan. Soc.* **28**, 125–152.

Kates, J. R., and Jones, R. F. (1964). The control of gametic differentiation in liquid cultures of *Chlamydomonas. J. Cellular Comp. Physiol.* **63**, 151–164.

Kies, L. (1962). Über die experimentelle Auslösung von Fortpflanzungsvorgängen bei Desmidiaceen. *Vorträge Gesamtgeb. Botanik Deut. Botan. Ges.* [N.F.] **1**, 65–70.

Kies, L. (1964). Über die experimentelle Auslösung von Fortpflanzungsvorgängen und die Zygotenkeimung bei *Closterium acerosum. Arch. Protistenk.* **107**, 331–350.

Kniep, H. (1928). Die Sexualität der niederen Pflanzen. Gustav Fischer, Jena.

Kochert, G. (1968). Differentiation of reproductive cells in *Volvox carteri. J. Protozool.* **15**, 438–452.

Köhler, K. (1956). Entwicklungsgeschichte, Geschlechtsbestimmung und Befruchtung bei *Chaetomorpha. Arch. Protistenk.* **101**, 224–268.

Köhler, K. (1957). Neue Untersuchungen zur Sexualität von *Dasycladus. Arch. Protistenk.* **102**, 209–218.

Köhler, K. (1967). Die chemischen Grundlagen der Befruchtung (Gamone). *In* "Handbuch der Pflanzenphysiologie" (Ruhland, W., ed.), Vol. 18, pp. 282–320. Springer, Berlin.

Kornmann, P. (1956). Zur Morphologie und Entwicklung von *Percursaria percursa. Helgoländer Wiss. Meeresuntersuch.* **5**, 259–272.

Kornmann, P. (1962). Zur Entwicklungsgeschichte von *Monostroma grevillei* und zur systematischen Stellung von *Gomontia polyrhiza. Vorträge Gesamtgeb. Botanik Deut. Botan. Ges.* [N.F.] **1**, 37–39.

Kotte, W. (1923). Zur Reizphysiologie der *Fucus*-Spermatozoiden. *Ber. Deut. Botan. Ges.* **41**, 24–32.

Krieger, W. (1937). Die Desmidiaceen Europas mit Berücksichtigung der aussereuropäischen Arten. *In* "Rabenhorst's Kryptogamenflora," Vol. 13, Abtl. 1, Teil 1. Akad. Verlagsgesellschaft, Leipzig.

Kristiansen, J. (1963). Sexual and asexual reproduction in *Kephyrion* and *Stenocalyx* (Chrysophyceae). *Botan. Tidsskr.* **59**, 244–254.

Kuhn, R. and Löw, I. (1960). Über Flavonolglycoside von *Forsythia* und über Inhaltsstoffe von *Chlamydomonas. Chem. Ber.* **93**, 1009–1010.

Kunieda, H., and Suto, S. (1938). The life-history of *Colpomenia sinuosa* (Scytosiphonaceae) with special reference to the conjugation of anisogametes. *Botan. Mag.* (*Tokyo*) **52**, 539–546.

Lerche, W. (1937). Untersuchungen über Entwicklung und Fortpflanzung in der Gattung *Dunaliella. Arch. Protistenk.* **88**, 236–268.

Levring, T. (1949). Fertilization experiments with *Hormosira Banksii* (Turn.). *Physiol. Plantarum* **2**, 45–55.

Levring, T. (1952). Remarks on the submicroscopical structure of eggs and spermatozoids of *Fucus* and related genera. *Physiol. Plantarum* **5**, 528–539.

Levring, T. (1955). Some remarks on the structure of gametes in *Ulva lactuca. Botan. Notiser* **108**, 40–45.

Lewin, R. A. (1952). Studies on the flagella of algae. I. General observations on *Chlamydomonas moewusii* Gerloff. *Biol. Bull.* **102**, 74–79.

Lewin, R. A. (1954a). Sex in unicellular algae. *In* "Sex in Microorganisms" (D. H. Wenrich, ed.), pp. 100–133. Am. Assoc. Advan. Sci., Washington, D. C.

Lewin, R. A. (1954b). Mutants of *Chlamydomonas* with impaired motility. *J. Gen. Microbiol.* **11**, 358–363.

Lewin, R. A., and Meinhart, J. O. (1953). Studies on the flagella of algae. III. Electron micrographs of *Chlamydomonas moewusii. Can. J. Botany* **31**, 711–717.

Lippert, B. E. (1967). Sexual reproduction in *Closterium moniliferum* and *Closterium ehrenbergii. J. Phycol.* **3**, 182–197.

Machlis, L., and Rawitscher-Kunkel, E. (1967). Mechanisms of gametic approach in plants. *In* "Fertilization: Comparative Morphology, Biochemistry, and Immunology" (C. B. Metz and A. Monroy, eds.), pp. 117–158. Academic Press, New York.

Machlis, L., Nutting, W. H., Williams, M. W., and Rapoport, H. (1966). Production, isolation and characterization of Sirenin. *Biochemistry* **5**, 2147–2157.

Magne-Simon, M. F. (1962). L'auxosporulation chez une Tabellariacée marine, *Grammatophora marina* (Lyngb.) Kütz. (Diatomée). *Cahiers Biol. Marine* **3**, 79–89.

Mainx, F. (1931). Gametenkopulation und Zygotenkeimung bei *Hydrodictyon reticulatum*. *Arch. Protistenk.* **75**, 502–516.

Manton, I. (1959). Observations on the internal structure of the spermatozoid of *Dictyota*. *J. Exptl. Botany* **10**, 448–461.

Manton, I., and Clarke, B. (1950). Electron microscope observations on the spermatozoid of *Fucus*. *Nature* **166**, 973–974.

Manton, I., and Clarke, B. (1951). An electron microscope study of the spermatozoid of *Fucus serratus*. *Ann. Botany (London)* **15** [N.S.], 461–471.

Manton, I., and Clarke, B. (1956). Observation with the electron microscope on the internal structure of the spermatozoid of *Fucus*. *J. Exptl. Botany* **7**, 416–432.

Manton, I., and Friedmann, I. (1960). Gametes, fertilization and zygote development in *Prasiola stipitata* Suhr. II. Electron microscopy. *Nova Hedwigia* **1**, 443–462.

Manton, I., and von Stosch, H. A. (1966). Observations on the fine structure of the male gamete of the marine centric diatom *Lithodesmium undulatum*. *J. Royal Microscop. Soc.* **85**, 119–134.

Manton, I., Clarke, B., and Greenwood, A. D. (1953). Further observations with the electron microscope on spermatozoids in the brown algae. *J. Exptl. Botany* **4**, 319–329.

Metz, C. B. (1967). Gamete surface components and their role in fertilization. *In* "Fertilization: Comparative Morphology, Biochemistry, and Immunology" (C. B. Metz and A. Monroy, eds.), pp. 163–224. Academic Press, New York.

Moewus, F. (1933). Untersuchungen über die Sexualität und Entwicklung von Chlorophyceen. *Arch. Protistenk.* **80**, 469–520.

Moewus, F. (1938). Carotinoide als Sexualstoffe von Algen. *Jahrb. Wiss. Botanik* **86**, 753–783.

Moewus, F. (1939). Über die Chemotaxis von Algengameten. *Arch. Protistenk.* **92**, 485–526.

Müller, D. G. (1965). Bemerkung zum Bau der Geisseln von Braunalgenschwärmern. *Naturwissenschaften* **52**, 311.

Müller, D. G. (1967). Generationswechsel, Kernphasenwechsel und Sexualität der Braunalge *Ectocarpus siliculosus* im Kulturversuch. *Planta* **75**, 39–54.

Müller, D. G. (1968). Versuche zur Charakterisierung eines Sexual-Lockstoffes bei der Braunalge *Ectocarpus siliculosus*. *Planta* **81**, 160–168.

Pascher, A. (1916). Über die Kreuzung einzelliger, haploider Organismen: *Chlamydomonas*. *Ber. Deut. Botan. Ges.* **34**, 228–242.

Pascher, A. (1931a). Über einen neuen einzelligen und einkernigen Organismus mit Eibefruchtung. *Botan. Centr. Beih.* **A48**, 466–480.

Pascher, A. (1931b). Über Gruppenbildung und "Geschlechtswechsel" bei den Gameten einer Chlamydomonadine (*Chlamydomonas paupera*). *Jahrb. Wiss. Botanik* **75**, 551–580.

Pascher, A. (1939). Über geisselbewegliche Eier, mehrköpfige Schwärmer und vollständigen Schwärmerverlust bei *Sphaeroplea*. *Botan. Centr. Beih.* **A59**, 188–213.

Pascher, A. (1943). Beiträge zur Morphologie der ungeschlechtlichen und geschlechtlichen Vermehrung der Gattung *Chlamydomonas*. *Botan. Centr. Beih.* **A62**, 197–220.

Pockock, M. A. (1953). Two multicellular motile green algae *Volvulina* Playfair and *Astrephomene*, a new genus. *Trans. Roy. Soc. S. Africa* **34**, 103–127.

Pockock, M. A. (1960). *Hydrodictyon*: a comparative biological study. *J. S. African Botany* **26**, 167–319.

Presscott, G. W. (1948). Desmids. *Botan. Rev.* **14**, 644–676.

Pringsheim, E. G. (1963). Farblose Algen. Gustav Fischer, Stuttgart.

Rawitscher-Kunkel, E., and Machlis, L. (1962). The hormonal integration of sexual reproduction in *Oedogonium. Am. J. Botany* **49**, 177–183.

Reimann, B. (1960). Bildung, Bau und Zusammenhang der Bacillariophyceenschalen. *Nova Hedwigia* **2**, 349–372.

Rieth, A. (1954). Beobachtungen zur Entwicklungsgeschichte einer *Vaucheria* der Section Woroninia. *Flora* **142**, 156–182.

Sager, R., and Granick, S. (1954). Nutritional control of sexuality in *Chlamydomonas reinhardti. J. Gen. Physiol.* **37**, 729–742.

Schreiber, E. (1925). Zur Kenntnis der Physiologie und der Sexualität höherer Volvocales. *Z. Botanik* **17**, 337–376.

Schreiber, E. (1931). Über die geschlechtliche Fortpflanzung der Sphacelariales. *Ber. Deut. Botan. Ges.* **49**, 235–240.

Schultz, M. E., and Trainor, F. R. (1968). Production of male gametes and auxospores in the centric diatoms *Cyclotella meneghiana* and *C. cryptica. J. Phycol.* **4**, 85–88.

Skuja, H. (1949). Drei Fälle von sexueller Reproduktion in der Gattung *Chlamydomonas* Ehrenberg. *Svensk Botan. Tidskr.* **43**, 586–602.

Starr, R. C. (1954a). Heterothallism in *Cosmarium botrytis* var. *subtumidum. Am. J. Botany* **41**, 601–607.

Starr, R. C. (1954b). Inheritance of mating type and a lethal factor in *Cosmarium botrytis* var. *subtumidum* Wittr. *Proc. Natl. Acad. Sci. U. S.* **40**, 1060–1063.

Starr, R. C. (1955). Isolation of sexual strains of placoderm desmids. *Bull. Torrey Botan. Club* **82**, 261–265.

Starr, R. C. (1959). Sexual reproduction in certain species of *Cosmarium. Arch. Protistenk.* **104** 155–164.

Starr, R. C. (1962). A new species of *Volvulina* Playfair. *Arch. Microbiol.* **42**, 130–137.

Starr, R. C. (1963). Homothallism in *Golenkinia minutissima. In* "Studies on microalgae and photosynthetic bacteria," pp. 3–6. Japan. Soc. Plant Physiologists. Univ. Tokyo Press, Tokyo.

Starr, R. C. (1968). Cellular differentiation in *Volvox. Proc. Natl. Acad. Sci. U. S.* **40**, 1060–1063.

Starr, R. C., and Rayburn, W. R. (1964). Sexual reproduction in *Mesotaenium kramstai. Phycologia* **4**, 23–27.

Stein, J. (1958a). A morphological study of *Astrephomene gubernaculifera* and *Volvulina steinii. Am. J. Botany* **45**, 388–396.

Stein, J. (1958b). A morphological and genetic study of *Gonium pectorale. Am. J. Botany* **45**, 664–672.

Stifter, I. (1959). Untersuchungen über einige Zusammenhänge zwischen Stoffwechsel und Sexualphysiologie an dem Flagellaten *Chlamydomonas eugametos. Arch. Protistenk.* **104**, 364–388.

Strehlow, K. (1929). Über die Sexualität einiger *Volvocales. Z. Botanik* **21**, 625–692.

Subrahmanyan, R. (1947). On somatic division, reproductive division, auxospore formation and sex differentiation in *Navicula halophila. J. Indian Botan. Soc., Iyengar Commem. Vol.,* pp. 239–266.

Trainor, F. R. (1959). A comparative study of sexual reproduction in four species of *Chlamydomonas. Am. J. Botany* **46**, 65–70.

Trainor, F. R., and Burg, C. A. (1965). *Scenedesmus obliquus* sexuality. *Science* **148**, 1094–1095.

Transeau, E. N. (1951). "The Zygnemataceae." Ohio State Univ. Press, Columbus, Ohio.

Tschermak-Woess, E. (1959). Extreme Anisogamie und ein bemerkenswerter Fall der Geschlechtsbestimmung bei einer neuen *Chlamydomonas* Art. *Planta* **52**, 606-622.

Tschermak-Woess, E. (1962). Zur Kenntnis von *Chlamydomonas suboogama*. *Planta* **59**, 68–76.

Tschermak-Woess, E. (1963). Das eigenartige Kopulationsverhalten von *Chloromonas saprophila*, einer neuen Chlamydomonacee, *Österr. Botan. Z.* **110**, 294–307.

Tsubo, Y. (1961a). Sexual reproduction of *Chlamydomonas* as affected by ionic balance in the medium. *Botan. Mag.* (*Tokyo*) **74**, 442–448.

Tsubo, Y. (1961b). Chemotaxis and sexual behavior in *Chlamydomonas*. *J. Protozool.* **8**, 114–121.

von Stosch, H. A. (1951). Entwicklungsgeschichtliche Untersuchungen an zentrischen Diatomeen. I. Die Auxosporenbildung von *Melosira varians*. *Arch. Mikrobiol.* **16**, 101–135.

von Stosch, H. A. (1954). Die Oogamie von *Biddulphia mobiliensis* und die bisher bekannten Auxosporenbildungen bei den Centrales. 8ᵉ *Congr. Intern. Botan. Paris*, Sect. 17, pp. 58–68.

von Stosch, H. A. (1956). Entwicklungsgeschichtliche Untersuchungen an zentrischen Diatomeen. II. Geschlechtszellenreifung, Befruchtung und Auxosporenbildung einiger grundbewohnender Biddulphiaceen der Nordsee. *Arch. Mikrobiol.* **23**, 327–365.

von Stosch, H. A. (1958). Kann die oogame Araphidee *Rhabdonema adriaticum* als Bindeglied zwischen den beiden grossen Diatomeengruppen angesehen werden? *Ber. Deut. Botan. Ges.* **71**, 241–249.

von Stosch, H. A. (1962). Über das Perizonium der Diatomeen. *Vorträge Gesamtgeb. Botanik* [N.F.] **1**, 43–52.

von Stosch, H. A. (1964). Zum Problem der sexuellen Fortpflanzung in der Peridineengattung *Ceratium*. *Helgolaender Wiss. Meeresuntersuch.* **10**, 140–152.

von Stosch, H. A. (1965). Sexualität bei *Ceratium cornutum* (Dinophyta). *Naturwissenschaften* **52**, 112–113.

von Stosch, H. A., and Drebes, G. (1964). Entwicklungsgeschichtliche Untersuchungen an zentrischen Diatomeen. IV. Die Planktondiatomee *Stephanopyxis turris*—ihre Behandlung und Entwicklungsgeschichte. *Helgolaender Wiss. Meersuntersuch.* **11**, 209–257.

von Witsch, H., Runkel, K. H., and Geranmayeh, R. (1966). Zur Kenntnis von *Gloeococcus bavaricus* Skuja. *Protoplasma* **61**, 114–126.

Wiese, L. (1961). Gamone. *Fortschr. Zool.* **13**, 119–145.

Wiese, L. (1965). On sexual agglutination and mating-type substances (gamones) in isogamous heterothallic Chlamydomonads. I. Evidence of the identity of the gamones with the surface components responsible for sexual flagellar contact. *J. Phycol.* **1**, 46–54.

Wiese, L. (1966). Geschlechtsbestimmung. *Fortschr. Zool.* **18**, 139–206.

Wiese, L., and Jones, R. F. (1963). Studies on gamete copulation in heterothallic Chlamydomonads. *J. Cellular Comp. Physiol.* **61**, 265–274.

Wiese, L., and Metz, C. B. (1969). On the trypsin sensitivity of fertilization as studied with living gametes in *Chlamydomonas*. *Biol. Bull.* **136**, 483–493.

Williams, J. L. (1904). Studies in the Dictyotaceae. II. The cytology of the gametophyte generation. *Ann. Botany* (*London*) **18**, 183–204.

Ziegler, J. R., and Kingsbury, J. M. (1964). Cultural studies on the marine green alga *Halicystis parvula-Derbesia tenuissima*. I. Normal and abnormal sexual and asexual reproduction. *Phycologia* **4**, 105–116.

CHAPTER 5

Fertilization Mechanisms in Higher Plants

H. F. Linskens

DEPARTMENT OF BOTANY, UNIVERSITY OF NIJMEGEN,
NIJMEGEN, THE NETHERLANDS

I. Introduction: Pollination and Fertilization in Spermatophytes

Fertilization or syngamy is the fusion of two sexually differentiated cells (gametes) resulting in the formation of a zygote and the processes

189

leading to it. Reproduction in plants is characterized by alternation of sexual and asexual generations.

In *phanerogamic* plants the sexual generation is much more reduced than in cryptogamic plants and is completely dependent on the metabolic supply of the asexual generation. Moreover the macrospore remains surrounded by sporophytic tissue. In higher plants a tendency to double fertlization is observed. In Coniferae this tendency appeared only as a sign of lability in differentiation, whereas it is normal in angiosperms.

Fertilization in higher plants is preceded by pollination, that is the transfer of the microspores (pollen grains) by means of ecological vectors to a position close to the macrospore (embryo sac). Before pollination can occur, sexually differentiated cells must be produced in the flower. The latter is a modified shoot, consisting of metamorphosed leaves which are specially adapted for reproduction, namely, the anthers bearing the microspores (male sex organs) and the pistil (female sex organ) containing the ovary, the style, and the receptive organ the stigma. The flower also has other differentiated parts including petals, sepals, nectaries, and receptacle. However, great variation of these structures having special functions in pollination occurs in different taxonomic groups.

Pollination depends on many ecological factors including adaptation to special conditions of the environment. *Fertilization* takes place within the tissue of the living plant. That means intensive interrelation, correlation, and exchange of genetic and biochemical processes. The latter processes are the subject of this treatise. The following literature deals with the anatomical and morphological specialization and differentiation in the various groups of the seed plants: Dahlgren, 1927; Schnarf, 1929; 1941; Johansen, 1950; Maheshwari, 1950, 1963; Vazart, 1958; Poddubnaya-Arnoldi, 1964.

In spermatophytes, as in animals, the following are primary topics of current research interest concerning problems of fertilization: (*a*) the pathway of the pollen tube through the female tissue and the biochemical exchange preceding the syngamic process *sensu strictu*; (*b*) the role of substances produced by the gametes in the fusion process; (*c*) the species specificity in fertilization. Although little is known concerning species barriers, many observations have been made concerning selective fertilization and the incompatibility barriers within a species.

II. Macrosporogenesis

A. MATURATION OF THE MACROSPORE

In angiosperms, macrospores develop inside the ovary—the enlarged basal part of the pistil—which becomes the fruit after fertilization. The

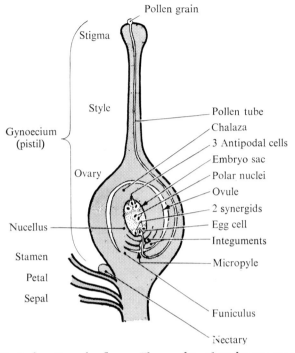

FIG. 1. Vertical section of a flower. The number of ovules per ovary depends on the species.

macrosporangium (ovule) consists of the nucellus and one or two surrounding integuments mounted on a short stalk (funiculus). The pore leading from the outer surface of the ovule between the edges of the integuments down to the surface of the nucellus is called the micropyle (Fig. 1).

The most common sequence of events in megaspore development is found in the monosporic type (Figs. 2 and 3). Here three out of four cells of the tetrad degenerate while the functional megaspore undergoes three mitotic divisions. The resulting eight nuclei organize into an embryo sac. The egg cell lies somewhat excentrically with its nucleus in the lower part, which is surrounded by strongly vacuolated cytoplasm. In the synergids, the nucleus is found above with a large basal vacuole (Figs. 4 and 14). Haustorial extension of the embryo sac has received much attention. Many examples of a high degree of polyploidy of the synergid and antipodal nuclei have also been observed in angiosperms.

In the gymnosperms (*Pinus*) (Fig. 2) the megaspores are also formed from megaspore mother cells by reduction divisions. Of the four megaspores formed from each mother cell, the one farthest from the micropyle

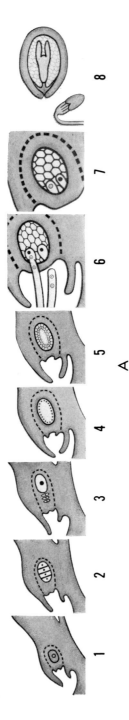

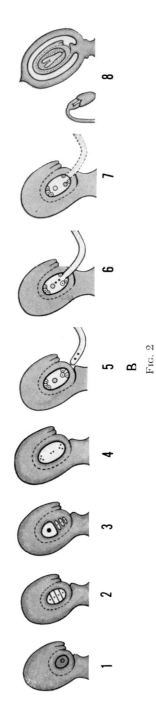

Fig. 2

enlarges at the expense of the other three, which are finally completely absorbed. At the time of pollination the nucellus of the megasporangium is surrounded by a single integument and encloses a single functional megaspore. During the year following pollination the megaspore develops into a considerable mass of cellular tissue, which differs from the surrounding nucellar cells in that the cells are haploid. Late in the spring of the following year, two or three archegonia normally develop in the micropylar end of the female gametophyte. When fully mature, each archegonium in gymnosperms consists of eight small neck cells in two tiers of four, the ventral canal cell which disorganizes before fertilization, and a very large cavity almost filled by the egg cell. The archegonium is not fully developed and ready for fertilization until about 1 year after pollination.

B. Metabolism in the Ovary

Although the importance of understanding the basic processes in fertilization in higher plants is appreciated, we still have practically no knowledge about the physiological conditions within the embryo sac (Johri, 1963). Localization of substances in embryonic organs of angiosperms was studied by Poddubnaya-Arnoldi and collaborators (Zinger and Poddubnaya-Arnoldi, 1961; Poddubnaya-Arnoldi, 1961, 1964; Poddubnaya-Arnoldi et al., 1964). High activity of peroxidase is found in the antipodal cells and in the embryonal suspensor, whereas cytochrome–oxidase activity was observed just before fertilization in the micropyle of

Fig. 2. Megasporogenesis and fertilization. A. Sequence of development in gymnosperms: (1) megaspore mother cell; (2) linear tetrad; (3) functional megaspore with three degenerating cells of the tetrad; (4) free nuclear gametophyte; (5) formation of cellular gametophyte by alveoli and laying down of cell plate; (6) development of archegonia at the micropylar region of the female gametophyte in the spring of the following year [the resting pollen tube at the tip of the nucellus grows rapidly, penetrates the neck of the archegonium, and discharge its contents (tube nucleus, stalk cell and the two sperm nuclei) into the egg—only the large functional male nucleus fuses with the egg]; (7) the nonfunctioning material disorganizes and forms part of the food store of the zygote; (8) each zygote gives rise to four or more proembryos—one out of these mature while others are arrested at various stages of development. B. Sequence of development in angiosperms: (1) Bitegmic ovule with megaspore mother cell; (2) megaspore tetrad formed by meiosis; (3) formation of functional megaspore and degeneration of the upper three cells of the tetrad; (4) eight-nucleate embryo sac embedded in the nucellar tissue; (5) mature embryo sac—the polar nuclei fuse at the time of fertilization; (6) arrival of the pollen tube, penetration through the micropyle and the nucellar tissue, and entry into the embryo sac; (7) one sperm nucleus fuses with the egg while the other fuses with the polar nuclei to form the triploid primary endosperm nucleus; (8) maturation of the ovule into a seed. This includes the formation of the embryo and the endosperm and, in most cases, the complete disappearance of the nucellus.

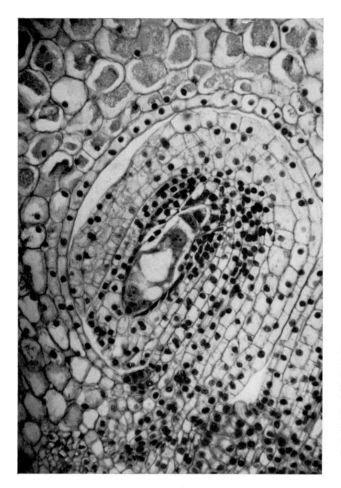

FIG. 3. Longitudinal section of a bitegmic anatropous ovule of *Lilium* showing a mature embryo sac. The number of ovules in an ovary ranges from one (as shown in the figure) to many thousands (orchids). (450 ×).

the ovule. High dehydrogenase activity is typical for the micropylar region of the embryo sac and for the developing embryo. Increasing concentration of sulfhydryl groups is found in the antipodals and in the cells of the young embryo. Plastids and mitochondria are the sites of synthesis in the endosperm mother cell. In the synergids, particularly, endoplasmatic reticulum is present and functional, whereas plastids degenerate.

Very different pathways of metabolism in the ovary have been observed in the different families of angiosperms. In orchids, e.g., with the absence of endosperm, antipodals and other haustorial organs are partly replaced by a suspensor haustorium.

In brief, there are two different ways for transport of metabolites in feeding the ovary (Fig. 5): (*1*) substances are transported from the integuments to the chalaza by diffusion and then to the embryo sac or

(2) from the integuments directly to the embryo sac, which is surrounded by an integumentary tapetum.

The cells of this layer frequently are specially differentiated. They serve as an intermediary for the transport of food material to the embryo sac. It is supposed that the integumentary tapetum contains

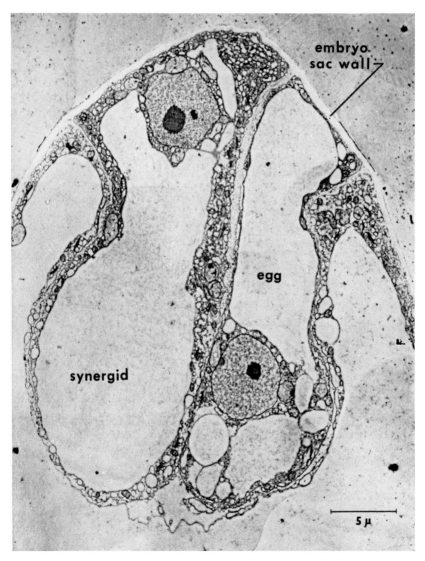

FIG. 4. Electron micrograph of a longitudinal section through the egg apparatus of the unfertilized embryo sac of *Torenia fournieri.* (Courtesy J. E. Van der Pluijm.)

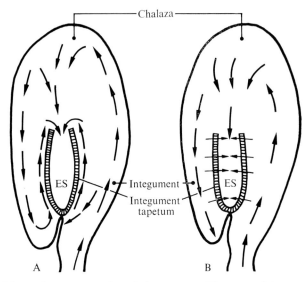

Fig. 5. Schematic representation of the two possible ways of transport of metabolic materials from the integuments into the embryo sac (ES). (A) The food material flows into the chalaza and then to the embryo sac; (B) soluble material is transported directly to the embryo sac from the integuments through the integumental tapetum. [After Zinger, 1958; taken from Poddubnaya-Arnoldi, 1964 (Fig. 71).]

diastase and other enzymes to convert the nutritive material into a suitable form for the use of the embryo sac.

Within the ovules, the central vacuole of the endosperm mother cell has special importance in synthesis and storage of proteins and starch, as well as the migration of soluble substances including sugars, amino acids, mineral salts, and water from the vegetative organs to the embryo sac (Ryczkowski, 1964). Decrease of viscosity and density of the central vacuolar sap in young ovules is associated with an intensive inflow of water into the ovule. Increase of these physicochemical factors is the result of an intensive absorption of water by the growing and developing embryo, causing an increase of protein content.

The contents [sugars, starch, proteins, lipids, and ribonucleic acid (RNA)] change with development. Starch grains and oil drops are not found within the embryo sac but are present in the surrounding tissues (integuments, funiculus, ovary wall). This fact indicates that these tissues act as energy reservoirs. The activity of enzymes (peroxidase, oxidase, dehydrogenase, phosphatase) depends on the stage of development (Miki-Hirosige, 1964b).

Metabolic changes in the ovary and the surrounding structures, in connection with succesful fertilization, are of great importance since they control subsequent development.

C. RECEPTIVITY OF THE STIGMA

The stigma appears to be a modified portion of the inner surface of the ovary wall (Thomas, 1934) and forms the expanded tip of the style to which the pollen grains adhere. Sometimes the epidermal cells of the stigma become papillate and covered by a cuticle (Kroh, 1964). In many cases the surface of the stigma produces a secretion. The chief function of the stigmatic fluid is to protect the stigma from desiccation and to play some role in receiving the pollen and giving optimal germination conditions for the pollen grains.

The *stigmatic liquid* contains enzymes which are able to digest starch to maltose and glucose (Iwanami, 1957), lipids (Bubar, 1958), sugars, and sometimes even resin (Baum, 1950). Two different inhibitory substances have been isolated from the stigma which prevent the entrance of bacteria and fungi, though the stigmatic liquid itself has no inhibiting effect (Jung, 1956; Jung and Stoll, 1957; Jung and Plempel, 1960). Duration of secretion depends on pollination. Although the unfertilized stigma continues to produce slime for 48 hours, it ceases after fertilization. Stigmatic slime promotes pollen germination (Tkachenko, 1959) and may work as a preventive against the germination of foreign pollen (Burck, 1901). The high content of boron, though nonspecific (Schmucker, 1935; Ehlers, 1951; Visser, 1955; Glenk, 1960), is important for normal development of pollen tubes.

In the Gramineae the stigma undergoes changes in the early stage of attachment of pollen grains: the stigmatic cells become more readily stainable, and the permeability changes; with the withering of the stigma, these cells change in size and nuclear structure. The physicochemical changes that follow in the stigma, due to the contact of the pollen (but not the growth of the pollen tube), have been termed the *stigma reaction* (Kato, 1953; Kato and Watanabe, 1957). It seems that when the stigma reaction takes place in a stigma cell it is not restricted within this cell but passes to the neighboring cells of the same feathery stigma.

Receptivity differs from species to species. Corn silks are receptive (but not uniformly) over a long period (about 24 days); the peak of receptivity was noticed on the eighth day (Jones and Newell, 1948). Pollination causes a decrease in the number of physiologically active substances in the stigmatic tissue (Pylnev, 1962). The age of the stigma has effects on fertilization (Rajki, 1961). There is a pronounced influence of the developmental stage of the stigma on the extent of seed setting (Rajki, 1962). Generally speaking, germination of pollen can also take place in the absence of the stigma (e.g., on the stylar stump after the stigma has been removed) or by injecting pollen suspensions directly into the stylar canal or the conducting tissue (Kroh, 1956).

III. Ripening of the Pollen Grain (Microsporogenesis)

Production of the microspores (pollen) takes place within the anther, which is borne at the top of a filament. Differentiation of the archesporial cells is followed by meiotic divisions. Development of pollen within the four longitudinal cavities (pollen sacs)—two in each of the lobes united by the connective—is more or less synchronized and linked with morphological and biochemical changes in the tapetum.

A. DEVELOPMENT OF TAPETUM AND SPOROGENOUS TISSUE

Figure 6 outlines anther development. In early stages when cells in the center of the young pollen sac enlarge to form pollen mother cells (PMC) the surrounding tissue forms the anther wall. This antherial wall consists of at least four layers: the epidermis, the fibrous layer (endothecium), the middle layer, and the tapetum. The fibrous layer, by its cohesion mechanism, is responsible for the opening of the ripe pollen sacs. The tapetum, with its richly cytoplasmic cells and (in many cases) polyploid nuclei, serves as a secretory tissue (*glandular* tapetum). It is of common occurrence in angiosperms. But in several families the walls of the tapetum cells break down, the cytoplasms remain intact, and enter the loculus to form a continuous mass (periplasmodium). This type of *ameboid* tapetum also serves for the nutrition of the developing microspores.

The sporogenous tissue, situated in the center of each sac, begins meiosis. Meiosis is induced by exogenous factors transmitted from the vegetative parts of the plant (Linskens, 1956, 1958b; Linskens and Schrauwen, 1963, 1968; Rodrigues-Pereira and Linskens, 1963; Neumann, 1963). The transport of the inducting principle depends on length and diameter of the parenchymatous and fasicular pathway. On entering the synapsis stage the protoplast contracts from the cell wall, and the cell wall thickens. When prophase of the first meiotic division is completed, the nuclei of the tapetal cells also start to divide, though cytokinesis does not take place. The result is two or more nucleated tapetal cells. After the formation of the phragmoplast the sporocytes reach the tetrad stage, and the middle layers of the anther wall degenerate. Vacuolization of the microspores is synchronized with the degeneration of the tapetum. The plasmatic material discharges into the cavity of the pollen sac and forms the nutritive substrate for the further development of the pollen.

During tapetum development many biochemical changes take place in the anther; these are summarized in Fig. 7. From these observations it can be concluded that the main function of the tapetum appears to be secretion of the material for building up of the microspores including their wall formation. Sterility of pollen can be caused by disturbed or retarded

Fig. 6. Longitudinal section through a locule of *Triticum vulgare* showing anther development. (Abbreviations: a, anther epidermis; b, fibrous layer or endothecium; c, middle layer; d, tapetum; e, archesporial tissue.) (1) Twenty-seven-day-old anther showing the central archesporial tissue surrounded by the anther epidermis; (2) and (3) 30-day-old anther—in the center is the sporogenous tissue surrounded by two or three undifferentiated wall layers; (4) the anther wall is differentiated into epidermis, endothecium, the middle layer, and the tapetum; (5) 33-day-old anther—all layers are fully developed and divide only anticlinally to add to the length of the anther (the amount of cytoplasm increases and the nuclei enlarge but still remain in the resting stage); (6) microspore mother cells nucleus enter meiosis while the cell wall remains angular; (7) protoplast of the spore mother cells separate from the walls, gets thickened, and further differentiated—the anther epidermis only undergoes stretching while the cells of the middle layer undergo divisions, the tapetal nuclei divide without cytokinesis resulting in bi- or multinucleate condition, and their fusion results in polyploidy; (8) dyad; (9) tetrad stage—middle layers have started to degenerate; (10) vacuolization of the microspore and the beginning of the degeneration of the tapetum; (11) developing microspore with large vacuole—the degenerating ameboid tapetum is used for pollen development; (12) ripening of the pollen grain is associated with increase in volume, thickening of the wall and accumulation of starch grains. [After Korobova, 1962 (Table 1 modified).] (Figures on pp. 200 and 201.)

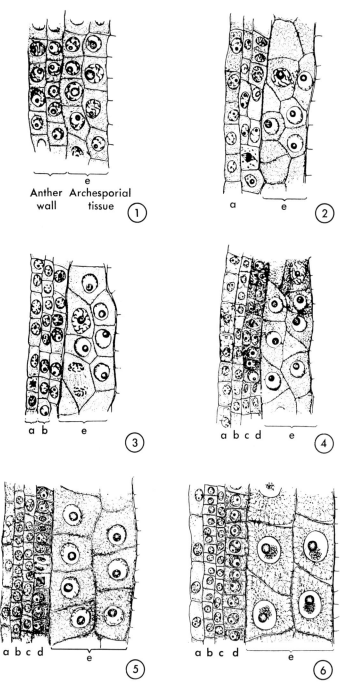

e
Anther Archesporial
wall   tissue   ①

a   e   ②

a b   e   ③

a b c d   e   ④

a b c d   e   ⑤

a b c d   e   ⑥

Fig. 6 [parts (1)-(6)]. See legend on p. 199.

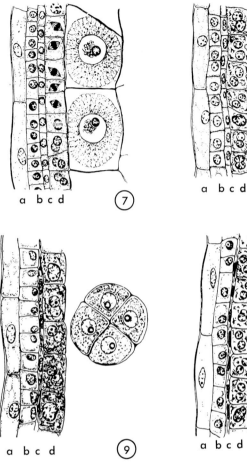

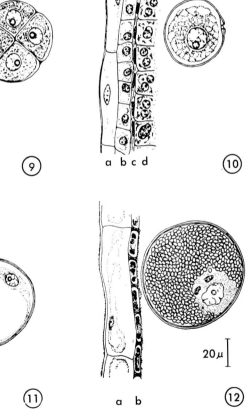

FIG. 6 [parts (7)-(12)]. See legend on p. 199.

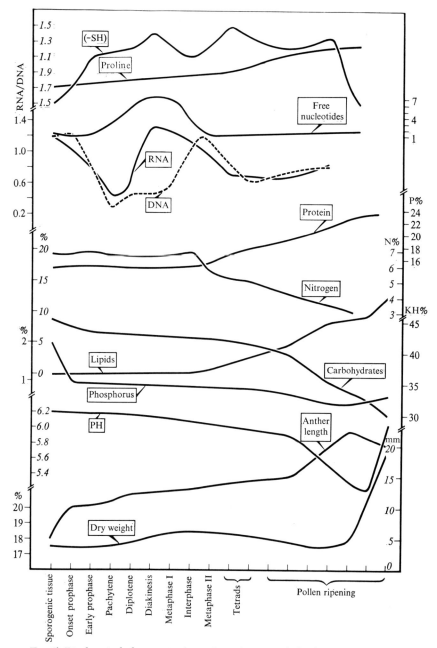

Fig. 7. Biochemical changes in the anther of *Petunia hybrida* during pollen meiosis and development. (Compiled from Linskens 1956, 1958b; Linskens and Schrauwen, 1963.)

activity of the tapetum. There is an intensive metabolic exchange between the growing microspore and the tapetum as demonstrated by pH changes, increase in auxin content, and transfer of carbohydrates, fats, proteins, and nucleic acids. All this is necessary not only for building up the pollen walls and storing reserve material but, also, for the precise synchronization and coordination of events in a locule and the whole anther. Autoradiographic studies have shown gradual changes in the amount of deoxyribonucleic acid (DNA) at the following two stages: (1) shortly before the division of the microspore nucleus and (2) before the division of the generative nucleus (Woodard, 1958).

In young anthers the cytoplasms of both the tapetal and the sporogenous cells are very rich in RNA. Protein synthesis occurs in the tapetal cells during meiosis but it is specially rapid during the maturation of the pollen grains. This is of significance, for the tapetal cells also secrete wall material for the maturing pollen grains (Vasil, 1962).

B. POLLEN WALL FORMATION

The ripe pollen grains are surrounded by the pollen wall which is differentiated into exine and intine and is made up of sporopollenin, cellulose, and a pectin layer. After completion of meiosis, certain materials appear in the tapetum which are of special significance in the formation of the exine. The presence of small granules (Ubisch bodies), closely "appressed" to the inner tangential wall of the tapetum, has been demonstrated. These granules gradually form a continuous layer which is considered homologous to the exine (Rowley, 1962). Although the tapetum delivers the metabolic elements in the form of sporopollenin-forming monomeres to the pollen wall and germpore formation, the wall morphology is largely under the biochemical and genetical control of the haploid spore nucleus after its isolation from the other cells of the anther. Heslop-Harrison (1963a,b) has shown that the differentiation of the pollen wall takes place in two steps: (1) In the first phase, while still at the tetrad stage and invested by the PMC membrane, the primexine is formed with a pore. (2) In the second phase, the initiated exine differentiates and intine forms.

Ultrastructural evidence shows that the protoplast of the spore is certainly involved in the initial establishment of wall pattern. The structure of the primexine is formed by the endoplasmic reticulum. The primexine and the differentiated pollen wall are, therefore, determined by the genotype of the haploid spore nucleus. The wall pattern is independent of the surrounding diploid tissue. Thus the whole exine grows like an organism. During the period of the most rapid growth of the exine, there are no protoplasmic continuities between the tapetal cells and the pollen surface.

The pollen grains float in a fluid medium which certainly contains organelles from the degenerating tapetum cells but no organized protoplast.
During the period of formation of the primexine and the establishment
of the patterning that ultimately characterizes the pollen wall, a special
mother cell wall persists composed of callose material. The function of
this wall may be interpreted as a molecular filter, permitting the passage
of basal nutrients into the spores from the tapetal fluid, but excluding
larger molecules the intrusion of which at this early stage might impair
the capacity of the haploid spore nucleus to establish autonomy within its
cytoplasm (Heslop-Harrison, 1963a,b, 1964). At a later stage this filter
and distribution function may be taken over by the exine (Rowley, 1964).

IV. On the Way to the Egg (Progamic Phase)

A. GERMINATION OF POLLEN

Germination of the pollen grains (Figs. 8 to 10) generally occurs on
the stigma. Innumerable publications on germination *in vitro* describe the
necessary conditions to form the pollen tube (see Linskens, 1964, 1967).
In the first place, pollen grains require water; the first function of any
medium is, therefore, to supply water. Germination capacity is influenced
by age and ripeness of the pollen and by temperature during and before
anthesis. As a rule the stigma exercises a considerable degree of specificity
on the kind of pollen that will germinate upon it. Pollination between
widely differing species is unsuccessful, either because the pollen grains

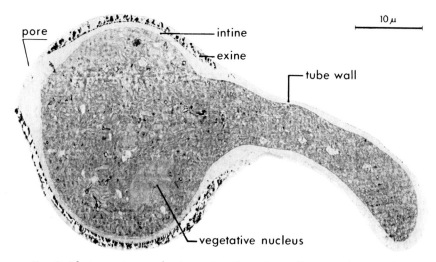

FIG. 8. Electron micrograph of a section through a pollen grain just germinating.
Only the unfolded vegetative nucleus can be seen. (Courtesy M. M. A. Sassen.)

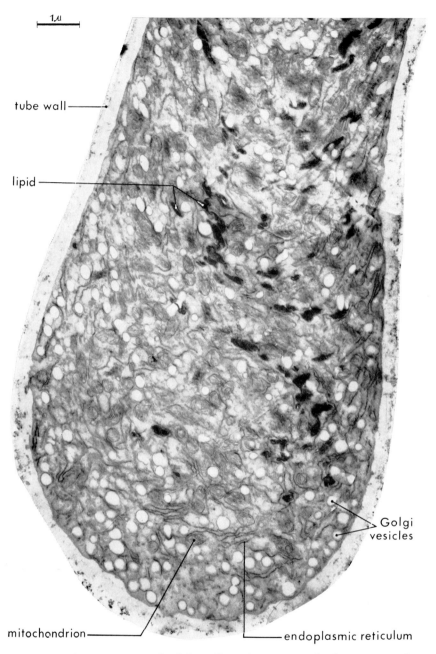

tube wall——

lipid

Golgi
vesicles

mitochondrion——

——endoplasmic reticulum

Fig. 9. Electron micrograph of the pollen tube tip. Note the flow pattern. There is visible accumulation of Golgi material and more or less globular vesicles, which possibly originate from the former. The contents of these vesicles seems to be pectic substances and are believed to be involved in cell wall growth. (Courtesy M. M. A. Sassen.)

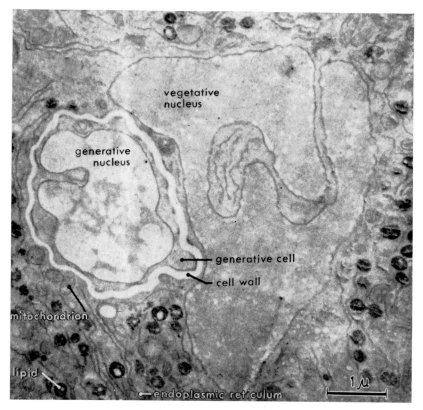

FIG. 10. Pollen tube cytoplasm showing the generative cell, surrounded by a "wall" and the folded vegetative nucleus as seen under an electron microscope. (Courtesy M. M. A. Sassen.)

do not germinate, or, if germination occurs, because the tubes fail to grow down the foreign style (Maheshwari, 1949). A more complex situation exists on the stigma of certain kinds of plants. If the flower is artificially pollinated, grains do not germinate; but if they are insect-pollinated, slight wounds are inflicted on the stigma by the visiting agent which enables the pollen to germinate. The same result can be achieved by artificial wounding. The time taken for germination is variable (a few seconds to 2 days) and is strongly influenced by external factors such as temperature, humidity, and the internal metabolic condition of the grains.

1. Boron Effect

Among the substances tested for effect on pollen germination, boron is the most surprising. It is essential for pollen tube formation as it reduces bursting of the pollen and increases the percentage of germination and the length of the tubes (Johri and Vasil, 1960; Linskens, 1964). It was suggested that the degree of boron sensitivity is related to the boron level of the pollen-producing plant and is affected by the osmotic properties of the medium of germination. The presence of boric acid is not required until the tube protudes but has to be continuously present thereafter (Visser, 1955).

a. Boron and Carbohydrate Metabolism. Boron can be seen to affect carbohydrate metabolism (Tupý, 1960; Vasil, 1964). It encourages sucrose absorption proportionate to the stimulation of pollen tube growth. In view of the fact that pollen tube growth is accelerated by boron at the time when the tubes are still depending to a great extent (or exclusively) on the reserve substances in the pollen grain, it follows that the primary factor is the influence of boron on the rate of metabolism. Boron may not be acting as a part of an enzyme but rather as part of a substrate. Possibly it has a protective effect in preventing excessive polymerization of sugars in sites of active sugar metabolism (Scott, 1960).

On the other hand, boron stimulates the absorption of some sugars during pollen germination and causes a corresponding increase in the rate of oxygen consumption. This suggests the hypothesis that boron modifies carbohydrate translocation by formation of a sugar–boron complex, and the activation of adenosine triphosphate (ATP) (Gauch and Duggar, 1953; Linskens, 1958a).

b. Boron and Pectin Synthesis. It seems possible that the borate ion is associated with the cell membrane and reacts chemically with the sugar molecule, facilitating its passage through the membrane. A relation between pectin synthesis and boron in germinating pollen has been suggested (Raghavan and Baruah, 1959). Pectin is present as a primary constituent of the pollen tube membrane in the tube tip—the region of greatest physiological activity (Fig. 9). From incorporation studies with *m*-inositol, which is a precursor in pectin synthesis (Stanley and Loewus, 1964), it can be concluded that boron plays a definitive role in the pectin synthesis of germinating pollen. Possibly it is related to the synthesis of D-galacturonosyl units to pectin. This hypothesis would also explain the well-established observation that addition of boron to a germination medium prevents the tube tip from bursting.

2. Group Effect

The density of pollen grains at the site of germination influences germination. Accordingly, there must be some interaction in interspecific pollen mixtures. Some authors believe in a kind of systematic and "sociological" relationship involving inhibition and stimulation. The mutual stimulation effect can also be observed among members of a species. It is universal that this population effect produces mutual stimulation whenever angiosperm pollen germinates. It results from the action of a stable, water-soluble growth factor in the pollen grains, which was shown to be calcium ions (Brewbaker and Kwack, 1964). The population effect is explained as a result of induced calcium deficiency. Calcium seems to be one of the primary regulators of pollen germination. High Ca content protects pollen from being killed by noxious gases and other substances, with the notable exception of pectinase. Methyl donors promote tube elongation under certain conditions. No other substance appears to enhance germination when calcium is limiting, provided borate and a suitable osmotic milieu is present. Cessation of growth *in vitro* and loss of viability in storage appears to be connected with oxidative lowering of pH and its probable effect on calcium-binding properties of the pollen wall.

B. Penetration into Female Tissue

After the tube has emerged from the pollen grain and found its way between the stigmatic papillae, it enters the tissue of the style (transmitting or conducting tissue). In some cases, certain thin areas in the walls of the stigmatic papillae, associated with starch and less resistant patches, serve as portals for entry. During the growth process through the style, only the distal part of the pollen tube contains cytoplasm. The mass of tubes grows at a more or less equal rate. At a certain regular distance behind the tip, callose partitions form, separating the cytoplasm contained in the front part from the rest of the tube, which subsequently collapses. Also the tube and the generative nucleus (which finally divides to form the two sperms cells) pass forward. The passage of the pollen tube is usually inter- and not intracellular (Fig. 11). The styles, which are extremely variable in length (a few millimeters to 30 cm and more) can be classified into three main types (Hanf, 1935; Vasil and Johri, 1964):

1. *Open* styles, with a wide stylar canal, without transmitting tissue. The inner, slimy epidermis itself assumes the function of conduction and nutrition of the pollen tubes.

2. In the *closed* conducting tissue (solid type) the tubes have to pass through a core of elongated, cytoplasm-rich cells, which sometimes have thickened middle lamellae.

3. The intermediate *half-closed* type has a canal with a rudimentary transmitting tissue which has two to three layers of glandular cells limited to one side of the stylar canal.

Usually the tubes make their way on the surfaces of the cells of the transmitting tissue (*endotropic growth*). They are conducted mechanically and oriented by the arrangement of the transmitting cells. The pollen tubes follow the path of least resistance. In *ectotropic* growth, the course of the pollen tubes is limited by the formation of a mucilagenous epidermis. The transmitting region from the para- to the coenocarpus parts of the ovary acts as a zone of distribution. Inside the loculi the growth of the tubes to the ovule is a matter of chance (Steffen, 1951, 1963).

C. METABOLIC RELATIONS DURING TUBE GROWTH

Germinating pollen is generally able to draw metabolic substrates from the reserve material which it contains. Nevertheless, it has been shown that growing tubes, especially during the passage in the middle and lower parts of the style, absorb stylar material and use it for reinforcing the tube wall.

1. *Metabolic Relations during Germination*

Within a few minutes after contact with the stigma surface, enzymes in the pollen grain become activated (Stanley and Linskens, 1964) and proteins diffuse into the medium (Stanley and Linskens, 1965). Aerobic

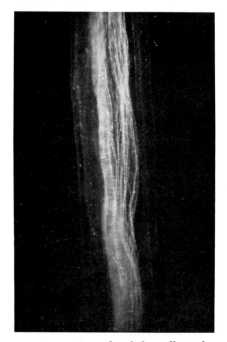

FIG. 11. Fluorescent microscopic study of the pollen tubes penetrating the style of *Petunia hybrida*. The contours of the style can be seen. The light spots are callose plugs in the tubes. The mass of tubes grow with equal velocity. (Preparations following Linskens and Esser, 1957.)

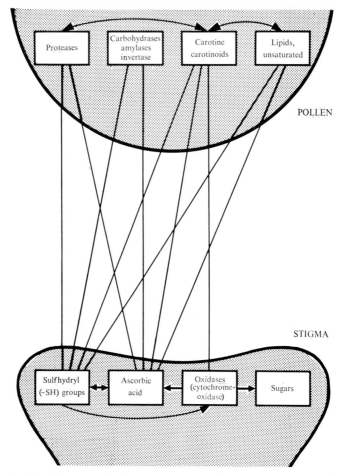

Fig. 12. Metabolic interrelations among enzymes and substrates in pollen grain and stigma cells. (After Britikov, 1954.)

enzyme activity associated with sucrose metabolism is found as soon as the tube emerges. Also the enzymic breakdown of cutin, if present, starts immediately after germination while the tube is still on the surface of the stigma. The metabolic relations between pollen and stigma surface have been summarized by Britikov (1954) (Fig. 12). Certainly this scheme is still incomplete and gives only a limited insight into the complex metabolic situation of the stigma surface during the first phase of the sexual contact.

2. Metabolism within the Style

The metabolic relations between pollen tube and transmitting tissue are complex and concern transfer of water to the tube, the preparation of substrate, and the opening of the passage. The view that pollen tubes must be nourished exclusively by the stylar tissue is contradicted by the fact that even *in vitro* pollen tubes from many species attain a tube length which is sufficient for bringing about fertilization without absorbing any food substance. Furthermore, there seems to be no relation between the style length and enzyme activity in the pollen tube (Haeckel, 1951). In any case, pollen tubes draw their nourishment at least partly from the transmitting tissue, as demonstrated by isotopic experiments (Linskens and Esser, 1959). Also, from the observations of Schoch-Bodmer and Huber (1945, 1947), it must be concluded that, by dissolution of the cell wall material with exoenzymes secreted by the pollen tube tips, the stylar material becomes available.

In orchids the scheme of reactions for pollen tube growth within the ovary is as follows (Hsiang, 1951): increase of catalase activity, stimulation of respiration, acceleration of water uptake, increase of salt resorption, mobilization of sugars, increase of hydrophylic colloids, increase of osmotic value, and increase of total nitrogen and phosphate content.

a. Enzymes of Protein Metabolism. Interaction of the pollen with tissue might activate components or induce modifications in the pistil tissue with the result that enzymes are released. Thus, if pollen tubes directly or indirectly release acid—and they have been shown to cause a shift in pH in germination and within the style (Schlösser, 1961)—a certain amount of loosely bound calcium in the style may be available for wall synthesis in growing pollen tubes. The role of cations in differentially modifying the metabolic pathway by which pollen metabolizes glucose was demonstrated by Stanley *et al.* (1958). The pollen tube is able to fix carbon dioxide, via the phosphoenolpyruvate–carboxylase route. Carbon dioxide can enter the organic acids which are subsequently transaminated and the resulting amino acids are then synthesized into proteins. A turnover of the proteins in pollen tubes is apparent, but net increase of protein has not been demonstrated (Pozsàr, 1960; Tupý, 1964).

A small, but significant, amount of RNA is synthesized during pollen tube elongation. Because the base ratio of the newly synthesized RNA is entirely different from the bulk RNA and resembles the DNA, it may be interpreted as a "messenger RNA" (Tano and Takahashi, 1964). Whether or not this RNA synthesis represents the formation of informational RNA in pollen tubes during penetration of the female tissue is still unknown.

b. Enzymes of Carbohydrate Metabolism. Growth of the tube through the style causes an increased inflow of carbohydrates into the entire pistil. The proportion of dextrose levulose in pollinated styles changes in favor of dextrose. This phenomenon is in agreement with the hypothesis (Tupý, 1961) that, in the respiratory process, pollen tubes consume mainly sucrose and of this primarily its fructofuranose component. During the growth through the style, an adaptation of the enzymic mechanism takes place. First, glucose is metabolized via the Embden-Meyerhoff pathway, afterward pollen tubes change to anaerobic respiration via the hexose–monophosphate shunt (Stanley, 1958). The strong increase of sugars in the style (Marré and Murneek, 1953) is followed by a decrease after the passage of the mass of the pollen tubes (Linskens, 1955).

c. Proline Metabolism. Proline is the major amino acid in the pollen of most plant species. It reaches up to about 1.6% of the total fresh weight of the pollen (Barthurst, 1954). On the other hand, pistil tissue contains only a small quantity of free proline, as a rule 0.001% or less of the total fresh weight. This pollen proline readily diffuses from the tube tip into pistil tissue and is rapidly incorporated into the protein fraction Britikov *et al.*, 1964). A high amount of CO_2 is produced from proline by pollinated styles. This indicates that proline is an active respiratory substrate (Tupý, 1964). Futhermore it seems that proline accelerates cell wall metabolism. This may permit integrating knowledge of protein and carbohydrate metabolism, when incorporation of proline in protein is associated with synthesis of primary cell wall pectins.

d. Growth Hormones. Growth hormone metabolism is switched on during the transmission of the pollen tubes in the style (Lund, 1956a,b). Although the content of indoleacetic acid in the different parts of the style is correlated with the growth of the tubes, the content of indole-acetonitrile decreases. A direct control of the synthesis of growth substances by the tubes seems to exist. The activity of the tryptophan-converting enzyme system increases with penetration of the pistil. It is possible that the growing pollen tubes transfer tryptophan-oxidizing enzymes into the style and the ovary, which are subsequently used for the growth of the embryo, endosperm, and the fruit as a whole.

D. DIRECTED TUBE GROWTH (CHEMOTROPISM)

The question of the oriented growth of pollen tubes is linked with tube entry into the micropyle. It is essential to distinguish between chemotropism, that is, the directed growth, and nonchemotropic growth stimulation. Solution of the problem depends on a suitable test method for detecting naturally occurring chemotropic substances in the pistils. Chemotropic reactions can be demonstrated in the ovules, placentae, the inner epidermis of the ovary, and the stigma. Apparently there are at least two different chemotropic systems. The active agents for these can be separated by dialyzing from an inhibiting system (Rosen, 1962; Mascarenhas and Machlis, 1962a). A chemotropic factor has been isolated from lilies. It is water-soluble, sensitive to heating, and has a molecular weight of about 600 (Noack, 1960). This factor may act by maintaining wall plasticity at the pollen tube tip (Rosen, 1961), since this is the region where the chemotropic response must be manifested by an effect on wall formation (Rosen, 1964). In this region, two types of vesicles are found: one is apparently rich in polysaccharides and is associated with plasmalemma and wall synthesis (Fig. 9). The other type appears to be rich in RNA. It is not quite clear whether the same tropic systems are responsible for the mechanism of selective fertilization (see Section VI,B). The other substance seems to be stable to heat and diffuses through both collodion and cellulose membranes.

It was concluded from *in vitro* studies that this second chemotropic substance is probably calcium (Mascarenhas and Machlis, 1962a). The growth response was correlated with a high level of Ca in the ovary. The ubiquitous distribution of calcium in plants may also be the reason why stylar tissues from taxonomically divergent plants induce a response to pollen (Miki-Hirosige, 1964b). The universality of this chemotropic factor explains the lack of specificity (Mascarenhas and Machlis, 1962a,b). But chemotropism does not seem to be the only hormonal control mechanism for the directed growth of pollen tubes. In certain cases pollen tubes also show hydrotropism and galvanotropism (Wulff, 1935; Zeijlemaker, 1956). According to Britikov (1952), there are other metabolic gradients in the upper part of the style which can direct tube growth, e.g., acidity (Fig. 13). In some parts of the style the conducting or transmitting tissue can form a predetermined slimy path. But, in any case, this oriented growth, directed by the anatomical structure, cannot explain phenomena such as selective fertilization and species barriers (Glenk, 1964) (see Section VI,B).

E. Mixed and Multiple Pollination

The effect of pollen from two different species germinating on the stigma at the same time has been discussed by Jones and Newell (1948). He advanced the hypothesis that, in mixed pollination (i.e., pollen from two or more different species), the heterologous pollen, although not involved in fertilization, can, nevertheless, contribute to zygote development. Isotope experiments with "hot" pollen (Eklund-Ehrenberg *et al.*, 1947) showed a single seed of *Alnus* contains radioactivity from the pollen of more than eight different parents. Compounds from pollen following repeated (multiple) pollinations are incorporated into the ovule or developing seed only after normal double fertilization has occurred and the process of embryonal development has already started. Various substances brought to the ovary by pollen tubes are involved in metabolism and assimilated by the developing ovule. This effect of supplementary pollen from other genera, species, or families is believed to result from action at a

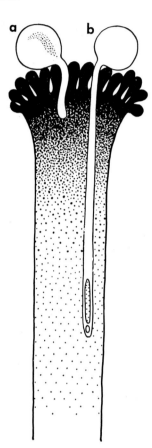

FIG. 13. pH gradient within the style (the degree of acidity is indicated by degree of blackening). At the germination stage (a) the tube has a higher pH than the stigma. The amount of acidity decreases within the style from 5.5 to 7.0. After penetrating the transmitting tissue (b) the cytoplasm in the tube changes to acid range. (Following Britikov, 1952.)

biochemical level involving cytoplasm but not extranuclear fertilization (Polyakov, 1964). This interpretation is in agreement with the recognized cytological pattern of zygote maturation. After fertilization the developing embryo utilizes nucleotides, amino acids, and other metabolites from the surrounding nucellar tissue. Once pollen tubes, including allogenic ones, have contributed their contents to the pistil or other female organs, diffusion or active transport can move the materials to the absorption sphere of the developing zygote. In this way ovular cytoplasm can be markedly influenced by additional, earlier or later pollinations, which contribute their biochemical compounds to influence the metabolic processes (Stanley, 1964).

Mixed pollination may also delay abscission of the style. This would mean that if compatible pollen is not initially present, metabolic activity stimulated by other pollen could extend the functional state of the pistil. Thus, when compatible pollen finally did germinate, it would find a functional egg apparatus to fertilize. Whether the sperm nuclei of non-compatible pollen or of excess pollen ever penetrates the egg still remains to be definitely answered. Further experiments must clarify whether supplemental DNA exerts a transforming effect on cell control or whether only DNase-released nucleotides are absorbed by the cells. This latter may be incorporated into the egg cell without transfer of information, but the possibility of assimilation of cytoplasmically localized information should be borne in mind.

V. The Fusion Process

A. Entry into the Embryo Sac

In most plants the pollen tubes enter the embryo sac through the micropyle, a process which is called *porogamy* (Fig. 1). Exceptionally they reach the embryo sac through chazala (*chalazogamy*) or through the lateral sides (*mesogamy*), and enter it at the apex.

1. *The Function of the Obturator*

Special structures that facilitate the entry of the pollen tube into the ovule are termed obturators. Usually this is a swelling of the placenta which grows toward the micropyle and serves as a sort of bridge for the pollen tubes (Maheshwari, 1950). In other cases, cells of the conducting tissue of the style or the inner integument elongate and form an apparently well-defined path for reaching the apex of the ovule (Kapil and Vasil, 1963). After fertilization, cells of the obturator shrink and soon disappear.

2. *The Function of the Synergids and the Filiform Apparatus*

The interval between the discharge of male gametes from the pollen tubes and their conjugation with the egg cell is very short. This interval can be extended for observation and experiment by cooling the material before and during fixation. The discharge under normal conditions takes place into one of the two synergids. In other cases the tube may enter the embryo sac between the two synergids without destroying any one of them, or the tube may destroy both synergids. In some cases they degenerate even before the pollen tube arrives (Vazart, 1958).

The pollen tube enters the embryo sac at the apex of the *filiform apparatus* at the point where the embryo sac wall is ruptured (Figs. 14, 15, and 16A). The cuticular layer seems to be destroyed by internal pressure. The synergids quickly absorb water and the filiform apparatus becomes a slimy, spongy mass. The filiform apparatus consists of the

1

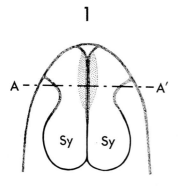

A -·- -·- A'

Sy   Sy

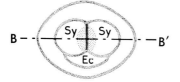

B -·- ─── -·- B'
Sy │ Sy
Ec

2

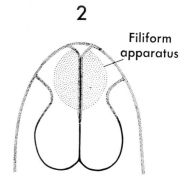

Filiform
apparatus

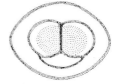

3

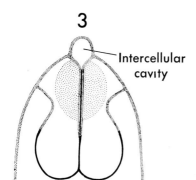

Intercellular
cavity

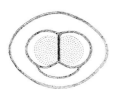

4

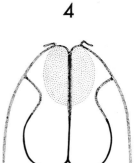

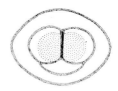

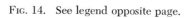

Fig. 14. See legend opposite page.

216

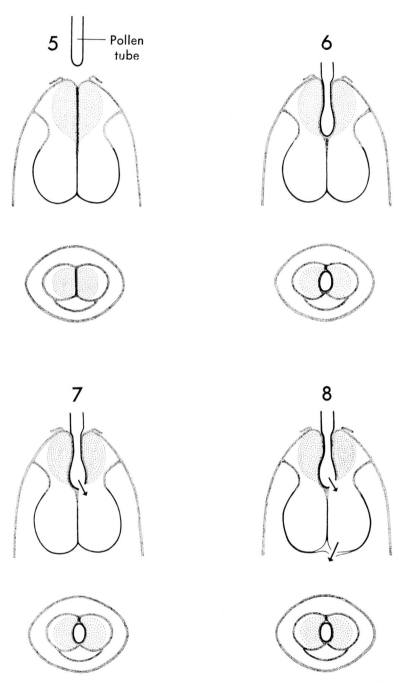

FIG. 14. Scheme of the processes associated with the ripening of the embryo sac and filiform apparatus, and the process during the entrance of the pollen tube into the embryo sac. Top, longitudinal sections; bottom, transverse sections. (Sy—synergids; Ec—egg cell.)

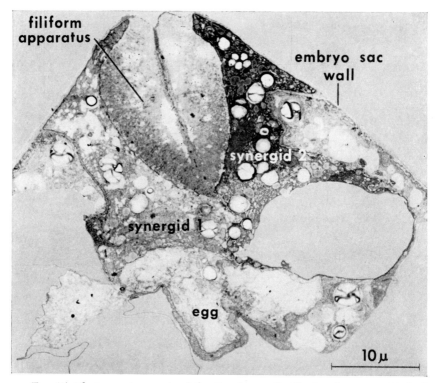

Fig. 15. Electron micrograph of the tip of an unfertilized embryo sac with both the synergids and the filiform apparatus. Under light microscope initially this structure shows longitudinal striations and consists of a large number of thin threads which later becomes more or less homogeneous.

thickened top walls of the synergids. Substances are deposited in this wall on both sides until the thickening becomes hemispherical.

The pollen tube usually forces its way through precisely where the middle lamella between the two synergids lies (Fig. 14). There appears to exist a causal relation between the quick bursting of one of the synergids, the degeneration and vacuolization of its cytoplasm, and discharge of the pollen tube contents (Fig. 16B). Tubes do not open into aborted or fertilized embryo sacs (Steffen, 1963). Ordinarily the other synergid remains unchanged during the first steps of the events that follow in the egg cell (Korobova, 1959; van Went and Linskens, 1967).

The *first* function of the synergids is the opening of the pollen tube tip (Schnarf, 1941). The disrupted tube is usually blocked by a callose plug at the place where it has penetrated the filiform apparatus with the result that the embryo sac is plugged off. The discharge of the tube contents does not require pressure.

In a few cases it has been observed that the top of the tube is bifurcated—one branch is directed to the egg while the other passes to the polar nuclei. This means that the two male gametes reach their mates through these bifurcations or by two subterminal openings through which they are discharged to the places of their destination (Mendes, 1941; Cooper, 1946).

The *second* function of the synergid is to bring the sperms and the discharged material into the right position; more or less deeply in the embryo sac. Normally this is between the egg nucleus and the secondary nucleus (Gerassimova-Navashina and Korobova, 1959; Gerassimova-Navashina, 1961). As soon as the tube reaches the bottom of the filiform apparatus, an open passage is made between the tube and one synergid. The contents of the pollen tube find their way into the synergid (Fig. 17). The synergid nuclei are probably polyploid (Tschermak-Woess, 1956; Hasitschka-Jenschke, 1957, 1959, 1962) and disappear as soon as the tube contents are released. At the same time the large vacuole in the lower part of the synergid is lost and the structure of the cytoplasm is changed. It becomes granular and the leucoplasts concentrate in the upper part (Van der Pluijm, 1964). The observed "x-bodies" are interpreted as degenerating synergid nuclei (Fig. 18).

A *third* function of synergids and the filiform apparatus may be the secretion of chemotropic substances. There are two possibilities: (*1*) the filiform apparatus itself may consist of chemotropic materials, which are leached from the embryo sac by the liquids secreted by the synergids, or (*2*) the filiform apparatus serves to convey the attracting substances produced by the synergids. The homogenous mass of the structure may serve as a depot maintaining the concentration gradient necessary for chemotropic reaction of the approximating tubes (Van der Pluijm, 1964).

The *fourth* function of the synergids seems to be to absorb, store, and transport compounds from the nucellus (Jensen, 1965). These highly organized, complex cells are characterized by a large number of mitochondria, plastids, spherosomes, Golgi material, and ribosomes, as well as large amounts of endoplasmic reticulum. The synergids act by providing material for the egg and the developing embryo and endosperm (Fig. 16C–F). When fertilization occurs the synergids change—both disintegrate some 6–7 days after fertilization (the one which was penetrated by the tube disintegrates immediately, the second one persists for some time, but undergoes marked changes). It is possible that the bulk of the material stored in the synergids is released and becomes part of the material surrounding the embryo. There is close relationship between the synergids and the cytoplasm of the embryo sac, particularly the cytoplasm associated with the polar nuclei.

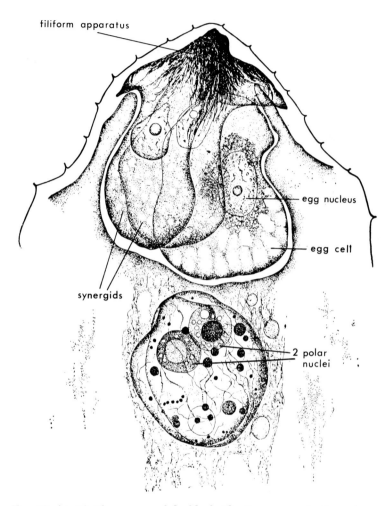

FIG. 16. (A–F) The process of double fertilization as seen in the embryo sac of *Triticum vulgare*. (After Batygina, 1962, modified by the author.) (A) The upper part of the embryo sac with the egg apparatus (egg cell and two synergids with their filiform apparatus) and the two polar nuclei.

3. *Discharged Material*

Generally speaking, the content of cytoplasmic organelles of both the uniting plasmas is similar: the ratio of the plastids to mitochondria plus microsomes in both sexes is about 1:4. The egg cell contributes 150–480 plastids and 1000–2500 mitochondria plus microsomes to the zygote of *Impatiens* (Richter-Landmann, 1959). This amount of cell organelles is the result of correlated growth of cytoplasm in two steps in the anther

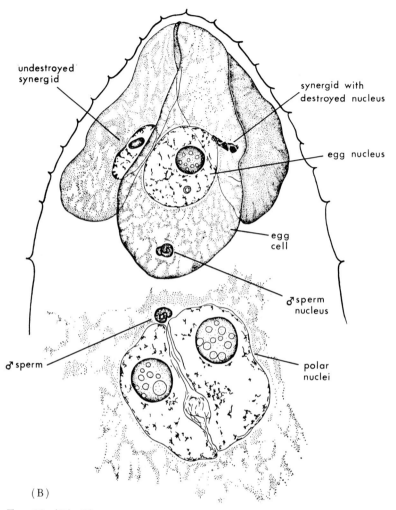

undestroyed synergid

synergid with destroyed nucleus

egg nucleus

egg cell

♂ sperm nucleus

♂ sperm

polar nuclei

(B)

FIG. 16. (B). The egg apparatus 30 minutes after pollination—both the male sperms entered the egg cell and the central cell. Within the egg cell the sperm nucleus is situated at the bottom. One synergid is destroyed due to the discharge of the pollen tube content. The distance between egg cell and polar nuclei is diminished.

during formation of PMC and differentiation of the vegetative cell and the sperm cells.

The question remains as to how the male cytoplasm is able to effect its influence on the egg nucleus in spite of the superiority of female plasma. Experiments, such as those carried out by Stubbe (1957) with genetically marked male plastids, may be able to clear up this question. What happens to the male cytoplasm in the egg cell remains to be

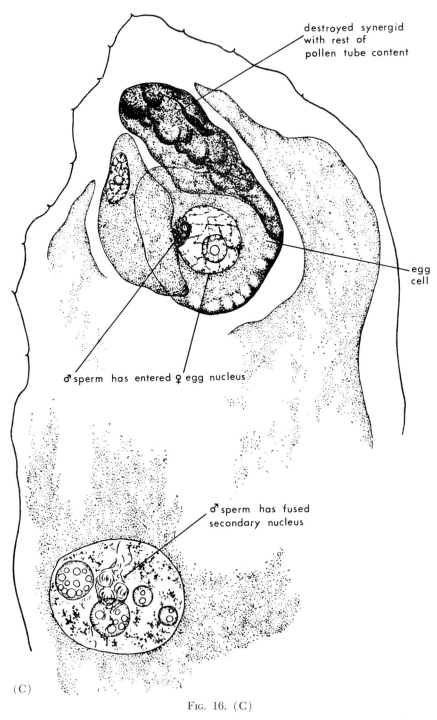

destroyed synergid
with rest of
pollen tube content

egg
cell

♂ sperm has entered ♀ egg nucleus

♂ sperm has fused
secondary nucleus

(C)

Fig. 16. (C)

determined. A thorough mixing does not take place before the first division of the zygote. This means that the homogenization of the cytoplasms is a process that depends on completion of the syngamy of the nuclei.

B. DOUBLE FERTILIZATION

In flowering plants, twin sperms are delivered presumably simultaneously into the embryo sac by the pollen tubes. They have to fertilize

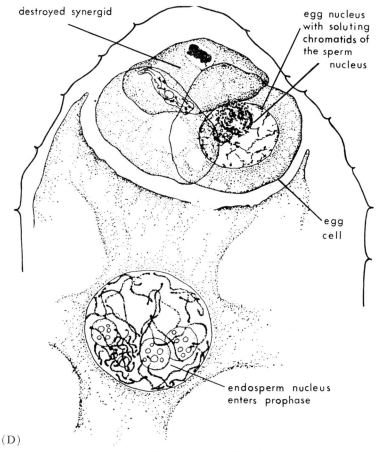

destroyed synergid

egg nucleus with soluting chromatids of the sperm nucleus

egg cell

endosperm nucleus enters prophase

(D)

FIG. 16. (C). Egg apparatus 2½ hours after pollination—the first nucleus is embedded in the egg nucleus. The second sperm has already fused with the polar fusion nucleus to give rise to the secondary nucleus. The nucleoli are still visible indicating that the nuclear material has not yet completely mixed. Rest of the pollen tube contents are visible in the destroyed synergid. The distance between egg cell and secondary nucleus is again extended. (D) The embryo sac 3 hours after pollination— the sperm nucleus lie adpressed to the egg nucleus (dissolution of the chromatic material and appearance of the nucleolus). The endosperm nucleus in the meanwhile enters prophase.

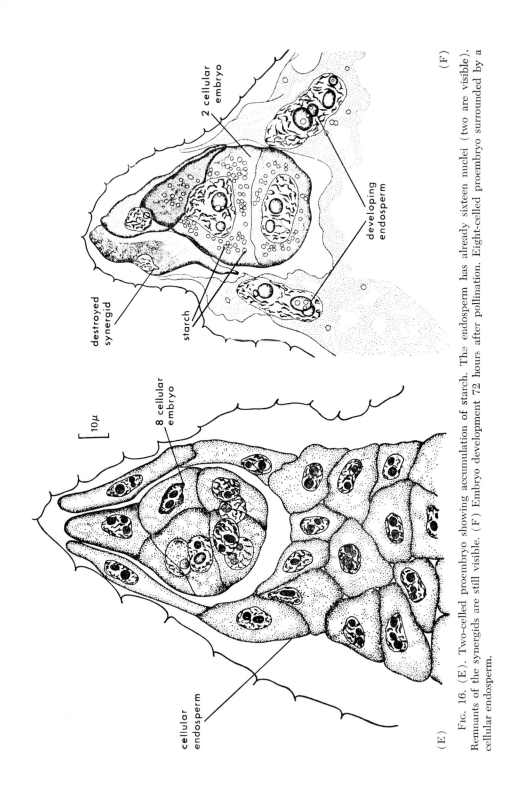

FIG. 16. (E). Two-celled proembryo showing accumulation of starch. The endosperm has already sixteen nuclei (two are visible). Remnants of the synergids are still visible. (F) Embryo development 72 hours after pollination. Eight-celled proembryo surrounded by a cellular endosperm.

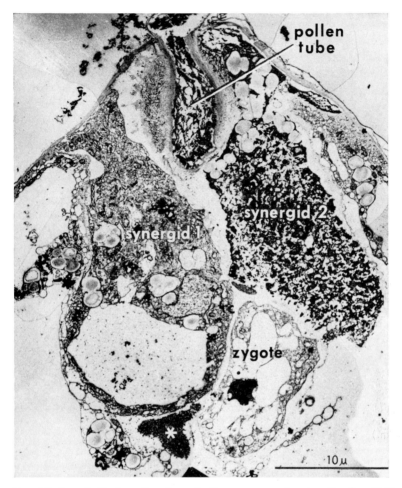

Fig. 17. Longitudinal section through the egg apparatus (on the line B-B' of Fig. 14) during fertilization process. The tube has entered the embryo sac at the apex of the filiform apparatus at the point where the wall is ruptured and forces its way between the two synergids. The tube contents are released into the right synergid which shows granular changed plasmatic structure. (Courtesy J. E. Van der Pluijm.)

the two different female elements: the egg nucleus and the secondary nucleus formed from the fusion of the polar nuclei, which is the progenator of the endosperm. This characteristic phenomenon of double fertilization was first discovered by Navashin in 1909. Many problems remain to be cleared up. Most observations are made in fixed and stained microscopic slides. Steffen (1953) introduced vital observation using phase contrast microscopy. Very few biochemical investigations are available

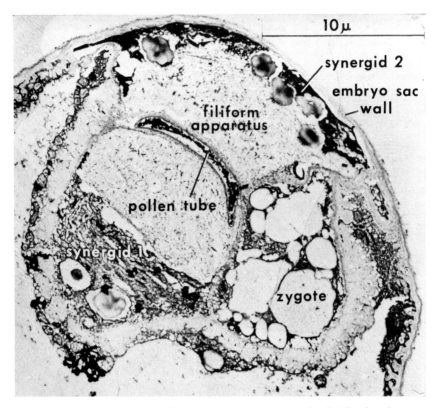

Fig. 18. Transverse section through the egg apparatus of a fertilized embryo sac (on the line A-A' of Fig. 14) 24 hours after fertilization. (Courtesy J. E. Van der Pluijm.)

regarding the complex processes that go on within the nucellar tissue. This is primarily because every treatment of analysis brings about changes in the general condition within the ovule.

The fundamental element of double fertilization consists of the directed divergence of the components of a pair of sperms, introduced by the pollen tube into the specifically organized cytoplasm of the two neighboring female nuclei.

1. Passage of the Sperms

The movement of the sperms inside the embryo sac is of great importance. Steffen (1953) demonstrated autonomous ameboid movement of isolated sperm nuclei which can be sufficient to approximate cytoplasmic streaming and at least passively enables the male elements to reach the female nuclei. According to Gerassimova-Navashina (1960, 1961), the direction of movement is a drift of the sperm to the spot where the female nucleus had been previously located, according to the general principles of nuclear location. This accounts for the fact that the male and female nuclei always come into mutual contact. Whereas nuclei, in the physiologically active state, mutually repel each other, they cease repulsion in aging and physiologically repressed cells. The sperms, therefore, exist in a state of "limited metabolic activity" surrounded by cytoplasm of the vegetative cell with its large metabolically active nucleus. Therefore all conditions are provided for the nonfunctioning of the repulsion forces resulting in contact of the nuclei and their subsequent fusion.

2. The Karyogamy

Double fertilization in the life cycle of the flowering plants consists of (*1*) the fusion of one of the two male nuclei with the egg nucleus and (*2*) the fusion of the second male nucleus with the combined polar nuclei.

a. Fusion of the Nuclei. Recent observations by electron microscopy (Jensen, 1964) elucidated the formerly vague, light optical observations on the actual fusion. The fusing nuclei migrate toward one another but are still separated by a layer of cytoplasm, including mitochondria, Golgi apparatus, vacuoles, spherosomes, and endoplasmic reticulum (ER). First the elements of the ER fuse, making the outer nuclear membranes of the two nuclei continuous. Next the inner membranes come into contact and merge, so that a complete bridge between the two nuclei is formed. The nuclear bridges are numerous. They enlarge and appear to coalesce. The data suggest that the membranes contract, possibly through some localized area of breakdown. The membranes of both nuclei contribute to the new nuclear membrane. This aspect of fusion between egg and sperm nuclei may prove to have some genetic and developmental implications in the case of the higher plants.

b. Types of Karyogamy. The whole variety of known aspects of double fertilization can be classified into three different types (Gerassimova-Navashina, 1960), depending on differences of the cyclic states of the sexual nuclei (Fig. 19). The female nuclei are in a state of deep mitotic rest whereas the male nuclei have not yet completed their mitotic cycle and are in telophase.

The *first* type, which corresponds to the so-called "sea urchin type" among animals—where mature eggs are inseminated—is characterized by the sperm nucleus fusing with the female nucleus immediately on coming in contact with the latter and reaching the ultimate phase of its mitotic cycle. It begins with dissolution of the contacting membranes (Korobova, 1959) which may occur by enzymic digestion (Morrison, 1955). In the large, secondary nucleus leveling of the sperm nucleus takes place at a

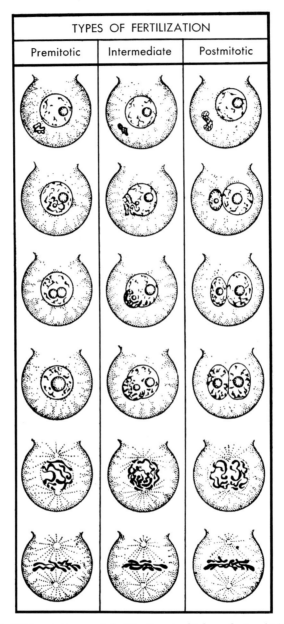

FIG. 19. Different types of fertilization in higher plants. (After Gerassimova-Navashina, 1960.)

rapid rate; the sperm sinks, so to speak, into the female nucleus and remains visible and easily recognizable (Steffen, 1951; Vazart, 1958). The male material ceases to show Feulgen reaction due to the loss or alteration of DNA and despiralization. At the same time a nucleolus is formed by it. This type is called *premitotic* because fusion takes place *before* the first zygote mitosis. This premitotic type of fertilization is found in Gramineae, Compositae, *Fucus,* etc.

The *second* type of karyogamy differs in the inability of the male nucleus immediately to *enter* the female one. Instead of completing its cycle within the female nucleus it undergoes a resting period while yet in contact with the latter. Union of the sexual nuclei occurs *after* the initiation of the first zygotic mitosis during prophase–metaphase. This type of fertilization may be compared with the so-called "*Ascaris* type" among animals where the sperm is incorporated into the immature egg and the sperm nucleus enters prophase still remaining outside the female nucleus. This is designated by the term *postmitotic*. This postmitotic fertilization occurs in a number of Liliaceae (*Lilium, Fritillaria, Gagea*) and in *Pinus*.

The *third* type of karyogamy differs from the previous one with respect to the duration of karyogamy. The sperm nucleus *enters* the female nucleus in a state of mitotic rest after completing its mitotic cycle, as in *Impatiens* (Steffen, 1951). The prophase is initiated in the fusion nucleus. The different areas of the sexual nuclei are clearly visible because of a delay in the female DNA. This *intermediate* type of fertilization is characterized by an incomplete mixing of the sexual nuclei. The chromosomes remain apart for varying lengths of time and can still be seen in the prophase of the zygotic nucleus. The spindle may show separate groups of chromosomes (Sax, 1918; Hoare, 1933) or separate spindles may be seen (Cooper, 1940).

3. *Fusion of Second Sperm*

In angiosperms, fertilization is a double process. The fusion of the second sperm with the secondary egg nucleus takes place in much the same way as the fusion between the egg nucleus and the first sperm. Usually the plasma of the second sperm is not seen during the movement to the two polar nuclei or in the secondary nucleus. From the paucity of stages observed in secondary fertilization, it can be concluded (Gerassimova, 1933; Steffen, 1951) that triple fusion is accomplished more quickly than syngamy. In some cases the fusion between the sperm and the polar nuclei occurs with the male gamete at some point on the line of juncture of the two polar nuclei such that the male nucleus lies in simultaneous contact with both female nuclei; thus all three nuclei are in mutual contact. The contents of the male nucleus mingle with the nucleoplasm of one of the polar nuclei. The fusion process is like the coalescence of oil droplets in an unstable oil–water emulsion (Bennet and Cope, 1959). In many cases the triple fusion resulting in a triploid secondary endosperm nucleus has already ended, while the male chromatin is still visible inside the egg nucleus.

4. Dynamics of the Nucleic Acids

Important changes have been observed in size, structure, and DNA content of the gametic nuclei before fusion. The egg nucleus shows a significant increase in its volume even before the arrival of the sperm cell. According to Gerassimova-Navashina (1960) this is a sign of aging or an inactive resting stage. During the development of the embryo sac and fertilization the female nuclei give positive Feulgen reactions which indicates the existence of DNA in their nuclei in all stages. In the mature egg the DNA-containing particles are scattered in the enlarged nucleus. The fact that the DNA content of the female nucleus shows marked cyclic changes (Wassilewa-Drenowska, 1959; Vassilcva Dryanovska and Tsoneva, 1959), as shown by histochemical studies, has to be interpreted as follows. During the process of development the Feulgen reactivity of the prefertilization female nucleus gradually decreases in intensity and concomitantly the size of the nucleus increases. The nucleus of the zygote reacts more strongly to Feulgen staining than the egg nucleus even though it is not much larger. Further, the egg nucleus is haploid and the zygote is diploid (Hu, 1964). Using microspectrophotometric methods it could be demonstrated (Nagaraj, 1954a,b) that the DNA content in the female gametophyte was half that of the nucellar cells. Therefore it can be concluded that the DNA content of the nucleus is consistent with the number of chromosomes.

The decrease in DNA content of the egg demonstrated by histochemical methods is linked with developmental processes. It can be explained by assuming a disperson of the DNA correlated with its molecular status and the degree of its polymerization. It is probable that during the process of fertilization there is not only a quantitative change in DNA, but also qualitative changes. All recent findings support the dispersion theory (Schnarf, 1941).

Only the male nucleus in contact with the female nucleus shows an increase in volume (Gerassimova, 1933) and at the same time a decrease in DNA. In the chromocentric nuclei the number of chromomeres correspond to the number of nuclear chromocenters (Steffen, 1951). It is suggested (Steffen, 1951; Vazart, 1958) that the DNA brought in by the sperm nucleus stimulates the egg nucleus to further division. In any case, the regeneration of the DNA in the female nucleus is a preliminary condition for the beginning of the next mitosis. Gerassimova-Navashina (1960) supposes that sperm and egg nuclei, lying in contact, become morphologically identical and do not differ in their DNA content. In her opinion the inactive female nucleus may be stimulated by the cytoplasm of the whole pollen tube content.

All elements in the embryo sac contain RNA. The RNA in the cytoplasm of the egg cell before and after fertilization shows definite differences. The young egg cell is rich in RNA, but with the growth of the embryo sac the RNA content gradually falls until the time of fertilization. After fertilization the RNA content increases again. These changes in RNA are correlated with the vacuolization of the cytoplasm and with the physiological status of the different developmental periods of the egg cell (Hu, 1964). Increase in RNA does not take place in all elements with the same velocity following fertilization. Generally the male and secondary nuclei show an early increase. It is probable that the RNA increase in the zygote is initiated by the sperm and the obligate supply of male RNA—an idea that was suggested by Zacharias in 1901. The increase occurs more rapidly in the endosperm nucleus than in the zygote, a fact which can explain the earlier division of the former.

C. Polyspermy

In some cases the entry of additional pollen tubes into the embryo sac is prevented by certain growth phenomena (Nemeč, 1931). Thus, usually only one pollen tube can enter the ovule. But in addition to mechanical contrivances, other barriers must exist that prevent multiple fertilization. Nevertheless, it seems that very often more than one pollen tube discharges cytoplasm, including the full complement of organelles, into one embryo sac. Such results were reported by Swedish scientists for the first time (Eklundh-Ehrenburg *et al.*, 1947) and later repeatedly confirmed by Soviet plant breeders and geneticists (Polyakov, 1955, 1964; Göring, 1960). This can also be proved by using labeled pollen. The present status of polyspermy can be summarized as follows: the fertilization of one egg by several male gametes has not been cytologically demonstrated except in a few plants. All these cases of polyspermy and fertilization of the several nuclei of the embryo sac must be regarded as abnormalities. Also the fertilization of the egg and the secondary nucleus by different pollen tubes (heterofertilization), as mentioned by Sprague (1932), should be considered an aberration. But it seems entirely possible that a pooled contribution of organelles from supernumary tubes may enter the cytoplasm of the zygote. In this sense a biochemical influence of "foreign" pollen after mixed pollination may be explained.

D. Special Situation in Coniferae

Pollination and fertilization in the gymnosperms differ in many respects from the situation in angiosperms. In Coniferae the wind-borne pollen arrives on the micropyle within the female strobilus and is transported by a special mechanism through the micropylar canal and onto the receptive surface of the unprotected nucellus. In Cycadales and Ginkgoales, autonomous motile spermatozoids develop in the haustorial microspore after reaching the pollen chamber within the nucellus. In all cases fertilization is achieved without water as a vector of the microspores. In Coniferae the fertilization process is preceded by the formation of a micropylar fluid, active closure of the micropyle after pollination, pollen germination and growth, development of the ovule by inductional effects of pollination, and syngamy.

1. Formation and Chemical Composition of the Micropylar Fluid

The integuments extend out in the form of a tubelike structure which covers the apex of the nucellus. This is termed the micropyle. The inner surfaces of the integuments are coated with a sticky film at the time of maximum receptivity, causing the pollen to adhere to this surface. Near the nucellus the canal broadens to form the pollen chamber in which a number of pollen grains can be accommodated quite readily. The appearance of the fluid in the micropylar canal depends on the water relations within the strobilus and the prevailing atmospheric humidity. The period for which ovular secretion will continue in unpollinated strobili probably depends on the species and the weather conditions. The presence of pollen induces fairly rapid, complete, and permanent withdrawal of the fluid. Unpollinated strobili retain the micropylar secretion (McWilliam, 1958). Thus the fluid drop in the micropylar canal can be compared to a "stigmatic" surface or a functional stigma. Sugars [saccharose, glucose, fructose up to 80% of the total dry weight (Fuji, 1903; Ziegler, 1959)] are present in the fluid with a total concentration of approximately 1.25% (McWilliam, 1958). Amino acids, peptides, malic, and citric acids have also been found (Ziegler, 1959). Secretion of the micropylar fluid is independent of respiration and may, therefore, be different from nectar secretion. The fluid is not secreted by the micropyle but probably by the saucerlike depression at the nucellar apex (pollen chamber). It resembles the process of guttation (Sarvas, 1962). The fact that the pollen comes to rest in this shallow depression after transversing the micropylar canal also supports the view that the fluid is absorbed through this region.

2. Closure of the Micropyle

After pollination the ovules develop rapidly. The prominent feature of the micropyle is the swelling of the neck cells. This effectively closes the micropylar canal and insures the protection of the pollen on the nucellus. The swelling occurs at a point in the neck, where the walls average three to four cells thick. The middle layer of cells elongates rapidly in a lateral direction, causing the cells on either side to bulge out. As this process continues, the inner cells meet and seal off the canal. Pollen grains adhering to the sides of the micropylar canal are often trapped in this position by the expanding neck cells. The force of the closing mechanism can be gauged by the manner in which such pollen can be crushed.

3. *Pollen Germination and Growth*

The role of the micropylar fluid is to serve as a transporting medium for caught pollen grains. The fluid is necessary for successful pollination. During the first month after pollination the pollen tubes grow very slowly and in an irregular pattern at the apex of the nucellus. Sometimes the tubes may show branching but the cytoplasm and nuclei remain in only one of the branches. In all such cases where the pollen grain has only one nucleus the division results in a typical composition: one spermatogen (= generative) nucleus, two free nuclei, tube nucleus, and stalk cell. The pollen tube quickly grows through the rest of the expanded nucellus and enters the neck of the archegonium and delivers its contents into the egg (Fig. 2). This final phase of growth of the pollen tube is completed in about 10 days and takes place approximately 12–13 months after pollination.

4. *Development of the Ovule*

The development of the female gametophyte is induced by the germinating pollen grains in Cycadales, Gingkoales, and in the Coniferae. Unpollinated ovules normally degenerate. Experimentally, pollination has not yet been replaced by the application of growth hormone to induce gametophyte development. Pollination of even a single ovule of a cone is sufficient to prevent fall off.

5. *Fusion of the Nuclei*

As the pollen tube approaches the archegonium the dense cytoplasmic contents are massed at its apex and contain the tube nucleus, the remains of the stalk cell, and the two prominent sperm nuclei. At this stage the sperm nuclei are in contact with one another and are surrounded by a delicate membrane. The larger nucleus generally precedes the smaller one. The pollen tube forces its way through the neck of the archegonium, rupturing the neck cells that seal the entrance to the egg, and delivers the tube contents into the egg. Shortly before the contents are released into the egg, a receptive spot in the form of a vacuole appears in the upper portion of the egg cytoplasm and persists during the subsequent develop-

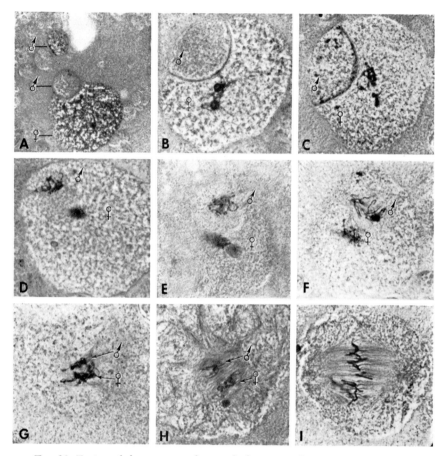

Fɪɢ. 20. Fusion of the sperm nucleus with the egg nucleus in *Pinus nigra* (Micrographs by courtesy of F. Mergen.) (A) One functional sperm nucleus in contact with the egg nucleus. The nonfunctional sperm nucleus lies above and adjacent to receptive vacuole. (B) Nuclei immediately after fusion, still separated by the double membrane. (C) Condensation of the chromatic material in the egg nucleus. (D) Separating membrane dissolved, the chromatic material of each nucleus condenses and forms a typical prophase configuration, consisting of an aggregation of slender, convoluted threads. (E) Chromatin material from each nucleus in prophase. (F) Appearance of chromosomes and formation of separate multipolar polyarch spindles. (G) Merging of the maternal and paternal chromosomes in a single polyarch spindle, on which two groups of chromosomes lie. (H) The two groups come together, the spindle gradually becomes diarch and broadens. (I) Metaphase of the first division. The maternal and paternal chromosomes arranged at the equator of a broad spindle, they are completely indistinguishable, showing 24 chromosomes.

ment of the egg. The larger, functional sperm nucleus approaches the egg nucleus. The smaller, nonfunctional nucleus usually does not move toward the egg nucleus but remains in the upper portion of the egg along with the tube nucleus and the remains of the stalk cell which gradually degenerate and disappear (Fig. 20A). After making contact, the functional sperm nucleus sinks into the egg nucleus until the outlines of the two nuclei form a continuous surface. The egg nucleus often contains prominent nucleoli. A double membrane separates the two nuclei where their surfaces are in contact and remains intact during the early stage of fertilization (Fig. 20B). As these membranes break down, the chromatic material of the nuclei condenses and forms a typical prophase configuration (Fig. 20C). As soon as the short and thick chromosomes appear, two multipolar polyarch spindles are formed (Fig. 20E). The spindles are initially separate but soon form a common polyarch spindle on which the two groups of chromosomes lie (Fig. 20G). As they come together the spindle gradually becomes diarch and broadens (Fig. 20H). The maternal and paternal chromosomes become arranged at the equator of this broad spindle and the chromosomes are no longer distinguishable as to parental origin (Fig. 20I). The metaphase of the first division shows the diploid chromosome set (McWilliam and Mergen, 1958).

E. INDUCTION OF THE EMBRYO DEVELOPMENT

After fertilization the zygote develops into the embryo while the secondary endosperm nucleus gives rise to the endosperm.

1. *The Embryo*

After syngamy the zygote nucleus generally undergoes a period of rest which varies from species to species and depends to some extent on environmental conditions; meanwhile the cytoplasm is augmented. The zygote is surrounded by a cellulose membrane. When it starts to divide the divisions follow a specific pattern (Fig. 16E and F). The developing zygote is a complex reacting system: it is the seat of autocatalytic reactions with patterned distribution of substances. The pattern of accumulation shows rapid movement of materials into the ovary suggesting activities in specific regions of the ovule (Coe, 1954). Next to the synergids, the greatest release of nutritive material into the embryo sac comes from the antipodal cells which have a glandular function (Beaudry, 1951). The nutrient materials accumulated in the nucellar cells pass directly into the embryo sac.

During the development of the embryo the nucellar cells undergo a continuous growth and breakdown. A considerable amount of food enters the embryo by the breakdown of these cells, which have a high metabolic

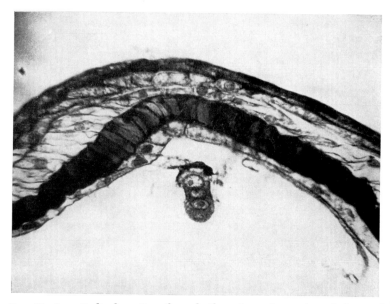

FIG. 21. Longitudinal section through the micropylar region of the ovule of *Papaver rhoeas*, 6 days after anthesis, showing four-celled proembryo. (Courtesy Kanta and Maheshwari, 1963.)

rate and concentration of synthesized foods. Fatty substances are brought into the embryo sac by pollen tubes and consumed at the time of fertilization. Further accumulation occurs in the embryo later on (Poddubnaya-Arnoldi *et al.*, 1961). The surrounding tissue is also responsible for inducing polarity. In a polarized zygote the apical or distal pole becomes the principal locus of protein synthesis, growth, and morphogenesis. The basal or proximal pole is characterized by the accumulation of osmotically active substances. Its cells become vacuolated and distended. The first division of the zygote typically results in a wall at right angles to the axis, i.e., to the principal direction of growth. Cell division is probably stimulated by the increase in size and the instability associated with a drift toward protoplasmic and metabolic heterogeneity.

The first division is always unequal (Wardlaw, 1955) resulting in a small terminal and a large basal cell. The former gives rise to the embryo proper whereas division products of the latter become the suspensors. The suspensors serve a feeding function and are highly polyploid (Nagl, 1962) (Fig. 21). The embryo cell divides to form two cells. The presumptive embryo cell has a larger amount of RNA (Jensen, 1963) and a greater number of plastids than the terminal cell. Continuous division of the terminal cell results in the formation of a globular mass of cells organized in a definite order with respect to planes of division. The further development of the embryo follows a definite pattern and is triggered off by factors in the surrounding tissue (Souèges, 1937; Wardlaw, 1955; Johansen, 1950). Metabolic gradients and biochemical patterns afford the basis for the morphological and histological pattern. As the embryo enlarges the reacting system changes in a characteristic manner (Wardlaw, 1963).

2. The Endosperm

The endosperm is important because it is the immediate environment and the main source of food for the embryo (Fig. 22). In angiosperms it is a new structure formed in most cases by the fusion of the two polar nuclei and the second sperm (also termed the second fertilization; see Section V,B). Generally, the three fusing nuclei are haploid, resulting in a triploid endosperm. In gymnosperms the female gametophyte itself functions at a later stage than the endosperm. Therefore, the endosperm in gymnosperms is haploid.

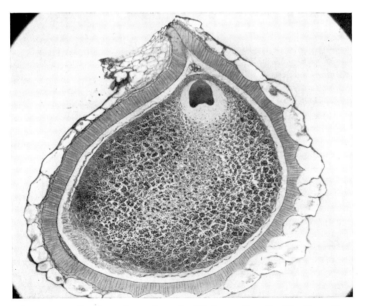

Fig. 22. Longitudinal section of ovule 15 days after fertilization—the embryo develops within the endosperm, surrounded by the seed coat (*Argemone mexicana*). (Courtesy Kanta and Maheshwari, 1963.)

Four different types of endosperm development have been distinguished (Schnarf, 1929; Maheshwari, 1950; Wunderlich, 1959; Chopra and Sachar, 1963):

1. The *nuclear* type—the first and subsequent divisions of the primary endosperm nucleus are not followed by wall formation. The division of this nucleus starts before the zygote divides. The nuclei distribute in a layer along the periphery of the embryo sac or show a special aggregation around the embryo. Later, the endosperm enters the cellular phase.

2. The *cellular* type—the first and subsequent divisions are followed by cell plate and subsequent wall formation.

3. The *helobial* type—the first division of the primary endosperm nucleus results in partitioning of the embryo sac into two unequal chambers; the micropylar chamber is usually larger and later becomes cellular, the chalazal chamber remains in a coenocytic condition and later also becomes cellular.

4. The *ruminate* type—characterized by some degree of irregularity and unevenness in its contour which, in turn, is conditioned by the irregular inner surface of the seed coat.

During the last few years much new information has been obtained on the synthesis of growth substances by the endosperm and their role in embryo development (Fig. 23). This has been made possible by *in vitro* culture of endosperm and embryo. The physicochemical properties of the central vacuolar sap in the developing ovule and embryo should be mentioned because it serves as a store of nutrients (proteins, carbohydrates, and growth hormones). The nutrient content in it increases or decreases depending on the developmental stage of the ovule (Ryczkowski, 1964).

VI. Fertilization Barriers

A. Gone Concurrence

In some breeding experiments the results are not in agreement with the basic hypothesis of the Mendelian law of free combination of genes. These were explained by assuming that a selection of gametes took place in the fertilization process. This was found to involve a variation in the rate of pollen tube growth (pollen tube concurrence)—the more rapidly growing tubes fertilize all the eggs present in the ovules, whereas the slower growing, late arriving ones, fail to fertilize.

In the case of hybrid plants, the different embryo sacs of one ovary do not have an equal chance to be fertilized since they depend on gametic constitution, and their further development is determined by random position (*embryo sac concurrence*). This unequal frequency of the various egg types can also be determined by the polarized position of the meiotic spindle. It follows that of certain chromosome combinations some are always directed toward the micropyle whereas others are directed to the chalaza. It seems that the position of the megaspore within the ovary determines the chance to be fertilized; therefore, certain embryo sacs are preferred. In all these cases, certain types of gametes are selected by processes *preceding* the events of fertilization. The breeding result is the phenomenon of gone concurrence, which is not linked immediately to syngamic events (Harte, 1952, 1953, 1958).

B. Selective Fertilization

During fertilization a selective relationship among the gametes can result in a disorder of random Mendelian segregation. In this case there exists unequal attraction force among the sexually differentiated gametes, the embryo sac, and pollen tube. This has been described by Renner (1929) and Schwemmle and his collaborators as *selective fertilization*

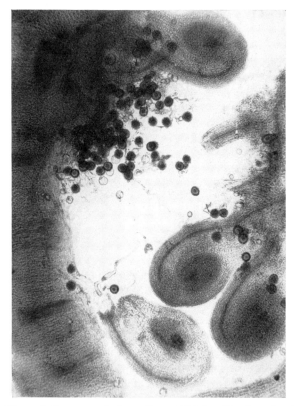

Fig. 23. View of the ovary, 3 days after artificial pollination showing the pollen grains germinating in the ovarian cavity. (Courtesy Kanta and Maheshwari, 1963.)

(reviews by Arnold, 1958; Haustein, 1955; Lamprecht, 1954). The phenomenon of selective fertilization has been definitively established by Schwemmle and his school in many species and races of *Oenothera*. The frequency of combination of two different types of gametes depends on their genetic constitution. The intensity of affinity and the velocity of reaction depend on phenotypic and genotypic factors. Selective fertilization also probably occurs in other species, as, e.g., *Pisum* (Lamprecht, 1954), *Salpiglossis* (Hiemenz, 1955), *Nicotiana* (Polyakov and Michailova, 1952), and *Pinus* (Barnes *et al.*, 1962).

Selective fertilization requires a new perspective of the fertilization process—instead of considering the individual act of one tube reaching one egg, here all the eggs in one bud, which are ready for fertilization, are considered. This requires a statistical evaluation of many processes taking place during the same time period.

The biochemical background of selective fertilization is not yet well understood. The growth of pollen tubes within the style and ovary may be largely a mechanical phenomenon. However, in order for the pollen tube to locate the micropyle, a chemotropical orientation would seem to be needed. In the case of selective fertilization, differences in manner and intensity of chemotropic attractivity of the various ovaries in one bud must exist.

If negative results in experiments of Kaienburg (1950) and, partly, Schneider (1956) are disregarded, the following general conclusions can be made:

1. *In vitro* differences in chemotropic activity were demonstrated concerning the affinity of different gametes correlated with positive-orientated pollen tubes.

2. Ovaries of different genetic constitution not only differ in concentration but also in the composition of released chemical substances that attract pollen tubes.

3. Optimal selective fertilization depends on specific maturity of the gametes.

4. From biochemical analyses of Schildknecht and collaborators (Schildknecht and Benoni, 1963a,b), it is apparent that a mixture of sugars and ninhydrin-positive substances (amino acids, peptides) is responsible for selective fertilization. The differences in affinity of tubes for various types of ovaries is determined by quantitative differences in qualitatively identical mixtures. There is no battery of specific attracting substances but rather a system of mixtures with optimal concentrations for genetically different tubes.

5. The capacity of selection by the pollen tubes, that is the reaction of the tubes, is species-specific (Glenk, 1964).

C. INCOMPATIBILITY

Incompatibility is a physiological barrier to fertilization. It prevents the fusion of the male and the female gametes although they are fully fertile. These barriers exist not only *between* species, genera, families, etc. (interspecific incompatibility) but also *within* a species (intraspecific incompatibility).

1. *Cross-Incompatibility*

Little is known concerning the biochemical mechanisms in cross-incompatibility, both intergenic and intragenic.

Blakeslee and his collaborators tried to analyze the nature of embryo inhibition by ovular tumors in interspecific *Datura* crosses. They found that hybrid embryos failed to develop as a result of disturbed growth hormone metabolism. The inhibition is caused by indoleacetic acid (IAA). It was possible to reproduce the typical inhibition by adding 0.4×10^{-3} mg IAA/1000 ovules to compatible pollinated ovaries. In intergeneric crosses between *Petunia* and *Salpiglossis* the respiration pattern is quite similar to that found in self-incompatible crossings. In the first hours after pollination an increase of CO_2 output can be observed in incompatible pollinated styles. The phenomenon is the same as in the case of self-pollination in self-incompatible species (Roggen and Linskens, 1967).

In other cases, Blakeslee and his collaborators were able to demonstrate that embryo abortions in incompatible crosses of *Datura* were caused by a water-soluble thermostable substance. This agent is unrelated to auxins, but ultraviolet absorption indicates that it is like nucleic acid. Embryo sac contents of ovules of selfed capsules contained about 6 times the amount of this substance present in the ovules of an incompatible cross. Although nucleic acids have the same inhibitory effect on embryo growth as ovular tumor extract, they probably are not alike in their action. The substance isolated from ovular tumors was effective in three successive subcultures, suggesting a self-duplication mechanism such as seen in viruses (Rappaport *et al.*, 1950; Rietsema *et al.*, 1954).

The few experimental findings from biochemical analyses of cross-incompatibility do not provide any general conclusion. However, it seems that there is more than one biochemical mechanism in cross-incompatibility.

2. *Self-Incompatibility*

Self-incompatibility reactions result only when identical genetic information (alleles) is present in both parental tissues. Thus, the same alleles act in different ways in pollen and megasporangial tissues

(Linskens, 1961, 1965a; Linskens and Kroh, 1967). There are different places and times during the process of fertilization when this physiological barrier operates.

a. The incompatibility reaction takes place at the *surface of the stigma* (Christ, 1959; Kroh 1956; Röbbelen, 1960). Either pollen tube germination is inhibited or the tube is unable to penetrate the stigma. The cuticle of the stigma is the incompatibility barrier and this is normally broken down enzymically. In all cases so far investigated the pollen of plants that have a stigma surface with a cuticular layer have a cutin-breaking enzyme system (Heinen and Linskens, 1961; Linskens and Heinen, 1962). The incompatibility barrier can be destroyed mechanically by injuring the stigma surface or by-passed by applying pollen directly into the conducting tissue of the style. Either of these procedures can result in normal fertilization and seed set. The incompatibility barriers, as represented by the families Cruciferae, Papilionaceae, Gramineae, Compositae, and others, is, therefore, linked with the cutinase enzyme system. During the contact of the pollen grain with a stigma, in the case of a compatible combination, the cutinase of the pollen is activated and the pollen tube can enter the stigma. Pollen coming from its own flower (with the identical S-allele in the case of an incompatible combination) is unable to germinate or to penetrate because the cutin-breaking enzyme system is not activated. Furthermore, no water is available from the stigmatic papillae for normal germination and, hence, the pollination fails. The reaction site is the point of surface contact between the stigmatic papillae and the coat of the pollen grain. The latter, according to recent observations of Zinger and Petrovskaja-Baranova (1961), can be regarded as a living, physiologically active structure. This pollen grain coat is rich in enzymes and plasmatic bridges which interconnect the living content of the pollen grain with the environment. Recently it has been demonstrated (Stanley and Linskens, 1965) that enzymes diffuse from the pollen grain within the first few minutes after contact between the pollen and the stigma.

b. The reaction of incompatibility takes place during the *growth to the pollen tubes through the style* (Straub, 1946, 1947; Linskens, 1955). This second type of incompatibility barrier is an inhibition of pollen tube growth within the conducting tissue of the style. It operates before the tube reaches the embryo sac. This has been thoroughly investigated. It results in abnormal behavior of the nuclei in the incompatible pollen tube and a disturbance in carbohydrate metabolism (strong deposit of cellulose fibrils in thickened walls, branched or swollen tube tips, increased number of callose plugs) and a higher density of fibrils in the tube walls (Linskens and Esser, 1957; Mühlethaler and Linskens, 1956; Schlösser, 1961; Tupý, 1961). The inhibited tubes have a high respiratory rate in the first few hours of growth (Linskens, 1953, 1955) and an increased resorption rate. Tubes from incompatible pollination show higher acid phosphatase and cytochrome oxidase activity (Schlösser, 1961). A changed protein pattern was demonstrated by several methods in extracts of pollinated styles (Linskens, 1955, 1961, 1965a,b). The newly formed compounds are complexes formed by both pollen and stylar proteins and carbohydrates in the pollinated style (Linskens, 1958, 1959). A protein specific to the S-allele is produced in the pollen and style having the same specificity as shown by serological tests (Lewis, 1952; Linskens, 1960). Increase in temperature augments and accelerates the incompatibility with a Q_{10} of about 2.4 (Lewis, 1952; Straub, 1958). The reaction is irreversible. The experimental observations of various authors support the hypothesis that incompatibility is a mechanism to prevent inbreeding and insure outbreeding.

3. *Incompatibility As a Gene Reaction*

Incompatibility genes are present in both female (ovule, ovary, style, stigma) and male tissue (pollen grains and pollen tubes). The special situation in all incompatibility reactions is that the information stored in the male and the female cells is responsible for (*a*) the specificity of the reaction, (*b*) the induction of the reaction, and (*c*) for the energetic and metabolic background of the reactions. Generally speaking, the incompatibility reaction can be interpreted as a sort of immunological reaction, in which the antibody production by nucleic acid template, and the synthesis of these templates, is inhibited by the products of an enzyme almost identical in configuration with the antibody. The antigen prevents the production of a repressor by combining with the enzyme (enzyme–antienzyme reaction). Both partners could be thought of as governed by the bipartite character of the S-allele, as shown by Lewis (1960).

The experimental data prompted Lewis (1965) to formulate his protein dimer hypothesis. According to him the S-gene complex specifies a different polypeptide for each allele. Each allelic polypeptide is an identical molecule in both the pollen and style. The polypeptide polymerizes into a dimer in both pollen and style.

The first step of the incompatibility reaction is that identical dimers in pollen and style combine to form a tetramer with an allosteric molecule (e.g., glucose). In the second step the tetramer acts as a genic regulator, either to induce the synthesis of a nonspecific inhibitor or to repress the synthesis of a growth principle essential for pollen tube growth.

Finally we come to the question, "Where is the seat of action of the S protein?" Staining tests and heat denaturation experiments on pollen extracts and exudates indicate that the antigenic substance is protein but also contains a small polysaccharide component (Mäkinen and Lewis, 1962). The tests showed that the S-protein is a substantial part of the total protein which can be readily dissociated in saline from macerated pollen cell fragments. This suggests that the S-protein diffuses out of the pollen grain. In agar–gel diffusion tests, it was found (Mäkinen and Lewis, 1962; Linskens, 1965) that the same precipitation lines of equal intensity were produced with macerated pollen. This indicates that the normal place of action of the S-protein—when the pollen tube is growing down the style—is either on the surface of the pollen tube or outside in the stylar tissue. The fact that incompatible and compatible tubes in the same style at the same time do not influence one another supports the view that the S-protein acts on the surface of the pollen tube and not outside in the style.

VII. Apomixis

Sexually differentiated higher plants generally reproduce by the act of fertilization and are termed *amphimictic* or mictic. In the normally amphimictic spermatophytes an individual (which may form a population) is produced without the act of fertilization, e.g., the process of fertilization has wholly or partially retrogressed. These are called *apomicts*. Apomixis can be considered as a stage in which the process of fertilization has been lost and no sexual fusion is involved (Winkler, 1934, Gustafsson, 1946, 1947a,b; Johansen, 1950; Battaglia, 1963). All types of apomixis (excluding only *vegetative propagation*) may be described as *agamospermy*.

Phanerogamic plants exhibit an alteration of generations i.e., sporophytic and gametophytic generations alternate, although the gametophytic phase can be very much reduced. This alternation fails to appear in certain agamospermous species. The embryos arise directly from the sporophyte (in which new reduction divisions take place) as an outgrowth of the nucellus or from the integument. This phenomenon is called *adventive* embryony. The gametophyte of the remaining agamospermous plants can arise as diploids from a macrospore mother cell (embryo sac mother cell), a process which is known as diplospory. The embryo when derived from somatic cells of the nucellus or chalaza is called *asporic*. Sometimes the egg cell can form an embryo without fertilization and this is termed *parthenogenesis*. If one of the other cells of the gametophyte develops into a plant, it is called *apogamety*.

Certain apomictic species need pollination for the induction of seed development, although no real fertilization of the egg cell takes place. The phenomenon is described as *pseudogamy*, a term which, in a broad sense, includes induced *adventitious embryony*.

All forms of agamospermy can be genetically classified in distinct families and species. The specific apomictic genes can be responsible for certain physiological constitutions, genotypically controlled, which must be present in species-hybridization. Information on the physiological basis of agamospermy is still very limited. It can be induced by various chemical and physical factors, including growth hormones (IAA, gibberellic acid), X-rays, temperature shocks, foreign or dead pollen, and pollen extracts. It is difficult to account for such a regular phenomenon as agamospermy in higher plants on a general physiological basis. The necrohormone theory first postulated by Haberlandt (1922) is still under discussion.

REFERENCES

Arnold, C. G. (1958). Selektive Befruchtung. *Ergeb. Biol.* **20**, 67–96.

Barnes, B. V., Bingham, R. T., and Squillace, A. E. (1962). Selective fertilization in *Pinus monticola. Silvae Genet.* **11**, 103–111.

Barthurst, N. O. (1954). The amino acids of grass pollen. *J. Exptl. Botan.* **5**, 253–256.

Battaglia, E. (1963). Apomixis. *In* "Recent Advances in Embryology of Angiosperms" (P. Maheshwari, ed.), pp. 221–264. Intern. Soc. Plant Morphologists, Delhi.

Batygina, T. B. (1962). Fertilization process in wheat (russ.) *Tr. Botan. Inst. Acad. Nauk. SSSR, Ser.* **9**, 260–293.

Baum, H. (1950). Das Narbensekret von *Koelreuteria paniculata. Österr. Botan. Z.* **97**, 517–519.

Beaudry, J. R. (1951). Seed development following the mating *Elymus virginicus* $\times$ *Agropyrum repens. Genetics* **36**, 109–135.

Bennet, M. C., and Cope, F. W. (1959). Nuclear fusion and nonfusion in *Theobroma cacao. Nature* **183**, 1540.

Brewbaker, J. L., and Kwack, B. H. (1964). The calcium ion and substances influencing pollen growth. *In* "Pollen Physiology and Fertilization" (H. F. Linskens, ed.), pp. 143–151. North Holland Publ. Co., Amsterdam.

Britikov, E. A. (1952). Über einige Besonderheiten der Pollenkeimung und das Wachstum der Pollenschläuche in den Fruchtblattgeweben. *Dokl. Akad. Nauk SSSR, Biol. Ser.* **1**, 121–134.

Britikov, E. A. (1954). On the physiological-biochemical analysis of pollen germination and the growth of pollen tubes in pistil tissues. *Tr. Timirjaseff Inst. Plant Physiol. Akad. Nauk SSSR* **8**(2), 3–58.

Britikov, E. A., Musatova, N. A., Vladimirtsbea, S. V., and Protsenko, M. A. (1964). Proline in the reproductive system of plants. *In* "Pollen Physiology and Fertilization" (H. F. Linskens, ed.), pp. 77–85. North Holland Publ. Co., Amsterdam.

Bubar, J. S. (1958). An association between variability in ovule development within ovaries and self-incompatibility in *Lotus. Can. J. Botan.* **36**, 63–72.

Burck, W. (1901). Preservatives on the stigma against the germination of foreign pollen. *Proc. Koninkl. Ned. Akad. Wetenschap* **3**, 264–274.

Chopra, R. N., and Sachar, R. C. (1963). Endosperm. *In* "Recent Advances in Embryology of Angiosperms" (P. Maheshwari, ed.), pp. 135–170. Intern. Soc. Plant Morphologists, Delhi.

Christ, B. (1959). Entwicklungsgeschichtliche und physiologische Untersuchungen über die Selbststerilität von *Cardamine pratensis* L. *Z. Botan.* **47**, 88–112.

Coe, G. E. (1954). Distribution of C-14 in ovules of *Zephyranthes drummondii*. *Botan. Gaz.* **115**, 342–346.

Cooper, D. C. (1940). Macrosporogenesis and embryology of the seed of *Phryma leptostachya*. *Am. J. Botany* **28**, 755–761.

Cooper, D. C. (1946). Double fertilization in *Petunia*. *Am. J. Botany* **33**, 54–57.

Dahlgren, K. R. O. (1927). Die Befruchtungserscheinungen der Angiospermen. *Hereditas* **10**, 169–229.

Ehlers, H. (1951). Untersuchungen zur Ernährungsphysiologie der Pollenschläuche. *Biol. Zentr.* **70**, 432–451.

Eklundh-Ehrenberg, C., von Euler, H., and Hevesy, G. (1947). Note on the number of pollen grain identified in the fruit of aspen. *Arkiv. Kemi, Mineral. Geol.* **23**(6), 1–5.

Fuji, K. (1903). Über den Bestäubungstropfen der Gymnospermen. *Ber. Deut. Botan. Ges.* **21**, 211–217.

Gauch, H. G., and Duggar, W. M. (1953). The role of boron in the translocation of sucrose. *Plant Physiol.* **28**, 457–466.

Gerassimova, H. (1933). Fertilization in *Crepis capillaris*. *Cellule Rec. Cytol. Histol.* **42**, 103–148.

Gerassimova-Navashina, H. (1960). A contribution to the cytology of fertilization in flowering plants. *Nucleus* (*Calcutta*) **3**, 111–120.

Gerassimova-Navashina, H. (1961). Fertilization and events leading up to fertilization, and their bearing on the origin of angiosperms. *Phytomorphology* **11**, 139–146.

Gerassimova-Navashina, H., and Korobova, S. N. (1959). On the role of synergids in fertilization. *Dokl. Akad. Nauk SSSR* **124**, 223–226.

Glenk, H. O. (1960). Keimversuche mit *Oenothera*-Pollen *in vitro*. *Flora* **148**, 378–433.

Glenk, H. O. (1964). Untersuchungen über die sexuelle Affinität bei Oenotheren. *In* "Pollen Physiology and Fertilization" (H. F. Linskens, ed.), pp. 170–179. North Holland Publ. Co., Amsterdam.

Göring, H. (1960). Untersuchungen zum Befruchtungsprozess bei *Zea* Mays mit radioaktivem Phosphor. *Naturwissenschaften* **47**, 142.

Gustafsson, A. (1946). Apomixis in higher plants, I. *Lunds Univ. Årsskr.* [N. F.] **42**(3), 1–67.

Gustafsson, A. (1947a). Apomixis in higher plants, II. *Lunds Univ. Årsskr.* [N. F.] **43**, (2), 71–178.

Gustafsson, A. (1947b). Apomixis in higher plants, III. *Lunds Univ. Årsskr.* [N. F.] **43**, (12), 183–370.

Haberlandt, W. (1922). Über Zellteilungshormone und ihre Beziehungen zur Wundheilung, Befruchtung, Parthenogenesis und Adventivembryogenie. *Biol. Zentr.* **42**, 145–172.

Haeckel, A. (1951). Beitrag zur Kenntnis der Pollenfermente. *Planta* **39**, 431–459.

Hanf, A. (1935). Vergleichende und entwicklungsgeschichtliche Untersuchungen über Morphologie und Anatomie der Griffel und Griffeläste. *Beih. Botan. Zentr.* **54A**, 99–141.

Harte, C. (1952). Untersuchungen über die Nachkommenschaft von Heterozygoten der gramini-folia Koppelungsgruppe von *Antirrhinum majus*. *Z. Vererbungslehre* **84**, 480–507.

Harte, C. (1953). Untersuchungen über Gonenkonkurrenz und crossing-over bei spaltenden *Oenothera-Bastarden*. *Z. Vererbungslehre* **85**, 97–117.

Harte, C. (1958). Untersuchungen über die Gonenkonkurrenz in der Samenanlage bei *Oenothera* unter Verwendung der Letalfaktoren als Markierungsgene. Z. *Vererbungslehre* **89**, 473–496, 497–507, 715–728.

Hasitschka-Jenschke, G. (1957). Die Entwicklung der Samenanlage von *Allium ursinum* mit besonderer Berücksichtigung der endopolyploiden Kerne in Synergiden und Antipoden. *Österr. Botan. Z.* **104**, 1–24.

Hasitschka-Jenschke, G. (1959). Vergleichende karyologische Untersuchungen an Antipoden. *Chromosoma* **12**, 229–267.

Hasitschka-Jenschke, G. (1962). Notizen über endopolyploide Kerne im Bereich der Samenanlagen von Angiospermen. *Österr. Botan. Z.* **109**, 125–137.

Haustein, E. (1955). Neuere Arbeiten zur Befruchtungsphysiologie. *Z. Botan.* **43**, 253–261.

Heinen, W., and Linskens, H. F. (1961). Enzymatic breakdown of stigmatic cuticula of flowers. *Nature* **191**, 1416.

Heslop-Harrison, J. (1963a). An ultrastructural study of pollen wall ontogeny in *Silene pendula*. *Grana Palynol.* **4**, 7–24.

Heslop-Harrison, J. (1963b). Ultrastructural aspects of differentiation in sporogenous tissue. *Symp. Soc. Exptl. Biol.* **17**, 315–340.

Heslop-Harrison, J. (1964). Cell walls, cell membranes and protoplasmatic connections during meiosis and pollen development. *In* "Pollen Physiology and Fertilization" (H. F. Linskens, ed.), pp. 39–47. North Holland Publ. Co., Amsterdam.

Hiemenz, G. (1955). Untersuchungen an Salpiglossis variabilis über Gonenkonkurrenz und selektive Befructung und ihre Nachwirkungen auf die Nachkommenschaft. *Biol. Zentr.* **74**, 337–370.

Hoare, G. V. (1933). Gametogenesis and fertilization in *Scilla non-scripta*. *Cellule Rec. Cytol. Histol.* **42**, 269–291.

Hsiang, T. H. T. (1951). Physiological and biochemical changes accompanying pollination in orchid flowers I, II. *Plant Physiol.* **26**, 441–455; 708–721.

Hu, S. Y. (1964). Morphological and cytological observations on the process of fertilization of the wheat. *Scientia (Peking)* **13** (6), 925–936; also in *Acta Botan. Sinica* **10**, 299–310 (1962).

Iwanami, Y. (1957). Physiological researches on pollen XI. Starch grains and sugars in stigma and pollen. *Botan. Mag.* **70**, 38–43.

Jensen, W. A. (1963). Cell development during plant embryogenesis. *Brookhaven Symp. Biol.* **16**, 179–200.

Jensen, W. A. (1964). Observations on the fusion of nuclei in plants. *J. Cell Biol.* **23**, 669–672.

Jensen, W. A. (1965). The ultrastructure and histochemistry of the synergids of cotton. *Am. J. Botany* **52**, 238–256.

Johansen, D. A. (1950). "Plant Embryology—Embryogeny of the Spermatophyta." Chronica Botanica, Waltham, Massachusetts.

Johri, B. M. (1963). Female gametophyte. *In* "Recent Advances in Embryology of Angiosperms" (P. Maheshwari, ed.), pp. 69–103. Intern. Soc. Plant Morphologists, Dehli.

Johri, B. M., and Vasil, I. K. (1960). The pollen and the pollen tube. *Ergeb. Biol.* **23**, 1–13.

Jones, M. D., and Newell, L. C. (1948). Longevity of pollen and stigmas in grasses. *J. Am. Soc. Agron.* **40**, 195–204.

Jung, J. (1956). Sind Narbe und Griffel Eintrittspforten für Pilzinfektionen. *Phytopathol. Z.* **27**, 405–426.

Jung, J., and Plempel, M. (1960). Über die Hemmwirkung des Gynäzeums bei *Primula obconica* II. *Phytopatholog. Z.* **38**, 245–249.

Jung, J., and Stoll, Chr. (1957). Über die Hemmwirkung des Gynäzeums bei *Primula obconica* auf Bakterien und Pilze I. *Phytopatholog. Z.* **31**, 180–184.

Kaienburg, A. (1950). Zur Kenntnis der Pollenplastiden und der Pollenschlauchleitung bei einigen Oenotheren. *Planta* **38**, 377–430.

Kanta, K., and Maheshwari, P. (1963). Intraovarian pollination in some Papaveraceae. *Phytomorphology* **13**, 215–229.

Kapil, R. N., and Vasil, I. K. (1963). Ovule. *In* "Recent Advances in Embryology of Angiosperms" (P. Maheshwari, ed.), pp. 41–67. Intern. Soc. Plant Morphologists, Delhi.

Kato, K. (1953). A phenomenon in the early stage of pollination in *Secale cereale:* the stigma reaction. *Mem. Coll. Sci., Univ. Kyoto* **B20**, 204–206.

Kato, K., and Watanabe, K. (1957). The stigma reaction II. *Botan. Mag. (Tokyo)* **70**, 96–101.

Korobova, S. N. (1959). On the course of fertilization in *Zea Mays* L. *Dokl. Akad. Nauk SSSR* **127**, 921–923.

Korobova, S. N. (1962). Embryology of corn. *Tr. Botan. Inst. Akad. Nauk SSSR Ser.* **7**, 294–314.

Kroh, M. (1956). Genetische und entwicklungsphysiologische Untersuchungen über die Selbststerilität von *Raphanus raphanistrum. Z. Induktive Abstammungs Vererbungslehre* **87**, 365–384.

Kroh, M. (1964). An electron microscopic study of the behavior of Cruciferae pollen after pollination. *In* "Pollen Physiology and Fertilization" (H. F. Linskens, ed.), pp. 221–224. North Holland Publ. Co., Amsterdam.

Lamprecht, H. (1954). Selektive Befruchtung im Lichte des Verhaltens interspezifischer Gene in Linien und Kreuzungen. *Agr. Hort. Genet.* **12**, 1–37.

Lewis, D. (1952). Serological reactions of incompatibility substances. *Proc. Roy. Soc.* **B140**, 127–135.

Lewis, D. (1960). Genetic control of specificity and activity of the S antigen in plants. *Proc. Roy. Soc.* **B151**, 468–477.

Lewis, D. (1965). A protein dimer hypothesis on incompatibility. *Genet. To-day* **3**, 657–663.

Linskens, H. F. (1953). Physiologische und chemische Unterschiede zwischen selbst- und fremd fremdbestäubten *Petunia*-Griffeln. *Naturwissenschaften* **40**, 28–29.

Linskens, H. F. (1955). Physiologische Untersuchungen der Pollenschlauch-Hemmung selbststeriler Petunien. *Z. Botan.* **43**, 1–44.

Linskens, H. F. (1956). Über die Änderung einiger physiologischer Zustandsgröben während der Pollenmeiose und Pollenentwicklung von *Lilium henryi. Ber. Deut. Botan. Ges.* **69**, 353–360.

Linskens, H. F. (1958a). Zur Frage der Entstehung der Abwehrkörper bei der Inkompatibilitätsreaktion von *Petunia I. Ber. Deut. Botan. Ges.* **71**, 3–10.

Linskens, H. F. (1958b). Über die Änderung des Nukleinsäurengehaltes während der Pollenmeiose und Pollenentwicklung von *Lilium henryi. Acta Botan. Neerl.* **7**, 61–68.

Linskens, H. F. (1959). Zur Frage der Entstehung der Abwehrkörper bei der Inkompatibilitätsreaktion von *Petunia II. Ber. Deut. Botan. Ges.* **72**, 84–92.

Linskens, H. F. (1960). Zur Frage der Entstehung der Abwehrkörper bei der Inkompatibilitätsreaktion von *Petunia III.* Mitteilungen serologische Teste mit leitgewebs- und Pollen-Extrakten. *Z. Botan.* **48**, 126–135.

Linskens, H. F. (1961). Biochemical aspects of incompatibility. *Recent Advan. Botan.* (*Toronto*) **2**, 1500–1503.

Linskens, H. F. (1964). Pollen physiology. *Ann. Rev. Plant Physiol.* **15**, 255–270.

Linskens, H. F. (1965a). Biochemistry of incompatibility. *Genet. To-day* **3**, 629–635.

Linskens, H. F. (1965b). Biochemische aspecten van incompatibiliteits-verschijnselen bij de bevruchting der bloemplanten. *Biol. Jaarboek Konink. Naturw. Genost. Dodonea Gent.* 33, 35–40.

Linskens, H. F. (1967). Pollen. In "Encyclopedia of Plant Physiology" (W. Ruhland, ed.), Vol. 18, pp. 368–406. Springer-Verlag, Berlin.

Linskens, H. F., and Esser, K. (1957). Über eine spezifische Anfärbung der Pollen-schläuche im Griffel und die Zahl der Kallosepfrofen nach Selbstung und Fremdung. *Naturwissenschaften* 44, 16.

Linskens, H. F., and Esser, K. (1959). Stoffaufnahme der Pollenschläuche aus dem Leitgewebe des Griffels. *Koninkl. Ned. Akad. Wetenschap., Proc.* C62, 150–154.

Linskens, H. F., and Heinen, W. (1962). Cutinase-Nachweis in Pollen. *Z. Botan.* 50, 338–347.

Linskens, H. F., and Kroh, M. (1967). Inkompatibilität der Phanerogamen. In "Encyclopedia of Plant Physiology" (W. Ruhland, ed.), Vol. 18, pp. 506–530. Springer-Verlag, Berlin.

Linskens, H. F., and Schrauwen, J. A. M. (1963). Änderungen des Gehaltes an Sulfhydryl-Gruppen während der Pollenmeiose und Pollenentwicklung. *Biol. Plant. Acad. Sci. Bohemoslov.* 5, 562–568.

Linskens, H. F., and Schrauwen, J. (1968). Quantitative nucleic acid determination in the microspore and tapetum fractions of lilly anthers. *Proc. Koninkl. Ned. Akad. Wet. Ser.* C71, 267–279.

Lund, H. A. (1956a). Growth hormones in the styles and ovaries of tobacco responsible for fruit development. *Am. J. Botany* 43, 562–568.

Lund, H. A. (1956b). The biosynthesis of IAA in the styles and ovaries of tobacco preliminary to the setting of fruit. *Plant Physiol.* 31, 334–339.

McWilliam, J. R. (1958). The role of the micropyle in the pollination of *Pinus. Botan. Gaz.* 120, 109–117.

McWilliam, J. R., and Mergen, F. (1958). Cytology of fertilization in *Pinus. Botan. Gaz.* 119, 246–249.

Mäkinen, Y. L. A., and Lewis, D. (1962). Immunological analysis of incompatibility proteins and of cross-reacting material in a self-compatible mutant of *Aenothera organensis. Genet. Res.* 3, 352–362.

Maheshwari, P. (1949). The male gametophyte of angiosperms. *Botan. Rev.* 15, 1–75.

Maheshwari, P. (1950). "Introduction to the Embryology of the Angiosperms." McGraw-Hill, New York.

Maheshwari, P. (ed.) (1963). "Recent Advances in the Embryology of Angiosperms." Intern. Soc. Plant Morphologists, Delhi.

Marré, E., and Murneek, S. E. (1953). Carbohydrate metabolism in the tomato fruit as affected by pollination, fertilizations and applications of growth regulators. *Plant Physiol.* 28, 255–266.

Mascarenhas, J. P., and Machlis, L. (1962a). Chemotropical response of *Antirrhinum majus* pollen to calcium. *Nature* 196, 292–293.

Mascarenhas, J. P., and Machlis, L. (1962b). The hormonal control of the directional growth of pollen tubes. *Vitamins Hormones* 20, 347–371.

Mendes, A. J. T. (1941). Cytological observations in *Coffea.* VI. Embryo and endosperm development in *Coffea arabica* L. *Am. J. Botany* 28, 784–789.

Miki-Hirosige, H. (1964a). Metabolism in the ovary of *Lilium langiflorum.* In "Pollen Physiology and Fertilization" (H. F. Linskens, ed.), pp. 26–33. North Holland Publ., Amsterdam.

Miki-Hirosige, H. (1964b). Tropism of pollen tubes to the pistils. In "Pollen Physiology and Fertilization" (H. F. Linskens, ed.), pp. 152–158. North Holland Publ. Co., Amsterdam.

Morrison, J. W. (1955). Fertilization and postfertilization development in wheat. *Can. J. Botan.* **33**, 168–176.

Mühlethaler, K., and Linskens, H. F. (1956). Elektronenmikroskopische Aufnahmen von Pollenschläuchen. *Experientia* **12**, 253–255.

Nagaraj, M. (1954a). Desoxyribonucleic acid (DNA) content in the female gametophyte and nucellus of *Rhoeo discolor*. *Proc. Indian Acad. Sci.* **B40**, 110–115.

Nagaraj, M. (1954b). DNA content in female gametophyte and nucellus of *Tradescantia paludosa*. *Current Sci.* (*India*) **23**, 300.

Nagl, W. (1962). Über Endopolyploide, Restitutionskernbildung und Kernstrukturen im Suspensor von Angiospermen und einer Gymnosperme. *Österr. Botan. Z.* **109**, 431–494.

Nemeč, B. (1931). Fecundation in *Gagea lutea*. *Preslia* **10**, 104–110.

Neumann, K. (1963). Meiotische Teilungswellen in Antheren und ihre morphologische Bedingtheit. *Biol. Zentr.* **82**, 665–719.

Noack, R. (1960). Die chemotropische Reaktionsfähigkeit der Pollenschläuche auf die Narbenstoffe der Blüten. *Z. Botan.* **48**, 463–487.

Poddubnaya-Arnoldi, V. A. (1961). The investigation of embryonic processes of some orchids on living material. *Publ. Main Botan. Garden Akad. Sci. SSSR, Moscow* **6**, 49–89.

Poddubnaya-Arnoldi, V. A. (1964). "General Embryology of the Angiosperms." Publ. House Nauka, Moscow.

Poddubnaya-Arnoldi, V. A., and Zinger, N. V. (1961). Application of histochemical technique to the study of embryonic processes in some orchids. *Recent Advan. Botan.* **1**, 711–714.

Poddubnaya-Arnoldi, V. A., Zinger, N. V., Petrovskaja, T. P., and Polunina, N. N. (1961). Histochemical study of the pollen grains and pollen tubes in angiosperms. *Recent Advan. Botan.* **1**, 682–685.

Poddubnaya-Arnoldi, V. A., Zinger, N. V., and Petrovskaja-Baranova, T. P. (1964). A histochemical investigation on the ovules, embryo sacs and seeds in some angiosperms. *In* "Pollen Physiology and Fertilization" (H. F. Linskens, ed.), pp. 3–7. Amsterdam.

Poljakov, I. M. (1955). The use of radioactive isotopes in studies of plant fertilization. *Sessiya Akad. Nauk SSSR po Mirnomu Ispol'z. At. Energii, Moscow, 1955, Zasedan. Otd. Biol. Nauk,* pp. 221–233.

Poljakov, I. M. (1964). New data on use of radioactive isotopes in studying fertilization in plants. *In* "Pollen Physiology and Fertilization" (H. F. Linskens, ed.), pp. 194–199, 215–218. North Holland Publ. Co., Amsterdam.

Poljakov, I. M., and Michailova, P. W. (1952). Das Wachstum der Pollenschläuche in den verschiedenen Teilen des Fruchtblattes und die Wahlbefruchtung. *In* "Über den Befruchtungsprozess bei Pflanzen und Tieren," pp. 201–231. Verlag, Berlin.

Pozsàr, B. I. (1960). The nitrogen metabolism of the pollen tube and its function in fertilization. *Acta Botan. Acad. Sci. Hung.* **6**, 389–395.

Pylnev, V. M. (1962). Effect of pollination on the histochemical nature of wheat pistils. *Dokl. Mosk. Sel'skokhoz. Akad.* **77**, 95–104.

Raghavan, V., and Baruah, H. K. (1959). Effect of time factor on the stimulation of pollen germination and pollen tube growth by certain auxins, vitamins and trace elements. *Physiol. Plantarum* **12**, 441–451.

Rajki, E. (1961). Effects of the age of the stigma on fertilization and on some basic characteristics of the progeny. *Novenytermeles* **10**, 51–58.

Rajki, E. (1962). Effect of pollination of the amount of pollen on the surface of stigma. *Novenytermeles* **11**, 35–44.

Rappaport, J., Satina, S., and Blakeslee, A. F. (1950). Extracts of ovular tumors and their inhibition of embryo growth in *Datura*. *Am. J. Botany* **37**, 586–595.

Renner, O. (1929). Störung durch selektive Befruchtung. *Handbuch Vererbungslehre* **2**, 47.

Richter-Landmann, W. (1959). Der Befruchtungsvorgang bei Impatiens glanduligera unter Berücksichtigung der plasmatischen Organelle von Spermazelle, Eizelle und Zygote. *Planta* **53**, 162–177.

Rietsema, J., Satina, S., and Blakeslee, A. F. (1954). On the nature of the embryo inhibitor in ovular tumors of *Datura*. *Proc. Natl. Acad. Sci. U. S.* **40**, 424–431.

Rodrigues-Pereira, A. S., and Linskens, H. F. (1963). The influence of glutathione and glutathione antagonists on meiosis in excised anthers of *Lilium henryi*. *Acta Botan. Neerl.* **12**, 302–314.

Röbbelen, G. (1960). Über die Kreuzungsunverträglichkeit verschiedener Brassica-Arten als Folge eines gehemmten Pollenschlauchwachstums. *Züchter* **30**, 300–312.

Roggen, H., and Linskens, H. F. (1967). Pollen tube growth and respiration in incompatible, intergenenic crosses between *Petunia* and *Salpiglossis*. *Naturwissenschaften* **54**, 542–543.

Rosen, W. G. (1961). Studies on the pollen tube chemotropism. *Am. J. Botany* **48**, 889–895.

Rosen, W. G. (1962). Cellular chemotropism and chemotaxis. *Quart. Rev. Biol.* **37**, 242–259.

Rosen, W. G. (1964). Chemotropism and fine structure of pollen tubes. *In* "Pollen Physiology and Fertilization" (H. F. Linskens, ed.), pp. 159–169. North Holland Publ. Co., Amsterdam.

Rowley, J. R. (1962). Stranded arrangement of sporopollenin in the exine of microspores of *Poa annua*. *Science* **137**, 526–528.

Rowley, J. R. (1964). Formation of the pore in pollen of *Poa annua*. *In* "Pollen Physiology and Fertilization" (H. F. Linskens, ed.), pp. 59–69. Amsterdam.

Ryczkowski, M. (1964). Physico-chemical properties of the central vacuolar sap in developing ovules. *In* "Pollen Physiology and Fertilization" (H. F. Linskens, ed.), pp. 17–25. North Holland Publ. Co., Amsterdam.

Sarvas, R. (1962). Investigations on the flowering and seed crop of *Pinus silvestris*. *Publ. Forest Res. Inst. Finland* **53** (4), 1–198.

Sax, K. (1918). The behaviour of the chromosomes in fertilization. *Genetics* **3**, 309–327.

Schildknecht, H., and Benoni, H. (1963a). Über die Chemie der Anziehung von Pollenschläuchen durch die Samenanlagen von Oenotheren. *Z. Naturforsch.* **18b**, 45–54.

Schildknecht, H., and Benoni, H. (1963b). Versuche zur Aufklärung des Pollenschlauch-Chemotropismus von Narcissen. *Z. Naturforsch.* **18b**, 656–661.

Schlösser, K. (1961). Cytologische und cytochemische Untersuchungen über das Pollenschlauchwachstum selbststeriler Petunien. *Z. Botan.* **49**, 266–288.

Schmucker, T. (1935). Über den Einfluss von Borsäure auf Pflanzen, insbesondere keimende Pollenkörner. *Planta* **23**, 264–283.

Schnarf, K. (1929). "Embryologie der Angiospermen." Bornträger, Berlin.

Schnarf, K. (1941). "Vergleichende Zytologie des Geschlechtsapparates der Kormophyten." Bornträger, Berlin.

Schneider, G. (1956). Wachstum und Chemotropismus von Pollenschläuchen. Z. Botan. **44**, 377–407.

Schoch-Bodmer, H., and Huber, P. (1945). Auflösung und Aufnahme von Leitgewebe-Substanz durch die Pollenschläuche. Verhandl. Schweiz. Naturforsch. Ges. **125**, 161–162.

Schoch-Bodmer, H., and Huber, P. (1947). Ernährung der Pollenschläuche durch des Leitgewebe. Vierteljahnesschr. Naturforsch. Ges. Zürich **92**, 43–48.

Scott, E. G. (1960). Effect of supra-optimal boron level on respiration and carbohydrate metabolism of Helianthus annuus. Plant Physiol. **35**, 653–661.

Souèges, R. (1937). "Les lois de developpement." Hermann, Paris.

Sprague, G. F. (1932). The nature and extent of hetero-fertilization in maize. Genetics **17**, 358–368.

Stanley, R. G. (1958). Methods and concepts applied to a study of flowering in pine. In "The Physiology of Forest Trees" (K. V. Thimann, ed.), pp. 583–599. Ronald Press, New York.

Stanley, R. G. (1964). Physiology of pollen and pistil. Sci. Progr. (London) **52**, 122–132.

Stanley, R. G., and Linskens, H. F. (1964). Enzyme activation in germinating Petunia pollen. Nature **203**, 542–544.

Stanley, R. G., and Linskens, H. F. (1965). Protein diffusion from germinating pollen. Physiol. Plantarum **18**, 47–53.

Stanley, R. G., and Loewus, F. A. (1964). Boron and myo-insitol in pollen pectin biosynthesis. In "Pollen Physiology and Fertilization" (H. F. Linskens, ed.), pp. 128–136. North Holland Publ. Co., Amsterdam.

Stanley, R. G., Young, L. C. T., and Graham, J. S. D. (1958). Carbon dioxide fixation in germinating pine pollen (Pinus ponderosa). Nature **182**, 1462–1463.

Steffen, K. (1951). Zur Kenntnis des Befruchtungsvorganges bei Impatiens glanduligera Lindl. Cytologische Studien am Embryosack der Balsamineen. Planta **39**, 175–244.

Steffen, K. (1953). Zytologische Untersuchungen an Pollenkorn und -schlauch. I. Phasenkontrast—optische Lebenduntersuchungen an Pollenschläuchen von Galanthus nivalis. Flora **140**, 140–174.

Steffen, K. (1963). Fertilization. In "Recent Advances in Embryology of Angiosperms" (P. Maheshwari, ed.), pp. 105–143. Delhi.

Straub, J. (1946). Zur Entwicklungsphysiologie der Selbststerilität I. Z. Naturforsch. **1**, 287.

Straub, J. (1947). Zur Entwicklungsphysiologie der Selbststerilität II. Z. Naturforsch. **2b**, 433.

Straub, J. (1958). Das Überwinden der Selbststerilität. Z. Botan. **46**, 98–111.

Stubbe, W. (1957). Entmischung von drei in der Zygote vereinigten Plastidomen. Ber. Deut. Botan. Ges. **70**, 21–226.

Tano, S., and Takahashi, H. (1964). Nucleic acid synthesis in growing pollen tubes. J. Biochem. **56**, 578–580.

Thomas, H. H. (1934). The nature and origin of the stigma. New Phytologist **33**, 173–198.

Tkachenko, G. V. (1959). The role of the substances secreted by the stigma in the pollination of grape. Botan. Zh. **44**, 963–967.

Tschermak-Woess, E. (1956). Notizen über die Riesenkerne und "Riesenchromosomen" in den Antipoden von Aconitum. Chromosoma **8**, 114–134.

Tupý, J. (1960). Sugar absorption callose formation and the growth rate of pollen tubes. Biol. Plant. Acad. Sci. Bohemslov. **2**, 169–180.

Tupý, J. (1961). Investigation on free amino-acids in cross-, self- and non-pollinated pistils of *Nicotiana alata*. *Biol. Plant. Acad. Sci. Bohemoslov.* **3**, 47–64.

Tupý, J. (1964). Metabolism of proline in styles and pollen tubes of *Nicotiana alata*. In "Pollen Physiology and Fertilization" (H. F. Linskens, ed.), pp. 86–94. Amsterdam.

Van der Pluijm, J. E. (1964). An electron microscopic investigation on the filiform apparatus in the embryo sac of *Torenia fournieri*. In "Pollen Physiology and Fertilization" (H. F. Linskens, ed.), pp. 8–16. North Holland Publ. Co., Amsterdam.

van Went, J., and Linskens, H. F. (1967). Die Entwicklung des sogenannten "Fadenapparates" in Embryosack von *Petunia hybrida*. *Genet. Breeding Res.* **37**, 51–55.

Vasil, I. K. (1962). Physiology of anthers. *Proc. Summer School Botany, Darjeeling, 1960*, pp. 477–487. (Ministry Sci. Res. and Cultural Affairs, New Delhi, India.)

Vasil, I. K. (1964). Effect of boron on pollen germination and pollen tube growth. In "Pollen Physiology and Fertilization" (H. F. Linskens, ed.), pp. 107–119. North Holland Publ. Co., Amsterdam.

Vasil, I. K., and Johri, B. M. (1964). The style, stigma and pollen tube. *Phytomorphology* **14**, 352–369.

Vassileva-Dryanovska, O. A., and Tsoneva, M. (1959). Dynamics of Desoxyribonucleic acid in lily's fertilization and embryogenesis. *Compt. Rend. Acad. Bulgare Sci.* **12** (6), 573–576.

Vazart, B. (1958). Differenciation des cellules sexuelles et fécondation chez les Phanérogames. *Protoplasmatologia* **7** (3a), 1–158.

Visser, T. (1955). Germination and storage of pollen. *Mededel. Landbouwhogeschool. Wageningen* **55** (1), 1–68.

Wardlaw, C. W. (1955). "Embryogenesis in Plants." Wiley, New York.

Wardlaw, C. W. (1963). Plant embryos as reaction systems. In "Recent Advances in Embryology of Angiosperms" (P. Maheshwari, ed.), pp. 355–360. Intern. Soc. Plant Morphologists, Delhi.

Wassilewa-Drenowska, O. (1959). Die Dynamik der Desoxyrhibonukleinsäure in dem weiblichen Gametophyten von *Lilium*. *Acta Histochem.* **7**, 74–87.

Winkler, H. (1934). Apomixis, Handwörterb. *Naturwissenschaften* **4**, 451–461.

Woodard, J. W. (1958). Intracellular amounts of nucleic acids and proteins during pollen grain growth in *Tradescantia*. *J. Biophys. Biochem. Cytol.* **4**, 383–389.

Wulff, H. D. (1935). Glavanotropismus bei Pollenschläuchen. *Planta* **24**, 602–608.

Wunderlich, R. (1959). Zur Frage der Phylogenie der Endospermtypen bei Angiospermen. *Österr. Botan. Z.* **106**, 203–293.

Wylie, R. B. (1941). Some aspects of fertilization in *Vallisneria*. *Am. J. Botany* **28**, 169–174.

Zacharias, E. (1901). Über Sexualzellen und Befruchtung. *Verhandl. Naturwiss. Ver. Hamburg* (cited by Steffen, 1951).

Zeijlemaker, F. C. J. (1956). Growth of pollen tubes *in vitro* and their reactions on potential differences. *Acta. Botan. Neerl.* **5**, 179–186.

Ziegler, H. (1959). Über die Zusammensetzung des "Bestäubungstropfens" und den Mechanismus seiner Sekretion. *Planta* **52**, 587–599.

Zinger, N. V., and Petrovskaja-Baranova, T. P. (1961). *Dokl. Akad. Nauk SSSR* **138**, 466–469.

Zinger, N. V., and Poddubnaya-Arnoldi, V. A. (1961). Application of histochemical technique to the study of embryonic processes in some orchids. *Publ. Main Botan. Garden Akad. Sci. SSSR Moscow* **6**, 90–169.

CHAPTER 6

Paramecium

Koichi Hiwatashi

BIOLOGICAL INSTITUTE, TOHOKU UNIVERSITY,

SENDAI, JAPAN

I. Introduction

The process of fertilization in *Paramecium,* usually called conjugation, is a biological phenomenon that has interested many investigators. The discovery by Sonneborn (1937) of mating types in this animal, a sort of sexual differentiation, opened the way to the modern work which showed that conjugation in *Paramecium* results in true fertilization—a process leading to genetic recombination.

Although specific features of the process of reproduction may differ in different organisms, the basic phenomena of fertilization are very much alike in all organisms including *Paramecium.* The process involves an initial adhesion of cells, meiotic reduction of the chromosome number with concomitant segregation of genes, and fusion of gametic nuclei. On the other hand, there are some specific features in the fertilization of *Paramecium.* These organisms do not produce special gamete cells as in higher organisms, although the paramecia must be sexually mature and reactive before fertilization can occur. Thus, vegetatively reproducing cells can conjugate at any time if they are sexually mature and environmental conditions are appropriate. Another specific feature is that the union of two paramecia in fertilization is not permanent; the mates separate again after exchange of the gametic nuclei and completion of the reciprocal cross-fertilization. These features, however, do not diminish the importance of the common basic phenomenon which *Paramecium* shares with other organisms.

Although one of the important problems of gamete physiology in *Paramecium* may be the nature of sexual maturity and mating reactivity, studies on these problems are rather limited (Cohen, 1964, 1965; Siegel, 1965). A comparable lack of available knowledge obtains for the mechanism of the transfer of gametic nuclei from one cell to its mate. On the other hand, beginning with the pioneer work of Metz (1946, 1947), much information has accumulated on the mechanism of the initial specific adhesion of the cells and its relation to activation. Extensive discussion on this problem has already been published (Metz, 1954).

In this chapter the general features of fertilization in *Paramecium* will first be described and then discussions in some detail will be given on the most important problems of the primary reaction, the so-called mating reaction, including the nature of the substances involved. Finally, an outline of the recently developed method of artificial induction of conjugation without mating differentiation will be presented.

II. Process of Conjugation in *Paramecium*

A. MORPHOLOGICAL CHANGES

Conjugation in *Paramecium* ordinarily involves three types of union occurring sequentially between the mates (Fig. 1d, e, and f). When paramecia of complementary mating type are brought together under appropriate conditions, the first step in the union is the agglutinative adhesion called the *mating reaction* (Sonneborn, 1939a). Under suitable conditions, tens or sometimes hundreds of paramecia stick together to form large agglutinates (Fig. 1b and c). In this reaction no evidence for oriented movement suggesting chemotactic attraction has been reported (Sonneborn, 1937). Thus the adhesion occurs upon random contact. The union of the animals in the mating reaction is highly specific (Sonneborn, 1937). If the paramecia of complementary mating types are marked differently by vital stains, it can be clearly observed that the animals in a clump always consist of two different mating types (Hiwatashi, 1951). Thus no clump consisting exclusively of the same mating type is observed, though more than two or three animals frequently adhere to a single animal of the complementary type. The same result has been confirmed in *Para-*

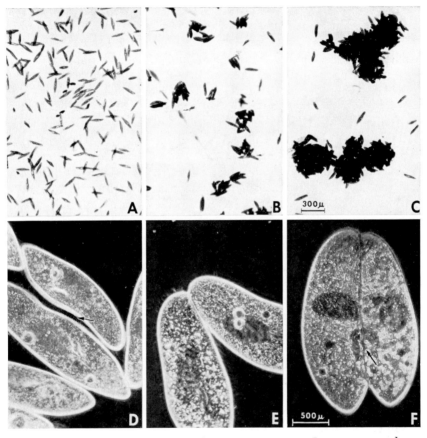

FIG. 1. (A–C) The time course of the mating reaction in *Paramecium caudatum.* (A) Animals of mating type V before mixing with mating type VI; (B) 1 minute after mixing; (C) 5 minutes after mixing. (D–F) Three types of union in the conjugation of *P. caudatum*; negative phase-contrast photomicrographs. (D) Mating reaction union—animals of complementary mating type are uniting by their cilia (arrow). (E) Hold-fast union—animals of complementary mating type are uniting at the cell surface of the anterior region. (F) Completion of the union of conjugation including paroral union—animals are uniting at the surface from anterior end to posterior regions including the paroral region (arrow).

mecia bursaria marked by the presence or absence of symbiotic algae (Jennings, 1938; Larison and Siegel, 1961). However, an animal which has first clumped with an animal of the complementary type can adhere briefly to an animal of the same mating type (Sonneborn, 1937, 1942).

All workers agree that the mating reaction involves the cilia of complementary mating types, and there seems to be little doubt that the paramecia in the reaction adhere at the tips of the cilia (Sonneborn, 1937;

Jennings, 1939a; Tartar and Chen, 1941; Metz, 1954; Vivier, 1960). This is readily confirmed by phase-contrast microscopic observation of the clumping paramecia. The role of the cilia in the mating reaction is further confirmed by the observation that detached cilia of one mating type adhere specifically to living paramecia of the complementary type (Metz, 1954; Cohen and Siegel, 1963; Miyake, 1964). From these observations, it is highly probable that the union of the animals in the mating reaction occurs between the cilia of complementary mating types. Unfortunately, however, these observations do not positively exclude other regions of the *Paramecium* surface, for example, the pellicle, from participation in the reaction, as stated by Metz (1954), although Cohen and Siegel (1963) showed that other fractions of the paramecia such as trichocysts do not carry mating reactivity.

Another problem concerns the location of the mating reactivity on the *Paramecium*. In *Paramecium bursaria*, Jennings (1939a) reported that in the agglutinative mating reaction any part of the body that comes in contact adheres to a potential mate. In *Paramecium caudatum*, however, Hiwatashi (1961) found that the ventral halves of bisected animals generally adhered to the complementary type but the dorsal halves failed to adhere except at the anterior tip. This observation was confirmed by Cohen and Siegel (1963) and by Miyake (1964) in *P. bursaria* and *Paramecium multimicronucleatum*, respectively, in their experiments on the mating reaction between detached cilia and whole animals, the details of which are given in Section III,B. There is some difference, however, between *P. caudatum* and *P. bursaria* in the most frequent region of the contact. It is the anteroventral region in *P. caudatum* but the posteroventral one in *P. bursaria*.

Electron microscopical observations have attempted to find structural differences in the cilia between mating-reactive and unreactive paramecia and also between the animals of different mating types. These observations, however, failed to detect any structural basis for the mating reactivity or mating-type difference (Metz, 1954). In view of the previously mentioned localization of mating-reactive cilia, this type of study should be done using only the ventral cilia or by carefully comparing the ventral-reactive with the dorsal-unreactive cilia.

The second step in the union of conjugation is called the *holdfast union*. This is the union of the paramecia, usually two, at the region near their anterior ends. At some time, usually from 45 to 60 minutes, after the beginning of the mating reaction, the clumps of paramecia break down releasing paired animals attached by the holdfast union as well as some single animals. The holdfast union is firmer and more intimate than the mating reaction union, though the pairs can still be separated without injury. Animal to animal contact in this union seems not to be mediated by

optics and highly synchronized conjugation (Hiwatashi, unpublished observations), or by use of a microcompression chamber (Wichterman, 1946b). Recent electron microscopical observations of the conjugation of ciliates reveals that there are many openings in the fused membrane between mates. These might permit cytoplasmic exchange (Elliott and Kennedy, 1962; Vivier and André, 1961; Schneider, 1963; Inaba *et al.*, 1966), though the actual meaning of the openings is not clear. Schneider (1963) showed that disintegration of the fused membranes is particularly extensive at the paroral region, reaches 10 μ in diameter, and may be large enough for the migration of the micronucleus (Fig. 2). On the other hand, Inaba *et al.* (1966) succeeded in observing the pronucleus at the exact moment of migration and indicated that the migration occurs in ameboid fashion through an opening only 1 μ in diameter (Figs. 3 and 4). Whether the difference between Schneider and Inaba's results is due to the difference in material or to the method of preparation remains for future study.

In ordinary conjugation the synkaryon is formed by the union of the migrated and stationary gametic nuclei and, since the third division is equational, the two synkarya formed in the pair should have the same genotype. However, some cases are known where the migration of gametic nuclei from one mate to the other does not occur. In these cases synkarya are formed by the union of the two gamete nuclei produced in the same animal (Diller, 1936; Wichterman, 1940; Sonneborn, 1947); this is called *cytogamy*. Essentially the same series of nuclear changes occurs in single paramecium. This is known as autogamy and frequently occurs in *Paramecium aurelia* (Hertwig, 1914; Diller, 1936; Sonneborn, 1939a) and in a few other species. Autogamy produces completely homozygous individuals and is useful for genetic studies. Unfortunately, we have little knowledge of the physiological mechanisms for the occurrence of autogamy. After the synkarya are formed in the conjugating pair, the pair mates separate. The synkarya divide several times and differentiate into new micro- and macronuclei. The old macronucleus fragments and disintegrates. The postzygotic process is a problem of differentiation rather than of fertilization and will not be considered here.

Fɪɢ. 3. Contact region of conjugating *Paramecium multimicronucleatum*, where one of the pronuclei (mp) is migrating and approaching the stationary pronucleus extending pseudopodlike extension. No extensive disintegration of fused membrane is observed as in Fig. 2. a—subpellicular alveolus; b—boundary membrane; c—cilium; i—interruption of the boundary; p—pellicle; r—ectoplasmic ridge; sp—stationary pronucleus; k—kinetosome; mp—migratory pronucleus; m—mitochondria; t—trichocyst. (Courtesy of F. Inaba *et al.*, 1966.)

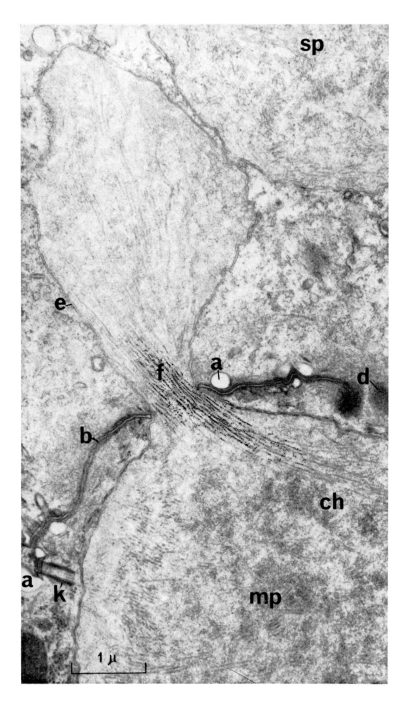

B. Initiation of Activation

As seen in the preceding section, many morphological changes occur in a precise and orderly sequence in conjugation. What, then, is the trigger for these changes? Are these various changes interrelated and initiated by a single primary reaction as the metazoan egg is activated by the entrance of the sperm? The first experimental approach to this problem was made by Metz (1947, 1948) with *Paramecium aurelia*. He found that formalin-killed reactive paramecia of one mating type clumped strongly and specifically with living animals of the complementary mating type. He also found that the living animals of one mating type, which had clumped with the formalin-killed animals, united to form pairs by a holdfast union. These selfing pairs united by the holdfast union are referred to as pseudoselfing pairs. The unions in pseudoselfing pairs involve only the holdfast regions of the animals. They never fuse at the paroral region though paroral cones are formed. Cytological observation revealed that the pseudoselfing pair members undergo meiosis and macronuclear breakdown and that these nuclear changes do not differ either morphologically or chronologically from those in ordinary conjugations except for the absence of migration of the gametic nucleus from one animal to its mate (Metz, 1947). These results clearly showed that paroral union, the third type of union in conjugation, is not necessary for the induction of the nuclear changes, although it may be prerequisite for the exchange of the gametic nuclei. Another fact which leads to the same conclusion concerns the conjugation of three animals. In this case the third animal unites only at the holdfast region but still undergoes meiosis and macronuclear breakdown (Chen, 1946a; Hiwatashi, 1955a).

The next question is whether the holdfast union, the second type of union, is necessary for nuclear activation. The answer was also given by Metz (1947). He mixed single, isolated, living paramecia of one mating type with formalin-killed animals of the complementary mating type. In this mixture the mating reaction occurs but further steps of union are prevented. The same nuclear changes occurred in these single living paramecia, which had engaged only in the mating reaction union with dead animals as occurred in the pseudoselfing pairs. This evidence, however, cannot entirely eliminate substances concerned with the holdfast union (holdfast substances) from participation in the activation of the nuclei because the dead animals had potentially the mechanism to form a hold-

Fig. 4. Enlarged figure of the framed portion of Fig. 3. Fibrils of chromonemata (f) are observed extending to the pseudopodlike extension. a—subpellicular alveolus; b—boundary membrane; ch—chromonema; d—dense material; e—nuclear envelope; k—kinetosome; mp—migratory pronucleus; sp—stationary pronucleus.

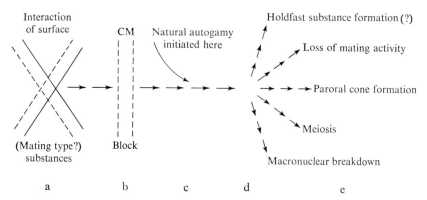

Fig. 5. Scheme for activation in *Paramecium* by Metz (1954). (From "Sex in Microorganisms," p. 302, with the permission of the American Association for the Advancement of Science.)

fast union. The conclusive experiment to eliminate this last possibility was carried out with a mutant stock of *Paramecium aurelia*. This stock gives a typical mating reaction but fails to form holdfast and paroral unions (Metz and Foley, 1949). It apparently has only the mechanism to give the mating reaction and seems to lack all mechanisms for other unions in conjugation. It is known as the "cannot mate" (CM) stock (Sonneborn, 1942). Finally, the CM paramecia never undergo meiosis and macronuclear breakdown except in the case of natural autogamy. In the experiment, single isolated living animals of a normal stock were mixed with formalin-killed CM animals of the complementary mating type. Meiosis and macronuclear breakdown occurred in the normal living paramecia just as in the case when normal dead animals were used. The result clearly shows that in *P. aurelia* the holdfast and paroral unions are not essential for activation and only the first or mating reaction union is required to initiate the nuclear changes. Since the CM stock paramecia, both formalin-killed and living, are not only capable of initiating the nuclear changes but also induce the holdfast union and loss of mating reactivity in the normal animals of the complementary type, Metz (1948) concluded that activation in the same sense as in the metazoan egg must result from the interaction of the surfaces of paramecia during the mating reaction (Fig. 5). Metz considered that holdfast substance formation could arise between b and c in Fig. 5, loss of mating reactivity between c and d, and, therefore, paroral cone formation, macronuclear breakdown, and meiosis could be sequential to the former two. Unfortunately, however, there is no evidence for the interdependence among those phenomena.

The same problem was studied in *Paramecium caudatum* partly by the same method employed by Metz (Hiwatashi, 1955a). Contrary to the result in *Paramecium aurelia,* nuclear reorganization did not occur in *P. caudatum* when single isolated living animals were mixed with formalin-killed reactive paramecia of the complementary mating type. The same negative result was also obtained when living animals which differ not only in mating type but also in serotype were mixed and then treated with an antiserum to one serotype. The antiserum treatment immobilizes animals of one mating type and prevents further interaction but has no demonstrable effect on the other mating type. Thus, it is easy to separate the animals of one mating type from the mixture at any time from the mating reaction to paroral union. The experiment confirmed that in *P. caudatum* neither the mating reaction nor the ordinary holdfast union was sufficient to induce nuclear reorganization. Additional evidence that the ordinary holdfast union does not induce nuclear change was obtained by preventing paroral union by treatment with KCl or the nonionic detergent Tween 80 (Hiwatashi, 1955b). However, even in *P. caudatum,* when three animals conjugate, the third animal undergoes nuclear reorganization even though united only at the holdfast region. This fact indicates either that the union at the holdfast region is sufficient to induce the nuclear change or that some change occurring in the first two mates is transferred to the third through the holdfast attachment. In any event, this fact still does not alter the conclusion that the mating reaction alone is insufficient to induce the nuclear changes in *P. caudatum.* The cause of this discrepancy between *P. aurelia* and *P. caudatum* remains unsolved. One of the important differences between the two species is the regular occurrence of autogamy in *P. aurelia* but not in *P. caudatum.* The question can be raised whether this difference has a relation to the above-mentioned discrepancy. Could the mating reaction induce antogamy in *P. aurelia?* Probably not, because the CM stock which does undergo natural autogamy cannot be induced to undergo nuclear changes by the mating reaction (Metz and Foley, 1949).

In *Paramecium bursaria,* Chen (1946b) reported that temporary pairs were formed between normal and some abnormal clones. In such pairs no conspicuous nuclear changes except a slight swelling of the micronuclei occurred. Since this mixture exhibits the typical agglutinative mating reaction, it is probable that the mating reaction does not initiate meiosis and macronuclear breakdown in this species. A useful means of attacking the problem of the activation-initiating mechanism may prove to be the reaction between detached cilia and whole animals as described in *P. bursaria* and *Paramecium multimicronucleatum* by Cohen and Siegel (1963) and by Miyake (1964), respectively. Miyake showed that isolated cilia from mating-reactive animals can induce paramecia to form selfing pairs. The selfing pair members are activated. Activation in single paramecia, however, has not been confirmed in these reports.

III. Mating Reaction and Mating-Type Substances

A. MATING REACTION WITH KILLED PARAMECIA

Since the mating reaction is a mating-type specific reaction and since it is the primary reaction followed by sequential changes in conjugation, studies on the chemical nature of the reaction and the substances involved should be of great importance for analyzing the problem of fertilization in *Paramecium*. The culture fluids or filtrates in which paramecia have lived produce no effect upon the mating reaction (Sonneborn, 1937; Metz, 1947). Accordingly, the mechanism of the reaction was sought for on the surface of the animals.

The first step in isolating the mechanism of the mating reaction was made with paramecia killed by various physical and chemical agents. A mating reaction between live and dead paramecia was first observed by Boell and Woodruff (1941). Metz (1946, 1947, 1948, 1954) regarded the reactive dead animals as a collection of specific substances built into the surfaces of the dead paramecia and developed a method to analyze the substances involved in the mating reaction. As already mentioned, when paramecia of one mating type are killed by appropriate agents and mixed with reactive living animals of the complementary mating type, the living and dead animals immediately adhere and, under favorable conditions, form large agglutinates as in the normal mating reaction. This has been reported to be successful in *Paramecium aurelia* (Metz, 1946, 1947), *Paramecium calkinsi* (Metz, 1948), and *Paramecium caudatum* (Hiwatashi, 1949b, 1950). The following are among the agents so far reported to be excellent killing agents for preserving the mating reactivity of the animals: formalin for *P. aurelia, P. calkinsi,* and *P. caudatum* (Metz, 1947, 1954; Metz and Butterfield, 1950; Hiwatashi, 1949b); picric acid and HgCl$_2$ for *P. calkinsi* (Metz and Butterfield, 1950); ammonium sulfate and glycerine for *P. caudatum* (Hiwatashi, 1950); lyophilization for *P. aurelia* and *P. calkinsi* (Metz and Fusco, 1949); and K$_2$Cr$_2$O$_7$ for *Paramecium multimicronucleatum* (Miyake, 1964) and *P. caudatum* (Hiwatashi, unpublished observations). *Paramecium caudatum* retains the mating reactivity for more than a week or even for months, if properly treated, in a saturated solution of ammonium sulfate or 50% glycerine (Hiwatashi, 1950 and unpublished observations). Properly lyophilized *P. calkinsi* also retains reactivity for several months at room temperature (Metz and Fusco, 1949).

On the other hand, a number of different physical and chemical agents are known to inactivate the mating reactivity of the living or dead paramecia. These have been reviewed extensively by Metz (1954) to whose article the reader is referred for details. Such agents include heating, extreme pH, grinding, freeze-thawing, X-ray and ultraviolet irradiations, detergents, organic solvents, and enzymic digestion. Among the inactivation experiments, the most meaningful result for characterizing the substances involved in the mating reaction was obtained with experiments on enzymic digestion (Metz and Butterfield, 1951). The enzyme preparations tested were crystalline trypsin, chymotrypsin, lysozyme, and ribonuclease; purified hyaluronidase and crude ptyalin and lecithinase (bee venom). Formalin- or picric acid-killed reactive paramecia of *P. calkinsi* were treated with the enzyme preparations for appropriate incubation periods and, after washing to remove the enzyme solution, they were tested for mating reactivity with living animals of the complementary mating type. Only the proteolytic enzymes, trypsin and chymotrypsin, destroyed the mating reactivity of the dead animals. The other enzymes had no detectable effect. Since loss of mating reactivity by trypsin and chymotrypsin did not necessarily accompany any serious structural change of the cilia, Metz and Butterfield (1951) concluded that the mating reactivity is dependent upon substances that are protein or intimately associated with protein.

Another important aspect of the mating reactivity is the differential action of some agents on dead animals of complementary mating types. Among the agents so far examined, only two, formalin and nitrous acid are known to show different action on complementary mating types. As stated previously, formalin-killed paramecia show strong mating reactivity. This is the case for mating type I of *P. calkinsi*. However, formalin-killed animals of type II, the complementary type to type I, never give strong mating reactions regardless of the concentration of formalin employed for killing (Metz and Butterfield, 1950; Metz, 1954). The same differential action of formalin is also seen in *P. caudatum*. Here one of the two complementary mating types in each of the four syngens (interbreeding groups) is unreactive after 20 minutes in 3% formalin, whereas the respective complementary type in each syngen gives a strong reaction after the same treatment (Hiwatashi, 1949b). Similar differential action of formalin is also observed in *P. multimicronucleatum*, syngen 2 (A. Miyake, personal communication). Finally, differential action of formalin on different mating types is due to the difference in stability of the mating reactivity of dead paramecia in formalin (Hiwatashi, 1950) (Fig. 6). The differential action of nitrous acid is known only in *P. calkinsi*. When formalin- or picric acid-killed type I animals were treated with 1 M nitrous

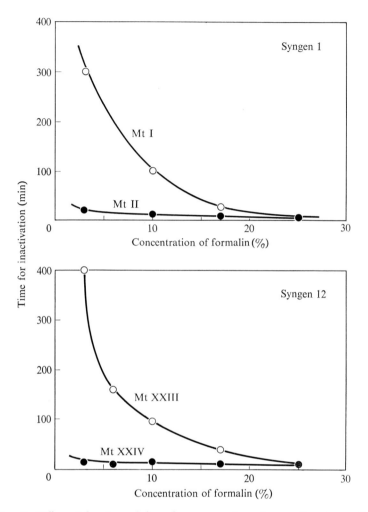

FIG. 6. Differential action of formalin on complementary mating types of *Paramecium caudatum*. Time for complete loss of the mating reactivity is plotted against concentration of formalin. Each plot is the mean of five sets of experiments. Mt— mating type. (Modified and redrawn from Hiwatashi, 1950.)

acid at 0°C, they were still reactive, whereas picric acid- or $HgCl_2$-killed reactive type II as well as formalin-killed type II animals lost reactivity after the nitrous acid treatment (Metz, 1954). A study on differential inactivation of mating reactivity would be an important key toward the chemical basis of mating-type differences, provided that the chemistry of action of the agents is known and is unambiguous and a quantitative description of mating reactivity is possible.

B. Mating Reaction with Detached Cilia

The second step in the isolation of the mating reaction mechanism would be not only to preserve the reactivity in the dead paramecia but also to isolate the reactive fraction of the animals. As stated previously, animals in the mating reaction adhere by the tips of their cilia. It is known that single isolated cilia from reactive formalin-killed paramecia can adhere to living animals of the complementary mating type (Metz, 1954). These observations suggested that an animal-free preparation of reactive cilia could be used for an analysis of the substances involved in the mating reaction.

The first successful preparation of detached mating reactive cilia was made with *Paramecium bursaria* by Cohen and Siegel (1963). They disrupted the mating reactive paramecia by repeated ejection through a medical syringe fitted with a 25-gauge needle and, subsequently, separated the trichocysts–cilia fraction from other cellular components in the brei by differential centrifugation. This fraction was mixed with mating-reactive intact paramecia of the complementary mating type and observed under the phase-contrast microscope. The observations revealed that the cilia, but not the trichocysts, in the preparation agglutinated to the cilia of the intact animals. Since this agglutinating ability of the free cilia was not decreased by washing, they concluded that no soluble cofactor for the agglutination was present in the brei. The agglutination of free cilia to intact paramecia occurs only when the animals providing the free cilia and also the intact tester animals are both mating-reactive and also complementary in mating types. The free cilia retained agglutinating ability for 24 hours at 4°C and for at least 2 weeks if quickly frozen and stored at −40°C. On the other hand, the free cilia lost reactivity upon heating or trypsin treatment but retained their microscopical structure. These results again suggest that the substances involved in the mating reaction are protein or intimately associated with protein.

Siegel and co-workers further confirmed Metz's hypothesis (Metz, 1954) that particular combinations of specific ciliary substances are responsible for mating-type specificity in syngen 1 of *P. bursaria*. *Paramecium bursaria* syngen 1 contains four complementary mating types (I, II, III, and IV). The differentiation of these types is controlled by two pairs of alleles at two different loci. Mating type I is determined by dominant alleles A and B, type III by the double recessive aa/bb, and types II and IV are determined, respectively, by aa/B- and A-/bb (Siegel and Larison, 1960). According to Metz's hypothesis, two pairs of ciliary substances are involved in the mating reaction among these types. The α reaction refers to the interaction between substances controlled by the A and a alleles; and the β reaction to interaction between substances con-

TABLE I

The Basis for the Specificity of Multiple Mating Types
in *Paramecium bursaria*, Syngen 1[a]

Mating type[b] (substances present)[c]	Mating type (substances present)[c]			
	I (AB)	II (Ab)	III (aB)	IV (ab)
I(AB)	—	β	α	α,β
II(Ab)	—	—	α,β	α
III(aB)	—	—	—	β
IV(ab)	—	—	—	—

[a] After Metz, 1954.

[b] Siegel's system for designating mating types in *Paramecium bursaria* is used here.

[c] Substance A reacts with substance a (reaction α); substance B reacts with substance b (reaction β).

trolled by the B and b alleles (Table I). Experiments on heat inactivation of the agglutinability of detached cilia revealed that the substances involved in the α reactions are more thermolabile than those in the β reactions (Cohen and Siegel, 1963). Thus two different kinds of reactions actually occur in syngen 1 of *P. bursaria* as suggested by Metz. However, the study did not make clear the difference between the two complementary substances in each reaction.

A comparable but technically different isolation of mating-reactive cilia was carried out with *Paramecium multimicronucleatum* by Miyake (1964). He treated mating reactive animals with 20 to 30 mM $K_2Cr_2O_7$ solution for a few hours at 25°C. This treatment detaches the cilia and discharges the trichocysts but does not disintegrate the bodies of the paramecia. The bodies and the trichocysts were removed, respectively, by mild centrifugation and filtration through filter paper. The animal body and discharged trichocyst-free preparations of cilia were washed or dialyzed against 40 mM KCl solution of pH 6.8 and stored in the same solution. Such isolated cilia not only showed strong and specific agglutination to living intact paramecia of the complementary mating type but they also induced selfing conjugations in the latter. The adherence of intact animals resulting in the formation of selfing conjugation is probably mediated by the detached cilia. Here again soluble factors responsible for the activity seem excluded by the fact that the activity of the free cilia does not decrease following repeated washing and the supernatant from the cilia preparation has no activity. The activity of the cilia preparation was retained for more than 10 days at 2° to 3°C and for a much longer time in glycerol at deep-freeze temperature (about —20°C), but was completely lost in 2 or 3 days at 25°C. It should be mentioned that according to

Miyake the activity of the cilia from one mating type (IV) is less stable than that from the complementary mating type (III) under the conditions he employed. Whether this is due to the differential action of $K_2Cr_2O_7$ on different mating types or to other conditions is still unknown. These studies, still in progress, should provide an important technical advance for extraction and characterization of active substances involved in the mating reaction of *Paramecium*.

C. NATURE OF THE MATING-TYPE SUBSTANCES

Although every effort to extract active substances involved in the mating reaction has so far been unsuccessful (Metz, 1948, 1954), it can be concluded with considerable assurance that the initial specific adhesion in the conjugation of *Paramecium*, the mating reaction, is dependent upon the interaction of substances at the surfaces, probably of the cilia, of the paramecia. The substances are called the "mating-type" substances (Metz, 1946, 1948). Such substances would be expected specifically to initiate conjugational effects upon the complementary mating-type paramecia or at least specifically inhibit the mating reaction of such animal. So far as the previously mentioned studies are concerned, the mating-type substances seem to be proteins or intimately associated with proteins and to be bound to or built into the cilia on the ventrolateral surface of the paramecium.

Metz (1954) tried to identify the active group(s) of the mating-type substances by applying various reagents, especially certain "protein group reagents," to reactive dead cells. Treatment with mercuric ion and iodoacetate failed to destroy the reactivity, but nitrous acid, dinitrofluorobenzene, diazonium compounds, iodine, benzoyl chloride, and formalin inactivated the reactivity when properly employed. From this he concluded that sulfhydryl and disulfide groups are not essential but amino and phenolic groups appear to be essential for the mating-type substance activity of *Paramecium calkinsi*. On the other hand, Kasuga (1964) reported that p-chloromercuribenzoate (PCMB) inactivated the mating reactivity of *Paramecium caudatum* in less than 40 minutes, but 6 hours were needed to block completely the SH groups detected cytochemically by 1-(4-chloromercuriphenylazo)-2-naphthol (RSR). He concluded that some SH group might have an essential role in the mating reaction. Since whole animals were employed in these studies, the possibility that the actions of the reagents on the mating-type substances might be indirect cannot be excluded. However, these results are a step toward a further analysis of the problem, i.e., in the case of the reactive components of isolated cilia or attempts to extract the mating-type substances.

In this situation where the mating-type substances as yet cannot be extracted, immunological methods might be useful to identify the substances because it would not necessarily require the extraction of the mating-type substances in pure form. Many antigenic types (serotypes) are found in paramecia (Sonneborn, 1948, 1950; Beale, 1954, 1957; Finger, 1957; Pringle, 1956; Koizumi, 1966; Hiwatashi, 1963), and the specific serotype antigens are known to be in the cilia (Preer and Preer, 1959). These antigens are also called "immobilization" antigens because the antigen–antibody reaction is easily assayed by the immobilization of the ciliary movement. Because of the probable location of the mating-type substances in the cilia, studies on the relation between the immobilization antigens and the mating-type substances should be rewarding. In all known cases, however, the serotypes are independent of mating types (Sonneborn, personal communication). Metz and Fusco (1948) tried to find a mating-type-specific action of antiserum. When mating-reactive formalin-killed *Paramecium aurelia* were treated with antiserum and then tested for mating reactivity, they failed to react with living animals of the complementary mating type. However, the inhibition by the antiserum treatment was not mating-type-specific. Furthermore, antiserum prepared against unreactive paramecia gave the same inhibiting effect, whereas antiserum against one race and mating type sometimes failed to give the effect to another race of the same mating type. From this they concluded that the inhibiting effect of the antiserum is not dependent upon the combination of the antibody with mating-type substances but prevents the mating reaction indirectly by combination with some "neighboring" antigen. Like Metz, Cohen (personal communication) and Hiwatashi (unpublished observations) also failed to find a mating-type-specific antibody in *Paramecium bursaria* and *Paramecium caudatum*, respectively. Why the mating-type-specific antibody cannot be obtained is still unknown. Possibly the antigen preparations used for the injection did not contain mating-type substances in amount sufficient to produce the specific antibody or the mating-type specificity is tightly bound to cellular structures and the immunizing procedure destroys them resulting in the loss of the specificity. Finally, the mating substances may not be antigenic in the rabbit. In any event, more extensive immunological work is desirable, especially because Metz and Fusco's work was done before the discovery of serotype systems in *Paramecium*.

In discussing the problems of the mating-type substances, it seems worth while to examine and compare the mating-type systems in various species of *Paramecium*. Two distinct groups of mating-type systems are known in *Paramecium*. In the first group, the species contain many sexually isolated syngens each of which consists exclusively of two mating types. In the second group, the species contain not only syngens of two mating types but also syngens of four, or even eight, interbreeding mating types. Those belonging to the first group are *P. aurelia* (Sonneborn, 1937, 1947, 1957; Rafalko and Sonneborn, 1959), *P. caudatum* (Gilman, 1941, 1950; Y. T. Chen, 1944; Hiwatashi, 1949a; Vivier, 1960; Ossipov, 1963), *P. calkinsi* (Sonneborn, 1939b; Wichterman, 1950), *Paramecium woodruffi* (Sonneborn, 1939b), *Paramecium jenningsi* (Sonneborn, 1958), and probably *P. multimicronucleatum* (Giese, 1941; Sonneborn, 1957, 1958). The two species in the second group are *P. bursaria* (Jennings, 1939a,b; Jennings and Opitz, 1944; Chen, 1946c; Bomford, 1966) and *Paramecium trichium* (Sonneborn, 1939b).

The mating reaction ordinarily occurs between different mating types within the same syngen. In some species of *Paramecium*, however, intersyngenic mating reactions are known to occur (Sonneborn, 1938; Sonneborn and Dippell, 1946; Gilman, 1949; Jennings and Opitz, 1944; Bomford, 1966). Intersyngenic mating reactions are usually weaker than intrasyngenic reactions but in some cases do lead to complete conjugation. The result of such intersyngenic mating is often lethal. In a group of syngens which show intersyngenic matings, mating types can be grouped into two homologous series of general types, plus and minus, according to the intersyngenic mating reaction. This is the case in the group of syngens 1, 3, 4, 5, 7, and 8 of *P. aurelia* (Sonneborn and Dippell, 1946; Sonneborn, 1957). A similar situation probably exists in *P. caudatum* where intersyngenic matings are also known (Gilman, 1949). In addition to the intersyngenic mating reactions other phenomena which probably indicate the general type ($+$ or $-$) are known in *P. aurelia* and *P. caudatum*. In *P. aurelia* the occurrence of even-numbered mating types after conjugation or autogamy increases with exposure to temperature during a sensitive postzygotic stage (Sonneborn, 1947). Also in *P. aurelia* only odd-numbered mating types can be genetically pure (Sonneborn, 1947; Butzel, 1955; Taub, 1963). In *P. caudatum* there is a mating type difference in stability to formalin as outlined in Section III,A. Another difference is the ability of paramecia to adhere to a cation-exchange resin. When living *P. caudatum* are mixed with Amberlite IR-120 (polystyrolsulfonic acid), they adhere to the resin usually by their cilia (K. Hiwatashi, unpublished observations). If the resin is pretreated with cation, adherence of paramecia to the resin is reduced in proportion to the amount of cation adsorbed to the resin. When Ca^{++} was adsorbed, the minimum amount of the adsorbed Ca^{++} necessary to prevent the adherence of the paramecia cells was the same in every stock of *P. caudatum* examined. However, when K^+ was adsorbed, the minimum amount to prevent the adherence was different for different mating types. The amount was always greater for one mating type than for the complementary type (Table II). Thus, at least

TABLE II

STICKING OF *Paramecium caudatum* TO AMBERLITE IR-120 WITH VARYING AMOUNTS OF ADSORBED POTASSIUM[a]

Potassium adsorbed (% saturation)	Syngen 3									Syngen 12		Syngen 13				
	V[b]						VI[b]			XXIII[b]	XXIV[b]	XXV[b]			XXVI[b]	
	Nd3[c]	288[c]	162[c]	Fe2[c]	Gal[c]	Ma1[c]	Kn6[c]	K4[c]	Ys3[c]	K_A6[c]	$K_3 2$[c]	Mo2[c]	Ku1[c]	Fw1[c]	Is4[c]	Tl4[c]
0	++	++	++	++	++	++	++	++	++	++	++	++	++	++	++	++
82	++	++	++	++	++	++	++	++	++	++	++	++	++	++	++	++
97	++	+	++	++	++	++	—	+	—	++	+	++	++	++	+	+
99	++	+	++	—	—	++	—	—	—	++	—	++	++	++	—	—
100	—	—	—	—	—	—	—	—	—	—	—	—	—	—	—	—

[a] Sticking intensive (++); sticking poor (+); no sticking (—).
[b] Mating-type numbers.
[c] Stock designations.

in syngens 3, 12, and 13 of *P. caudatum,* two homologous groups, mating types V, XXIII, XXV and VI, XXIV, XXVI, are defined by the difference in the reaction to the resin. These homologous series of mating types correspond perfectly with those found by the stability of mating reactivity to formalin. Unfortunately, the relation of these results to Gilman's data on intersyngenic matings are still unknown. Thus the establishment of general mating types in this species is incomplete.

According to Metz's scheme (Metz, 1954) the adherence of paramecia in the mating reaction is due to the combination of complementary mating-type substances in an antigen–antibody-like reaction. This means that the interaction of the complementary substances is rather a type of steric fitting than an enzymically mediated reaction. This interpretation seems to be supported by the fact that the adhesion of animals in the mating reaction is instantaneous* and also temperature independent over a wide temperature range (Cohen and Siegel, 1963; K. Hiwatashi, unpublished observations). On the basis of the nature of the mating reaction as well as on the general type systems, Metz (1954) postulated a general hypothesis for mating-type specificity. According to his scheme the number of interbreeding mating types in a syngen is represented as 2^n, where n indicates the number of pairs of complementary substances involved in mating-type specificity. Thus, mating reactions within a two-type syngen such as *Paramecium aurelia* would result from the interaction of a pair of complementary substances, A and a, which combine in antigen–antibody-like fashion. According to the concept of general types, pairs of substances in different syngens should be represented as A^1 and a^1, A^3 and a^3, etc., where superscripts designate the syngen. The complementariness would be less perfect when the superscript are different, e.g., A^1 and a^3, and other combinations of intersyngenic matings. In the case of a four-type syngen, such as syngen 1 of *Paramecium bursaria,* two pairs of substances would be involved in mating-type specificity. Each mating type would have two substances, one from each of two different pairs of substances. If the two pairs of substances involved in the syngen are designated as A and a, and B and b, the four different mating types would have AB, Ab, aB, and ab, respectively. Mating reactions in this syngen would result from the interactions of A and a, B and b, or both (Table I). The interaction between A and a is referred to as the α reaction and that of B and b as the β reaction. The validity of this scheme is supported by the work of Cohen and Siegel (1963) as outlined in Section III,B. This scheme can

---

* In *Euplotes,* mating reactions do not occur instantaneously but only after a considerable period. Probably new biosynthesis of mating-type substances occurs between initial mixing and the beginning of mating behavior (Heckmann and Siegel, 1964).

be extended to an eight-type syngen such as syngen 2 of *P. bursaria*. Here there would be three independent pairs of substances involved in mating-type specificity and three kinds of reactions, α, β, and γ. Although a case of intersyngenic mating is reported in *P. bursaria* that suggests the presence of homologous types in some syngens (Jennings and Opitz, 1944; Bomford, 1966) available data are insufficient to establish a general type system among the syngens in this species and, hence, the exact relation between the pairs of substances among the known six syngens is still not clear. To establish general type systems in this species, methods other than intersyngenic matings, as in the cases of *P. aurelia* and *Paramecium caudatum*, might be worthy of investigation.

Because of the possible protein nature of mating-type substances, studies on the biosynthesis and its control mechanism should be of great interest. Mating reactivity appears only at a certain stage of the life cycle and also under appropriate environmental conditions. The life cycle of *Paramecium* begins with fertilization and proceeds successively through immaturity, maturity, and senility (Sonneborn, 1954b, 1957). Paramecia have no mating reactivity in immaturity, strong reactivity in maturity, and weak or no reactivity at senility. Even in the mature period of the life cycle, paramecia express no mating reactivity if they are supplied with excess food or subjected to extreme starvation. Thus, more or less depletion of food or so-called "relative hunger" is necessary for the expression of mating reactivity. In addition to nutrition, temperature and light conditions are important for mating. The temperature optima are often different for different syngens (Sonneborn, 1939a, 1957; Gilman, 1941). Diurnal periodicities of mating reactivity exist in some syngens of *P. aurelia* (Sonneborn, 1939a, 1957) and *P. bursaria* (Jennings, 1939b; Ehret, 1953). Such differential expression of mating reactivity with different stages and conditions might reflect any of the following phenomena: biosynthesis, activation and inactivation, or transportation of the mating-type substances from inside of the animals to the cilia. In *P. bursaria*, which shows diurnal periodicity of mating reactivity, Cohen (1965) demonstrated that the change of the reactivity actually reflects the biosynthetic process. Treatment of the intact animals during the mating-reactive period with puromycin or actinomycin D resulted in the loss of mating reactivity, whereas reactive detached cilia retain reactivity throughout the same treatment. From the well-known actions of puromycin and actinomycin D, Cohen concluded that the manifestation of mating reactivity in living paramecia requires protein synthesis. Furthermore, he showed that the daily loss of mating ability in the diurnal cycle is also an active process requiring metabolism since, when detached cilia and the intact animals were maintained side by side, the latter lost the mating ability much faster than the former. Thus, one possibility is that mating-type substances or their precursors are constantly being synthesized and degraded during the period of mating competence. However, another possibility is that some substance(s) activating or masking the mating-type substance is responsible for the phenomena.

Much less is known about the mechanism determining mating-type differences. In *Paramecium aurelia,* most stocks have the potentiality to yield both mating types but a few are pure for one mating type and never yield the complementary type. This difference is controlled by a pair of alleles (Sonneborn, 1937, 1939a; Butzel, 1955; Taub, 1963). Of particular interest is the fact that the pure-type stocks are confined to one and the same mating type. Thus, in syngen 1 of *P. aurelia* the pure-type stocks are always mating type I, in syngen 7 always mating type XIII, and in syngen 4 mating type VII (Sonneborn, personal communication). These facts led Butzel (1955) to formulate an hypothesis that the pure-type stocks lack a terminal reaction necessary for the development of their complementary mating type. In other words, one of the mating-type substances would be a precursor of the other (Taub, 1963). From this point of view, particular interest is directed to the cases where mating types can change reversibly with change of some environmental factors. In some stocks of *Paramecium caudatum* the mating type changes reversibly with change of fission rate which, in turn, is affected by nutrition (Hiwatashi, 1960a). In *Paramecium multimicronucleatum* the cells of syngen 2 change mating type with the daily cycle of light and darkness (Sonneborn and Barnett, 1958; Sonneborn and Sonneborn, 1958; Barnett, 1966). From the above-mentioned hypothesis, one explanation for these phenomena might be the switching on and off of the terminal reaction in the synthetic pathway of the mating-type substances by environmental factors. In any case, these stocks of paramecia may be excellent materials for the study of biosynthetic processes that produce mating-type differences.

IV. Chemical Induction of Conjugation

A. Induction of Conjugation by Inorganic Salts

Since mating-type specificity is the particular feature of sexual differentiation in *Paramecium* and the interaction of mating-type substances is the primary reaction in the whole sequence of changes in fertilization, artificial induction of conjugation without the presence of mating-type differentiation should provide valuable information on the mechanism of fertilization in *Paramecium.*

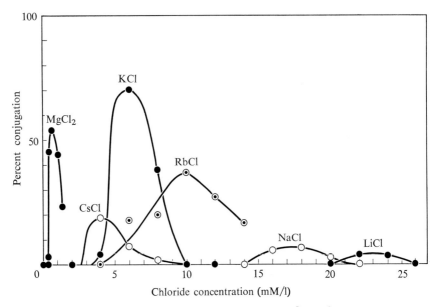

FIG. 7. Induction of conjugation in *Paramecium caudatum* by various inorganic chlorides. (From Miyake, 1958, with permission.)

There are two kinds of fertilization in *Paramecium*, conjugation and autogamy. Autogamy occurs regularly at a certain stage in the life cycle and under proper conditions in some species. It is a natural type of parthenogenesis in the sense that it occurs in single individuals and the presence of two different sexes is not necessary. It results in self-fertilization through union of the two gametic nuclei in single animals. Hertwig (1914) actually called the process parthenogenesis. Studies on the induction of autogamy by physical or chemical agents are limited. The only agent so far known to induce nuclear changes similar to autogamy in single animals is killer fluid which contains killing particles (P particle) secreted from killer stocks of *Paramecium* (Sonneborn, 1939a). Killer fluid from stock Ru22 of *Paramecium bursaria* induces nuclear changes, probably including meiosis, in paramecia of several syngens of the same species (Chen, 1945). Fluid from stock G (stock 7) killer of *Paramecium aurelia*, syngen 2, induces macronuclear breakdown in the cells of stock P (stock 16) of syngen 1 (Preer, 1948; Jacobson, 1948). However, the mechanism of the action of these killer fluids remains for future study.

On the other hand, excellent techniques for inducing conjugation by chemical agents have recently been developed by Miyake (1956, 1957, 1958, 1959, 1960, 1961). He treated mating-reactive animals of one mating type in *Paramecium caudatum* with 4 to 12 mM KCl in Ca-poor medium and induced conjugating pairs among animals of the same mating type. Various inorganic salts were examined for the conjugation-inducing action. The most effective proved to be K and Mg salts. Cesium and rubidium were less effective; Na, Li, and NH_4 were slightly effective; and Be, Sr, and Hg salts were ineffective (Fig. 7). Some organic reagents including acetamide, biuret, and urea, promote the conjugation-inducing effect of the inorganic salts, though these organic reagents alone have no inducing action. An important factor for the chemical induction of conjugation is a Ca-poor condition of the inducing as well as the culture medium. Likewise, the addition of a small amount of Ca inhibits the inducing action. Nevertheless, Ca itself has a weak inducing action in high concentration. In the chemical induction of conjugation, holdfast unions appear directly without prior occurrence of agglutinative mating reactions. Other processes of normal conjugation, including degeneration of cilia on the oral side, paroral cone formation, paroral union, meiosis, macronuclear breakdown, and exchange of gametic nuclei, are all observed in the in-

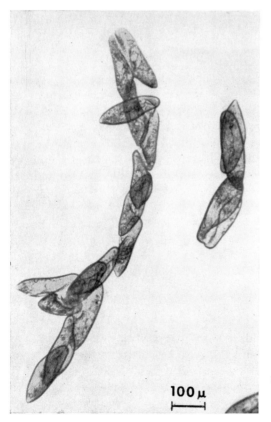

100 μ

Fig. 8. Aberrant unions of many *Paramecium caudatum* induced by K and acetamide. (Courtesy of A. Miyake.)

duced conjugation. In some cases, especially when a promoting agent such as acetamide or biuret is used, irregularities in the relative positions of the conjugants occur. In extreme cases, the aberrant unions of many animals, even more than twenty, are observed (Fig. 8).

Miyake (1960, 1961) extended his study to other species of *Paramecium* and succeeded in obtaining chemical induction of conjugation in *P. aurelia, Paramecium multimicronucleatum, Paramecium jenningsi,* and *Paramecium traunsteineri* but failed in *P. bursaria, Paramecium trichium, Paramecium polycaryum,* and *Paramecium calkinsi.* The method is slightly different for certain species. In addition to the K in Ca-poor medium, acriflavine or proflavine should be added to obtain the induction in *P. aurelia, P. multimicronucleatum,* and *P. jenningsi.* In the chemically induced conjugation of *P. aurelia,* syngen 4, the occurrence of micronuclear transfer and synkaryon formation was confirmed with serotypic alleles as genetic markers (Miyake, 1960). The induction failed in the mutant CM stock of *P. aurelia* which gives mating reactions but lacks the mechanism to form the holdfast union as well as other conjugation effects (Miyake, 1961; see also Section II,B). An additional confirmation is the successful induction of conjugation in stocks of *P. aurelia* pure for one mating type. Such stocks are incapable of differentiating the complementary mating type. This would seem to exclude the possibility of conjugation due to change of mating type by the action of the chemical agents. Another important aspect of this study is the formation of nonspecific pairs by the chemical agents. Miyake could induce not only intersyngenic conjugations between the cells of different syngens in *P. aurelia* and *P. caudatum* but also interspecific pairings between any two of *P. aurelia, P. multimicronucleatum,* and *P. caudatum* (Fig. 9). This is strong evidence that the substances involved in the holdfast union are nonspecific and that chemical induction of conjugation may occur directly through the formation of the holdfast substances.

Artificial induction of conjugation is also produced by salts of various heavy metals (Hiwatashi and Kasuga, 1960; Kasuga, 1962 and unpublished observations). Among the chlorides of various heavy metals so far examined for *P. caudatum,* the most effective for the induction were Mn and Co, less effective were Ni and Al (though this is not a heavy metal), whereas Hg, Cu, Cd, and Zn were ineffective. When strongly mating-reactive animals are treated with $MnCl_2$ or $CoCl_2$, aberrant unions of many paramecia result. Often more than ten paramecia are joined with irregularities in the relative positions of the conjugants. This is very much like the induction by K or Mg with the promoting agents, acetamide or biuret, as previously mentioned. Meiosis and macronuclear breakdown

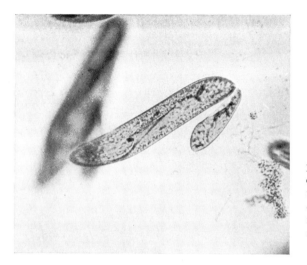

Fig. 9. Interspecific pairing between *Paramecium multimicronucleatum* (large) and *Paramecium aurelia* (small) induced by K and acriflavin. (Courtesy of A. Miyake.)

are observed in the conjugants induced by the heavy metals. Calcium inhibits the induction. It is of interest that the salts of metals that are not effective for the induction, namely Hg, Cu, Cd, and Zn, are known to be strongly bound to SH groups forming mercaptides with low dissociation constants (Cecil, 1963).

The effect of various inorganic salts on the conjugation of *Paramecium caudatum* was studied by Zweibaum (1912) long before the discovery of mating types. He showed that various inorganic salts including $CaCl_2$ favor the occurrence of conjugation. In the stock of *P. caudatum* he used, however, natural selfing pairs, probably due to temporary self-differentiation of mating types, occurred. When nonselfing stocks were used the effect of the inorganic salts was never observed (Hopkins, 1921). Thus, Zweibaum's results should be interpreted as an increase in the occurrence of natural selfing pairs produced by the inorganic salts—a process entirely different from the chemical induction of conjugation.

Concerning the mechanism of the conjugation-inducing action of the inorganic salts, Miyake presented an hypothesis that the agents might make the surfaces of paramecia sticky by liquefying the cortical protoplasma and this results in the formation of holdfast unions (Miyake, 1958). This view seems to be supported by the fact that K, Mg, or Na decreases the rigidity of the cortical gel in *Amoeba* and Ca inhibits this effect (Heilbrunn and Daugherty, 1934). However, this interpretation requires an explanation of why only mating-reactive paramecia respond to the inducing agents and also why only the ventral sides of the animals take part in the union. Perhaps, as Miyake suggested, quantitative study of the cationic exchange between paramecia and the inducing medium might give an important clue toward elucidation of the mechanism.

B. INDUCTION OF CONJUGATION BY OTHER AGENTS

In addition to the inorganic salts mentioned above, some organic compounds are known to induce conjugation. Miyake found that heparin and acriflavine have conjugation-inducing action (Miyake, 1958). The action of heparin is as efficient as that of K or Mg but seems not to be promoted by the addition of acetamide. The action of acriflavine is less effective than heparin. Heparin works cooperatively with K and acriflavine with Mg. Both of their actions are inhibited by Ca in the same way as the action of the inorganic salts. Heparin is a well-known anticoagulant for blood. Various biological activities of heparin other than anticoagulating action are also reported (see Stacey and Barker, 1962). However, no evidence has ever been obtained to suggest a possible mechanism for induction of conjugation by heparin. Acridine dyes are known to interact with nucleoproteins (McIlwain, 1941) and probably sandwich between purine bases of the deoxyribonucleic acid (DNA) double helix (Lerman, 1961). The dyes interfere with the self-reproduction of cellular particles containing nucleic acid (Lwoff, 1949; Ephrussi, 1953; Hirota, 1960). Elimination of the bacterial sex factor by acridines as well as salts of heavy metals such as Co and Ni (Hirota, 1956) is of particular interest because all of these agents can induce conjugation. However, nothing is known concerning the mechanism of the conjugation-inducing action of acriflavine.

Since the Ca-poor condition is necessary for the induction of conjugation by chemical agents, the action of agents that are known to bind Ca would be of great interest. Hiwatashi (1959) treated mating-reactive *Paramecium caudatum* with a strong Ca-chelating agent, ethylenediaminetetraacetic acid-2Na (EDTA) and succeeded in the induction of conjugation. Here again, only mating-reactive animals respond to the action of EDTA. An important fact in this study is that the Ca-poor condition is not necessary for the induction of conjugation. When the Ca is added to the medium, the induction of conjugation is inhibited but this is easily reversed by adding more EDTA into the medium. Thus the conjugation-inducing effect of EDTA and its inhibition by Ca are competitive over a wide range of concentrations. No such competitive effect was observed between EDTA and Mg, though EDTA is known to chelate Mg as well as Ca. On the other hand, the cooperative effect of EDTA and K was remarkable and multiple unions of conjugation, such as those observed in the inductions by K and acetamide or by Mn, were often observed. A less extensive cooperation action appears to exist between EDTA and Mg.

The question arises whether EDTA acts directly to induce conjugation. It might be that the EDTA changes the cationic balance, especially between Ca and other cations in the medium and that this imbalance of salts in the medium induces conjugation. However, this possibility is eliminated by the fact that EDTA dissolved in pure deionized water can induce conjugation in paramecia that have been washed repeatedly with deionized water. These results strongly suggest that removal of Ca from paramecia induces the formation of holdfast unions. From this point of view, the induction of conjugation by the inorganic salts might be interpreted as caused by depriving the cell of Ca by cationic exchange. The possibility of the exchange of Ca within the cell with K in the medium across the cell membrane of *Paramecium* was reported (Kamada, 1940). However, much additional evidence is needed to validate such an interpretation.

In addition to the artificial agents thus far described, a natural product of a stock of *Paramecium* is known to induce conjugation. The killer fluid of stock Rs22 of *Paramecium bursaria* not only induces the nuclear changes in single animals as stated earlier but also induces clotting and pair formation in various stocks of the same species, though the clotting induced by the killer fluid differs somewhat from the agglutination which occurs when paramecia of different mating types are mixed (Chen, 1945). Similar action of the killer fluid is known in stocks 7 and 5 of *Paramecium aurelia* (Sonneborn; see Chen, 1945). Unfortunately, however, the nature of the active substance(s) in this fluid has not been studied.

C. IMPLICATION OF THE STUDY

The foregoing results show that the treatment of paramecia with conjugation-inducing agents leads to the same results as the interaction of complementary mating types, namely formation of holdfast union, degeneration of cilia on the oral side, paroral cone formation and paroral union, meiosis, macronuclear breakdown, and exchange of gametic nuclei. Both chemical agents and animals of complementary mating type failed to induce these effects in the mutant CM stock of *Paramecium aurelia*. Furthermore, the time from the beginning of the treatment with the chemical agents to the appearance of holdfast unions is almost the same as that from the onset of the mating reaction to the formation of holdfast unions in normal conjugation. Though the mating reaction normally results in conjugating pairs consisting of different mating types, the reaction can also induce selfing pairs consisting of animals of the same mating type as mentioned earlier. Thus, the only difference between the features of the chemically induced conjugation and the normal conjugation induced by the interaction of mating types seems to be the lack of a typical mating reaction in the former. The first question of concern is whether the action of the chemical agents involves mechanisms in common with the interaction of mating-type substances. Miyake (1966) reported that degeneration of cilia at the anterior region including the holdfast region was observed before the formation of holdfast unions in both inductions of conjugation by the mating reaction and by conjugation-inducing chemicals.

The fact that only animals in the mating-reactive condition can respond to the conjugation-inducing action of the chemical agents suggests two alternative interpretations concerning the mechanism involved in the chemical induction of conjugation: (1) the chemical agents activate formation of the holdfast substance(s) through an action on the mating-type substances or (2) the mechanism necessary to form the holdfast substances appears only in the mating-reactive condition of the animals and the chemical agents act upon this mechanism through a different route and site than the mating-type substances. The facts that mating reactivity is not seriously affected by treatment with the inducing agents and that the addition of Ca, which inhibits the chemical induction of conjugation, does not interfere with the normal mating reaction, seem to favor the latter interpretation. Since there is a time lag between the mating reaction or the treatment with the conjugation-inducing agents and the formation of the holdfast union, the holdfast substance(s) may be synthesized by enzymically mediated reactions which are activated either by the mating reaction or the conjugation-inducing agents. Then, according to the hypothesis suggested in the previous section (i.e., the chemical induction of conjugation is due to depriving the cell of Ca), Ca might well be masking or repressing the reaction necessary to synthesize the holdfast substance(s). This hypothesis, however, cannot easily explain the induction of conjugation by heparin and acriflavine. Furthermore, there not only is lack of evidences for applying this hypothesis to natural conjugation but also some difficulty arises from the fact that the addition of Ca does not interfere with the formation of holdfast unions induced by the mating reaction. More diverse physical and chemical agents should be examined for the induction of conjugation and its inhibition before the proposal of any general concept to relate the artificial induction of conjugation to natural conjugation. Finally, the technique of inducing conjugation by chemical agents is very helpful for genetic studies of *Paramecium*, because it makes conjugation possible among animals in which natural conjugation is impossible.

V. Conclusion

Fertilization of *Paramecium* is a process that occurs in a very precise and orderly manner comparable to fertilizations in other organisms. The primary reaction in this orderly sequence of fertilization events is the mating reaction. It brings about the initial adhesion of paramecia, controls fertilization specificity, and initiates activation. The mating reaction depends upon the interaction of complementary surface substances, the mating-type substances. Partial isolation of the system involved in the mating reaction was achieved first by killing paramecia and then by isolating the cilia without destroying their ability to react with and activate living paramecia. The available evidence so far obtained shows that the mating-type substances are protein or intimately associated with the protein of the cilia. However, attempts at isolation and purification of the mating-type substances have not yet been successful.

The mating reaction activates the formation of holdfast substance(s) independent of the cilia and by which the animals unite more firmly. The holdfast substances(s) seem to be a nonspecific type of cementing substance(s) though the chemical nature of this substance(s) is largely unknown. The union of cells in conjugation is completed by the formation of the paroral union. The substance(s) involved in this union may not be much different in nature from the holdfast substance(s). The trigger mechanism inducing meiosis and macronuclear breakdown appears to reside in the mating reaction, in some species, and in other reactions brought about by the holdfast union, in other species.

Recently developed techniques have made it possible to induce conjugation without the presence of sexually different cells. This method induces holdfast unions directly without the mating reaction. Whether this type of study will elucidate some essential mechanism in the fertilization of *Paramecium* or only demonstrate that the animal contains within itself every essential mechanism for activation, as artificial parthenogenesis has done for metazoan fertilization, however, depends upon further investigations.

In *Paramecium*, mating types or potentiality to differentiate mating types are known to be controlled by one or more pairs of alleles (Sonneborn, 1937; Siegel and Larison, 1960; Taub, 1963; Hiwatashi, 1964). Mating-type specificity in *Paramecium* depends upon the specificity of mating-type substances. Some evidences have already been found to show that synthesis of the mating-type substances involves DNA-dependent protein synthesis. Many aspects of fertilization in *Paramecium* favor the view that *Paramecium* is one of the best materials for the study of the genetic control of the specific substances involved in fertilization.

ACKNOWLEDGMENTS

The author is indebted to Drs. T. M. Sonneborn, R. W. Siegel, C. B. Metz, and A. Miyake for kindly going through the manuscript and making valuable comments. The author is also indebted to Drs. F. Inaba, L. Schneider, and A. Miyake for their contribution of electron- and photomicrographs to illustrate this paper.

REFERENCES

Barnett, A. (1966). A circadian rhythm of mating type reversals in *Paramecium multimicronucleatum*, syngen 2, and its genetic control. *J. Cellular Comp. Physiol.* **67**, 239.

Beale, G. H. (1954). "The Genetics of *Paramecium aurelia*." Cambridge Univ. Press, London and New York.

Beale, G. H. (1957). The antigen system of *Paramecium aurelia*. *Intern. Rev. Cytol.* **6**, 1.

Boell, E. J., and Woodruff, L. L. (1941). Respiratory metabolism of mating types in *Paramecium calkinsi*. *J. Exptl. Zool.* **87**, 385.

Bomford, R. (1966). The syngens of *Paramecium bursaria:* New mating types and intersyngenic mating reaction. *J. Protozool.* **13**, 497.

Butzel, H. M., Jr. (1955). Mating type mutations in variety 1 of *Paramecium aurelia*, and their bearing upon the problem of mating type determination. *Genetics* **40**, 321.

Cecil, R. (1963). Intramolecular bonds in proteins: The role of sulfur in proteins. *In* "The Proteins" (H. Neurath, ed.), Vol. I, pp. 380–477. Academic Press, New York.

Chen, T. T. (1945). Induction of conjugation in *Paramecium bursaria* among animals of one mating type by fluid from another mating type. *Proc. Natl. Acad. Sci. U. S.* **31**, 404.

Chen, T. T. (1946a). Conjugation in *Paramecium bursaria*. I. Conjugation of three animals. *J. Morphol.* **78**, 353.

Chen, T. T. (1946b). Temporary pair formation in *Paramecium bursaria*. *Biol. Bull.* **91**, 112.

Chen, T. T. (1946c). Varieties and mating types in *Paramecium bursaria*. I. New variety and types, from England, Ireland, and Czechoslovakia. *Proc. Natl. Acad. Sci. U. S.* **32**, 173.

Chen, Y. T. (1944). Mating types in *Paramecium caudatum*. *Am. Naturalist* **78**, 334.

Cohen, L. W. (1964). Diurnal intracellular differentiation in *Paramecium bursaria*. *Exptl. Cell Res.* **36**, 398.

Cohen, L. W. (1965). The basis for the circadian rhythm of mating in *Paramecium bursaria*. *Exptl. Cell Res.* **37**, 360.

Cohen, L. W., and Siegel, R. W. (1963). The mating-type substances of *Paramecium bursaria*. *Genet. Res.* **4**, 143.

Diller, W. F. (1936). Nuclear reorganization process in *Paramecium aurelia*, with description of autogamy and "hemixis." *J. Morphol.* **59**, 11.

Ehret, C. F. (1953). An analysis of the role of electromagnetic radiations in the mating reaction of *Paramecium bursaria*. *Physiol. Zool.* **26**, 274.

Elliot, A. M., and Kennedy, J. R., Jr. (1962). The morphology and breeding system of variety 9, *Tetrahymena pyriformis*. *Trans. Am. Microscop. Soc.* **81**, 300.

Ephrussi, B. (1953). "Nucleo-cytoplasmic Relation in Microorganisms." Oxford Univ. Press, London and New York.

Finger, I. (1957). Immunological studies of the immobilization antigens of *Paramecium aurelia*, variety 2. *J. Gen. Microbiol.* **16**, 350.

Giese, A. C. (1941). Mating types in *Paramecium multimicronucleatum*. *Anat. Record* **81**, Suppl., 131.

Gilman, L. C. (1941). Mating types in diverse races of *Paramecium caudatum*. *Biol. Bull.* **80**, 384.

Gilman, L. C. (1949). Intervarietal mating reactions in *Paramecium caudatum*. *Biol. Bull.* **97**, 239 (abstract).

Gilman, L. C. (1950). The position of Japanese varieties of *Paramecium caudatum* with respect to American varieties. *Biol. Bull.* **99**, 348 (abstract).

Heckmann, K., and Siegel, R. W. (1964). Evidence for the induction of mating-type substances by cell to cell contacts. *Exptl. Cell Res.* **36**, 688.

Heilbrunn, L. V., and Daugherty, K. (1934). A further study of the action of potassium on *Amoeba* protoplasm. *J. Cellular Comp. Physiol.* **5**, 207.

Hertwig, R. (1914). Über Parthenogenesis der Infusorien und die Depressionzustände der Protozoen. *Biol. Zentr.* **34**, 557.

Hirota, Y. (1956). Artificial elimination of the F factor in *Bact. coli* K12. *Nature* **178**, 92.

Hirota, Y. (1960). The effect of acridine dyes on mating type factors in *E. coli. Proc. Natl. Acad. Sci. U. S.* **46**, 57.

Hiwatashi, K. (1949a). Studies on the conjugation of *Paramecium caudatum*. I. Mating types and groups in the races obtained in Japan. *Sci. Rept. Tohoku Univ., Fourth Ser.* **18**, 137.

Hiwatashi, K. (1949b). Studies on the conjugation of *Paramecium caudatum*. II. Induction of pseudoselfing pairs by formalin killed animals. *Sci. Rept. Tohoku Univ., Fourth Ser.* **18**, 141.

Hiwatashi, K. (1950). Studies on the conjugation of *Paramecium caudatum*. III. Some properties of the mating type substances. *Sci. Rept. Tohoku Univ., Fourth Ser.* **18**, 270.

Hiwatashi, K. (1951). Studies on the conjugation of *Paramecium caudatum*. IV. Conjugating behavior of individuals of two mating types marked by a vital staining method. *Sci. Rept. Tohoku Univ., Fourth Ser.* **19**, 95.

Hiwatashi, K. (1955a). Studies on the conjugation of *Paramecium caudatum*. V. The time of the initiation of nuclear activation. *Sci. Rept. Tohoku Univ., Fourth Ser.* **21**, 199.

Hiwatashi, K. (1955b). Studies on the conjugation of *Paramecium caudatum*. VI. On the nature of the union of conjugation. *Sci. Rept. Tohoku Univ., Fourth Ser.* **21**, 207.

Hiwatashi, K. (1959). Induction of conjugation by ethylenediamine tetraacetic acid (EDTA) in *Paramecium caudatum*. *Sci. Rept. Tohoku Univ., Fourth Ser.* **25**, 81.

Hiwatashi, K. (1960a). Analyses of the change of mating type during vegetative reproduction in *Paramecium caudatum*. *Japan. J. Genet.* **35**, 213.

Hiwatashi, K. (1960b). An aberrant selfing strain of *Paramecium caudatum* which shows multiple unions of conjugation. *J. Protozool.* **7**, Suppl., 20 (abstract).

Hiwatashi, K. (1961). Locality of mating reactivity on the surface of *Paramecium caudatum*. *Sci. Rept. Tohoku Univ., Fourth Ser.* **27**, 93.

Hiwatashi, K. (1963). Serotype inheritance in *Paramecium caudatum*. *Genetics* **48**, 892 (abstract).

Hiwatashi, K. (1964). Mating type inheritance in *Paramecium caudatum* syngen 3. *Genetics* **50**, 255 (abstract).

Hiwatashi, K., and Kasuga, T. (1960). Artificial induction of conjugation by manganese ion in *Paramecium caudatum*. *J. Protozool.* **7**, Suppl., 20 (abstract).

Hopkins, H. S. (1921). The conditions for conjugation in diverse races of *Paramecium*. *J. Exptl. Zool.* **34**, 338.

Inaba, F., Imamoto, K., and Suganuma, Y. (1966). Electron-microscopic observations on nuclear exchange during conjugation in *Paramecium multimicronucleatum*. *Proc. Japan Acad.* **42**, 394.

Jacobson, W. E. (1948). Non-mating-type-specific "activation" in *Paramecium aurelia*: induction of macronuclear breakdown in sensitive stock P of variety 1 by killer stock G of variety 2. *Anat. Record* **101**, 708 (abstract).

Jennings, H. S. (1938). Sex reaction types and their interrelation in *Paramecium bursaria*. I and II. *Proc. Natl. Acad. Sci. U. S.* **24**, 112.

Jennings, H. S. (1939a). Genetics of *Paramecium bursaria*. I. Mating types and groups, their interrelations and distribution: mating behavior and self-sterility. *Genetics* **24**, 202.

Jennings, H. S. (1939b). *Paramecium bursaria*: mating types and groups, mating behavior, self-sterility; their development and inheritance. *Am. Naturalist* **73**, 414.

Jennings, H. S., and Opitz, P. (1944). Genetics of *Paramecium bursaria*. IV. A fourth variety from Russia. Lethal crosses with American variety. *Genetics* **29**, 576.

Kamada, T. (1940). Ciliary reversal of *Paramecium*. *Proc. Imp. Acad. Tokyo* **16**, 241.

Kasuga, T. (1962). Nuclear changes in *Paramecium caudatum* treated with $MnCl_2$. *Japan. Zool. Mag.* **71**, 401 (abstract in Japanese).

Kasuga, T. (1964). SH-groups of *Paramecium caudatum*. *Japan. Zool. Mag.* **73**, 375 (abstract in Japanese).

Koizumi, S. (1966). Serotypes and immobilization antigens in *Paramecium caudatum*. *J. Protozool.* **13**, 73.

Larison, L. L., and Siegel, R. W. (1961). Illegitimate mating in *Paramecium bursaria* and the basis for cell union. *J. Gen. Microbiol.* **26**, 499.

Lerman, L. S. (1961). Structural considerations in the interaction of DNA and acridines. *J. Mol. Biol.* **3**, 18.

Lwoff, A. (1949). Les organites doués de continuité génétique chez les protistes. *In* "Unités biologiques douées de continuité génétique," pp. 7–23. Centre Natl. Rech. Sci., Paris.

McIlwain, H. (1941). A nutritional investigation of the antibacterial action of acriflavin. *Biochem. J.* **35**, 1311.

Metz, C. B. (1946). Effect of various agents on the mating type substance of *Paramecium aurelia* variety 4. *Anat. Record* **94**, 347 (abstract).

Metz, C. B. (1947). Induction of "pseudoselfing" and meiosis in *Paramecium aurelia* by formalin killed animals of opposite mating type. *J. Exptl. Zool.* **105**, 115.

Metz, C. B. (1948). The nature and mode of action of the mating type substances. *Am. Naturalist* **82**, 85.

Metz, C. B. (1954). Mating substances and the physiology of fertilization in ciliates. *In* "Sex in Microorganisms" (D. H. Wenrich, ed.), pp. 284–334. Am. Assoc. Advan. Sci., Washington, D. C.

Metz, C. B., and Butterfield, W. (1950). Extraction of a mating reaction inhibiting agent from *Paramecium calkinsi*. *Proc. Natl. Acad. Sci. U. S.* **36**, 268.

Metz, C. B., and Butterfield, W. (1951). Action of various enzymes on the mating type substances of *Paramecium calkinsi*. *Biol. Bull.* **101**, 99.

Metz, C. B., and Foley, M. T. (1949). Fertilization studies on *Paramecium aurelia*: An experimental analysis of a non-conjugating stock. *J. Exptl. Zool.* **112**, 505.

Metz, C. B., and Fusco, E. M. (1948). Inhibition of the mating reaction in *Paramecium aurelia* with antiserum. *Anat. Record* **101**, 654 (abstract).

Metz, C. B., and Fusco, E. M. (1949). Mating reactions between living and lyophilized paramecia of opposite mating type. *Biol. Bull.* **97**, 245 (abstract).

Miyake, A. (1956). Physiological analysis of the life cycle of the Protozoa. III. Artificial induction of selfing conjugation by chemical agents in *Paramecium caudatum*. *Physiol. Ecol.* (*Kyoto*) **7**, 14 (in Japanese; English summary).

Miyake, A. (1957). Aberrant conjugation induced by chemical agents in amicronucleate *Paramecium caudatum. J. Inst. Polytech. Osaka City Univ., Ser. D* **8**, 1.

Miyake, A. (1958). Induction of conjugation by chemical agents in *Paramecium caudatum. J. Inst. Polytech. Osaka City Univ., Ser. D* **9**, 251.

Miyake, A. (1959). Chemically induced mating without mating type difference in *Paramecium caudatum. Science* **130**, 1432 (abstract).

Miyake, A. (1960). Artificial induction of conjugation by chemical agents in *Paramecium aurelia, Paramecium multimicronucleatum, Paramecium caudatum* and between them. *J. Protozool.* **7**, Suppl., 15 (abstract).

Miyake, A. (1961). Artificial induction of conjugation by chemical agents in *Paramecium* of the "aurelia group" and some of its applications to genetic work. *Am. Zoologist* **1**, 373.

Miyake, A. (1964). Induction of conjugation by cell-free preparations in *Paramecium multimicronucleatum. Science* **146**, 1583.

Miyake, A. (1966). Local disappearance of cilia before the formation of holdfast union in conjugation of *Paramecium multimicronucleatum. J. Protozool.* **13**, Suppl., 28. (abstract).

Ossipov, D. V. (1963). Mating types of *Paramecium caudatum* clones from the basins in some regions of the Soviet Union. *Rept. Leningrad Univ.* **21**, Biol. 4, 106 (in Russian; English summary).

Preer, J. R., Jr. (1948). A study of some properties of the cytoplasmic factor, "kappa", in *Paramecium aurelia*, variety 2. *Genetics* **33**, 349.

Preer, J. R., Jr., and Preer, L. B. (1959). Gel diffusion studies on the antigens of isolated cellular components of *Paramecium. J. Protozool.* **6**, 88.

Pringle, C. R. (1956). Antigenic variation in *Paramecium aurelia*, variety 9. *Z. Induktive Abstammungs Vererbungslehre* **87**, 421.

Rafalko, M., and Sonneborn, T. M. (1959). A new syngen (13) of *Paramecium aurelia* consisting of stocks from Mexico, France and Madagascar. *J. Protozool.* **6**, Suppl., 30 (abstract).

Schneider, L. (1963). Elektronenmikroskopische Untersuchungen der Konjugation von *Paramecium*. 1. Die Auflösung und Neubildung der Zellmembran bei den Konjuganten. (Zugleich ein Beitrag zur Morphogenese cytoplasmatischer Membranen.) *Protoplasma* **56**, 109.

Siegel, R. W. (1954). Mate-killing in *Paramecium aurelia*, variety 8. *Physiol. Zool.* **27**, 89.

Siegel, R. W. (1965). Hereditary factors controlling development in *Paramecium. In* "Genetic Control of Differentiation," pp. 55–65. Brookhaven Natl. Lab. Upton, New York.

Siegel, R. W., and Larison, L. L. (1960). The genetic control of mating types in *Paramecium bursaria. Proc. Natl. Acad. Sci. U. S.* **46**, 344.

Sonneborn, T. M. (1937). Sex, sex inheritance and sex determination in *Paramecium aurelia. Proc. Natl. Acad. Sci. U. S.* **23**, 378.

Sonneborn, T. M. (1938). Mating types in *Paramecium aurelia*: diverse conditions for mating in different stocks; occurrence, number and interrelations of the types. *Proc. Am. Phil. Soc.* **79**, 411.

Sonneborn, T. M. (1939a). *Paramecium aurelia*: Mating types and groups; lethal interactions; determination and inheritance. *Am. Naturalist* **73**, 390.

Sonneborn, T. M. (1939b). Sexuality and related problems in *Paramecium. Collecting Net* **14**, 1.

Sonneborn, T. M. (1942). Evidence for two distinct mechanisms in the mating reaction of *Paramecium aurelia. Anat. Record* **84**, 92 (abstract).

Sonneborn, T. M. (1947). Recent advances in the genetics of *Paramecium* and *Euplotes*. *Advan. Genet.* **1**, 263.

Sonneborn, T. M. (1948). The determination of hereditary antigenic differences in genically identical *Paramecium* cells. *Proc. Natl. Acad. Sci. U. S.* **34**, 413.

Sonneborn, T. M. (1950). The cytoplasm in heredity. *Heredity* **4**, 11.

Sonneborn, T. M. (1954a). Patterns of nucleocytoplasmic integration in *Paramecium*. *Caryologia*, Suppl. **1**, 307.

Sonneborn, T. M. (1954b). The relation of autogamy to senescence and rejuvenescence in *Paramecium aurelia*. *J. Protozool.* **1**, 38.

Sonneborn, T. M. (1955). A third point of attachment between conjugants in *Paramecium aurelia* and its significance. *J. Protozool.* **2**, Suppl., 13 (abstract).

Sonneborn, T. M. (1957). Breeding systems, reproductive methods, and species problems in Protozoa. *In* "The Species Problem" (E. Mayr, ed.), pp. 155–324. Am. Assoc. Advan. Sci., Washington, D. C.

Sonneborn, T. M. (1958). Classification of syngens of the *Paramecium aurelia-multimicronucleatum* complex. *J. Protozool.* **5**, Suppl., 17 (abstract).

Sonneborn, T. M., and Barnett, A. (1958). The mating type system in syngen 2 of *Paramecium multimicronucleatum*. *J. Protozool.* **5**, Suppl., 18 (abstract).

Sonneborn, T. M., and Dippell, R. V. (1946). Mating reactions and conjugation between varieties of *Paramecium aurelia* in relation to conceptions of mating type and variety. *Physiol. Zool.* **19**, 1.

Sonneborn, T. M., and Sonneborn, D. R. (1958). Some effects of light on the rhythm of mating type changes in stock 232-6 of syngen 2 of *P. multimicronucleatum*. *Anat. Record* **131**, 601 (abstract).

Stacey, M., and Barker, S. A. (1962). "Carbohydrates of Living Tissues." Van Nostrand, Princeton, New Jersey.

Tartar, V., and Chen, T. T. (1941). Mating reactions of enucleate fragments in *Paramecium bursaria*. *Biol. Bull.* **80**, 130.

Taub, S. R. (1963). The genetic control of mating type differentiation in *Paramecium*. *Genetics* **48**, 815.

Vivier, E. (1960). Contribution a l'étude de la conjugaison chez *Paramecium caudatum*. *Ann. Sci. Nat., Zool. Biol. Animale* [12] **2**, 287.

Vivier, E., and André, J. (1961). Nonnées structurales et ultrastructurales nouvelles sur la conjugaison de *Paramecium caudatum*. *J. Protozool.* **8**, 416.

Wichterman, R. (1940). Cytogamy: a sexual process occurring in living joined pairs of *Paramecium caudatum* and its relation to other sexual phenomena. *J. Morphol.* **66**, 423.

Wichterman, R. (1946a). The behavior of cytoplasmic structures in living conjugants of *Paramecium bursaria*. *Anat. Record* **94**, 93 (abstract).

Wichterman, R. (1964b). Direct observation of the transfer of pronuclei in living conjugants of *Paramecium bursaria*. *Science* **104**, 505.

Wichterman, R. (1950). The occurrence of a new variety containing two opposite mating types of *Paramecium calkinsi* as found in sea water of high salinity content. *Biol. Bull.* **99**, 366 (abstract).

Zweibaum, J. (1912). Les conditions nécessaires et suffisantes pour la conjugaison du *Paramecium caudatum*. *Arch. Protistenk.* **26**, 275.

Addendum to References

Since this article was written several reports have appeared that are relevant to the problems of fertilization in *Paramecium* and related ciliates. These are listed below. The important evidences reported in these papers extend, but do not alter, the essentials of the chapter.

Beisson, J., and Capdeville, Y. (1966). Sur la nature possible des étapes de différenciation conduisant à l'autogamie chez *Paramecium aurelia. Compt. Rend. Acad. Sci.* **263**, 1258.

Cronkite, D. (1968). The mating substances of *Paramecium. Proc. 12th Intern. Congr. Genet.* **2**, 265.

Hiwatashi, K., and Takahashi, M. (1967). Inhibition of mating reaction by antisera without ciliary immobilization in *Paramecium. Sci. Rept. Tohoku Univ., Biol.* **33**, 281.

Miyake, A. (1968). Induction of conjugation by chemical agents in *Paramecium. J. Exptl. Zool.* **167**, 359.

Miyake, A. (1968). Mechanism of initiation of sexual reproduction in *Paramecium multimicronucleatum*, syngen 2. *Proc. 12th Intern. Congr. Genet.* **2**, 263.

Miyake, A. (1968). Chemical induction of nuclear reorganization without conjugation union in *Paramecium multimicronucleatum*, syngen 2. *Proc. 12th Intern. Congr. Genet.* **1**, 72.

Miyake, A. (1968). Induction of conjugating union by cell-free fluid in the ciliate *Blepharisma. Proc. Japan Acad.* **44**, 837.

Nobili, R. (1967). Ultrastructure of the fusion region of conjugating *Euplotes* (Ciliata Hypotrichida). *Monitore Zool. Ital.* **1**, [N.S.] 73.

Siegel, R. W., and Cole, J. (1967). The nature and origin of mutations which block a temporal sequence for genic expression in *Paramecium. Genetics* **55**, 607.

CHAPTER 7

Fishes

Eizo Nakano

BIOLOGICAL INSTITUTE, NAGOYA UNIVERSITY, NAGOYA, JAPAN

I. Introduction

Fertilization is a step in the development of the egg cell. The cell originates from a primordial germ cell, undergoes repeated oogonial divisions, and becomes an oocyte. After a growth phase, the oocyte enters an inert phase, in which both growth and cell division are blocked. Fertilization not only results in the fusion of two gametes but also a release from the block, resulting in the initiation of development. At fertilization, some prominent cortical reactions occur which have attracted much attention in the past, but fertilization concerns much more than surface phenomena between the two gametes. As pointed out by Runnström (1949), fertilization involves the activation of an egg by a spermatozoon; that is, the removal of the block which is formed during the development of the egg cell. For the complete understanding of fertilization, therefore, it is desirable to investigate the process of oogenesis, in which the block is established. As described below, some fish eggs have advantages for such a study, since their oogenesis takes place within a short period. Further, fish eggs are much larger in size than those of marine invertebrates and they afford good materials for study of fertilization. Transparent eggs, such as those of *Oryzias* and *Fundulus,* are especially advantageous for the observation of structural changes at the time of fertilization.

Considerable work has been done on fertilization of fish eggs, but one should note the following special properties which fish eggs do not share with most marine invertebrates: (1) fish eggs are surrounded by a rigid egg membrane (chorion) even before fertilization; (2) spermatozoa can penetrate into the egg only through the micropyle in the egg membrane; (3) fish eggs have a large amount of yolk. At fertilization, ooplasmic segregation takes place, resulting in the formation of a blastodisc.

Since several aspects of fertilization of fish eggs have been reviewed by others (Rothschild, 1958; T. Yamamoto, 1961), the main part of this account will be devoted to the metabolic changes which underlie the initiation of development.

II. Oogenesis

A. STRUCTURAL CHANGES DURING OOGENESIS

The fish oocyte in the early phase has a large nucleus and a small amount of cytoplasm. As oogenesis proceeds, the cytoplasm increases in volume and the nucleus becomes enlarged to form the germinal vesicle. In a later phase of oogenesis, the formation of yolk begins with the appearance of yolk globules in the cytoplasm. These globules gradually increase in size and number and may fuse into a continuous yolk mass, e.g., *Fundulus* (Marza *et al.*, 1937) and *Liopsetta* (K. Yamamoto, 1957). In other cases, e.g., *Cyprinus* and *Gobio* (Konopacka, 1935), *Perca* (Mas, 1952), *Leufa* (Osanai, 1956), and *Clupea* (K. Yamamoto, 1958), yolk globules persist even after maturation and compose a nonmassed yolk. Besides yolk globules, yolk vesicles containing mucopolysaccharide are found in the cytoplasm. Marza *et al.* (1937) assumed that yolk vesicles transform into yolk globules which subsequently grow into yolk. However, according to T. S. Yamamoto (1955) and K. Yamamoto (1956b,c,d), yolk vesicles may give rise to cortical alveoli by migration toward the periphery of the cytoplasm during oogenesis. From the nature of the polysaccharides occurring in yolk vesicles, K. Yamamoto (1956d) classified fish eggs into three types: (1) the yolk vesicles containing only neutral mucopolysaccharides (e.g., *Clupea* and *Liopsetta*); (2) vesicles having at first neutral, but later also acid mucopolysaccharides (e.g., *Hypomesus* and *Salmo*); (3) vesicles containing only acid mucopolysaccharides throughout oogenesis (e.g., *Oryzias*). K. Yamamoto suggested that these types may be related to the habitat of the fishes, as the first type comprises marine, the second anadromous, and the third freshwater fishes.

The growing oocyte is surrounded by follicle cells, which may play a part in the transfer of the nutrient substances into the oocyte. Between the follicle cells and the cortical cytoplasm the zona radiata appears at first as a vacuolar layer. Later radial striations become visible with the thickening of the zona. In the early phase of oogenesis of the flounder, *Liopsetta*, the zona radiata consists mainly of polysaccharides but later its polysaccharide content diminishes because the protein component predominates (K. Yamamoto, 1956a). In cyprinoids, however, the zona radiata contains only protein throughout oogenesis, and a thin hyaline layer outside the zona radiata contains polysaccharides (Arndt, 1960).

As described below, at fertilization the zona radiata is lifted from the egg surface to become the egg membrane. This membrane is often called the chorion, although it is believed to arise from the oocyte itself. On the other hand, villi or threads attached to the egg membrane are known to originate from a substance secreted by the follicle cells outside the zona radiata, e.g., *Salarias* (Eggert, 1929), *Cyprinoids* (Arndt, 1960), and *Oryzias* (T. S. Yamamoto, 1955).

In *Oryzias*, Nakano (1953) observed that underripe eggs are still unfertilizable at the beginning of the first maturation division, even if their follicular walls have been removed. It seems that cytoplasmic maturity is related to the fertilizability of this egg as in eggs generally. Yanagimachi (1958) also reported that herring eggs are not fertilizable before the first maturation division is completed. Recent work by Iwamatsu (1965) has confirmed these observations. He has shown that fertilizability of preovulation eggs increases gradually after the breakdown of the germinal vesicle. In preovulation eggs the cortical change is comparatively slow and the breakdown of the cortical alveoli is incomplete. When eggs are fertilized before ovulation, polyploidy is found in the embryos that develop. Although there is no definite explanation for such polyploidy, one possibility is that the sperm nucleus unites with the diploid egg nucleus before the maturation division.

B. METABOLIC CHANGES DURING OOGENESIS

Rapid growth of the oocyte takes place in the later phase of oogenesis, mainly due to the accumulation of the yolk. This phase is called vitellogenesis. Some data indicate that intense metabolic activity is associated with vitellogenesis. In *Oryzias* oocytes, the rate of oxygen uptake increases several fold during the growth phase. This rise is followed by a rapid drop after accumulation of the yolk has ceased with the result that the level of the respiratory activity observed in ripe unfertilized eggs is lower than that in oocytes during vitellogenesis (Nakano, 1953). Carbohydrate and protein accumulate in growing oocytes in parallel with the increase of the respiratory activity (Ando, 1960).

Concerning phosphorus metabolism, Tsusaka and Nakano (1965) reported that the incorporation rate of phosphorus in oocytes increases with

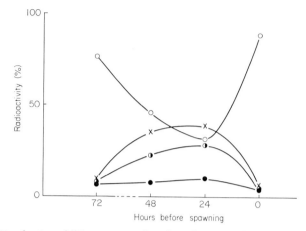

FIG. 1. Distribution of ^{32}P in various phosphate fractions of *Oryzias* oocytes during growth. Open circles—total acid-soluble P; filled-in circles—phospholipid P; half-filled circles—total nucleic acid P; crosses—phosphoprotein P. (From Tsusaka and Nakano, 1965.)

the growth of the oocyte. After 8 hours incubation, ^{32}P is incorporated most strongly in large oocytes, less in small oocytes. The distribution of ^{32}P changes considerably with oocytes of different stages. In small oocytes, the radioactivity is mainly distributed in the acid-soluble fraction and much less in the phosphoprotein and nucleic acid fractions. On the other hand, in large oocytes the radioactivity of both phosphoprotein and nucleic acid fractions shows a maximum level, whereas that of the acid-soluble fraction remains relatively low (Fig. 1). After prolonged incubation (8 days), the distribution pattern of radioactivity in fractions becomes uniform among oocytes of different stages. This period may coincide with that of vitellogenesis. The same holds for ^{35}S- or ^{14}C-labeled amino acids which are incorporated about 3 times more rapidly in large oocytes than in small oocytes (Nakano and Ishida-Yamamoto, 1968). With the growth of the oocyte, the yolk protein is synthesized at the expense of amino acids in the pool fraction. Experiments with isolated oocytes showed that the maximum incorporation of inorganic phosphorus appears in the smallest oocytes, and the minimum amount is incorporated in unfertilized eggs. On the other hand, the incorporation of labeled amino acids increases to a maximum at the large oocyte stage, and about 80 to 90% of the amino acids was found in the protein fraction. These authors concluded that some of the reserve substances may be synthesized *in ovo* during oogenesis, although some maternally synthesized materials may also contribute to the large size-increase in the oocyte. The relationship between these syntheses requires investigation. In oocytes of

Oryzias, Ando (1959) reported that an increase in the amount of ribonucleic acid (RNA) and deoxyribonucleic acid (DNA) precedes the intense accumulation of protein and other substances. It seems likely that a large amount of DNA or its precursors exist not in the nucleus but in the cytoplasm as a reserve substance. This view is supported by the fact that in the developed embryo there is only 3 times as much DNA as in the mature oocyte (Ohi, 1961). Recently, Shmerling (1965) isolated the cytoplasmic DNA from sturgeon oocytes and demonstrated the ability of such DNA preparations to act as primers in the RNA polymerase system.

III. Fertilization

A. Structural Changes following Fertilization

1. *Structure of the Unfertilized Egg*

In fishes the ripe unfertilized egg is surrounded by the egg membrane, which is formed of at least two layers with different properties. Since both layers are sensitive to proteolytic enzymes (Ishida, 1944a,b; Kanoh and T. S. Yamamoto, 1957; Sakai, 1961), the membrane is made up largely of proteins. At the animal pole, the egg membrane is pierced by the micro-

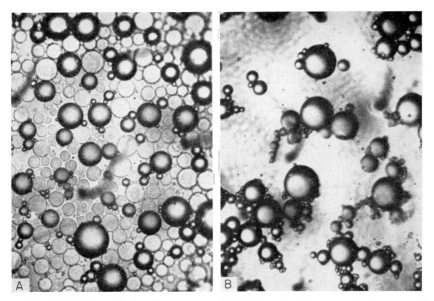

Fig. 2. Surface of the *Oryzias* egg. (A) Before fertilization, showing cortical alveoli and oil droplets; (B) after fertilization—note the disappearance of cortical alveoli. Magnification: × 150.

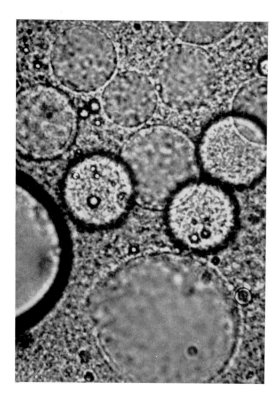

Fig. 3. Isolated cortical alveoli from the *Oryzias* egg. Magnification: × 1000.

pylar canal through which the spermatozoon passes to fertilize the egg. Beneath the egg membrane, there is a cortical layer of cytoplasm containing alveoli and oil droplets, except in a small area at the animal pole. The cortical alveoli vary in size (10–40 μ in *Oryzias* eggs; 4–25 μ in *Pungitius* eggs) and are distinguishable from the oil droplets and the yolk globules by a difference in refractive index (Fig. 2A). Histochemical tests showed that the cortical alveoli contained mucopolysaccharide (summarized by Rothschild, 1958). It is possible to isolate the alveoli from the cortical layer in isotonic saline solution. The isolated alveoli appear as smooth hyaline spheres and are fairly stable for a considerable length of time (Fig. 3). If, however, they are suspended in distilled water, they swell rapidly and burst. Aqueous solutions of burst alveoli show colloidal properties and give polysaccharide reactions (Kusa, 1956, 1958; Nakano, 1956). T. Yamamoto (1956, 1962) presented some data on the properties of isolated cortical alveoli of *Oryzias* eggs. The isolated alveoli dissolve in lipid solvents, as well as in venoms of snakes and bees which contain phospholipase. They are also dissolved by traces of ferrous ions. He suggested that ferrous ions may activate an endogenous esterase, resulting in the breakdown of the alveoli. Attempts to detect such an esterase in the cortical layer of fish eggs should be made.

In addition to cortical alveoli, Kusa (1956) found numerous particles in the cortical layer of stickleback eggs particularly surrounding the cortical alveoli. These particles resemble mitochondria in histochemical properties and do not disappear in the process of cortical alveolar breakdown. In *Oryzias* eggs, T. Yamamoto (1961, 1962) described a- and b-granules distributed in the cortical cytoplasm. According to him, a-granules, which surround the cortical alveoli, start to dissolve before the breakdown of the cortical alveoli, whereas b-granules do not disappear after fertilization. Aketa (1962) histochemically demonstrated that RNA-rich granules are embedded in the cortical cytoplasm of *Oryzias* eggs and disappear upon fertilization. He assumed that these granules correspond to the so-called a-granules. T. Yamamoto (1962) believed that a-granules may contain some enzymes that dissolve the membrane of the cortical alveoli. This, however, remains to be proven. In the cortical cytoplasm of *Oryzias* eggs, Ohtsuka (1964) detected another type of granule which remains unchanged after fertilization. According to him, these granules are different from both a- and b-granules.

There is no agreement in the literature about the presence of a lipid membrane surrounding the cortical alveoli. In *Oryzias* eggs, T. Yamamoto (1956) suggested that the cortical alveoli are surrounded by a lipid membrane. Histochemical work by Aketa (1954) demonstrated the presence of lipid in the surface of the cortical alveoli. However, T. S. Yamamoto (1955) could not detect the lipid membrane of the cortical alveoli in *Oryzias* eggs. In other fishes, the presence of such a lipid membrane was also denied (Kusa, 1956; Dettlaff, 1957). T. Yamamoto (1962) suggested that lipid in the membrane of the cortical alveoli may be present but is bound to protein or other substances, so that ordinary lipid staining cannot demonstrate its presence.

2. *Process of Sperm Entry*

Following insemination the spermatozoa are located in a single row in the micropylar canal; this results from the fact that the canal is as narrow as the head of one spermatozoon (Fig. 4). When the first spermatozoon reaches the egg surface, the cortical alveoli begin to break down and at the same time the supernumerary spermatozoa are pushed out from the canal (K. Yamamoto, 1952; Sakai, 1961; Ginsburg, 1961). In *Limanda* eggs, K. Yamamoto (1952) observed the formation of a pluglike deposit at the entrance of the micropyle immediately after fertilization. According to Sakai (1961) the discharge of extra spermatozoa is accomplished by the secretion of colloidal material into the canal from the inside. This

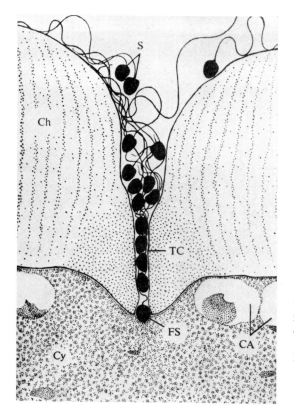

FIG. 4. Section of the trout egg in the region of the micropylar canal. CA, cortical alveoli; Ch, chorion (egg membrane); Cy, cytoplasm; FS, fertilizing spermatozoon, S, spermatozoa; TC, terminal canal of micropyle. (From Ginsburg, 1963.)

material occasionally bulges out from the canal, forming a plug. Ginsburg (1961) presented data showing a relationship between the block to polyspermy and the discharge of the cortical alveoli. In fishes, the mechanism responsible for monospermic fertilization seems to be attributable to the fact that a fertilizing spermatozoon can reach the egg surface only through the micropylar canal, which undergoes some change making it impermeable to supernumerary spermatozoa following fertilization.

There seems to be no block to polyspermy in the cortical structure of the egg which prevents the penetration of more than one spermatozoon. Following egg membrane removal by proteolytic enzyme treatment, spermatozoa can enter any part of the egg. Insemination of such denuded eggs usually results in polyspermy (Kanoh and Yamamoto, 1957). Yanagimachi (1957) also reported that in the denuded herring egg refertilization occurs if reinseminated within 20 to 30 minutes after the initial fertilization. In *Oryzias* eggs, Sakai (1961) confirmed these observations and concluded that the block to polyspermy in the denuded fish egg does not take effect immediately after fertilization. However, by the time of bipolar differentiation spermatozoa are no longer able to enter the egg.

In fishes, it is well known that both eggs and spermatozoa lose their fertilizability within a short period when they are immersed in freshwater. However, if ripe eggs are kept in a saline solution, isotonic with ovarian eggs, they retain their fertilizability for a long period. This fact was discovered by T. Yamamoto (1939) in *Oryzias* eggs and later applied to other fish eggs (Kusa, 1951, 1953b; Yanagimachi, 1957), although some fish eggs undergo parthenogenetic activation in an isotonic solution and lose their fertilizability. The main reason for the loss of fertilizability in freshwater may be due to the change of the egg membrane, particularly to the closure of the micropyle. On standing in freshwater, the egg membrane swells or hardens by parthenogenetic activation and the closure of the micropyle takes place. Since spermatozoa can penetrate only through the micropyle, its closure causes a loss of fertilizability.

According to Ishida (1948), addition of a small amount of salts is sufficient to maintain fertilizability of *Oryzias* eggs, since the micropyle remains open to allow sperm entry. There are some curious observations in salmon eggs indicating that fertilization does not occur in an isotonic solution, even though the eggs retain their fertilizability for a long period. Kusa (1950) observed that a spermatozoon enters the egg in isotonic solution, but no cortical change occurs in the solution. However, if the eggs are put into water after the spermatozoa have lost their fertilizing capacity, they develop in the normal way. Recent cytological observations by Ginsburg (1963) have shown that in isotonic solution the spermatozoon reaches the cortical layer of the egg along the micropylar canal and the anterior part of its head becomes embedded in the cytoplasm.

3. Breakdown of the Cortical Alveoli

In fish eggs the first visible change at fertilization is the breakdown of the cortical alveoli; this has many features in common with the cortical change occurring in the eggs of other animals (Fig. 2).

In *Fundulus* eggs, Kagan (1935) described the disappearance of the platelets embedded in the cortical layer of unfertilized eggs after fertilization or artificial activation. The platelets in *Fundulus* eggs seems to be identical to the cortical alveoli of other fish eggs (T. Yamamoto, 1961). Tchou and Chen (1936) first discovered the progressive breakdown of the cortical alveoli at fertilization in the egg of the goldfish. They suggested that the breakdown of the cortical alveoli is followed by the separation of the egg membrane. In *Oryzias* eggs, T. Yamamoto (1939) observed the same phenomenon and published a series of papers on the cortical change under various conditions (T. Yamamoto, 1944a,b, 1949a,b, 1954). He established the fact that when *Oryzias* eggs are fertilized, the breakdown of the cortical alveoli begins near the animal pole and ends at the vegetal pole. Similar cortical changes are also observed following treatment with parthenogenetic agents. Progressive breakdown of the cortical alveoli from the animal pole, as observed in *Oryzias* eggs, has been described in other species. A detailed review has been devoted to this subject by T. Yamamoto (1961).

The process of the alveolar breakdown was observed by Kusa (1953b) in *Pungitius* eggs. At the beginning of the breakdown, the outline of the alveoli becomes indistinct and the alveolar contents are forced out through a wide aperture into the space between the egg membrane and the surface of egg proper. In *Oryzias* eggs (T. Yamamoto, 1961), the contents of the cortical alveoli are squeezed out through a narrow aperture, and form an aggregate of colloidal substance between the egg membrane and the egg surface; this soon becomes a homogeneous solution.

There is a possibility that some sort of invisible change propagates at the egg surface prior to the breakdown of the cortical alveoli (T. Yamamoto, 1944a,b). In *Oryzias* eggs, the cortical alveoli can be displaced by centrifugation and accumulate at the centrifugal side. Even when the cortical alveoli accumulate at the vegetal pole, the eggs can still be activated by normal fertilization or by pricking at the animal pole. After activation, the cortical alveoli embedded in the cortical layer break down, whereas those in the yolk do not. This indicates that an invisible change, which causes the breakdown of the cortical alveoli, is propagated through the cortical layer. This change is called the fertilization wave (T. Yamamoto, 1944a,b) or impulse (Rothschild, 1958), but its nature is still obscure. In acipenserid eggs, according to Dettlaff (1962), the fertilization impulse brings about nuclear stimulation in addition to progressive disintegration of the cortical granules (cortical alveoli).

4. Separation of the Egg Membrane

The breakdown of the cortical alveoli is followed by the release of the colloid alveolar contents into the space between the egg membrane and the protoplasmic surface. This, in turn, results in the formation of the perivitelline space which widens rapidly. It seems likely that the colloidal contents of the cortical alveoli, which cannot pass through the egg membrane, cause influx of water with consequent formation of the perivitelline space (T. Yamamoto, 1944a). In *Fundulus* eggs, Kao and Chambers (1954) also noticed that a colloidal material derived from the platelets in the cortical layer causes the formation of the perivitelline space. According to Kao (1956), the volume of the egg proper is controlled by the hydrostatic pressure developed in the perivitelline space.

Nakano (1956) observed the process of separation of the egg membrane in *Oryzias* eggs. At fertilization, the egg membrane contracts immediately after the breakdown of the cortical alveoli and its thickness diminishes rapidly. This change takes place mainly in the inner layer of the egg membrane. Subsequently the membrane hardens.

In salmon eggs, Kusa (1949a,b) emphasized an important role of Ca ions in egg membrane hardening following immersion in water. Nakano (1956) reported that egg membrane hardening is inhibited by cysteine and thioglycollate, if the eggs are treated within 10 minutes after fertilization. On the other hand, colloidal substances, such as gum arabic, cause egg membrane hardening without activation. The colloidal substance derived from the cortical alveoli may play a part in egg membrane hardening. In salmon eggs, Zotin (1958) found that the hardening of the egg membrane is linked to the release of a "hardening enzyme" from the egg surface. The release of this enzyme is blocked by salt solutions and is stimulated by water. An assumption was made that egg membrane hardening may result from initiation of chain polymerization of membrane substances. However, in X-ray diffraction patterns from trout egg membranes, Kusa (1959) found no evidence for the changes in the molecular organization of the egg membrane expected to occur upon activation. According to Ohtsuka (1957), $CdCl_2$ inhibits the hardening of the egg membrane, whereas some fatty acids cause membrane hardening. He suggested that phospholipid of the cortical layer may play a role in egg membrane hardening. In other experiments, Ohtsuka (1960) reported that oxidizing agents cause the hardening of the egg membrane, whereas in the presence of reducing reagents, the egg membrane fails to harden. The SH-reagents also inhibit hardening, and aldehydes produce hardening. He concluded that oxidation is involved in the hardening of the egg membrane. Recent work by Ohtsuka (1964) has shown that the hardening of the egg membrane occurs without the breakdown of the cortical alveoli when eggs are exposed to ion-binding agents, such as nitroso-R-salt and α,α'-dipyridyl. The hardening of the egg membrane without alveolar breakdown was also observed following treatment with ethylenediaminetetraacetate (EDTA) and subsequent immersion in isotonic solution. He finally concluded that the release of some metallic ions from the cortical layer brings about an invisible change which induces egg membrane hardening. However, there is little doubt that such an invisible change plays a role in the egg membrane hardening as well as in the incitement of alveolar breakdown.

In contrast to these observations, Dettlaff (1962) found in acipenserid eggs that the materials of the cortical granules (cortical alveoli) are expelled from the cortical layer at fertilization and merge with the inner surface of the egg membrane, as in sea urchin eggs. It is worth noting that in these eggs the formation of the perivitelline space does not directly follow the extrusion of the cortical granules but is related to the swelling of the colloidal substance excreted from the animal pole region of the egg. This colloidal substance differs from the cortical granule material and does not contain sulfated acid mucopolysaccharide. Probably, the mechanism of the separation and hardening of the egg membrane varies among fish eggs. The whole situation still remains obscure.

5. Ooplasmic Segregation

When alveolar breakdown is complete, ooplasmic segregation or bipolar differentiation follows. In this process the protoplasm accumulates at the animal pole, forming the blastodisc, with the oil droplets at the vegetal pole. Sakai (1964a,b) studied the mechanism of ooplasmic segregation in *Oryzias* eggs. She found that if the egg is separated into two fragments and both are activated by pricking, ooplasmic segregation proceeds in each fragment. In the animal fragment, a large blastodisc is formed at the animal pole, while a small protoplasmic accumulation is observed at the activated point. In the vegetal fragment, on the other hand, protoplasm accumulates at the activated point and the oil droplets migrate toward the opposite side. These observations suggest that the animal–vegetal axis, which is presumably responsible for ooplasmic segregation, can be altered, at least in the vegetal fragment. In partially activated eggs, obtained by blocking the propagation of the alveolar breakdown by heating, ooplasmic segregation is restricted to the activated part at first, but later it spreads into the unactivated part. The segregation proceeds along the egg axis regardless of the position of the partially activated part in the egg. It seems likely that a cytoplasmic activation can be induced under the influence of already activated cytoplasm without the propagation of the breakdown of the cortical alveoli.

In this connection, it is of interest to note that some fish eggs are activated without the breakdown of the cortical alveoli by immersion in isotonic or hypertonic solutions of $CaCl_2$ or $MgCl_2$ (Kanoh, 1952; Kusa, 1953a). Activation without alveolar breakdown is also induced in eggs subjected to double treatment with acidified isotonic saline solution (pH 1.8) and 0.2% pancreatin (Kanoh and T. S. Yamamoto, 1957). Recently T. S. Yamamoto (1962) found that salmon eggs can be activated without alveolar breakdown by immersion in isotonic solutions containing heavy metals, such as Zn, Cd, Ni, Co, and Mn. From these observations, Kusa (1953a) concluded that the alveolar breakdown is not a necessary factor for the initiation of development, but seems only to be an independent phenomenon associated with or accessory to it. As Rothschild (1956) pointed out, it would be dangerous to ascribe too important or dominating a role to visible cortical change in fertilization. More intrinsic changes at fertilization may occur in the egg itself, particularly in the metabolic process.

B. Physical Changes following Fertilization

Attempts to detect an invisible electric change have been made in *Oryzias* eggs. Maeno *et al.* (1956) first reported that the unfertilized egg has a positive membrane potential and that it is transiently reversed simultaneously with or shortly before the visible cortical change at fertilization. Such a transient change in the potential was also observed by Hori (1958) together with the gradual decrease during the breakdown of the cortical alveoli. He emphasized that the change is recorded prior to the breakdown of the cortical alveoli. In these cases, however, Ito and Maeno (1960) pointed out that the electrodes may not have penetrated into the egg proper. According to these authors, by successful insertion of electrodes into the unfertilized egg, the membrane potential should be negative and its magnitude is about -10 mV in isotonic saline solution. Recent work by Ito (1962) showed that no transient change occurs in the membrane potential after fertilization, but there is a gradual increase in the membrane potential, which reaches a plateau of -20 to -30 mV within 1 minute after initiation of the potential change. This potential change (called "activation potential") begins before the initiation of the breakdown of the cortical alveoli and ends before the completion of the breakdown. It was suggested that the potential change is not attributable to the alveolar breakdown but to the change of ionic permeability, since the primary response may take place in the protoplasmic membrane of the egg. More recently, Ito (1963) has presented some data on the relationship between the potential change and ionic concentrations of external media. When the concentration of external monovalent cations (K, Na) increases, the magnitude of potential change upon activation decreases and finally reverses its sign (Fig. 5). On the other hand, the concentration of Ca ions has no effect on the potential change. He suggested that in the *Oryzias* egg permeability to K ions transiently increases during activation. In *Fundulus* eggs, Kao (1955) failed to observe any change in the membrane potential upon activation, though the membrane resistance increases 2–7 times in the fully activated state. The changes in membrane resistance are interpreted to result from alterations in the effective pore size in the plasma membrane.

Concerning the permeability of fish eggs, there is some evidence that both the egg membrane and the plasma membrane are water-permeable before fertilization. Upon fertilization, however, the plasma membrane

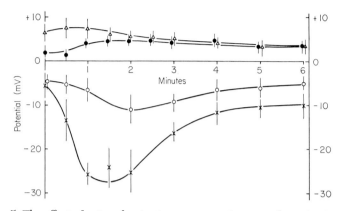

Fig. 5. The effect of external potassium concentration upon the activation potential in *Oryzias* eggs. Triangles—in 0.133 *M* K-phosphate buffer solution; filled-in circles—in 0.133 *M* KCl; open circles—in 0.013 *M* KCl; crosses—in 0.001 *M* KCl. (From Ito, 1963.)

becomes impermeable to water, whereas the egg membrane retains its permeability after separation (Kusa, 1951; Kanoh, 1951, 1952). This fact was confirmed by Kao and Chambers (1954) in *Fundulus* eggs and by Prescott (1955) in salmon eggs. Recent work by Ito (1960) also showed that the permeability of the *Oryzias* egg to water decreases after fertilization. In contrast to the sea urchin egg, the cortical change does not increase the permeability of the fish egg, but decreases it. This is also shown in the uptake of ^{32}P in *Oryzias* eggs, which does not change after fertilization.

C. Sperm–Egg Interacting Substances

At fertilization swarms of spermatozoa enter the micropyle by means of their active movement. Yanagimachi and Kanoh (1953) and Yanagimachi (1957) reported that herring spermatozoa become more active in the vicinity of the micropyle. In other fishes, e.g., *Rhodeus, Sarcocheilichthys,* and *Acheilognathus* (Suzuki, 1958, 1961), activation of spermatozoa is also observed around the micropyle area and, in addition, aggregation also occurs. The factors responsible for sperm activation and aggregation is located around the micropyle and can be extracted with distilled water. This factor is not species-specific but acts on heterologous spermatozoa of fishes (Suzuki, 1959). According to Ginsburg (1963), agglutination of spermatozoa takes place when semen is diluted with a saline solution that previously contained eggs with egg membrane extract or with perivitelline fluid. The agglutination differs somewhat from that of sea urchin spermatozoa. The agglutinated spermatozoa are stuck together by their heads and tails, and in isolated spermatozoa the tails form loops. Some information is available on sperm–egg interacting substances extracted from fish semen (Hartmann, 1944; Runnström *et al.*, 1944). A detailed discussion on the nature of these substances was given by Rothschild (1958) in his review.

It is well established in marine invertebrates that the spermatozoa undergo the acrosome reaction when they come into contact with unfertilized eggs (reviewed by Dan, 1956; Colwin and Colwin, 1964). The existence of a similar phenomenon has been shown in several species of mammals (Austin and Bishop, 1958; Austin, 1963). The acrosome reaction may have a role in initiating the fertilization reaction. Further, it may function in preventing polyspermic fertilization since spermatozoa may not be able to fertilize eggs once they have reacted.

In teleostean fishes, however, such a prominent change in spermatozoan structure has not been reported. In fact, electron-microscopic analysis does not reveal an acrosomelike structure at the anterior part of the head (Fig. 6). In keeping with this, trout spermatozoa do not extrude an acrosomal filament, when they are suspended in media containing egg secretion products (Ginsburg, 1963). Apparently in fishes the fusion of gametes

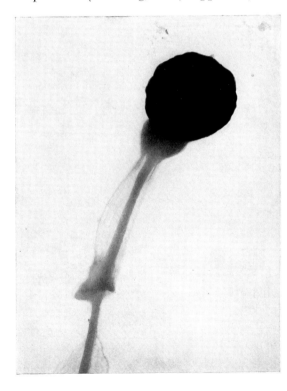

FIG. 6. Electron micrograph of a spermatozoon of *Oryzias latipes*. Magnification: × 18,000.

may take place without the acrosome reaction. Ginsburg concluded that the fertilizing spermatozoon passes directly to the cytoplasm through the micropylar canal, and the acrosome reaction loses its role in fusion of gametes. Although this conclusion of Ginsburg (1963) is probably correct, a reservation should be made because sperm entry has not yet been observed with the electron microscope. It should be noted that spermatozoa of acipenserid fishes have a long acrosome filament, even though their eggs have a membrane equipped with a micropyle similar to that of teleostean fish eggs (Dettlaff and Ginsburg, 1963).

D. Metabolic Changes following Fertilization

1. Respiratory Metabolism

In fish eggs only a few studies have been made on the changes in respiratory metabolism at the time of fertilization. Boyd (1928) found a marked rise in the oxygen uptake after fertilization in *Fundulus* eggs. On the contrary, Philips (1940) stated that such a transient rise in the respiration does not take place after fertilization. In *Oryzias* eggs, no immediate rise in the oxygen uptake is observed at fertilization, but there is a gradual rise followed by an exponential increase at later stages (Nakano, 1953). As pointed out by Whitaker (1933) in marine invertebrates, fertilization is not necessarily accompanied by a sudden increase in respiration, but it regulates the metabolic exchange of the eggs for further development. Since fish eggs contain a large amount of yolk, one would not expect a sudden change in the respiratory metabolism. It is of interest to note in this connection that the rate of oxygen uptake in *Oryzias* eggs begins to increase after the completion of ooplasmic segregation in which the cytoplasm accumulates at the animal pole. Although there is no sudden increase in the oxygen uptake at the time of fertilization, the metabolic process may be changed with the initiation of development. The evidences for such a metabolic change are obtained mainly in *Oryzias* eggs. The respiratory quotient (R.Q.) of *Oryzias* eggs has a low value (0.75) at the beginning of development, but this value soon begins to increase and reaches the maximum value (0.92) 24 hours after fertilization. This suggests that some changes in metabolic pathways may occur in the egg as a consequence of fertilization. Hishida and Nakano (1954) found that addition of intermediates of the tricarboxylic acid cycle, such as citrate, succinate, malate, glutamate, and pyruvate, stimulates the oxygen uptake of egg homogenates. These authors further showed that the endogenous oxygen uptake by homogenates increases during early development with or without the addition of substrate to the homogenates. It may be assumed, then, that a synthesis of the respiratory enzymes begins with the commencement of development. This view was confirmed by the measurement of activities of cytochrome oxidase and succinoxidase. Activities of these two enzymes increase even in the early stages of development. In the course of the study on the isozymic pattern of dehydrogenases in *Oryzias* eggs, it was found that unfertilized eggs possess one isozyme of malate dehydrogenase and that its activity intensifies during development. After hatching, some additional isozymes appear (Nakano and Whiteley, 1965).

Concerning the increase in respiratory metabolism, it is possible that at the beginning of development the repiratory enzymes in the eggs are not saturated by their substrates, but at a later stage the increase in oxidation results from the saturation of respiratory enzymes without synthesis (cf. Runnström, 1949; Spiegelman and Steinbach, 1945). In *Oryzias* eggs the oxygen uptake by cytochrome oxidase amounts to about 2 times that of intact eggs at the beginning of development. This indicates that the fertilized eggs contain an amount of cytochrome oxidase sufficient to catalyze such higher rate of oxidation. Normally the enzyme is probably far from being saturated by its substrate. As described above, however, the synthesis of respiratory enzymes also takes place even in the early stages of development. It may be safely concluded that fertilization brings about the saturation of respiratory enzymes by their substrates, together with the synthesis of these enzymes.

It is of interest to determine if coenzymes are responsible for respiratory rise in fish eggs. Hishida and Nakano (1954) found that the amount of diphosphopyridine nucleotide (DPN) of *Oryzias* eggs is rather constant throughout the development. This suggests that DPN may play a minor role in the respiratory rise. The concentration of DPN is 20 μg/100 eggs and of the same order as that found in the sea urchin egg (Jandorf and Krahl, 1942; Fujii and Ohnishi, 1963). The flavin contents of *Oryzias* eggs does not change during development, though the amount of flavin nucleotide increases after fertilization at the expense of free flavin. The amount of flavin nucleotide of *Oryzias* eggs is lower than that of sea urchin eggs (Krahl *et al.*, 1940), which suggests that *Oryzias* eggs are not dependent upon the activity of flavin enzymes, such as amino acid oxidase, to get energy for development.

There is no information about the presence of an inhibitor of respiratory enzymes in the unfertilized fish egg, as reported by Maggio and monroy (1959), and Maggio *et al.* (1960) in the sea urchin egg.

2. Carbohydrate Metabolism

In *Oryzias* eggs, the glycogen content gradually decreases during the first 24 hours of development, instead of a sudden drop immediately after

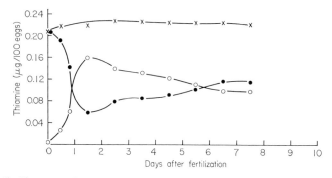

Fig. 7. Change in free and phosophothiamine in *Oryzias* eggs following fertilization and during early development. Crosses—total thiamine; filled-in circles—free thiamine; open circles—phosphothiamine. (From Hishida and Nakano, 1954.)

fertilization (Hishida and Nakano, 1954; Ando, 1960). In accordance with the breakdown of glycogen, the activity of anaerobic glycolysis increases gradually. This glycolysis is inhibited by *p*-chloromercuribenzoate and iodoacetamide. Hishida and Nakano (1954) reported that there is no phosphothiamine in the unfertilized egg. After fertilization, however, the amount of phosphothiamine increases sharply, while that of free thiamine decreases (Fig. 7). This result indicates that *Oryzias* eggs can synthesize phosphothiamine during the early stages of development. In view of the role of phosphothiamine in carbohydrate metabolism, it is of interest to note that there is a comparable increase in glycogenolysis and synthesis of phosphothiamine at the beginning of development.

Very little is known concerning the occurrence of the pentose phosphate shunt of carbohydrate metabolism in fish eggs. The activity of glucose-6-phosphate dehydrogenase, which is a terminal enzyme for this shunt, is very weak both in unfertilized and in fertilized eggs of *Oryzias*. The activity begins to increase at much later stages of development (Nakano and Whiteley, 1965). From this and the fact that glycolysis is inhibited by iodoacetamide and *p*-chloromercuribenzoate, it may be postulated that in *Oryzias* eggs the usual pathway of glycolysis is mainly responsible for carbohydrate breakdown at the early stages of development.

In summary, fertilization stimulates the oxidation of carbohydrates, as revealed by the rise of R.Q. value, the decrease in the amount of glycogen, and the increase of glycolytic activity.

3. Phosphorus Metabolism

The phosphorus metabolism of *Oryzias* eggs was first described by Ohi and Nakano (1954). With the initiation of development, the amounts of acid-soluble and nucleic acid phosphorus begin to increase at the expense of phosphoprotein phosphorus.

In later stages of development, acid-soluble phosphorus maintains a constant level until hatching, whereas nucleic acid phosphorus increases almost linearly. These findings were confirmed by Yamagami (1960) and were extended by Ohi (1961) who used ^{32}P as a tracer. According to Ohi, the increase of acid-soluble phosphorus precedes that of total nucleic acid phosphorus. It was suggested that phosphoprotein phosphorus is transferred into nucleic acid phosphorus through the pool of acid-soluble phosphorus. The experiments with ^{32}P clearly showed that the radioactivity is transferred from phosphoprotein to the acid-soluble fraction at an early stage of development (Fig. 8). These observations are quite the reverse of those in the oocytes during oogenesis, in which phosphoprotein is synthesized at the expense of acid-soluble phosphorus (Tsusaka and Nakano, 1965). In contrast to other phosphorus fractions, phospholipid showed no marked change during early development. In the rainbow trout, Yamagami and Mohri (1962) reported that phospholipid of unfertilized eggs can be fractionated into seven components, one cephalin, four lecithins, one sphingomyelin, and one unknown component (Fig. 9). According to these authors, there is no significant change in the relative and absolute amounts before and after fertilization, and utilization of phospholipid is a much later process in development.

The content of adenosine triphosphate (ATP) remains at a constant level during the whole course of development from fertilization to hatching (Ishida *et al.*, 1959; Taguchi, 1962). On the contrary, the content of adenosine diphosphate (ADP) undergoes a marked change at later stages of development (Taguchi, 1962). Since ADP is believed to be a main

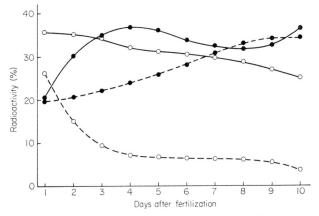

FIG. 8. Distribution of ^{32}P in various phosphate fractions of *Oryzias* eggs following fertilization and during early development. Filled-in circles (solid line)—total acid-soluble P; open circles (solid line)—phospholipid P; filled-in circles (dashed line)—total nucleic acid P; open circles (dashed line)— phosphoprotein P. (From Ohi, 1961.)

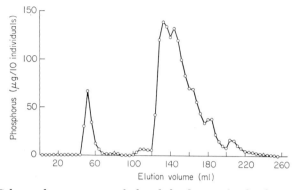

FIG. 9. Column chromatograms of phospholipids in unfertilized eggs of the trout. (From Yamagami and Mohri, 1962.)

factor controlling the respiratory metabolism in the cell (Slater and Hülsmann, 1959), the rise in the ADP content may be related to the increase of the oxygen uptake.

4. Protein Metabolism

It has been well established in the sea urchin egg that egg proteins undergo some sort of rearrangement after fertilization (summarized by Monroy, 1957, 1965). Similar changes were reported for fish eggs by Hamano (1957) with electrophoretic methods (Fig. 10). In *Hypomesus* eggs, the transient appearance of a new component was observed in the lipovitellin fraction. However, this new component does not appear in fertilized eggs. These results were confirmed with the eggs of other salmonid fishes, *Salmo irideus* and *Oncorhynchus keta*. Hamano (1957) explained these results as follows. The appearance of a new component may be concerned with denaturation of proteins, and the degree of de-

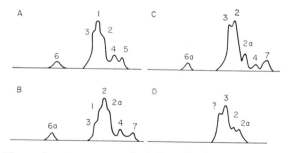

FIG. 10. Electrophoretic patterns of whole yolk of *Hypomesus* eggs. (A) unfertilized eggs; (B) eggs 30 seconds after activation; (C) eggs 2 minutes after activation; (D) eggs 15 minutes after activation. (From Hamano, 1957.)

naturation is stronger in activated eggs than in fertilized eggs. In this connection, it is of interest to note that if the proteins of unfertilized eggs are denatured by heat, the electrophoretic diagrams become similar to those of activated eggs. Using paper electrophoresis, Ohi (1962a) detected five components in water-soluble proteins of *Oryzias* eggs. Although the transient appearance of a new component immediately after fertilization was not mentioned by this author, one component decreases after the beginning of development, while a new component, probably lipoprotein, appears at much later stages after fertilization. Experiments with ^{32}P showed that the radioactivity is distributed in all the components. The radioactivity in two lipoprotein components could be removed by treatment with alcohol–ether. Besides these components, two protein bands containing acid phosphatase and inorganic pyrophosphatase were detected. Immunoelectrophoretic analysis of water-soluble proteins of *Oryzias* eggs confirmed these results, though there were some differences between the protein pattern obtained by paper electrophoresis and that by immunoelectrophoresis (Ohi, 1962b). He assumed that some bands found in paper electrophoresis are artificial products derived from the protein of one component in the agar gel pattern, and that these are antigenically similar.

Evidence for rearrangement of egg proteins at fertilization in the sea urchin egg has been provided by Lundblad (1949, 1950, 1954) in studies on the activation of proteolytic enzymes at fertilization. Likewise Hamano (1957) reported a transient rise of proteolytic activity at fertilization of some fish eggs. In salmonid fishes, the activity of livetin and vitellin-splitting enzymes increases within a few minutes after fertilization, followed by a decrease 15 minutes after fertilization. The mechanism of activation of proteolytic enzymes at fertilization is still unknown. Hamano (1957) suggested that some sort of protease inhibitor, such as fatty acids, antitrypsin, and Ca ions, may block the enzymes in unfertilized eggs.

The yolk proteins of fish eggs contain three components: lipovitellin; livetin; and phosphoprotein (Young and Phinney, 1951; Hamano, 1957). Ito *et al.* (1963) reported that phosphoprotein of the trout egg contains 4.0% of phosphorus, an amount appreciably lower than in phosvitin but significantly higher than in casein or milk phosphoprotein. Amino acid composition of phosphoprotein is as follows: glycine; alanine; serine; aspartic acid; glutamic acid; and an unidentified one. Serine exists in unusually large amount. The sedimentation constant of phosphoprotein is 1.9 S, a value significantly lower than that for phosvitin (2.9–3.1 S). Fujii (1960) carried out comparative studies on the yolk proteins, particularly on the lipoproteins of yolk from various animal species. In the trout egg, he found that the lipoprotein fraction is more soluble in lower NaCl con-

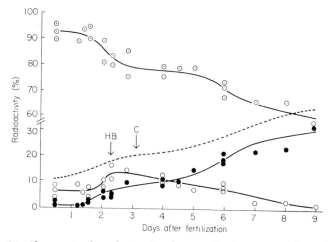

Fig. 11. Changes in the relative distribution of radioactivity (^{35}S-methionine) in yolk (dotted circles), pool (open circles), and embryo (filled-in circles) during the development of *Oryzias latipes*. (From Monroy *et al.*, 1961.) The dotted line indicates the pattern of oxygen uptake according to Hishida and Nakano, 1954. HB, beginning of heart beat; C, beginning of circulation.

centrations and more difficult to precipitate by ammonium sulfate. Both $S_{20,w}$ and the intrinsic viscosity (10.1 and 0.46) are smaller than those of any lipoproteins from other animal species. Monroy *et al.*, (1961) succeeded in labeling the unfertilized egg of *Oryzias latipes* with ^{35}S-methionine and ^{14}C-leucine by injection into the egg-laying female. In the unfertilized eggs, most radioactivity was found in the yolk fraction. Once development begins, the radioactivity is transferred from the yolk to the embryo through the low molecular weight pool (see Fig. 11). This seems to indicate that the embryonic proteins begin to be synthesized at the expense of the material derived from the yolk. The exponential rise of the respiratory activity of the egg (Hishida and Nakano, 1954) may be causally related to this yolk breakdown. According to Boell (1955) the synthesis of embryonic material from yolk should result in an increased oxygen uptake. These results are also in good agreement with the breakdown of phosphoproteins (Ohi, 1961). The nature of the low molecular weight pool is not clear, though the radioactivity is mainly concentrated in peptides.

Corti (1950) analyzed the amino acid composition of the trout egg. The percentages found in the fresh egg material are, arginine 1.77, histidine 1.07, leucine 3.20, methionine 0.91, phenylalanine 1.72, threonine 2.88, tyrosine 0.74, and valine 3.64.

In *Oryzias* eggs, Ohi (1961) detected acid phosphatase activity and found it to increase sharply during the first 24 hours after fertilization. In addition, he found another phosphatase that can hydrolyze inorganic pyrophosphate. Yamagami (1961) also demonstrated acid phosphatase activity in *Oryzias* eggs and suggested that the majority of phosphoprotein is directly dephosphorylated by acid phosphatase. Recent work by Yamagami (1963) confirmed that the activity of acid phosphatase increases during development. According to this author, alkaline phosphatase is also present in *Oryzias* eggs, though its activity is very weak at the beginning of development. The occurrence of phosphoprotein phosphatase, reported by Harris (1946) in frog eggs, is not detected in fish eggs at early stages of development (Ohi, 1961). At later stages, however, *Oryzias* embryos can hydrolyze casein, which suggests the presence of phosphoprotein phosphatase (Tsusaka, 1967). These results all suggest that the character of protein metabolism in the egg changes following fertilization. The synthesis of yolk protein ceases, and its degradation begins, together with the synthesis of new proteins.

IV. Conclusion

Fertilization is often considered as a rapid process beginning with the contact of two gametes and ending with the separation of the egg membrane. Considerable work has been devoted to elucidation of the mechanism underlying the initial events of fertilization; that is, the interaction between gametes and the separation of the egg membrane. In fishes, the presence of a sperm–egg interacting substance which activates spermatozoa has been reported. The cortical changes of the egg at the time of fertilization have been extensively investigated. However, fertilization has additional biological significance; namely, the initiation of development. After oogenesis, the egg enters an inert phase and fertilization brings about the release from inertness. Our knowledge is rather scanty concerning the mechanism that induces development. Some evidence suggests that metabolic processes of the egg undergo gradual changes following fertilization. Particularly in fish eggs, these changes proceed slowly. This may be owing to the fact that fish eggs contain a large amount of inert yolk. It seems likely that these changes may be more or less related to the mechanism that induces the initiation of development, but the available data are too few to draw any conclusion.

REFERENCES

Aketa, K. (1954). The chemical nature and the origin of the cortical alveoli in the egg of the medaka, *Oryzias latipes*. Embryologia **2**, 63.

Aketa, K. (1962). Cytochemical observations on disappearance of ribonucleic acid-rich granules upon fertilization in the egg of the medaka (*Oryzias latipes*). *Exptl. Cell Res.* **28**, 245.

Ando, S. (1959). Changes in the content of nucleic acids during oogenesis of *Oryzias* eggs. *Zool. Mag.* (*Tokyo*) **68**, 42 (in Japanese).

Ando, S. (1960). Physiological study on egg formation of the fish. I. Accumulation of carbohydrates and proteins during oogenesis. *Embryologia* **5**, 239.

Arndt, E. A. (1960). Untersuchungen über die Eihüllen von Cypriniden. *Z. Zellforsch.* **52**, 315.

Austin, C. R. (1963). Acrosome loss from the rabbit spermatozoon in relation to entry into the egg. *J. Reproduction Fertility* **6**, 313.

Austin, C. R., and Bishop, M. W. H. (1958). Some features of the acrosome and perforatorium in mammalian spermatozoa. *Proc. Roy. Soc.* **B148**, 234.

Boell, E. J. (1955). Energy exchange and enzyme development during embryogenesis. *In* "Analysis of Development" (B. H. Willier, P. S. Weiss, and V. Hamburger, eds.), pp. 520–555, Saunders, Philadelphia, Pennsylvania.

Boyd, M. (1928). A comparison of the oxygen consumption of unfertilized and fertilized eggs of *Fundulus heteroclitus*. *Biol. Bull.* **55**, 92.

Colwin, A. L., and Colwin, L. H. (1964). Role of the gamete membranes in fertilization. *In* "Cellular Membranes in Development" (M. Locke, ed.), pp. 233–279. Academic Press, New York.

Corti, U. A. (1950). Die Matrix der Fische X. Notizen zur postembryonalen Ontogenese von *Salmo irideus*. (W. Gibbs). *Schweiz. Z. Hydrol.* **12**, 288.

Dan, J. C. (1956). The acrosome reaction. *Intern. Rev. Cytol.* **5**, 365.

Dettlaff, T. A. (1957). Cortical granules and substances secreted from the animal portion of the egg during the period of activation in Acipenseridae. *Dokl. Acad. Nauk, SSSR* **116**, 341.

Dettlaff, T. A. (1962). Cortical changes in acipenserid eggs during fertilization and artificial activation. *J. Embryol. Exptl. Morphol.* **10**, 1.

Dettlaff, T. A., and Ginsburg, A. S. (1963). Acrosome reaction in sturgeons and the role of Ca-ions in sperm–egg association. *Dokl. Acad. Nauk, SSSR* **153**, 1461.

Eggert, B. (1929). Entwicklung und Bau der Eier von *Salarias flavoumbrinus Rupp.* *Zool. Anz.* **83**, 241.

Fujii, T. (1960). Comparative biochemical studies on the egg-yolk proteins of various animal species. *Acta Embryol. Morphol. Exptl.* **3**, 260.

Fujii, T., and Ohnishi, T. (1963). Inhibition of acid production at fertilization by nicotinamide and other inhibitors of diphosphopyridine nucleotidase (DPNase) in the sea urchin. *J. Fac. Sci. Univ. Tokyo, Sect. IV*, **9**, 333.

Ginsburg, A. S. (1961). The block to polyspermy in sturgeon and trout with special references to the role of cortical granules (alveoli). *J. Embryol. Exptl. Morphol.* **9**, 173.

Ginsburg, A. S. (1963). Sperm–egg association and its relationship to the activation of the egg in salmonid fishes. *J. Embryol. Exptl. Morphol.* **11**, 13.

Hamano, S. (1957). Physico-chemical studies on the activation and fertilization of fish egg. *Mem. Fac. Fisheries Hokkaido Univ.* **5**, 91.

Harris, D. L. (1946). Phosphoprotein phosphatase, a new enzyme from the frog egg. *J. Biol. Chem.* **165**, 541.

Hartmann, M. (1944). Befruchtungsstoffe (Gamone) bei Fischen (Regenbogenforelle). *Naturwissenschaften* **32**, 231.

Hishida, T., and Nakano, E. (1954). Respiratory metabolism during fish development. *Embryologia* **2**, 67.

Hori, R. (1958). On the membrane potential of the unfertilized egg of the medaka, *Oryzias latipes,* and changes accompanying activation. *Embryologia* **4**, 79.

Ishida, J. (1944a). Hatching enzyme in the fresh-water fish, *Oryzias latipes*. *Annotationes Zool. Japon.* **22**, 137.

Ishida, J. (1944b). Further studies on the hatching enzyme of the fresh-water fish, *Oryzias latipes*. *Annotationes Zool. Japon.* **22**, 155.

Ishida, J. (1948). The maintenance of the fertilizability in unfertilized eggs of *Oryzias latipes* by means of a small quantity of some salts and its mechanism. "Oguma Commemorative Volume on Cytology and Genetics," pp. 48–52.

Ishida, J., Taguchi, S., and Maruyama, K. (1959). ATP content and ATPase activity in developing embryos of the teleost, *Oryzias latipes*. *Annotationes Zool. Japon.* **32**, 1.

Ito, S. (1960). The osmotic property of the unfertilized egg of fresh water fish, *Oryzias latipes*. *Kumamoto J. Sci. Ser. B, Sect. 2*, **5**, 61.

Ito, S. (1962). Resting potential and activation potential of the *Oryzias* egg. II. Change of membrane potential and resistance during fertilization. *Embryologia* **7**, 47.

Ito, S. (1963). Resting potential and activation potential of the *Oryzias* egg. IV. The effect of monovalent cations. *Embryologia* **7**, 344.

Ito, S., and Maeno, T. (1960). Resting potential and activation potential of the *Oryzias* egg. I. Response to electrical stimulation. *Kumamoto. J. Sci. Ser. B, Sect. 2*, **5**, 100.

Ito, Y., Fujii, T., and Yoshioka, R. (1963). On a phosphoprotein isolated from trout egg. *J. Biochem.* **53**, 242.

Iwamatsu, T. (1965). On fertilizability of pre-ovulation eggs in the medaka, *Oryzias latipes*. *Embryologia* **8**, 327.

Jandorf, B. T., and Krahl, M. E. (1942). Studies on cell metabolism and cell division. VIII. The diphosphopyridine nucleotide (coenzyme) content of egg of *Arbacia punctulata*. *J. Gen. Physiol.* **25**, 749.

Kagan, B. M. (1935). The fertilizable period of the eggs of *Fundulus heteroclitus* and associated phenomena. *Biol. Bull.* **69**, 185.

Kanoh, Y. (1951). Über Wasseraufnahme und Aktivierung der Lachseier. II. Die Wirkung der hypertonischen Salzlösung. *J. Fac. Sci. Hokkaido Univ., Sect. VI*, **10**, 260.

Kanoh, Y. (1952). Über die Beziehung zwischen dem Zerfallen der Kortikalalveoli und der Entwicklung bei Fischeiern. *Japan. J. Ichthyol.* **2**, 99.

Kanoh, Y., and Yamamoto, T. S. (1957). Removal of the membrane of the dog salmon egg by means of proteolytic enzymes. *Bull. Japan. Soc. Sci. Fisheries* **23**, 166.

Kao, C. Y. (1955). Changing electrical constants of the *Fundulus* egg surface. *Biol. Bull.* **109**, 361.

Kao, C. Y. (1956). Pressure-volume relationship of the *Fundulus* egg in sea water and in sucrose. *J. Gen. Physiol.* **40**, 91.

Kao, C. Y., and Chambers, R. (1954). Internal hydrostatic pressure of the *Fundulus* egg. *J. Exptl. Biol.* **31**, 139.

Konopacka, B. (1935). Recherches histochimiques sur le développement des poissons, I. La vitellogénèse chez lagoujon et la carpe. *Bull. Acad. Polon. Sci., B, Classe II*, **62**, 163.

Krahl, M. E., Keltch, A. K., and Clowes, G. H. A. (1940). Flavin dinucleotide in eggs of sea urchin, *Arbacia punctulata*. *Proc. Soc. Exptl. Biol. Med.* **45**, 719.

Kusa, M. (1949a). Hardening of the chorion of salmon egg. *Cytologia* **15**, 131.

Kusa, M. (1949b). Further notes on the hardening of the chorion of salmon eggs. *Cytologia* **15**, 145.

Kusa, M. (1950). Physiological analysis of fertilization in the egg of the salmon, Oncorhynchus keta. I. Why are the eggs not fertilized in isotonic Ringer solution? Annotationes Zool. Japon. 24, 22.

Kusa, M. (1951). A brief note on the permeation of heavy water into the unactivated eggs of the rainbow trout. J. Fac. Sci. Hokkaido Univ. 6, 10, 271.

Kusa, M. (1953a). Significance of the cortical change in the initiation of development of the salmon egg. (Physiological analysis of fertilization in the egg of the salmon, Oncorhynchus keta, II.) Annotationes Zool. Japon. 26, 73.

Kusa, M. (1953b). On some properties of the cortical alveoli in the egg of the stickle back. Annotationes Zool. Japon. 26, 138.

Kusa, M. (1956). Studies on cortical alveoli in some teleostean eggs. Embryologia 3, 105.

Kusa, M. (1958). Explosion of isolated cortical alveoli of the stickleback. Annotationes Zool. Japon. 31, 212.

Kusa, M. (1959). X-ray diffraction patterns from the egg membrane of the egg of the brook trout. Nature 183, 410.

Lundblad, G. (1949). Proteolytic activity in eggs and sperms from sea urchins. Nature 163, 643.

Lundblad, G. (1950). Proteolytic activity in sea urchin gametes. Exptl. Cell Res. 1, 264.

Lundblad, G. (1954). "Proteolytic Activity in Sea Urchin Gametes." Almqvist & Wiksells, Uppsala.

Maeno, T., Morita, H., and Kuwabara, M. (1956). Potential measurements on the eggs of Japanese killi-fish, Oryzias latipes. Mem. Fac. Sci. Kyushu Univ., Ser. B, 2, 87.

Maggio, R., and Monroy, A. (1959). An inhibitor of cytochrome oxidase activity in the sea urchin egg. Nature 184, 68.

Maggio, R., Ajello, F., and Monroy, A. (1960). Inhibitor of the cytochrome oxidase of unfertilized sea urchin eggs. Nature 188, 1195.

Marza, V. D., Marza, E. V., and Guthrie, M. J. (1937). Histochemistry of the ovary of Fundulus heteroclitus with special reference to the differentiating oöcytes. Biol. Bull. 73, 67.

Mas, F. (1952). Contribution a l'histologie de l'ovogénèse chez un teleostéen, Perca fluviatilis, L. Bull. Soc. Zool. France 77, 414.

Monroy, A. (1957). Studies of proteins of sea urchin egg and of their changes following fertilization. In "The Beginnings of Embryonic Development" (A. Tyler, R. C. von Borstel, and Ch. B. Metz, eds.), pp. 169–174. Am. Assoc. Advance. Sci., Washington, D. C.

Monroy, A. (1965). "Chemistry and Physiology of Fertilization." Holt, New York.

Monroy, A., Ishida, M., and Nakano, E. (1961). The pattern of transfer of the yolk material to the embryo during the development of the teleostern fish, Oryzias latipes. Embryologia 6, 151.

Nakano, E. (1953). Respiration during maturation and at fertilization of fish eggs. Embryologia 2, 21.

Nakano, E. (1956). Changes in the egg membrane of the fish egg during fertilization. Embryologia 3, 89.

Nakano, E., and Ishida-Yamamoto, M. (1968). Uptake and incorporation of labeled amino acids in fish oocytes. Acta Embryol. Morphol. 10, 109.

Nakano, E., and Whiteley, A. H. (1965). Differentiation of multiple molecular forms of four dehydrogenases in the teleosts, Oryzias latipes, studied by disc electrophoresis. J. Exptl. Zool. 159, 167.

Ohi, Y. (1961). Studies on the phosphate metabolism during the development of *Oryzias latipes. Japan. J. Zool.* **13**, 199.

Ohi, Y. (1962a). Immuno-electrophoretic analysis of water-soluble proteins of fish eggs. *Japan. J. Zool.* **13**, 383.

Ohi, Y. (1962b). Water-soluble proteins in the fish egg and their changes during early development. *Embryologia* **7**, 208.

Ohi, Y., and Nakano, E. (1954). Studies on the metabolism of fish egg. *Zool. Mag. (Tokyo)* **63**, 167.

Ohtsuka, E. (1957). On the hardening of the chorion in the fish egg after fertilization I. Role of the cortical substance in chorion hardening of the egg of *Oryzias latipes. Sieboldia.* **2**, 19.

Ohtsuka, E. (1960). On the hardening of the chorion of the fish egg after fertilization III. The mechanism of chorion hardening in *Oryzias latipes. Biol. Bull.* **118**, 120.

Ohtsuka, E. (1964). Studies on the invisible cortical change of the fish egg. *Embryologia* **8**, 101.

Osanai, K. (1956). On the ovarian eggs of the loach, *Lefua echigonia,* with special reference to the formation of the cortical alveoli. *Sci. Rept. Tohoku Univ. Fourth Ser.* **12**, 181.

Philips, F. S. (1940). Oxygen consumption and its inhibition in the development of *Fundulus* and various pelagic fish eggs. *Biol. Bull.* **78**, 256.

Prescott, D. M. (1955). Effect of activation on the water permeability of salmon eggs. *J. Cellular Comp. Physiol.* **45**, 1.

Rothschild, Lord (1956). "Fertilization." Methuen, London.

Rothschild, Lord (1958). Fertilization in fish and lampreys. *Biol. Rev.* **33**, 372.

Runnström, J. (1949). The mechanism of fertilization in Metazoa. *Advan. Enzymol.* **9**, 241.

Runnström, J., Lindvall, S., and Tiselius, A. (1944). Gamones from the sperm of sea-urchin and salmon. *Nature* **153**, 285.

Sakai, Y. T. (1961). Method for removal of chorion and fertilization of the naked egg in *Oryzias latipes. Embryologia* **5**, 357.

Sakai, Y. T. (1964a). Studies on the ooplasmic segregation in the egg of the fish, *Oryzias latipes.* I. Ooplasmic segregation in egg fragments. *Embryologia* **8**, 129.

Sakai, Y. T. (1964b). Studies on the ooplasmic segregation in the egg of the fish, *Oryzias latipes.* II. Ooplasmic segregation of the partially activated egg. *Embryologia* **8**, 135.

Shmerling, J. G. (1965). Isolation and properties of DNA of sturgeon oocytes. *Biokhimiya* **30**, 113.

Slater, E. C., and Hülsmann, W. C. (1959). Control of rate of intracellular respiration. *Ciba Found. Symp. Regulation Cell Metab.,* pp. 58–83.

Spiegelman, S., and Steinbach, H. B. (1945). Substrate-enzyme orientation during embryonic development. *Biol. Bull.* **88**, 254.

Suzuki, R. (1958). Sperm activation and aggregation during fertilization in some fishes. I. Behavior of spermatozoa around the micropyle. *Embryologia* **4**, 93.

Suzuki, R. (1959). Sperm activation and aggregation during fertilization in some fishes. III. Non species-specificity of stimulating factor. *Annotationes Zool. Japon.* **32**, 105.

Suzuki, R. (1961). Sperm activation and aggregation during fertilization in some fishes. VII. Separation of the sperm-stimulating factor and its chemical nature. *Japan. J. Zool.* **13**, 79.

Taguchi, S. (1962). Changes in adenosine nucleotide content during embryonic development of the teleost, *Oryzias latipes,* as determined by an improved ion exchange resin chromatography method. *Annotationes Zool. Japon.* 35, 51.

Tchou, S., and Chen, C. H. (1936). Fertilization in goldfish. *Contrib. Inst. Zool. Natl. Acad. Peiping* 3, 35.

Tsusaka, A. (1967). Activity changes of phosphoprotein phosphatase, acid and alkaline phosphomonoesterase and proteinase during oogenesis and embryogenesis in the teleostean fish, *Oryzias latipes. Acta Embryol. Morphol.* 10, 44.

Tsusaka, A., and Nakano, E. (1965). The metabolic pattern during oogenesis in the fish, *Oryzias latipes. Acta Embryol. Morphol.* 8, 1.

Whitaker, D. M. (1933). On the rate of oxygen consumption by fertilized and unfertilized eggs. V. Comparisons and interpretation. *J. Gen. Physiol.* 16, 497.

Yamagami, K. (1960). Phosphorus metabolism in fish eggs. I. Changes in contents of some phosphorus compounds during early development of *Oryzias latipes. Sci. Papers Coll. Gen. Educ., Univ. Tokyo* 10, 99.

Yamagami, K. (1961). Phosphorus metabolism in fish eggs. III. Enzymatic dephosphorylation of endogenous phosphoprotein in *Oryzias* eggs. *Sci. Papers Coll. Gen. Educ., Univ. Tokyo* 11, 153.

Yamagami, K. (1963). Phosphorus metabolism in fish eggs. V. Acid and alkaline phosphomonoesterases in the whole egg of *Oryzias latipes. Sci. Papers Coll. Gen. Educ., Univ. Tokyo* 13, 223.

Yamagami, K., and Mohri, H. (1962). Phosphorus metabolism in fish eggs. IV. Column chromatographic separation of phospholipids in the rainbow trout during embryonic and postembryonic development. *Sci. Papers Coll. Gen. Educ., Univ. Tokyo* 12, 233.

Yamamoto, K. (1952). Studies of fertilization in the dog-salmon. I. The morphology of the normal fertilization. *J. Fac. Sci., Hokkaido Univ., Ser. VI,* 11, 81.

Yamamoto, K. (1956a). Studies on the formation of fish eggs. III. Localization of polysaccharides in oocytes of *Liopsetta obscura. J. Fac. Sci., Hokkaido Univ., Ser. VI,* 12, 391.

Yamamoto, K. (1956b). Studies on the formation of fish eggs. IV. The chemical nature and the origin of the yolk vesicles in the oocyte of the flounder, *Liopsetta obscura. Japan. J. Zool.* 11, 567.

Yamamoto, K. (1956c). Studies on the formation of fish eggs. VII. The fate of the yolk vesicle in the oocytes of the herring, *Clupea pallasii,* during vitellogenesis. *Annotationes Zool. Japon.* 29, 91.

Yamamoto, K. (1956d). Studies on the formation of fish eggs. VIII. The fate of the yolk vesicle in the oocyte of the smelt, *Hypomesus japonicus,* during vitellogenesis. *Embryologia* 3, 131.

Yamamoto, K. (1957). Studies on the formation of fish eggs. XI. The formation of a continuous mass of yolk and the chemical nature of lipids contained in it in the oocyte to the flounder, *Liopsetta obscura. J. Fac. Sci., Hokkaido Univ., Ser. VI,* 13, 344.

Yamamoto, K. (1958). Studies on the formation of fish eggs. XII. On the non-massed yolk in the egg of the herring, *Clupea pallasii. Bull. Fac. Fisheries, Hokkaido Univ.* 8, 270.

Yamamoto, T. (1939). Changes of the cortical layer of the egg of *Oryzias latipes* at the time of fertilization. *Proc. Imp. Acad. Tokyo* 15, 269.

Yamamoto, T. (1944a). Physiological studies on fertilization and activation of fish eggs. I. Response of the cortical layer of the egg of *Oryzias latipes* to insemination and to artificial stimulation. *Annotationes Zool. Japon.* 22, 109.

Yamamoto, T. (1944b). Physiological studies on fertilization and activation of fish eggs. II. The conduction of the "fertilization-wave" in the egg of *Oryzias latipes*. *Annotationes Zool. Japon.* **22**, 126.

Yamamoto, T. (1949a). Physiological studies on fertilization and activation of fish eggs. III. The activation of the unfertilized egg with electric current. *Cytologia* **14**, 219.

Yamamoto, T. (1949b). Physiological studies on fertilization and activation of fish eggs. IV. Fertilization and activation in narcotized eggs. *Cytologia* **15**, 1.

Yamamoto, T. (1954). Physiological studies on fertilization and activation of fish eggs. V. The role of calcium ions in activation of *Oryzias* eggs. *Exptl. Cell Res.* **6**, 56.

Yamamoto, T. (1956). The physiology of fertilization in the medaka (*Oryzias latipes*). *Exptl. Cell Res.* **10**, 387.

Yamamoto, T. (1961). Physiology of fertilization in fish eggs. *Intern. Rev. Cytol.* **12**, 361.

Yamamoto, T. (1962). Mechanism of breakdown of cortical alveoli during fertilization in the medaka, *Oryzias latipes*. *Embryologia* **7**, 288.

Yamamoto, T. S. (1955). Morphological and cytological studies on the oogenesis of the fresh-water fish, medaka (*Oryzias latipes*). *Japan. J. Ichthyol.* **4**, 170.

Yamamoto, T. S. (1962). Activation of dog salmon egg by heavy metal ions. *J. Fac. Sci. Hokkaido Univ., Ser. VI*, **15**, 148.

Yanagimachi, R. (1957). Studies of fertilization in *Clupea pallasii*. VI. Fertilization of the egg deprived of the membrane. *Japan. J. Ichthyol.* **6**, 41.

Yanagimachi, R. (1958). Studies of fertilization in *Clupea pallasii*. VIII. On the fertilization reaction of the under-ripe eggs. *Japan. J. Ichthyol.* **7**, 61.

Yanagimachi, R., and Kanoh, Y. (1953). Manner of sperm entry in herring egg, with special reference to the role of calcium ions in fertilization. *J. Fac. Sci. Hokkaido Univ. Ser. VI*, **11**, 145.

Young, E. G., and Phinney, J. I. (1951). On the fraction of the proteins of egg yolk. *J. Biol. Chem.* **193**, 73.

Zotin, A. I. (1958). The mechanism of hardening of the salmonid egg membrane after fertilization or spontaneous activation. *J. Embryol. Exptl. Morphol.* **6**, 546.

Addendum

The growth of goldfish oocytes has been studied by the electron microscope (Yamamoto and Onozawa, 1965). As in the oocytes of other animals, rapid multiplication of mitochondria occurs during the early phase of oogenesis, while no marked change occurs in the form of the nucleus. Early oocytes have smooth surfaces closely applied to the follicle cells. As oogenesis proceeds, both oocytes and follicle cells develop microvilli and the microvilli of the cell types are in contact. These microvilli are considered to be the main site of pinocytosis.

There is no difference in the ATP level between activated and nonactivated eggs of the loach, *Misgurnus fossilis*. During late stages of development, respiration increases while the ATP level in the embryo drops (Zotin *et al.*, 1967).

Oryzias oocytes actively incorporate amino acids into proteins and this incorporation is inhibited by puromycin. The follicle cells accelerate the uptake of amino acids by the oocyte (Nakano and Ishida-Yamamoto, 1968).

In the egg of the river lamprey, *Lampetra fluviatilis*, electron microscopy has shown that the cortical alveoli have homogeneous contents which are expelled into the perivitelline space after fertilization (Afzelius *et al.*, 1968). The peripheral layer of fertilized eggs has long protrusions containing a dense cytoplasmic matrix. These often adhere to the chorion. The fertilizing sperm nucleus enters the egg through one of the peripheral protrusions (Nicander *et al.*, 1968).

The hypothesis of the masked mRNA store in unfertilized eggs has been proposed in the loach, *Misgurnus fossilis* (Spirin *et al.*, 1965; Spirin, 1966). After fertilization, "unmasking" of stored mRNA takes place and protein synthesis is switched on. At the same time, newly synthesized mRNA appears in the form of inactive ribonucleoprotein particles (informosomes) and will be incorporated into active polyribosomes during later stages of development.

REFERENCES TO ADDENDUM

Afzelius, B. A., Nicander, L., and Sjödén, J. (1968). Fine structure of egg envelopes and the activation changes of cortical alveoli in the river lamprey, *Lampetra fluviatilis*. *J. Embryol. Exptl. Morphol.* 19, 311–318.

Nakano, E., and Ishida-Yamamoto, M. (1968). Uptake and incorporation of labeled amino acids in fish oocytes. *Acta Embryol. Morphol. Exptl.* 10, 109–116.

Nicander, L., Afzelius, B. A., and Sjödén, I. (1968). Fine structure and early fertilization changes of the animal pole in eggs of the river lamprey, *Lampetra fluviatilis*. *J. Embryol. Exptl. Morphol.* 19, 319–326.

Spirin, A. S., Belitsina, N. V., and Ajtkhozhin, M. A. (1965). Messenger RNA in early embryogenesis. *Federation Proc.* 24, T907-915.

Spirin, A. S. (1966). On "masked" forms of messenger RNA in early embryogenesis and in other differentiating systems. *In* "Current Topics in Developmental Biology" (A. Monroy and A. A. Moscona, eds.), Vol. I, pp. 1–38. Academic Press, New York.

Yamamoto, K., and Onozawa, H. (1965). Electron microscope study on the growing oocyte of the goldfish during the first phase. *Mem. Fac. Fisheries Hokkaido Univ.* 18, 79–106.

Zotin, A. I., Faustov, V. S., Radzinskaja, L. I., and Ozernyuk, N. D. (1967). ATP level and respiration of embryos. *J. Embryol. Exptl. Morphol.* 18, 1–12.

CHAPTER 8

Gamete Structure and Sperm Entry in Mammals*

Lajos Pikó†

DIVISION OF BIOLOGY, CALIFORNIA INSTITUTE OF TECHNOLOGY,
PASADENA, CALIFORNIA

I. Introduction

The biological function of the male gamete is twofold: to initiate development of the egg and to transfer paternal genetic material. To

* This chapter is dedicated to the memory of Albert Tyler.
† *Present address:* Developmental Biology Laboratory, Veterans Administration Hospital, Sepulveda, California.

achieve these functions mammalian spermatozoa, like those of most animals, have evolved certain characteristic properties that may be summarized as follows: (1) independent motility enabling the sperm, along with other transport mechanisms in the genital tracts, to encounter the egg; (2) ability to recognize the egg as distinct from other tissue cells, and to recognize eggs of the same or closely related species (although available evidence of the species specificity of fertilization in mammals is limited at the cellular level, it can be assumed on general grounds); (3) ability to penetrate the egg's extraneous coats, interact with the egg's plasma membrane, and fuse with the egg.

For the property of motility, the sperm tail or flagellum is of primary interest. Recognition of the egg depends upon the properties of the sperm surface and may involve specific receptor sites on the sperm plasma membrane and, perhaps, materials adsorbed by the sperm from genital secretions (e.g., "coating antigens"). Penetration of the outer coats of the egg is aided by a specialized organelle containing enzymic material: the acrosome. Finally, the ability to interact with the egg plasma membrane and to fuse with the egg involves specific receptors and probably depends also on various specialized structures of the sperm.

The structure of the sperm flagellum and the mechanism of flagellar movement have been investigated intensively and will not be reviewed here (Fawcett, 1961; D. W. Bishop, 1962a,b; and Nelson in Chapter 2 of Volume I of this treatise). This chapter summarizes recent progress in our understanding of the structural and chemical bases of those properties of the gametes that relate to sperm entry as outlined above under (2) and (3). Recent fine-structural observations on the structure of the gametes and on the changes accompanying sperm entry are emphasized. An attempt is made to correlate these newer findings with some of the known features of fertilization and to elucidate the cellular mechanisms operating during sperm entry. The interaction of the gametes with the environment in the genital tracts is considered only inasmuch as it concerns some cellular aspect of sperm entry, for instance in the problem of sperm capacitation. Work relating to nonmammalian material is discussed only briefly and for comparative purposes; for a more detailed treatment the reader is generally referred to other chapters of this treatise.

II. Structure and Properties of the Sperm Head

Spermatozoa of mammals, like those of other animals, vary greatly in their microscopic appearance according to the species. Species differences also occur at the fine-structural level and may be used for taxonomic purposes (André, 1962, 1963; Hughes, 1965). The possible adaptive and evolutive significance of this variability is obscure (Rothschild, 1962). Even interstrain differences can be detected and appear to be under genetic control by the organism in which the gametes are produced and, perhaps, even by the haploid genome of the spermatozoa themselves (Beatty, 1961; Braden, 1958, 1960; Braden and Weiler, 1964). But the basic structural plan of the sperm of various mammals is remarkably similar and agrees in its essential features with that known for flagellate animal spermatozoa in general. This structural similarity suggests that the fundamental mechanisms of sperm entry are common in most of these species.

The fine structure of the sperm head has been studied recently in several species of mammals, e.g., man (Ånberg, 1957), bull (Kojima, 1962, 1966; Saacke and Almquist, 1964; Blom and Birch-Andersen, 1965), boar (Nicander and Bane, 1962), rabbit (Moricard, 1960; Hadek, 1963a; Bedford, 1964a), guinea pig (Fawcett and Hollenberg, 1963; Fawcett, 1965), rat (Pikó and Tyler, 1964b), opossum (Holstein, 1965), bat (Fawcett and Ito, 1965), sheep, horse, cat, dog, and hare (Nicander and Bane, 1966), and several primates (Bedford, 1967). There are several reviews on the structure and development of mammalian sperm—Nath (1956, 1966), M. W. H. Bishop and Austin (1957), Fawcett (1958, 1965), M. W. H. Bishop and Walton (1960), Roosen-Runge (1962), Leblond *et al.* (1963), André (1963), and Hancock (1966). The biology of mammalian spermatozoa has been reviewed by Parkes (1960) and D. W. Bishop (1961); Mann's (1964) comprehensive treatise deals with the biochemistry of semen. The terminology adopted here follows largely that of M. W. H. Bishop and Walton (1960). The structure of the rat sperm head, with the nomenclature used, is illustrated diagrammatically in Fig. 1 (modified after Pikó and Tyler, 1964b).

A. Nucleus

The nuclei of mature mammalian spermatozoa are in general highly flattened, compact structures that appear as homogeneous dense bodies in electron microscope pictures (Fig. 2). They contain the paternal genetic material consisting of a haploid set of somatic chromosomes and, in addition, one sex chromosome (X or Y). The deoxyribonucleic acid (DNA) content of mammalian sperm nuclei is about 3×10^{-12} gm— about 40% of the dry weight of the nucleus; the remaining portion of the dry material is made up mainly of basic nuclear protein (cf. M. W. H. Bishop and Walton, 1960; Mann, 1964).

The aggregation of chromatin fibers and their coalescence during spermiogenesis have been described in several species: cat (Burgos and

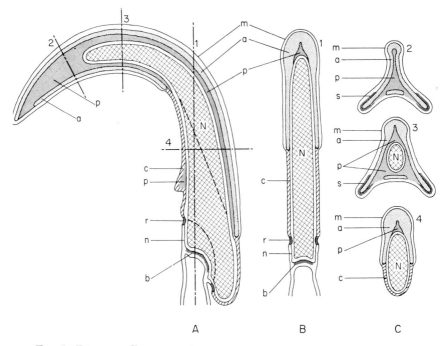

A   B   C

Fig. 1. Diagrams illustrating the structure of the rat sperm head at different sectional planes: A, median; B, frontal; and C, transverse sections. Key: (a) acrosome; (b) basal plate; (c) postnuclear cap (layer); (m) plasma membrane; (N) nucleus; (n) outer nuclear membrane (omitted under the acrosome); (p) perforatorium; (r) posterior ring; (s) surface coating of the outer acrosomal membrane (exaggerated in thickness).

Fawcett, 1955); man (Fawcett and Burgos, 1956); mouse (Prokofieva-Belgovskaya and Chai, 1961); rat (Brökelmann, 1963). Small, irregular cavities of various sizes, so-called "nuclear vacuoles," are frequently seen and believed to be due to accidents of chromatin condensation (cf. Fawcett, 1958). The nuclei of mature spermatozoa are unusually stable as shown by their resistance to deoxyribonuclease treatment (Daoust and Clermont, 1955); to extraction by concentrated salt solutions at neutral pH (Dallam and Thomas, 1953); and to physical stress, e.g., ultrasonic treatment (Henle et al., 1938) and mechanical shear (Mann, 1949, p. 340). The nucleus is covered by a double-layered nuclear envelope in which the usual pores appear to be lacking (André, 1963; Blom and Birch-Andersen, 1965). The two membranes often show considerable curling, particularly in the area covered by the postnuclear cap, making assessment of their structure difficult. At the base of the head, two specialized structures are closely associated with the nuclear envelope: the basal plate and the lamellar bodies. The basal plate is a disclike thickening of the outer nuclear membrane that connects the head with the "connecting piece" of the tail (Fig. 3). The lamellar bodies consist of intricately folded membranes in the area of the sperm neck (Fig. 4). Nicander and Bane (1962, 1966) consider them as folded extensions of the nuclear membrane that arise as a result of nuclear shrinkage during spermiogenesis. Similar structures were described in the spermatids of several species, e.g. crayfish (G. I. Kaye *et al.*, 1961), man (Horstmann, 1961), rat (Brökelmann, 1963) and salamander (Picheral, 1967). The lamellar bodies could be analogous to the argentophil bodies, named "accessory bodies," of light microscopic studies (Gatenby and Beams, 1935; Gresson and Zlotnik, 1945; cf. M. W. H. Bishop and Walton, 1960). Their function is not known.

B. ACROSOME

The term "acrosome" is used here to denote the caplike membrane-bound structure which covers the anterior and, in part, the lateral surfaces of the mammalian sperm nucleus. The name acrosome will thus

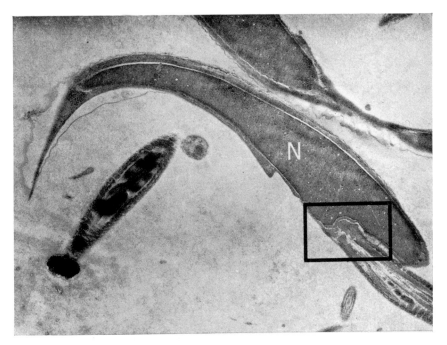

FIG. 2. Electron micrograph of a nearly median section through the rat sperm head showing the shape and general appearance of the nucleus (N). The area within the rectangle is enlarged in Fig. 3. Magnification: × 10,500.

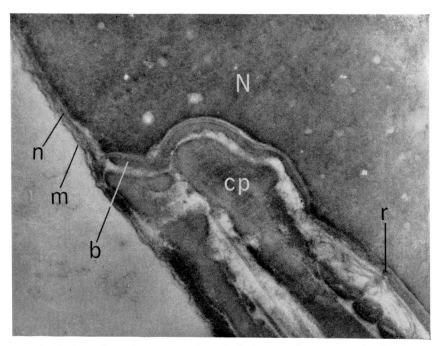

Fig. 3. A nearly median section through the base of the rat sperm head (cf. Fig. 2). Key: (N) nucleus; (n) nuclear envelope; (m) plasma membrane; (b) basal plate; (cp) connecting piece of the tail; (r) posterior ring (cf. Section II,D). Magnification: × 57,000.

include structures often designated as "head cap," "galea capitis," and "equatorial segment." Electron microscopic studies have demonstrated clearly that all these are but parts of a single unit surrounded by a limiting membrane of its own, here called the "acrosomal membrane." Since the acrosomal membrane is continuous it can be referred to as the "acrosomal sac" and its contents as the "acrosomal material." The acrosomal material may be subdivided further into two regions of somewhat differing properties: a bulkier apical region and a thinner lateral region (roughly corresponding to the "acrosome" and "head cap" of, e.g., Fawcett, 1958); however, the distribution of the two regions is not always clear-cut in the mature acrosome, and considerable differences exist according to species (Figs. 5, 6, and 7).

The acrosome is one of the most characteristic sperm organelles, inasmuch as it is of nearly universal occurrence in flagellated animal spermatozoa (cf. Chapters 6 and 7 of Volume I of this treatise). There is good reason to believe that its function is essentially the same in all these species, namely, it contains the enzymes and various other components that enable the sperm to penetrate the outer egg envelopes and to contact the surface of the egg plasma membrane proper.

The acrosome of mammals, as that of all other species investigated, is formed during spermatid differentiation as a secretory product of the Golgi apparatus (a view clearly expressed by Bowen, 1922, 1924). Recently new techniques, such as periodic acid–Schiff (PAS) staining which reacts selectively with acrosomal material (Wislocki, 1949; Leblond and Clermont, 1952; Clermont and Leblond, 1955) and electron microscopy (Burgos and Fawcett, 1955; Fawcett and Burgos, 1956), have yielded

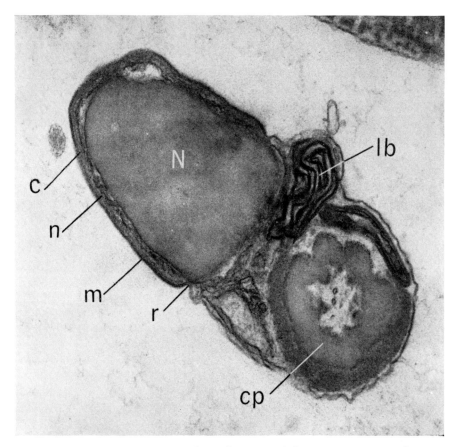

Fig. 4. Transverse section in the neck region of the rat spermatozoon showing the connecting piece (cp) of the tail and the caudal portion of the nucleus (N). (lb) Lamellar body; (n) nuclear envelope; (c) postnuclear cap (layer); (m) plasma membrane; (r) posterior ring. Note that the plasma membrane appears thicker and denser in the area of the postnuclear cap than in other areas of the sperm (cf. Section II,E). Lead citrate staining. Magnification: × 47,000.

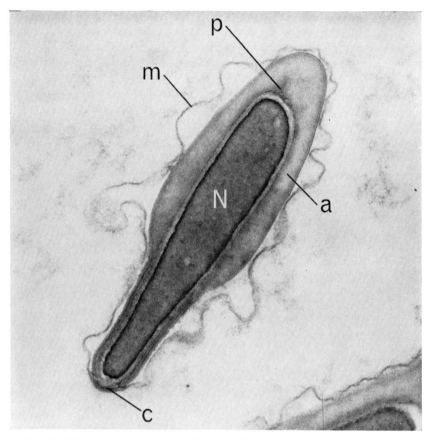

Fig. 5. Transverse section through the midregion of the golden hamster sperm head. Key: (N) nucleus; (a) acrosome; (p) perforatorium; (c) postnuclear cap (layer); (m) plasma membrane. Note division of the acrosome into a bulkier dorsal region (upper half of the picture) and a thinner ventrolateral region (lower half of the picture). Magnification: × 54,000.

new details on the process of acrosome development in a number of mammalian species (see reviews by Fawcett, 1958; M. W. H. Bishop and Walton, 1960; Leblond et al., 1963). These accounts show that, in connection with the Golgi apparatus, a round acrosomal vesicle arises at the future anterior pole of the spermatid nucleus. This vesicle contains a dense, strongly PAS-positive granule, the so-called acrosomal granule, surrounded by a clear zone that gives a faint PAS-positive reaction. In later stages of acrosome development, the acrosomal vesicle flattens and its clear zone extends laterally around the anterior half of the nucleus, forming a caplike structure (sometimes called "head cap") filled with a thin layer of material; the origin of the latter is not clearly understood. The material of the acrosomal granule seems to be confined mainly to the apical region in the mature acrosome which reaches a considerable size in the sperm of certain species (e.g., guinea pig, Chinese hamster).

The maturation of the acrosome continues during sperm descent through the epididymis. In the rabbit the acrosome decreases in size and its margins become more attenuated (Bedford, 1963a, 1964b). The large apical region of the guinea pig acrosome undergoes particularly marked

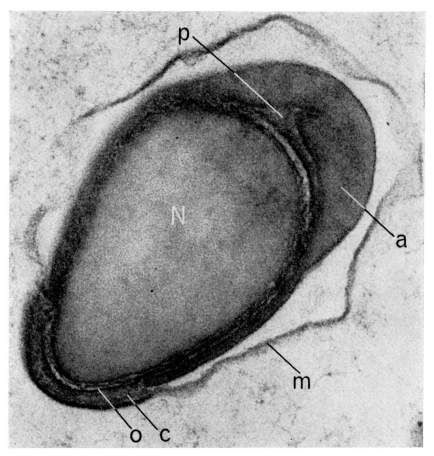

Fig. 6. Transverse section (somewhat oblique) through the midregion of the rat sperm head. Key: (N) nucleus; (a) acrosome—note the bulkier dorsal and thin lateral regions; (p) perforatorium; (o) outer nuclear membrane; (c) postnuclear cap, which appears to consist of two layers in this area; (m) plasma membrane. Note that the plasma membrane is detached and swollen over the acrosome but is firmly adherent to the postnuclear cap material. Magnification: × 110,000.

changes in shape and in internal structure as judged by the distribution of zones of differing electron density (Fawcett and Hollenberg, 1963; Fig. 8). The functional significance of these changes is obscure but they could be related somehow to the known progressive increase in fertilizing capacity of the spermatozoa as they pass through the epididymal tract (Young, 1931). A change in the character of swimming movements during epididymal transport has been described in rat sperm (Blandau and Rumery, 1964), but it is not known whether this is in any way connected

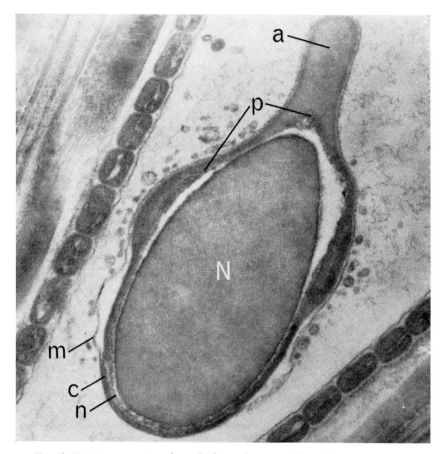

Fig. 7. Transverse section through the midregion of the Chinese hamster sperm head. Key: (N) nucleus; (a) acrosome, showing a more complex distribution of acrosomal material than in golden hamster and rat spermatozoa (cf. Figs. 5 and 6); (p) perforatorium, consisting of dorsal and lateral portions; (n) nuclear membrane; (c) postnuclear cap (layer); (m) plasma membrane. Note the accumulation of small subcellular particles on the surface of the plasma membrane in the acrosomal region (cf. Section II,F and Fig. 16). Magnification: × 54,000.

with changes in the size and shape of the sperm head. Testicular and epididymal spermatozoa of the guinea pig also differed in the antigenic behavior of the acrosome in immunofluorescence tests (Willson and Katsh, 1965).

The acrosomal material of mature sperm of most mammals appears rather homogeneous under the electron microscope. However, the thin lateral portion of the acrosome is often more electron-dense and relatively more resistant to swelling and breakdown than the anterior portion, e.g., in bull sperm (Saacke and Almquist, 1964; Blom and Birch-Andersen, 1965; Nicander and Bane, 1966). The mature acrosome does not normally contain other cytoplasmic organelles or elements of any kind. This would argue against the possibility that it could carry out protein biosynthesis, as has been suggested (Abraham and Bhargava, 1963; Iype *et al.*, 1963). Available evidence also indicates that no new synthesis of hyaluronidase, a major acrosomal component (see below), occurs in mature sperm (Swyer, 1947a).

The acrosomal membrane exhibits, at least after certain fixing and staining procedures, a geometrically asymmetric triple-layered structure with its thicker layer facing to the outside, i.e., toward the overlying sperm plasma membrane and the outer nuclear membrane, respectively (Fig. 6). The acrosomal membrane is then structurally similar to plasma membrane (Sjöstrand, 1963). An additional thickening of the surface of the outer acrosomal membrane (facing the sperm plasma membrane) is visible in certain areas of rodent acrosomes (see Figs. 1, 12, and 13). This coating seems to have the function of delimiting the area undergoing fusion and vesiculation during acrosome reaction and sperm penetration (see Figs. 25 and 26). In most mammals the acrosome is a single unit. In rat and mouse sperm, however, a small separate vesicle resembling the thin lateral portion of the acrosome is located on the ventral surface of the perforatorium (Figs. 9, 11, and 13). It seems to arise through a pinching-off process from the lateral edge of the acrosomal vesicle during spermiogenesis (Brökelmann, 1963; Nicander, personal communication).

The chemical composition of the acrosome is complex and not fully known (see the extensive review by Mann, 1964). Its characteristic PAS-positive reaction has been generally attributed, on histochemical evidence, to the presence of neutral polysaccharides (Wislocki, 1949; Leblond, 1950; Leuchtenberger and Schrader, 1950; Schrader and Leuchtenberger, 1951). However, Onuma and Nishikawa (1963) and Onuma (1963) were able to remove much of the PAS-reactive material from boar, bull, and stallion spermatozoa with lipid solvents and concluded that glycolipid is a major acrosomal component.

Clermont *et al.* (1955a) were the first to prepare an "acrosomal ex-

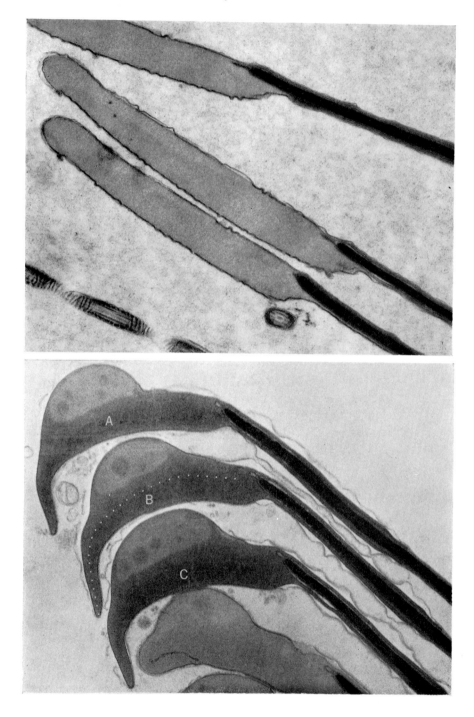

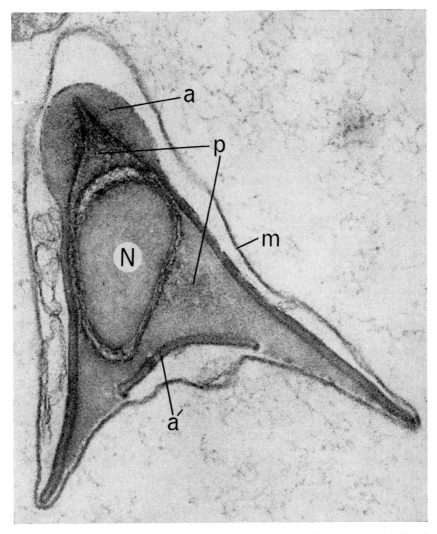

Fɪɢ. 9. A transverse section of the anterior portion of the rat sperm head (cf. diagram 3 of Fig. 1C). Key: (N) nucleus; (a) acrosome; (a′) small vesicle on the ventral surface of the perforatorium (p)—it is probably part of the acrosomal system; (m) plasma membrane. Magnification: × 96,000.

Fɪɢ. 8. Sagittal sections of the anterior portions of the heads of guinea pig spermatozoa showing changes in the shape and internal structure of the acrosome during passage through the epididymis. Upper: spermatozoa from a proximal segment of the epididymis. Lower: spermatozoa from a more distal segment of the epididymis; note also the aggregation ("stacking") of the spermatozoa (cf. Section II,F). Magnification: × 18,000. (From Fawcett and Hollenberg, 1963.)

tract" from guinea pig spermatozoa with 0.1 N sodium hydroxide; they isolated an alcohol-precipitable fraction consisting of a carbohydrate–protein complex, later named "acrosomin" (Clermont and Leblond, 1955). Hartree and Srivastava (1965) obtained roughly identical lipid and glycoprotein fractions from ram spermatozoa extracted either with 0.0125 N sodium hydroxide or with an anionic detergent. Treatment with detergent results in a detachment *in toto* of many acrosomes which can be collected by differential centrifugation. The acrosomal lipids had a high content of phospholipid of a composition similar to that of whole spermatozoa, i.e., consisting mainly of plasmalogen and lecithin; however, contamination from nonacrosomal parts of the sperm cell could not be ruled out. The glycoprotein fraction contained glutamic acid as the predominant amino acid. The following sugars were present: mannose, galactose, fucose, glucosamine, and sialic acid. All the sialic acid content of rat sperm was removed by the detergent treatment; but since sialic acid is known to be present on the plasma membrane, dissolved by the detergent, the contribution of the acrosomal material itself remains to be determined.

Very little is known at the present time about the distribution and molecular arrangement of the various components within the acrosome. Mann (1964, p. 132) has recently discussed evidence indicating the presence of large molecular carbohydrate–protein–lipid conjugates. The polysaccharides themselves could be of several kinds. The latter view is corroborated by the apparent antigenic complexity of the acrosome. "Acrosomal extracts" (with 0.01 N NaOH) of Chinese hamster epididymal spermatozoa gave six precipitation lines in immunodiffusion tests with rabbit antiserum against Chinese hamster epididymal sperm (Pikó and Tyler, 1962; Fig. 10); similar tests with ram spermatozoa yielded four precipitation lines (Hathaway and Hartree, 1963). The structural and functional role of these various components is not clear; they may be involved in the enzymic function of the acrosome during sperm penetration, e.g., by a progressive release of acrosomal enzymes (see Section IV,B and C).

Present evidence indicates that the acrosome contains at least two enzymes which operate during sperm entry—hyaluronidase and zona lysin. The first evidence that an egg membrane lysin originates from the acrosome was presented by Tyler (1949) in the mollusk *Megathura crenulata*.[*] In the case of hyaluronidase of mammalian sperm, evidence

---

[*] Parat (1933a) showed that isolated *Discoglossus* acrosomes, separated from the rest of the sperm head by microdissection, contain a lytic agent capable of digesting the jelly around the egg. But he concluded (Parat, 1933b) that the acrosomal enzyme functions in the activation of the egg as suggested earlier by Bowen (1924).

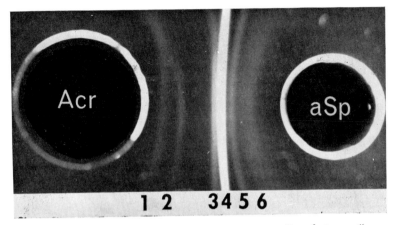

Fig. 10. Immunodiffusion plate showing precipitation lines between "acrosomal extracts" of Chinese hamster sperm (Acr)—obtained after extraction with 0.01 N sodium hydroxide—and rabbit antiserum against Chinese hamster sperm (aSp). One very strong line (No. 4) and five weaker precipitation lines are visible.

of its acrosomal origin was indicated by Leuchtenberger and Schrader (1950) on the following grounds: (1) the PAS-positive staining of the acrosome correlates with enzyme activity; (2) the enzyme appears in the developing testis when acrosomal material begins to accumulate; and (3) a PAS-positive reaction was also obtained with potent hyaluronidase preparations *in vitro*. The mucoprotein nature of highly purified bovine testicular hyaluronidase has been confirmed (Brunish and Högberg, 1960). When anti-hyaluronidase antibody was used in immunofluorescence techniques, specific fluorescence was observed only in the acrosome of mature bull sperm and in the perinuclear area of spermatids (Mancini *et al.*, 1964). Removal of the acrosome results in hyaluronidase release *in vitro* (Austin, 1960, 1961c; Hathaway and Hartree, 1963); on the other hand, sperm passage through the cumulus oophorus *in vivo*, which has long been thought to require the release of hyaluronidase by individual spermatozoa (cf. Austin, 1948), is accompanied by a change in the properties, and eventual "loss," of the acrosome (Austin and Bishop, 1958b). More recent electron microscopic observations make it clear that the acrosomal material is emptied during sperm entrance through the outer egg envelopes (Moricard, 1960, 1964; Hadek, 1963b; Austin, 1963b; Pikó and Tyler, 1964a,b; Barros *et al.*, 1967; see Section IV,B and C).

Since sperm penetration through the zona pellucida seems to involve a lytic action and since hyaluronidase itself does not dissolve the zona, a separate "zona lysin" was postulated with the perforatorium as its possible source (Austin and Bishop, 1958b). Recently, Srivastava *et al.* (1965a,b) have obtained an acrosomal extract from ram, bull, and rabbit sperm that was found effective in dispersing the corona radiata and dissolving the zona pellucida of rabbit eggs. The extract contained a lipoglycoprotein complex and also had proteolytic activity. This observation is the first direct evidence of the existence of a zona lysin in mammalian sperm. It also indicates that the acrosome, rather than the perforatorium, could be the source of this enzyme—a proposition supported by fine-structural observations of sperm entry (Pikó, 1964; see Section IV,C). According to Stambaugh and Buckley (1968) the rabbit acrosomal enzyme responsible for the dissolution of the zona pellucida is similar to trypsin.

The relative distribution of hyaluronidase and zona lysin within the acrosome cannot be ascertained at present. Since hyaluronidase is first to function during sperm entry, it might be expected to occupy the apical region (or the leading edge) of the acrosome. This would be consistent with the probable origin of this region from the PAS-positive acrosomal granule. On the other hand, the zona lysin might be located mainly in the thin lateral portion of the acrosome that appears to originate from the "clear zone" of the acrosomal vesicle. This portion of the acrosome seems to be preserved during sperm passage through the cumulus oophorus (see Section IV,C). Immunochemical studies at the fine-structural level (Sternberger, 1967) may provide information on the intra-acrosomal localization of acrosomal enzymes.

C. PERFORATORIUM

The name perforatorium is used here to denote the fibrous or electron-opaque material which, in certain areas of the sperm head, fills the interspace between the acrosomal sac and the outer nuclear membrane. The perforatorium is probably present in most or all mammalian spermatozoa but its size and shape vary greatly: in the more usual "bat-shaped" sperm head it is a rather inconspicuous structure visible only under the electron microscope; but it can reach great proportions in the crescent-shaped sperm heads of some rodents. However, the human sperm appears to lack any structure comparable to the subacrosomal space or perforatorium (Bedford, 1967).*

---

* A note may be added here on the origin and history of the term "perforatorium." It was introduced by Waldeyer (1906, p. 105 *et seq.*) to designate the whole apical complex of the head of flagellate sperm (in allusion to its supposed function as a means of boring into the egg during sperm entry); the same author refers to the "acrosome" (a term introduced by Von Lenhossék, 1898, p. 282) as the tough inner component of certain perforatoria, e.g., in rat sperm. These two terms

Structures probably analogous to the perforatorium have been described in the spermatozoa of many invertebrate and lower vertebrate species. Afzelius (1955) was the first to demonstrate electron microscopically the presence of two different components within the intact acrosomal region of sea urchin sperm and to show (Afzelius, 1956; Afzelius and Murray, 1957) that, upon eversion of the acrosome, the deeper-located component elongates to form a fibrous filament. The formation of this so-called "acrosome filament"—a characteristic feature of sperm penetration in many species—has been studied in considerable detail (Dan, 1956; Colwin and Colwin, 1957, 1963a; and Chapters 6 and 7 of Volume I of this treatise). These studies suggest that the fibrous core seen in thin sections of the acrosomal process has the role of a supporting element. Although no further elongation of the mammalian perforatorium is evident during sperm entry, it appears likely that its role is also structural: (1) providing support for the acrosome and (2) serving as an attachment device between the inner acrosomal membrane, which is exposed during sperm entry, and the nuclear envelope.

The perforatorium of rat sperm is particularly large and has been repeatedly described in light microscopic studies (M. W. H. Bishop and Walton, 1960). It projects apically to a considerable extent and has the shape of a three-pointed star in cross sections (Pikó and Tyler, 1964b). The dorsal prong runs along most of the convex (leading) edge of the sickle-shaped nucleus forming a crest in the median plane of the sperm head. The two ventral prongs extend posteriorly to about one-third of the length of the nucleus (Figs. 1, 9, and 11). Clermont *et al.* (1955b) studied some of the histochemical properties of the rat perforatorium and found it PAS- and Feulgen-negative, moderately acidophilic, strongly staining with iron hematoxylin, and extremely resistant to alkaline hydrolysis.

Among other murine and cricetine rodents, the spermatozoa of the mouse, cotton rat, golden hamster, and Chinese hamster have fairly large perforatoria (Austin and Bishop, 1958a; Figs. 12 and 13). In the sperm of rabbit, bull, ram, boar, dog, and guinea pig, an accumulation of

---

have often been used synonymously by later authors (Wilson, 1928, p. 281). In recent years, the term perforatorium has been adopted, by several authors, to describe the rodlike structure at the apex of the head of certain rodent spermatozoa (M. W. H. Bishop and Walton, 1960); with the advent of the electron microscope, the term has been extended to include analogous structures in the spermatozoa of other mammalian and vertebrate species, e.g., toad (Burgos and Fawcett, 1956), chicken (Nagano, 1962), various mammals (Bane and Nicander, 1963). It is retained here in this wider sense. The functional implications of the word are, of course, open to criticism. Perhaps "subacrosomal material" would be a more appropriate name.

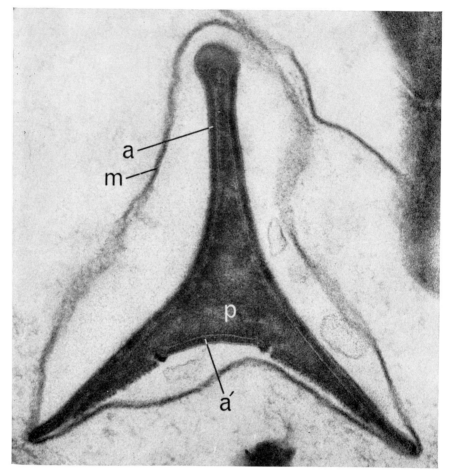

FIG. 11. Transverse section through the tip of the rat sperm head (cf. diagram 2 of Fig. 1C). The perforatorium (p) has the shape of a three-pointed star. Most of its surface is covered by the thin portion of the acrosomal sac (a, a'). The detachment of the plasma membrane (m) is probably a fixation artifact. Magnification: × 105,000.

electron-dense material, most likely equivalent to the perforatorium of the above rodents, is seen along the anterior edge of the nucleus, where it forms a thin crest, and laterally in the so-called "equatorial" (middle) region of the sperm head (Nicander and Bane, 1962, 1966; Bane and Nicander, 1963; Blom and Birch-Andersen, 1965; Fig. 14).

In the rat the development of the perforatorium can be reconstructed from the description of Brökelmann (1963). As nuclear condensation advances, a widening space is formed between the acrosomal vesicle and the nuclear membrane. This "subacrosomal space" appears as a clear area

with some filamentous contents but becomes quite electron-dense later.
A similar course of events has been observed in several mammalian
species, indicating that the perforatorium of the rat and that of other
species are, indeed, related structures (Bane and Nicander, 1963; Hopsu
and Arstila, 1965). An essentially similar process for the accumulation of
subacrosomal material has been described in more distant animal species:
toad (Burgos and Fawcett, 1956), chicken (Nagano, 1962), and house
cricket (J. S. Kaye, 1962). The origin of the material forming the perforatorium is not clear. Clermont *et al.* (1955b) regarded the rat perforatorium, on the basis of histochemical evidence, as an extension of the
nuclear membrane; according to Burgos and Fawcett (1956), however,
the fibrous substance of the perforatorium of toad spermatids is formed
outside the nucleus and independently of the nuclear membrane. J. S.
Kaye (1962) suggested that the "inner cone" of the acrosome of the house
cricket (presumably analogous to the perforatorium) could derive, at

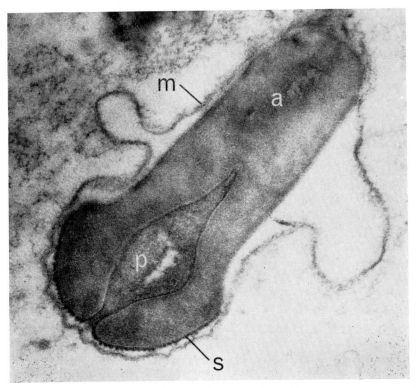

Fig. 12. Transverse section through the tip of the golden hamster sperm head
(anterior to the nucleus). Key: (p) perforatorium; (a) acrosome; (s) material
coating the surface of the outer acrosomal membrane; (m) plasma membrane.
Magnification: × 90,000.

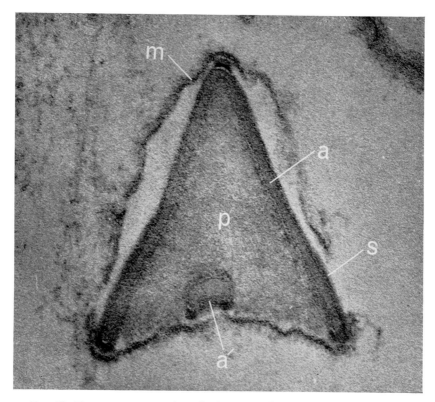

FIG. 13. Transverse section through the tip of the mouse sperm head (anterior to the nucleus). Key: (p) perforatorium; (a) acrosome; (a′) small ventral vesicle, probably part of the acrosomal system; (s) surface coating of the outer acrosomal membrane; (m) plasma membrane. Magnification: × 180,000.

least in part, from the acrosomal granule. The perforatorium of the salamander *Pleurodeles* shows a complex tripartite structure; but the origin of the various components is not known (Picheral, 1967).

The chemical composition of the perforatorium has not been explored aside from the above-mentioned study of Clermont *et al.* (1955b). Burgos and Fawcett (1956) suggest, on the basis of the fine-structural appearance of the toad perforatorium, that it may be comprised of a scleroprotein. The presence of highly insoluble keratinlike proteins, rich in sulfur, has been reported in the sperm head of several mammalian species (cf. Mann, 1964; Henricks and Mayer, 1965).

D. POSTNUCLEAR CAP

A strongly argentophil layer covering the posterior half of the nucleus was first described by Gatenby and Wigoder (1929) in guinea pig sperm

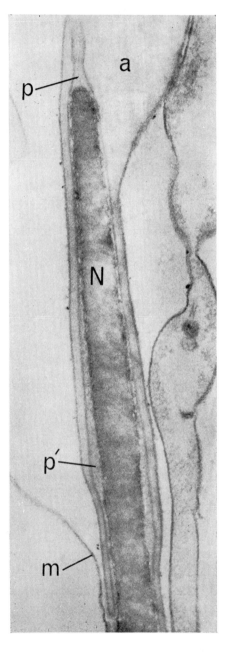

FIG. 14. Sagittal section of the guinea pig sperm head showing the apical and "equatorial" portions of the perforatorium (p and p', respectively). Key: (N) nucleus; (a) acrosome; (m) plasma membrane. Magnification: × 55,000. (Courtesy of Drs. L. Nicander and A. Bane.)

and later demonstrated in human spermatids (Gatenby and Beams, 1935). A similar layer, usually referred to as the "postnuclear cap," has been observed in the sperm of a variety of mammalian species (e.g., Fig. 10 of M. W. H. Bishop and Walton, 1960). But whether a comparable structure occurs in all flagellate spermatozoa as suggested by Gatenby and Wigoder (1929) appears doubtful. In invertebrate sperm, in general, no accumulation of material is seen in the space between the plasma membrane and the nuclear envelope in the area posterior to the acrosome.

In electron micrographs of thin sections of mammalian sperm the postnuclear cap appears as a thin intermediate layer between the sperm plasma membrane and the outer nuclear membrane (cf. Figs. 4, 6, and 7). Its fine structure has not been resolved clearly and may vary in different species; it has been described as fibrous (Randall and Friedlaender, 1950; Pikó and Tyler, 1964b), microtubular (Blom and Birch-Andersen, 1965), and membranous (Fawcett and Ito, 1965). This layer is always tightly attached to the overlying plasma membrane and, to a lesser extent, to the nuclear membrane, indicating a possible structural role in binding these membranes together. Various intercellular attachment devices are known to exist between the plasma membranes of epithelial cells that are exposed to mechanical stress (e.g., Farquhar and Palade, 1963).

Anteriorly, the postnuclear cap is limited by the posterior edge of the acrosomal sac which often shows a local thickening in this area (see Figs. 5 and 7). The posterior border of the postnuclear cap is demarcated by a narrow band of dense material that forms a tight connection between the plasma membrane and the outer nuclear membrane (cf. Figs. 1, 3, 4, and 15); it could be identical with the "posterior ring" described by Gresson and Zlotnik (1945).

The origin and formation of the postnuclear cap are not fully understood. An argentophil nuclear ring appears in spermatids at the posterior border of the acrosome and spreads progressively in a caudal direction (Zlotnik, 1943; Gresson and Zlotnik, 1945; Hancock, 1957; Hancock and Trevan, 1957). According to Burgos and Fawcett (1955) the nuclear ring probably arises through a local differentiation of the cytoplasm or the sperm plasma membrane. As the nuclear ring moves posteriorly some of its material may be left behind to form the postnuclear cap (Nicander, personal communication; cf. Fig. 8 of Hadek, 1963a). It is not clear whether the microtubular system of the manchette, thought to be of endoplasmic reticular origin, actually contributes to the formation of the postnuclear cap as suggested by Das (1962).

The chemical composition of the postnuclear cap is not known. Its strongly argentophil nature could be due to the presence of phospholipids (Alsterberg, 1941); this would also be consistent with its suggested origin from membranous structures. The strong staining with phosphotungstic acid noted by Nicander and Bane (1962, 1966) may indicate the presence of proteins rich in basic amino acids, especially arginine (Kühn *et al.*, 1958). A faint acid phosphatase reaction in the postnuclear cap area was reported in deer sperm (Wislocki, 1949). The postnuclear layer may have a role in the rapid breakdown of surrounding membranes during sperm entry (see Section IV,D). Alternatively, it could be a

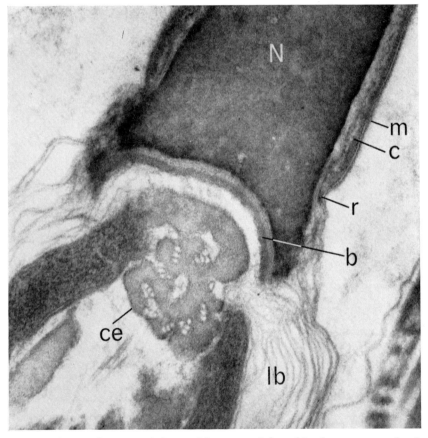

Fɪɢ. 15. Frontal section of the caudal portion of the golden hamster sperm head. Key: (N) nucleus; (c) postnuclear cap (layer); (r) posterior ring (see text); (m) plasma membrane—note a layer of coating material on the outer surface of the plasma membrane in the postnuclear cap area (cf. Section II,E); (b) basal plate; (lb) lamellar body; (ce) centriole. Magnification: × 85,000.

purely structural device for maintaining firm cohesion between the
plasma membrane and the nuclear envelope in this area (Fawcett and
Ito, 1965).

E. PLASMA MEMBRANE

The mature spermatozoon is covered by a continuous plasma membrane deriving from the original plasma membrane of the spermatid.
Besides serving as a permeability barrier, the sperm plasma membrane
most likely contains the specific receptor sites that interact with complementary egg substances. The plasma membrane covering the sperm head
is of particular interest because it is in this area where the major events
of sperm penetration are initiated—the acrosome reaction and the incorporation of the sperm by the egg cytoplasm (see Section IV,B and D).
A general survey of gamete membrane structure and of interacting substances is presented by Metz (Chapter 5 of Volume I of this treatise).

A special differentiation of the spermatid plasma membrane takes
place toward the end of spermiogenesis. The Sertoli cytoplasm surrounding the sperm head shows considerable metabolic activity as indicated by
the presence of numerous vesicles and an accumulation of electron-dense
material (man—Horstmann, 1961; rat—Brökelmann, 1963). This period
coincides with intense nucleoside triphosphatase activity at the interface
between the plasma membranes of the two cells suggesting a transport
of materials from the Sertoli cell to the spermatid (Tice and Barrnett,
1963). The spermatid plasma membrane gradually becomes thicker and
more electron-dense. André (1963) has suggested that the surface coating of the late spermatid might be made of mucopolysaccharides; its role
would be to reduce the adhesiveness of the cell surface and, thus, to
facilitate the release of the sperm into the lumen. The polysaccharide
nature of the coating would be in agreement with the observation that
the heads of spermatozoa from the caput epididymis of rabbit carry a
high negative surface charge and show little tendency to head-to-head
autoagglutination (Bedford, 1963b). Sialic acid appears to be one of the
surface constituents since the negative charge of bull spermatozoa
diminishes after treatment with neuraminidase (Fuhrmann et al., 1963).
A special coating consisting of regularly spaced small granules was observed on the plasma membrane overlying the acrosome (except for its
apical region) in testicular spermatozoa of the salamander *Pleurodeles
waltlii* (Picheral, 1967).

The extra coating of the late spermatid seems to persist at least
during sperm passage through the epididymis. In rat and golden hamster
sperm from the cauda epididymis, the plasma membrane of the sperm
head appears thicker and denser than that of the sperm tail. This is

particularly well seen over the postnuclear cap where the plasma membrane is tightly attached and where, therefore, it is easier to obtain a favorable section (Figs. 4 and 15).

The plasma membrane covering the acrosome is usually detached and more or less distorted in fixed and sectioned specimens (e.g., Figs. 5, 6, 9, and 11). This is probably due to an artifact of fixation. The degree of detachment depends, to some extent, on the osmolarity of the fixative and the physiological state of the sperm; e.g., the plasma membrane separates more easily in uterine spermatozoa than in epididymal and ejaculate sperm, probably because of a change in membrane permeability (Bedford, 1964a,c).

F. COATING SUBSTANCES; CAPACITATION

Recent observations indicate that spermatozoa readily adsorb materials from the secretions of the epididymis and the accessory glands. Some of the adsorbed substances become quite tightly bound to the sperm surface. For example, they may not be removed by repeated washings in physiological saline, although, in some instances, they can be detached electrophoretically (e.g., Searcy *et al.*, 1964). "Spermatozoa-coating antigens" of seminal vesicle origin have been reported in rabbit and man (Weil, 1960, 1965; Weil and Rodenburg, 1960, 1962) as well as in bull (Hunter and Hafs, 1964; Matousek, 1964). The human A and B blood group antigens were found to be present on the sperm of secretors only (Edwards *et al.*, 1964), and these same antigens were also adsorbed by spermatozoa *in vitro* (Boettcher, 1965). Stone (1964, p. 99) maintains that, among the blood group antigens of cattle, solely the so-called J substance (where all J-positive individuals are secretors) can be detected with certainty on spermatozoa. The major spermatozoa-coating antigen in human seminal plasma appears to be antigenically related to lactoferrin, an iron-binding protein which occurs in milk and various other external secretions (Hekman and Rümke, 1969).

Investigations of the electrophoretic mobility of spermatozoa also indicate that the sperm surface may change upon contact with the secretions of the male tract. In rabbit the initially high negative charge of the sperm head, as compared to that of the sperm tail, decreases as the spermatozoa migrate through the epididymis (Bedford, 1963b). In ejaculated spermatozoa of the bull (Nevo *et al.*, 1961) and the ram (Bangham, 1961) the tail also carries a higher negative charge than the head. Fuhrmann *et al.* (1963) noted that bull spermatozoa from whole semen had a lower negative charge than epididymal spermatozoa; the negative charge rose when the sperm were washed repeatedly but fell again when the washed cells were incubated with seminal plasma. Hartree and

Srivastava (1965) found the sialic acid content of ejaculated bull sperm about the same as that of epididymal sperm. This observation suggests that the change in surface charge is not due to a loss of sialic acid residues but rather to the binding of positively charged components of accessory gland secretions.

Some of the adsorbed materials are relatively large, for they sediment readily in the ultracentrifuge. Thus, the antiagglutinating factor of seminal plasma, "sperm antagglutin" (cf. Lindahl, 1960), was found to be associated with subcellular particles (Lindahl and Brattsand, 1962). The so-called "decapacitation factor" of seminal plasma and epididymal fluid [which reversibly inhibits the fertilizing capacity of previously "capacitated" sperm as described originally by Chang (1957)] also has a fast rate of sedimentation (Bedford and Chang, 1962; Pinsker and Williams, 1967). The secretory function of the epididymis is well established, including the presence of holocrine secretory cells (Risley, 1963; Martan and Risley, 1963; Martan and Allen, 1964); these cells may be the source of some of the large aggregates found in seminal plasma.

An array of small tubules and vesicles can be seen attached to the surface of the plasma membrane overlying the acrosome in Chinese hamster spermatozoa from the cauda epididymis (Figs. 7 and 16; Pikó, 1967). According to Fawcett and Phillips (1967) this special coating is acquired toward the end of spermatid differentiation and is apparently peculiar to Chinese hamster spermatozoa.

A somewhat different kind of association may be mentioned in this context, namely, the aggregation ("stacking") of guinea pig spermatozoa in the epididymis (Fawcett and Hollenberg, 1963) and the so-called conjugating pairs of opossum spermatozoa (Biggers and Creed, 1962). In both instances, the associated spermatozoa are firmly held together in an area of the plasma membrane overlying the acrosome (Fawcett and Hollenberg, 1963). Although the mechanism of attachment is not known, one likely explanation is that the plasma membrane has special properties that result in a selective cohesion of the membrane surfaces at this site.

Little is known about the physiological significance of the various coating substances. The presence of some substances, such as some of the blood group antigens, seems incidental; nevertheless, once present, they could conceivably interfere with the normal process of sperm transport and fertilization (cf. Tyler, 1961; Tyler and Bishop, 1963; Stone, 1964). The secretory activity of the epididymis has long been implicated in the physiological maturation of spermatozoa during their descent in this organ, but its exact role is still unknown (e.g., Gaddum and Glover, 1965). For example, the fertilizing capacity of rabbit spermatozoa increases greatly upon passage through the distal portion of the corpus epididymis (Orgebin-Crist, 1967). It is conceivable that some of the coating substances might have a protective action on spermatozoa in one or another of several ways, e.g., by (*1*) preventing their agglutination in the ejaculate and in the female tract; (*2*) regulating membrane permeability and thus possibly affecting the metabolism of spermatozoa; (*3*) stabilizing the plasma membrane over the acrosome, by virtue of covering receptor sites at its surface, and thus preventing the premature occurrence of the acrosome reaction; and (*4*) preventing the early uptake of spermatozoa by phagocytes. It has been reported recently that rabbit spermatozoa that have resided in the estrous uterus for some hours undergo a surface change which renders them vulnerable to leukocytes (Bedford, 1965). The plasma membrane of uterine spermatozoa becomes

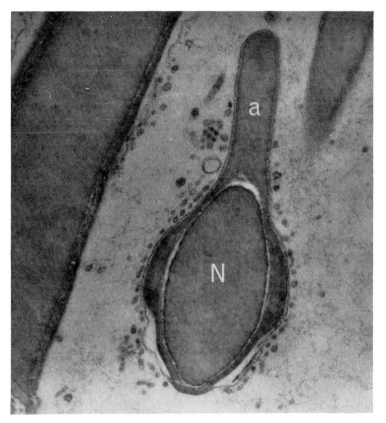

FIG. 16. Transverse section through the anterior portion of the head of Chinese hamster epididymal sperm showing aggregation of small particles on the surface of the plasma membrane in the acrosomal region (cf. also Fig. 7). (N) Nucleus; (a) acrosome. Magnification: ✕ 43,000.

more permeable and generally more adhesive over the acrosome (Bedford, 1964a,c). It seems likely that these changes are concomitant with the so-called "capacitation" process, that is, the phenomenon that epididymal or normally ejaculated spermatozoa must spend a certain length of time in the female tract in order to be able to penetrate the egg (Austin, 1951a; Chang, 1951; Noyes, 1959; Austin, 1963a, 1964).

The nature and mechanism of sperm capacitation are still incompletely understood. But in the light of the above considerations it seems reasonable to suppose that capacitation consists, at least in part, in the gradual removal of coating material from the sperm surface, especially in the acrosome region. This would result in an exposure of sperm receptor sites allowing spermatozoa to interact specifically with egg receptors, or with substances emanating from the egg, and to undergo the acrosome reaction upon contact with the outer coats of the egg (see Section IV,B). In rat sperm that have recently penetrated through the zona pellucida, the plasma membrane in the postnuclear cap area appears thinner and less electron-dense than in epididymal sperm (compare Figs. 4 and 6 with Figs. 22 and 24). This observation suggests that the sperm plasma membrane loses all its coating materials, including those deposited by the Sertoli cells (see preceding section), by the time it contacts the plasma membrane of the egg. Capacitation may involve enzymic action by secretions of the female tract as suggested by recent evidence regarding *in vitro* capacitation of rabbit spermatozoa when incubated in the presence of β-amylase (Kirton and Hafs, 1965). Similarly the "decapacitation factor" of rabbit seminal plasma was inactivated by β-amylase but not by lysozyme or Pronase (Dukelow *et al.*, 1966a,b; Pinsker and Williams, 1967). High levels of amylase activity have been reported in the fallopian tubes of several species (McGeachin *et al.*, 1958), in cysts of the human fallopian tube (Green, 1957), and in human cervical mucus (Gregoire *et al.*, 1967). The involvement of large molecular species in the capacitation process is also indicated by the observation that capacitation does not occur when rabbit spermatozoa are enclosed in a cellophane dialysis bag within the uterus of an estrous female (Noyes *et al.*, 1958; Noyes, 1959). However, these observations do not necessarily rule out the possibility that small molecular components may also play a role in sperm capacitation. In general, spermatozoa seem to require a relatively long interval to attain a capacitated state: aproximately 6 hours in the rabbit (Chang, 1955) and the sow (Thibault, 1959); 2–4 hours in the rat (Austin and Braden, 1954) and the golden hamster (Chang and Sheaffer, 1957); at least 1½ hours in the ewe (Mattner, 1963); but not more than 1 hour in the mouse (Braden and Austin, 1954a).

Thus far no clear-cut evidence of capacitation, comparable to the process in mammals, has been described in other animal groups. In species with external fertilization, the acrosome reaction in general occurs readily upon contact of spermatozoa with the egg envelopes (cf. Dan, 1956, and Chapter 6 of Volume I of this treatise). In a wider sense, if capacitation is regarded as an exposure of specific receptor sites that had been previously unavailable, some analogy may be found with the so-called "activation" phenomenon of T-even bacteriophages. Activation requires the presence of certain cofactors of which tryptophan is the most effective. In the absence of cofactor molecules the tail fibers, which form the adsorption organelle of the phage, are tightly wound around the tail sheath; upon reaction with tryptophan the fibers are released and are able to attach to the receptor sites of the bacterial cell surfaces (Stent, 1963; Kellenberger *et al.*, 1965). Further analogies may be found with the inhibition of viral hemagglutination by various mucoproteins, and the reversal of this inhibition (e.g., in the case of influenza virus). The inhibition is due to a coating of the viral surface by soluble mucoprotein molecules that are, presumably, similar to the receptors present on red blood cells. The influenza virus can progressively destroy the inhibitory activity of the mucoprotein, and regain its hemagglutinating capacity, by means of an enzyme, neuraminidase, of the viral surface (cf. Buzzel and Hanig, 1958; Fazekas de St. Groth, 1962; Bayer, 1964).

The proposition that capacitation of mammalian spermatozoa involves a loss of surface coatings is amenable to experimental verification. It would be interesting to see, for example, whether the "spermatozoa-coating antigen" of Weil is still present on capacitated sperm. The behavior of other antigenic components, e.g., blood group antigens and specific antigens of the sperm surface, could also be studied. However, a complete elucidation of the mechanism of capacitation will require a reliable method of *in vitro* fertilization of mammalian eggs under precisely controlled conditions (cf. Thibault in Chapter 9 of this volume).

III. Structure and Properties of the Ovulated Egg

In most mammals, the egg undergoes the first meiotic division shortly before ovulation and remains arrested in second meiotic metaphase until sperm penetration occurs (for variations from this general pattern see Austin in Chapter 10 of this volume). Typically, the cytoplasm of the freshly ovulated egg is surrounded by its plasma membrane proper and by two extracellular coats: the zona pellucida and the cumulus oophorus (Fig. 17). Both of these coats arise in the ovarian follicle where they are closely associated with the growth and differentiation, and with eventual

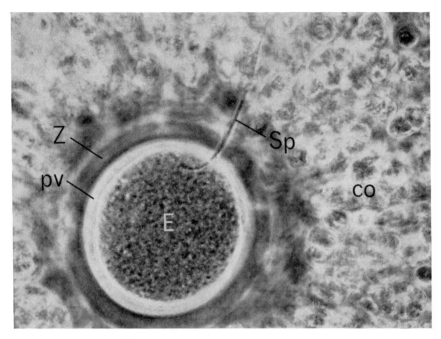

FIG. 17. A recently ovulated Chinese hamster egg surrounded by cumulus oophorus (co) and penetrated by a spermatozoon (Sp). The sperm head is attached to the surface of the egg cytoplasm (E) while most of the tail is still located within the cumulus. (Z) Zona pellucida; (pv) perivitelline space. Phase contrast microscopy. Magnification: × 460.

release, of the oocyte. Recently the fine structural aspects of follicle development and oocyte differentiation have been studied extensively in several mammalian species: rat (Sotelo and Porter, 1959; Franchi, 1960; Odor, 1960; Björkman, 1962); mouse (Yamada *et al.*, 1957; Chiquoine, 1960; Hadek, 1963c); rabbit (Trujillo-Cenóz and Sotelo, 1959; Blanchette, 1961; Merker, 1961; Zamboni and Mastroianni, 1966a,b); guinea pig (Anderson and Beams, 1960; Adams and Hertig, 1964); golden hamster (Odor, 1965; Weakley, 1966, 1967); rhesus monkey (Hope, 1965); and man (Wartenberg and Stegner, 1960; Stegner and Wartenberg, 1961; Tardini *et al.*, 1960, 1961; Lanzavecchia and Mangioni, 1964; Hertig and Adams, 1967). In the following survey the structure and properties of the egg coats and of the cytoplasmic surface of the freshly ovulated egg will be discussed primarily from the point of view of the mechanisms of sperm entry. For earlier literature the reader is referred to reviews by Austin (1961a, 1965) and by Blandau (1961). More details on oogenesis and the ovarian follicle can be found in reviews by Brambell (1956), Raven (1961), Franchi *et al.* (1962), Harrison (1962), and Jacoby (1962) and

Srivastava (1965). The fine structure of the oocyte has been reviewed by Beams (1964) and Hadek (1965).

A. Cumulus Oophorus

The newly ovulated egg is usually surrounded by the cumulus oophorus, consisting of several layers of granulosa cells interspersed in a jelly-like ground substance. The cells closest to the zona pellucida are more tightly packed and radially arranged, forming the corona radiata. The size, compactness, and the persistence of the cumulus in tubal eggs vary according to species (Austin, 1961a; Blandau, 1961). There is considerable evidence showing that the granulosa cells of the growing follicle carry out intensive synthetic and secretory activities (Björkman, 1962; Hadek, 1963c). They are attached to each other partly by cellular adhesion and partly by desmosome-like connections (Björkman, 1962). As the follicle approaches maturity, the cells of the cumulus become separated (except for the corona radiata) and undergo degenerative changes as shown by their microscopic and fine-structural appearance (Björkman, 1962) and their reduced ability to grow in tissue culture (Blandau and Rumery, 1962).

Data on the chemical composition of the intercellular matrix are limited but indicate the presence of several components. Hyaluronic acid is considered a major constituent because the matrix is dissolved rapidly by hyaluronidase preparations from various sources: spermatozoa, testes, snake venom, leech, and bacteria (McClean and Rowlands, 1942; Fekete and Duran-Reynals, 1943). The cumulus reacts strongly in histochemical tests employed for the demonstration of acid mucopolysaccharides, e.g., it gives strong metachromasia with toluidine blue (Braden, 1952, 1955). The fact that there is no significant uptake of ^{35}S-sulfate in the intercellular cement of growing follicles indicates an absence of sulfated mucopolysaccharides (cf. Zachariae, 1959). The presence of hyaluronic acid is also consistent with the physical properties of the matrix, such as its highly viscous nature. (The physiological functions of hyaluronic acid in connective tissue have been discussed recently by Fitton Jackson, 1964). The cumulus also contains protein, as it is readily disintegrated by proteolytic enzymes, and some PAS-positive material other than hyaluronic acid (Braden, 1952, 1955).

The chemical structure of hyaluronic acid and the mechanism of action of testicular hyaluronidase are reasonably well established. Hyaluronic acid is probably an unbranched polymer composed of equimolar amounts of D-glucuronic acid and N-acetyl-D-glucosamine units attached through alternating $\beta(1 \to 3)$-glucuronidic and $\beta(1 \to 4)$-glucosaminidic bonds (Meyer, 1958). Estimates of its molecular weight vary depending

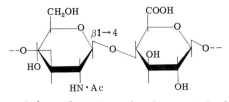

FIG. 18. Structure of the endo-β-N-acetyl-D-glucosaminidic bond that is acted upon by testicular hyaluronidase.

on the source and the method of preparation but in most samples it is on the order of several million (Balazs, 1958). Assuming a distance of 5 Å between two glucosidic bonds, an extended chain of hyaluronic acid with 2×10^6 molecular weight (composed of about 10^4 monosaccharide units) would be about 5μ long, but the actual configuration may be randomly coiled. Fessler and Fessler (1966) found the average molecular weight of hyaluronic acid isolated from human knee joint fluid to be about 2.5×10^6 by sedimentation and viscosity measurements; an electron microscope examination of the same preparation showed unbranched molecules $0.2–7 \mu$ long. Testicular hyaluronidase degrades hyaluronic acid by splitting the endo-β-glucosaminidic bond specifically (cf. Weissmann et al., 1964; Fig. 18).

The role of the cumulus in the fertilization process is not clear. In some species (e.g., sheep, cow, sow) it disintegrates rapidly after ovulation. However, experiments with rabbit eggs indicate that in this species the presence of the cumulus (and especially that of the corona radiata) may improve the chances of fertilization and prolong the fertilizable life of the egg. Removal of most of the cumulus, although leaving the corona radiata intact, significantly lowered the number of spermatozoa penetrating through the zona pellucida but did not affect the ferilizability of the eggs (Dickmann, 1964a); but dispersal of the corona cells (by shaking) reduced the fertilization rate considerably (Chang and Bedford, 1962; see Table I). Austin (1961a) suggested that the cumulus may facilitate penetration of the egg by offering a larger target for the spermatozoa to encounter and by guiding the spermatozoa toward the egg, through the radial arrangement of the follicle cells. As a highly viscous, slightly permeable layer (Austin and Lovelock, 1958; Glass, 1963) it may also contribute to the maintenance of a favorable microenvironment in the vicinity of the egg.

B. ZONA PELLUCIDA

The zona pellucida is a transparent layer that envelops the plasma membrane of the freshly ovulated egg. Its thickness ranges in different

TABLE I

FERTILIZATION OF RABBIT OVA AFTER REMOVAL OF THE CORONA RADIATA[a]

Age of ova (hr after ovulation)	No. of recipient rabbits	Ova without corona radiata				Ova with corona radiata			
		No. transferred	No. examined	Fertilized No.	(%)	No. transferred	No. examined	Fertilized No.	(%)
2	8	80	68	31	46	46	41	37	90
4	5	45	44	24	55	44	39	33	85
6	8	71	50	6	12	99	88	53	60

[a] From Chang and Bedford (1962).

species from about 3 to 15 μ. It persists through cleavage until the blastocyst stage. It is a structure analogous to the extracellular coats (zona radiata, vitelline membrane) surrounding the eggs of other vertebrate and invertebrate species (cf. Raven, 1961).

The zona pellucida is formed early during follicular development (Odor, 1965) by a gradual accumulation of material in the interspace between the oocyte and the surrounding follicle cells. The old controversy whether the zona derives from secretions of the follicle cells or of the egg remains unsettled (cf. Hope, 1965). The zona is penetrated by numerous egg microvilli and granulosa cell processes and is the site of active metabolic exchange during vitellogenesis. With the formation of the first polar body, the egg microvilli retract and a perivitelline space develops between the cytoplasmic surface and the zona (Odor, 1960; Weakley, 1966). At the same time the follicle cell processes begin to withdraw, but complete retraction seems to require a tubal factor (cf. Zamboni et al., 1965; Zamboni and Mastroianni, 1966a).

The zona pellucida of the ovulated egg appears as an intricate filamentous network under the electron microscope (Fig. 19). It is rich in carbohydrate and probably contains glycoprotein as a major structural component (Braden, 1952, 1955; cf. Jacoby, 1962). Variable amounts of sialic acid residues may be present (Soupart and Noyes, 1964) and may contribute to some of the structural features of the zona, e.g., its elasticity (Soupart and Clewe, 1965). The zona is dissolved by several proteolytic enzymes of which the enzyme Pronase (a *Streptomyces griseus* protease) was found to be the most effective (Mintz, 1962; Gwatkin, 1964).

There is evidence that the zona pellucida of at least some mammals consists of two, and perhaps more, layers that are morphologically and chemically different. Wartenberg and Stegner (1960) and Stegner and Wartenberg (1961) showed two layers histochemically in the zona of the human oocyte: the inner layer stained as a neutral mucopolysac-

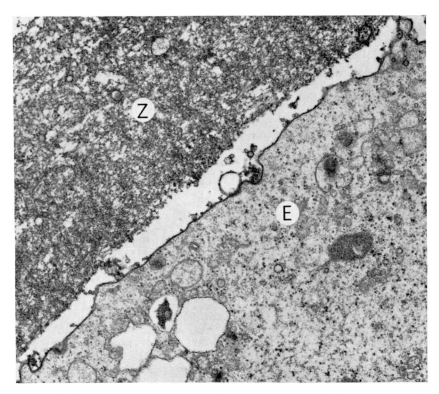

Fig. 19. Electron micrograph of a section of a recently fertilized rat egg. Note the filamentous structure of the zona pellucida (Z) and the relative smoothness of the surface of the egg cytoplasm (E). Ruthenium red staining, which stains mucopolysaccharides preferentially, was used. The zona and the surface of the plasma membrane are strongly stained. Magnification: × 17,000. (Courtesy of Dr. Daniel Szollosi.)

charide and the outer (thinner) layer as an acid mucopolysaccharide. These workers also presented evidence that the acid mucopolysaccharide layer is secreted by the follicle cells. Odor (1960) described an outer thin granular layer interposed between the zona pellucida proper and the adjacent corona cells; and Hope (1965) noted an electron-dense flocculent material in the outer region of the zona of the rhesus egg. In light microscope studies, Dickmann (1963b) described two distinct concentric layers in the zona in recently ovulated rabbit eggs, and Dickmann and Dziuk (1964) observed two to three layers in the zona of pig eggs. Seshachar and Bagga (1963) suggested that the outer layer of the zona pellucida in oocytes of two primate species (*Loris tardigradus lydekkerianus* and *Macaca mulatta mulatta*) contained acid mucopolysac-

charides with ester-bound sulfate groups. This would be in accordance
with the observation (Moricard and Gothié, 1955; Gothié, 1958) that, in
the mature follicle of the rabbit, there is a marked accumulation of ^{35}S-
sulfate between the corona cells and the zona pellucida (but not within
the zona or the egg itself). It is tempting to speculate that the outer acid
layer of the zona may be analogous to the gelatinous coat of the sea
urchin egg (cf. Tyler and Tyler, 1965) and may be the source of the
fertilizin-like substance reported to occur in rabbit and other mammalian
eggs (D. W. Bishop and Tyler, 1956; Thibault and Dauzier, 1960, 1961).
The experiments of D. W. Bishop and Tyler (1956) pointed, in fact, to
the zona pellucida as the most likely source of the mammalian "fertilizin"
(sperm-agglutinating factor).

The zona pellucida of tubal eggs probably functions both by providing mechanical protection for the egg (Gwatkin 1963a,b) and by regulating the chemical environment in the immediate vicinity of the egg.
(Regarding the nature and function of extracellular coatings in general,
see the discussion by Bennett, 1963). Data on the permeability properties
of the zona pellucida are limited. There is evidence that relatively large
particles, such as a virus, may pass through the zona of the mouse egg
(Gwatkin, 1963b) as does heterologous serum albumin (Glass, 1963).
But the zona pellucida of rat and rabbit eggs is apparently impermeable
to heparin (Austin and Lovelock, 1958), a strongly acidic molecule. That
the chemical composition of the perivitelline fluid is different from that
of the luminal secretions is indicated by the observation of Odor and
Blandau (1949), and of others, that supplementary spermatozoa within
the perivitelline space appear to be well preserved morphologically even
until the time of implantation, while those suspended in the tubal fluids
(and not taken up by phagocytes) disintegrate at an early stage.

In many species the zona pellucida plays a role in the mechanism of
preventing polyspermic fertilization in a manner analogous to the "fertilization membrane" of echinoderm eggs (see Section IV,E).

C. CYTOPLASMIC SURFACE

In the freshly ovulated egg of the rat the microvilli are retracted from
the zona pellucida, and the plasma membrane shows an irregular, undulating outline (Odor, 1960). According to Hadek (1964, 1965) the
cytoplasmic surface of the ovulated egg of the rabbit changes with the
passage of time. At about the time of ovulation, relatively smooth villi,
of approximately equal length, are seen. At a somewhat later stage the
surface shows evidence of considerable activity: irregular projections and
folds appear that seem to combine or flow together rather freely, often
entrapping extraneous material from the perivitelline space. This surface

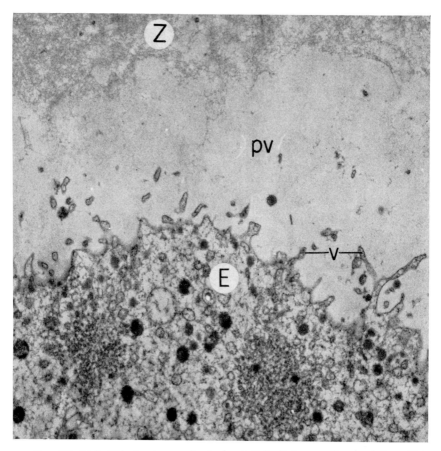

Fig. 20. Unfertilized rat egg fixed about 2 to 3 hours after ovulation. Note irregularity of the cytoplasmic surface at this stage. Key: (E) egg cytoplasm; (Z) zona pellucida; (v) microvilli; (pv) perivitelline space. Magnification: × 12,500.

appears to be characteristic of fertilizable ova. The surface of aged eggs is covered by small villi that are more or less similar in size. Irregular cytoplasmic folds are also seen in unfertilized tubal eggs of the rat recovered a few hours after ovulation (Fig. 20). In contrast, the surface of the penetrated egg appears considerably smoother (cf. Fig. 19). The reasons for the apparent increase in surface activity in the ovulated egg are not known, e.g., whether it is induced by tubal secretions. A certain analogy presents itself with the surface transformation observed in the feeding cells of *Hydra* after stimulation for nutritive phagocytosis (Afzelius and Rosén, 1965).

The salient feature of sperm incorporation into the egg cytoplasm is the intimate, and apparently specific, interaction of the plasma membrane of the egg with that of the sperm (cf. Section IV,D). This is presumably due to the presence of specific complementary receptor groups on the respective plasma membranes of the two gametes. In the *Mytilus* egg, numerous delicate fibrils, possibly bearing receptor sites for sperm, extend from the plasma membrane of the microvilli into the surroundings of the egg (Dan, 1962). It is not known whether, in mammals, there are similar processes extending from the microvilli.

The permeability properties, electric charge, and antigenic structure of the egg surface have been studied in some detail in the sea urchin and some other species (cf. Monroy, 1965) but little is known about the corresponding characteristics of mammalian eggs. A conspicuous change upon sperm penetration is the loss of receptivity of the egg plasma membrane to additional spermatozoa. The possible mechanism of this reaction is discussed in Section IV,E.

D. Cortical Granules

The presence of cortical granules (vesicles) near the egg surface is of widespread occurrence in animal eggs (see Pasteels, 1961, 1965a; Monroy, 1965). In mammals, cortical granules were first described by Austin (1956, 1961a) in the eggs of the golden hamster. Szollosi (1962) observed them in unfertilized tubal eggs of rabbits, guinea pigs, rats, mice, golden hamsters, coypus, and pigs. The cortical granules range from 160 to 350 mμ in diameter and are usually located at or near the plasma membrane, although some are scattered throughout the cytoplasm (Fig. 21). Cortical granules were also described in ovarian oocytes of the human (Tardini *et al.*, 1961), the guinea pig (Adams and Hertig, 1964), and the rhesus monkey (Hope, 1965). According to Adams and Hertig (1964) the cortical granules arise in connection with the so-called vesicular aggregates in the egg cytoplasm. In the rabbit, cortical granule formation may continue after ovulation since the number of these granules appears to increase in aged eggs (Hadek, 1963d). In unfertilized golden hamster eggs, however, the cortical vesicles appear to undergo spontaneous breakdown. Their contents are released into the perivitelline space and may contribute to the loss of fertilizability in these eggs (Yanagimachi and Chang, 1961). The cortical granules of the golden hamster give a strong PAS-positive reaction, resistant to diastase, indicating the presence of mucopolysaccharides. The cortical granules of mammalian eggs would then be chemically related to those of echinoderm, fish, and amphibian eggs (cf. Monroy, 1965). Cortical granules appear to be absent in the region of the plasma membrane where it is elevated over the second meiotic spindle (Yanagimachi and Chang, 1961; Szollosi,

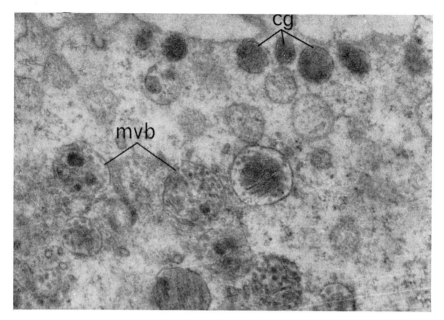

FIG. 21. Cortical granules (cg) in the rat egg. The cortical granules may be produced in multivesicular bodies (mvb) which often contain dense granules of various sizes. Magnification: × 40,000.

1962, 1967); but they may persist in the first polar body of the human egg even after fertilization (Zamboni *et al.*, 1966). Their possible role in the mechanism of the block to polyspermy in mammalian eggs is discussed in Section IV,E.

IV. Mechanisms of Sperm Entry

A. The Environment of Fertilization

In most mammalian species that have been studied, sperm penetration occurs in a dilated part of the upper segment of the fallopian tube, the ampulla. The ova are rapidly transported to this site after rupture of the follicles, partly by ciliary activity, and partly by muscular contractions of the tube (cf. Blandau, 1961; Winterberger-Torrès, 1961; Austin, 1963a). There are conflicting reports in the literature as to the rapidity with which spermatozoa are able to reach the site of fertilization (cf. Dauzier, 1958; D. W. Bishop, 1961; Hartman, 1962; Restall, 1967). There is no doubt, however, that under normal conditions of fertilization the actual number of spermatozoa present in the ampulla at the time of sperm penetration is exceedingly low—of the order of tens or hundreds, depending on the species (Braden and Austin, 1954b). Motion picture studies of the rat oviduct (Blandau, 1965) reveal that the freshly ovulated eggs (which are surrounded by masses of cumulus cells) are moved about vigorously within the ampulla by contractions of the tubal wall. This mechanical agitation undoubtedly plays a role in the loosening and eventual dispersal of follicle cells around the eggs, as suggested by Swyer (1947b); however, a tubal factor has also been implicated in the detachment of corona cells from the zona pellucida of rabbit eggs (cf. Austin, 1961b; Dickmann, 1963a; Korolev and Petrov, 1964; Zamboni *et al.,* 1965). According to Thibault and Dauzier (1960, 1961, and Chapter 9 of this volume), continued agitation facilitates the removal of a substance (fertilizin) that, if present in excessive concentration around the egg, could effectively block sperm penetration. If the mixing action continues until the time of sperm entry (in the rat 2–4 hours after ovulation; Austin and Braden, 1954), it may perhaps assist the spermatozoa in encountering the egg.

It appears likely that the initial contact of spermatozoa with the surface of the egg coats (usually the cumulus oophorus) is the result of a random collision rather than of any chemotactic attraction by the egg (cf. Austin, 1963a). In plants there is good evidence for chemotactic and chemotropic mechanisms in gametic approach (Rothschild, 1956a; Machlis and Rawitscher-Kunkel, 1963), but claims of chemotaxis in animal spermatozoa are equivocal (cf. Monroy, 1965; Machlis and Rawitscher-Kunkel in Chapter 4 of Volume I of this treatise). However, the recent studies of Miller (1966) strongly suggest that chemotaxis does occur in some of the coelenterates. The spermatozoa of the hydroids *Campanularia flexuosa* and *C. calceolifera* were both activated and attracted by a substance emanating from the female gonangium. Chemotaxis was concluded on the basis of the following criteria: (*1*) the spermatozoa showed oriented movements toward the source of the attractant; (*2*) the effect was concentration-dependent and species-specific; and (*3*) the active material, a low molecular weight organic substance, could be extracted and it induced similar sperm behavior *in vitro.*

The secretory activity of the fallopian tube and the composition of the luminal fluids have been studied mainly in the rabbit and in some of the larger mammalian species (cf. D. W. Bishop, 1961; Mastroianni, 1962; Hamner and Williams, 1965). Some of the chemical constituents present in secretions of the rabbit fallopian tube are given in Table II. These may not necessarily reflect conditions prevailing in the ampulla at the time of fertilization. The tubal secretions may contain special factors necessary for the process of fertilization and the initiation of development. These and other related problems are discussed in connection with

TABLE II

CONCENTRATIONS OF VARIOUS COMPONENTS OF RABBIT OVIDUCT FLUID[a]

| Component | Mean | | Standard error |
	(mg/ml)	(μg/ml)	
Sodium	2.94	—	0.044
Chloride	4.10	—	0.390
Calcium	0.32	—	0.066
Potassium	0.22	—	0.006
Bicarbonate	1.76	—	0.102
Polysaccharide	0.37	—	0.029
Total protein	2.73	—	0.293
Dry matter	8.28	—	1.120
Phosphate	—	5.98	1.298
Magnesium	—	3.43	0.338
Zinc	—	6.48	0.180
Lactic acid	—	31.35	7.566

[a] From Hamner and Williams (1965).

the *in vitro* fertilization of mammalian eggs (cf. Thibault in Chapter 9 of this volume).

B. SPERM PASSAGE THROUGH THE CUMULUS OOPHORUS

The enzyme, hyaluronidase, is probably widespread in mammalian tissues (Bollet *et al.*, 1963) but its richest sources are the testes and spermatozoa. The relation of spermatozoal hyaluronidase to fertilization came to light through the discovery by McClean and Rowlands (1942) and Fekete and Duran-Reynals (1943) that the cumuli of rat and mouse ova can be dispersed by hyaluronidase preparations from various sources (besides mammalian testes and spermatozoa). The cumulus-dispersing effect of sperm suspensions had been known before (Schenk, 1878; Yamane, 1935; Pincus, 1936) but it had been ascribed to the presence of a proteolytic enzyme in the spermatozoa (Yamane, 1930). It is now well known that, contrary to earlier belief, actual denudation of the eggs is not a necessary condition for sperm entry but that the individual spermatozoon carries enough of the enzyme to make a path for itself through the intact cumulus (cf. reviews by Tyler, 1948; Austin and Walton, 1960; Blandau, 1961; Mann, 1964).

It appears likely from the evidence discussed in Section II,B that most or all of the testicular hyaluronidase is derived from the acrosomal material or its precursors and that in the mature spermatozoon the enzyme is confined in the acrosome. If the acrosomal hyaluronidase is to function during sperm passage through the cumulus, one would expect that (1) the acrosome would remain intact until the spermatozoa arrive in the vicinity of the egg, and (2) the acrosomal material (or at least that part containing hyaluronidase) would be released during sperm entry. In fact, recent phase-microscopic and fine-structural studies show that the acrosome remains unchanged in the majority of "capacitated" spermatozoa recovered from uterine and tubal fluids in the rabbit (Adams and Chang, 1962; Moricard, 1961, 1964; Bedford, 1963c, 1964a; Austin, 1963b) and in several rodent species (Austin and Bishop, 1958b; Yanagimachi and Chang, 1964). On the other hand, the acrosomal material (or most of it) was found to be invariably absent in spermatozoa observed in the thickness of the zona pellucida and within the perivitelline space (Austin and Bishop, 1958b; Moricard, 1960, 1964; Moricard *et al.*, 1960; Hadek, 1963b; Austin, 1963b; Pikó and Tyler, 1964b). These observations indicate that the acrosomal contents are released during, or immediately preceding, sperm passage through the outer coats of the egg. A significant observation is that the inner acrosomal membrane remains intact during this process. This shows that the acrosome does not detach *in toto* but rather that it opens up, thereby releasing its contents. There is evidence indicating that the opening up of the acrosome involves fusion and vesiculation of the outer acrosomal membrane and the overlying sperm plasma membrane (Pikó and Tyler, 1964a,b; Barros *et al.*, 1967; Bedford, 1968; Fig. 22). Further fine-structural studies are needed, however, in order to elucidate the sequence of events in the mammalian acrosome reaction. (For the acrosome reaction of invertebrate spermatozoa, see Dan in Chapter 6 of Volume I of this treatise.)

The nature of the stimulus that results in the release of hyaluronidase from the penetrating spermatozoon is not known. Swyer (1947a) found that the presence of cumulus masses, with or without saline extracts of the fallopian tube, had no effect on the rate of liberation of hyaluronidase from a suspension of rabbit spermatozoa. He, therefore, concluded that no specific mechanism existed ensuring selective liberation of the enzyme in the fallopian tube. The experiments of Austin (1960, 1961c) help to clarify this point. When small numbers of epididymal or ejaculated spermatozoa of rabbit, rat, and mouse were brought together with intact cumuli, *in vitro*, the spermatozoa were seen to attach readily to the surface of the cumulus but were unable to penetrate it. On the other hand, capacitated spermatozoa were able to move through the cumulus and also exhibited a typical change in the acrosome. These observations can be interpreted to show that capacitation (cf. Section II,F) is a prerequisite for the specific release of hyaluronidase, presumably by enabling spermatozoa to undergo the acrosome reaction. (Note: Hyaluronidase can be released nonspecifically from moribund or damaged spermatozoa.)

As a working hypothesis, the following mechanism may be proposed for sperm passage through the cumulus. In capacitated spermatozoa the

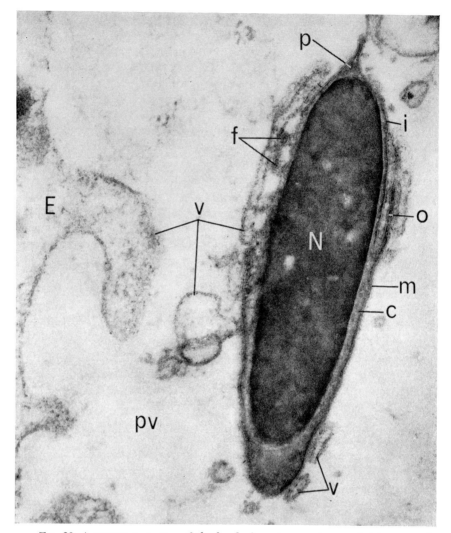

FIG. 22. A transverse section of the head of a rat spermatozoon that has recently penetrated through the zona pellucida and has become attached to the surface of the egg cytoplasm (E). The acrosomal sac is open and its contents are absent (compare with diagram 4 of Fig. 1C and with Fig. 6). The inner acrosomal membrane (i) is intact. The lateral portion of the acrosomal sac (o), covered by the sperm plasma membrane (m), is mostly present on one side of the sperm head, whereas on the other side small vesicles (f) are seen which have probably arisen by the fusion of the outer acrosomal membrane and the sperm's plasma membrane. The egg microvilli (v) are attached to the sperm plasma membrane and are spread out into a thin cytoplasmic sheet at several places. (N) Sperm nucleus; (p) perforatorium; (c) postnuclear cap; (pv) perivitelline space. Magnification: × 80,000.

specific receptors of the sperm plasma membrane are exposed and able to interact with a substance present in the cumulus or emanating therefrom. This would elicit the acrosome reaction, i.e., the opening up of the acrosomal sac through a process of membrane fusion and vesiculation, and subsequent release of acrosomal contents.

The interacting substance in the cumulus may well be identical with the "fertilizin" of D. W. Bishop and Tyler (1956) and Thibault and Dauzier (1960, 1961). The observation (D. W. Bishop and Tyler, 1956) that the fertilizin would be diffusing slowly from the zona pellucida toward the periphery of the cumulus is of interest. The concentration gradient resulting from such a diffusion may be instrumental in the progressive release of hyaluronidase that presumably occurs when individual spermatozoa pass through the intact cumulus.

Little can be said at the present time about the mechanism of hyaluronidase action on the cumulus matrix except for what can be inferred from studies in other systems. One feature that requires explanation is the rapidity of liquefaction of the matrix during sperm passage. Available kinetic data on the action of hyaluronidase from testes, as well as from snake venom, do indicate a very rapid reaction rate on the highly polymerized substrate, but the reaction rate falls off with the degradation of the substrate into smaller molecular weight products (Meyer and Rapport, 1952; Weissmann, 1955). Since the breakage of relatively few bonds gives rise to marked changes in the physicochemical properties of native hyaluronic acid (decrease in viscosity, loss of mucin clot formation), this would explain the rapid "spreading" action of the enzyme from the above sources (Meyer and Rapport, 1952). According to Fessler (1960) the initial action of hyaluronidase in connective tissue is to reduce the interaction of hyaluronic acid with the confining structures (e.g., collagen); this would allow free movement of fluid, together with polysaccharide, through the tissue spaces.

The pH optimum of human and bovine testicular hyaluronidases falls within a broad range between pH 4.5 and 6.0, but enzyme activity can be detected up to pH 7.6. This differs markedly from the pH optima of hyaluronidases from other tissues and serum that are maximally active below pH 4 and are inactive above pH 4.9. Therefore, the testicular enzyme appears to be intrinsically different from the hyaluronidases found in other mammalian tissues (De Salegui *et al.*, 1967).

C. Sperm Passage through the Zona Pellucida

When the sperm passes through the zona it leaves a "slit" or channel, suggesting lytic action (Austin, 1951a; Fig. 23). However, only recently has direct evidence been produced that a lytic agent other than hyalu-

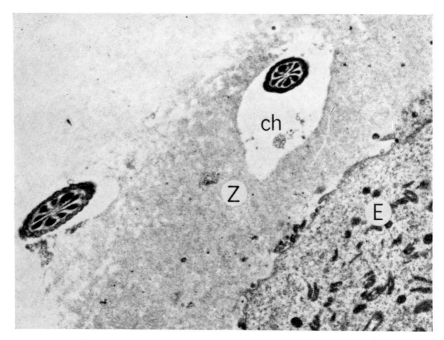

Fig. 23. Electron micrograph of a section of a recently penetrated Chinese hamster egg. The penetration channel (ch) in the zona pellucida (Z) is seen, with sections of the tail of the fertilizing sperm. (E) Egg cytoplasm. Magnification: × 8000.

ronidase may be carried by mammalian spermatozoa (Srivastava *et al.*, 1965a,b; Stambaugh and Buckley, 1968).

The presently available evidence indicates that the "zona lysin" is contained in the acrosome. A possible location could be in the thin lateral portion of the acrosome (cf. Section II,B) and, perhaps, on the surface of the inner acrosomal membrane. The alternative possibility (Austin and Bishop, 1958b) that the perforatorium would be the source of the zona lysin appears less likely since this organelle remains unchanged during sperm entry (Fig. 24), and its fine-structural appearance and chemical properties suggest a structural rather than enzymic function (Section II,C). The location of the zona lysin in the lateral portion of the acrosome would not necessarily contradict earlier phase-microscopic observations (Austin and Bishop, 1958b; Yanagimachi and Chang, 1964) according to which "loss" of the acrosome precedes sperm penetration through the zona. It seems likely that the loss of acrosomal material involves initially only the apical region of the acrosome—presumably carrying hyaluronidase—thus exposing the more deeply located lysin. Electron micrographs of rabbit spermatozoa found in the material of the zona (Moricard, 1960; Hadek, 1963b) show remains of the lateral portion of the acrosomal sac; such remnants can also be seen in rat spermatozoa that have recently passed through the zona (Pikó and Tyler, 1964b; Fig. 22). The observations of Barros *et al.* (1967) and Bedford (1968) indi-

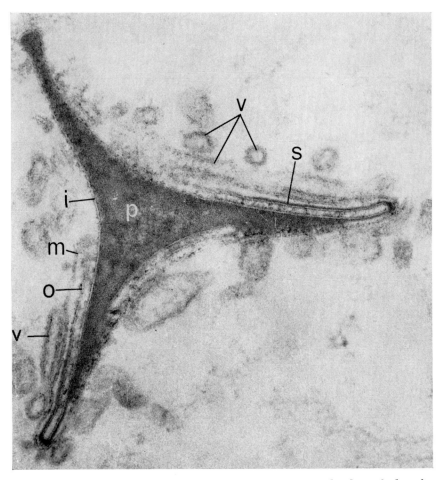

Fɪɢ. 24. Transverse section through the tip of a rat sperm head attached to the surface of the egg cytoplasm (sections of the same spermatozoon are shown in Figs. 27 and 28). The acrosomal sac is open and its contents are absent (compare with diagram 2 of Fig. 1C and with Fig. 11). The perforatorium (p) is unchanged and remains covered by the inner acrosomal membrane (i). That portion of the outer acrosomal membrane (o) covered by a surface coating (s) is also present and is now continuous with the sperm plasma membrane (m). Several egg microvilli (v) are attached to the sperm plasma membrane. (For interpretation, see also Fig. 25A.) Magnification: × 102,000.

cate that the thin posterior part of the acrosome ("equatorial segments") remains intact in golden hamster and rabbit spermatozoa during passage through the cumulus oophorus, while the anterior part breaks down and its contents are released. These findings are consistent with the hypothesis that the contents of the lateral region of the acrosome are liberated during or shortly before penetration of the zona.

The mechanism of release of the zona lysin and its mode of action are not known. It is reasonable to suppose that the outer acid mucopolysaccharidic layer of the zona, which shows a certain analogy with the "jelly layer" of the sea urchin egg (Section III,B) may be in some way connected with the liberation of the lysin, possibly through a further opening up of the acrosomal sac. The penetration channel is rather narrow, suggesting that little diffusion of the lysin takes place. The channel often follows a curved, oblique path entering the zona at a small angle and turning gradually toward the cytoplasmic surface, e.g., in the guinea pig and the Libyan jird (Austin and Bishop, 1958b), in the rabbit (Moricard, 1960; Dickmann, 1964b), and in the pig (Dickmann and Dziuk, 1964). This indicates that in these species the sperm head is usually positioned nearly parallel to the zona surface when it first enters. Dickmann (1964b) and Dickmann and Dziuk (1964) noted a fine filamentlike structure protruding from the apex of rabbit and pig spermatozoa that were found embedded in the zona. The reality of the filament is uncertain, and it is unlikely that it could be analogous to the acrosomal filament of invertebrate spermatozoa. A similar structure has not been seen in electron micrographs.

The recent studies of Soupart and Clewe (1965) suggest that sialic acid residues in the zona of rabbit eggs may have a functional significance in sperm penetration. Treatment of unfertilized eggs with the enzyme, neuraminidase, diminished significantly the rate of fertilization and the number of spermatozoa penetrating through the zona pellucida.

Elucidation of the mechanism of action of the zona lysin will require a more detailed knowledge of the properties of this enzyme and of the zona pellucida itself. It is interesting to note, however, that the preparation of Srivastava et al. (1965a,b) was non-species-specific in its action, since acrosomal extracts from ram and bull spermatozoa were effective in dissolving the zona of the rabbit egg.

D. SPERM INCORPORATION INTO THE EGG CYTOPLASM

Phase-microscopic observations of rat, mouse, and hamster eggs (Austin 1951b; Austin and Braden, 1956; cf. Austin and Bishop, 1957a,b) show that the head of the fertilizing spermatozoon becomes attached to the egg's surface very shortly after traversing the zona pellucida. The head is not dislodged by the residual movements of the sperm tail which persist for some time and may actually aid in bringing the gamete surfaces into close apposition. The sperm head comes to lie flat on the surface of the egg and enters the cytoplasm from this position. It takes about half an hour from early attachment to complete incorporation of the sperm head by which time transformation of the nucleus (swelling and loss of contrast) is evident. An intriguing finding is that changes first appear in the caudal portion of the nucleus and then extend gradually toward the tip (see, also, Odor and Blandau, 1951). Absorption of the head is followed by a progressive incorporation of the midpiece and the tail of the spermatozoon—a process that may take several hours.

Fine-structural observations of sperm entry in mammals are limited, due mainly to the technical difficulties involved in locating the sperm head during fine sectioning and in obtaining sections cut at an angle favorable for interpretation. Szollosi and Ris (1961), studying the rat, found that the head of a sperm that had recently entered the egg cytoplasm was denuded of membranes over a large area and that the sperm plasma membrane, where it remained, was continuous with the egg plasma membrane. They postulated that after the sperm had come to lie on the plasma membrane of the egg, the egg and sperm plasma membranes ruptured in the area of contact and then fused with each other, first starting at the tip and then proceeding posteriorly along the entire length of the spermatozoon. As a result the sperm would come to lie inside the cytoplasm, leaving its own plasma membrane incorporated into the egg plasma membrane at the surface of the egg. However, other available evidence conflicts with this hypothesis. Austin (1961a) published an electron micrograph showing a section of the entering sperm tail in a golden hamster egg; the tail appeared to be enclosed within a vesicle, suggesting a phagocytotic mechanism of uptake. Hadek (1963e) described a strong localized cortical reaction, resembling a "cytoplasmic wave," that surrounded the tail of the (possibly fertilizing) spermatozoon in the rabbit. Pikó and Tyler (1964b) presented evidence showing that in the rat the spermatozoon is engulfed progressively by the egg cytoplasm through a process resembling phagocytosis.

More extensive studies have been made on sperm entry in several invertebrate species: *Hydroides* and *Saccoglossus* (Colwin and Colwin, 1961, 1963b); sea urchin (Rothschild, 1956b; Afzelius and Murray, 1957; Takashima and Takashima, 1960; Franklin, 1965); the mollusk *Barnea candida* (Pasteels, 1965b,c); and the gephyrean worm *Urechis caupo* (Tyler, 1965). But the opinions differ as to the interpretation of the fine-structural observations and the mechanisms involved in sperm incorporation. Colwin and Colwin propose a hypothesis of sperm entry in *Hy-*

droides and *Saccoglossus* similar to that of Szollosi and Ris (1961), namely that the sperm and egg membranes fuse at the point of initial contact, which on the part of the sperm involves the everted inner acrosomal membrane at the tip of the head. Thus, a funnel-like connection is established between the two gametes through which the sperm nucleus and other sperm structures move into the egg cytoplasm. The sperm plasma membrane is considered to remain at the surface as part of the plasma membrane of the fertilized egg. However, Tyler (1965) suggests that in *Urechis*, progressive lateral joining of the egg and sperm membranes continues also for the nonacrosomal part of the sperm plasma membrane and that this lateral joining of the membranes provides the mechanism for drawing the sperm nucleus into the egg cytoplasm. The conjoined membranes disintegrate, and the plasma membrane that is

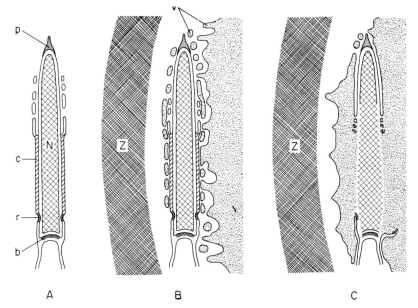

A   B   C

FIG. 25. Diagrammatic representation of sperm entry into the egg cytoplasm in the rat. A: Frontal section of a sperm head that has recently passed through the zona pellucida illustrating fusion and vesiculation of plasma and outer acrosomal membranes and absence of acrosomal contents (compare with Fig. 1B). (N) Nucleus; (p) perforatorium; (c) postnuclear cap; (r) posterior ring; (b) basal plate. B: An attached sperm head surrounded by microvilli (v) of the egg; (Z) zona pellucida. C: A sperm head at an intermediary stage of incorporation into the egg cytoplasm. All membranes have disintegrated in the postnuclear cap area and here the nucleus is exposed to the egg cytoplasm. The egg plasma membrane and cytoplasm have spread anteriorly around the inner acrosomal membrane and posteriorly around the plasma membrane of the sperm tail. The remains of the posterior ring mark the point where the plasma membrane of the egg unites with that of the sperm.

left above the sperm is that of the egg. Pasteels (1965b,c) also concludes that in *Barnea* the sperm is completely engulfed by the egg cytoplasm, for the characteristic microvilli of the egg cortex form a continuous layer above the sperm, and no "cicatrix" is left behind.

The following account of sperm entry in the rat is based on observations of serial sections of spermatozoa in various early stages of penetration into the egg. The chances of locating the entering sperm head were increased by using a double-embedding procedure in agar and plastic (Pikó and Tyler, 1964b). The eggs were fixed in 3% glutaraldehyde, post-osmicated, and sectioned serially with an LKB-Ultrotome. The sections were stained with lead citrate and uranyl acetate and examined in a Philips 200 electron microscope.

The fine-structural changes during sperm entry are summarized diagrammatically in Figs. 25 and 26. The process of sperm incorporation can be divided into three phases: (*1*) specific attachment; (*2*) membrane fusion and breakdown; and (*3*) progressive engulfment.

In the attachment phase, numerous egg microvilli surround the sperm head which is covered anteriorly by the inner portion of the acrosomal membrane and posteriorly by the sperm plasma membrane. The villi

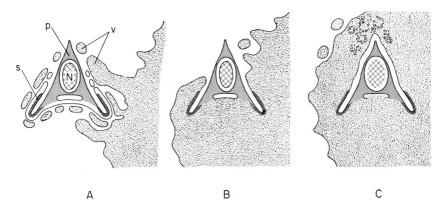

A   B   C

FIG. 26. Diagrammatic representation of sperm entry into the egg cytoplasm in rat as shown in a transverse section through the anterior portion of the sperm head. A: The acrosomal sac is open, and the remaining portion of the outer acrosomal membrane is now continuous with the sperm plasma membrane (compare with diagram 3 of Fig. 1C). Egg microvilli (v) are projected around the sperm head and are attached to the sperm plasma membrane. (N) Sperm nucleus; (p) perforatorium; (s) surface coating of the outer acrosomal membrane. B: The conjoined egg and sperm plasma membranes have disintegrated and the microvilli have coalesced into a continuous layer in this area. The outer acrosomal membrane is now continuous with the egg plasma membrane. C: The egg plasma membrane and cytoplasm completely surround the anterior region of the sperm head. The vesiclelike structure is composed of the remaining portion of the acrosomal sac and of egg plasma membrane that became internalized in the engulfment process.

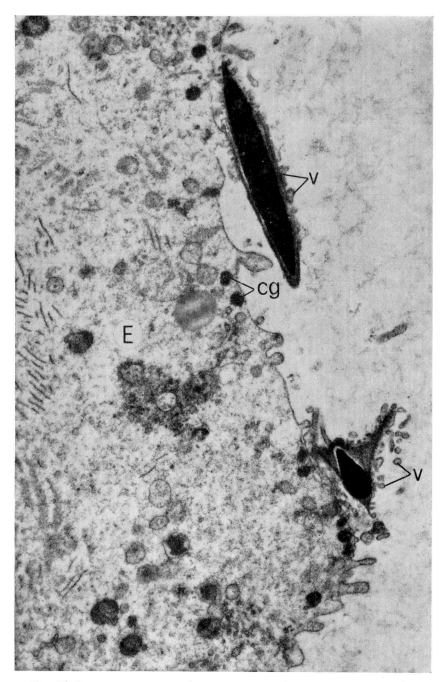

Fig. 27. Section cutting across the anterior region (in lower part of the picture) and across the base (in upper part of the picture) of a rat sperm head attached to the surface of the egg cytoplasm (E). Note microvilli (v) around the sperm head and elevation of the egg cytoplasm. The cortical granules (cg) are still present.

appear to be particularly firmly bound to the plasma membrane proper of the sperm head where they often spread out into a thin cytoplasmic sheet (Figs. 22, 24, 27, and 28).

The next phase is characterized by the further joining and breakdown of the apposed sperm and egg plasma membranes in the region of the postnuclear cap; as a result, the villi coalesce into a continuous cytoplasmic layer in this area. The underlying postnuclear layer and nuclear membranes also disintegrate, and the resulting exposure of the nucleus to the egg cytoplasm appears to trigger nuclear transformation (Fig. 29). These events probably occur relatively early after initial attachment— perhaps within 10 minutes as judged from the appearance of the first changes in the caudal portion of the sperm nucleus (Austin, 1951b).

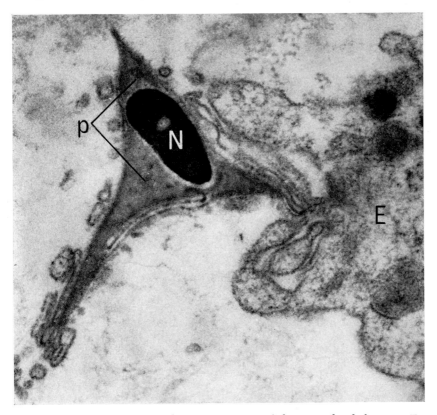

Fig. 28. A section through the anterior region of the sperm head shown in Fig. 27. (E) Egg cytoplasm; (N) sperm nucleus; (p) perforatorium. (For interpretation, see Fig. 26A.) Magnification: × 63,000.

(Sections of the same spermatozoon are shown in Figs. 24 and 28.) Magnification: × 18,000.

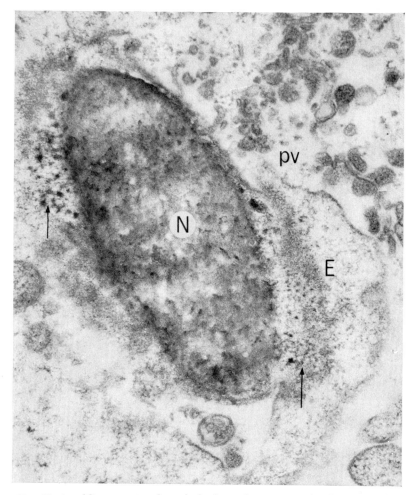

Fɪɢ. 29. An oblique section through the base of a rat sperm head entering the egg cytoplasm. The egg microvilli have coalesced into a continuous cytoplasmic layer (E) around part of the sperm head. In the area of the postnuclear cap, products of disintegrated membranes are seen (arrows). The perivitelline (pv) contains material extruded from the egg cytoplasm—probably the products of the cortical reaction which occurs at this time. (N) Sperm nucleus. Magnification: × 47,000.

In the engulfment phase the egg plasma membrane and cytoplasm progressively envelop the anterior region of the head (covered by the inner acrosomal membrane) and spread posteriorly around the plasma membrane of the sperm tail. The apposed egg and sperm membranes in these areas apparently disintegrate slowly. In sections at this stage the posterior part of the nucleus lies denuded in the egg cytoplasm and shows

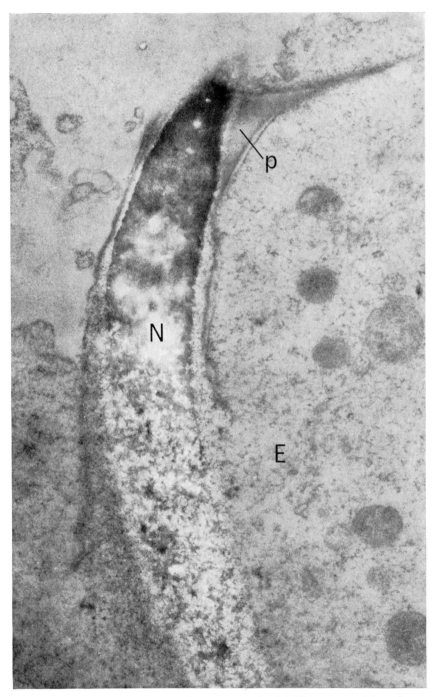

FIG. 30. Longitudinal section of a rat sperm head half-incorporated into the egg cytoplasm (E). The posterior region of the nucleus (N) is denuded of membranes.

(*Legend continued p. 378.*)

considerable transformation, whereas the inner acrosomal membrane is still only partially engulfed and the anterior part of the nucleus remains unchanged (Figs. 30, 31, 32).

The nature of the engulfment process makes it unlikely that any part of the sperm plasma membrane would be left behind at the surface of the egg once incorporation is completed. This is particularly well seen in the apical region of the sperm head which becomes enclosed in a vesicle-like structure. The vesicle is a composite of the remaining portion of the acrosomal sac and that portion of the egg plasma membrane that became internalized in the engulfment process (Figs. 26 and 33).

There is evidence that sperm entry in another rodent, the Chinese hamster, follows that pattern described above for the rat egg (Figs. 34 and 35). Also, the recent observations of Barros and Franklin (1968) reveal similar fine-structural changes during sperm entry in the golden hamster. The middle portion of the sperm head enters the egg first. The membranes surrounding this region (which corresponds to the post-nuclear cap area in this species) disintegrate early. Incorporation then proceeds toward the anterior and posterior regions of the sperm head.

Sperm incorporation in the rat, Chinese hamster, and golden hamster appears to be, then, essentially a process akin to phagocytosis—the sper-matozoon is engulfed by a progressive spreading of the egg's plasma membrane and cytoplasm around it. The following distinctive features are apparent, however: (1) incorporation is initiated by what appears to be a highly specific interaction between the plasma membrane of the sperm head and the egg plasma membrane; and (2) the apposed gamete membranes disintegrate early around the sperm head (in the species studied so far this first occurs in the postnuclear cap area), thus exposing the sperm nucleus to the egg cytoplasm at an early stage of incorporation.

It is reasonable to interpret these observations in terms of the exist-ence of specific complementary receptor substances on the plasma mem-branes of the two gametes as postulated in the fertilizin–antifertilizin theory of fertilization (cf. Tyler, 1948, 1949, 1956, 1959, 1960; Metz, 1957, and Chapter 5 of Volume I of this treatise). In rat spermatozoa the re-ceptors appear to be located primarily on the plasma membrane of the sperm head, whereas in the unfertilized egg they are distributed over the whole surface except for the region above the second meiotic spindle that is normally nonreceptive to spermatozoa (e.g., Blandau and Odor,

Nuclear transformation advances from this region toward the tip. Note that double membranes arise as the egg plasma membrane and cytoplasm envelop the inner acrosomal membrane around the anterior region of the sperm head. (p) Perforatorium. Magnification: × 32,000.

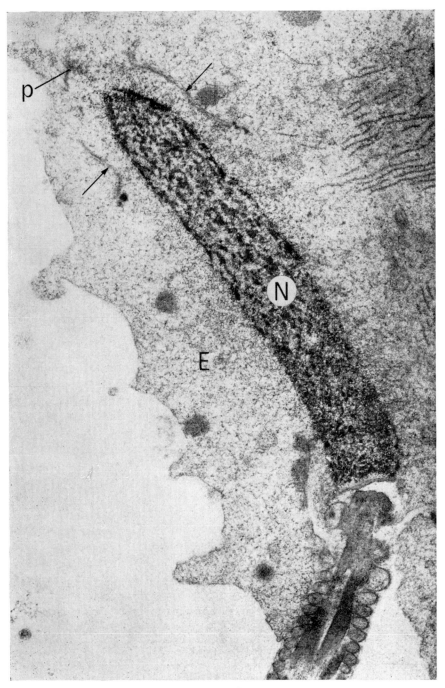

Fig. 31. For legend see p. 380.

1952). It is noteworthy that this region also lacks cortical granules (see Section III,D) and microvilli (Pikó, unpublished observation).

E. The Block to Polyspermy

In mammals, as in most animal groups, fertilization is normally monospermic, i.e., only one spermatozoon gains entry into the egg cytoplasm. However, simultaneous penetration of two spermatozoa may occur accidentally in a small proportion of the eggs. Dispermic mammalian eggs characteristically divide by a bipolar first cleavage spindle and go through a normal early development. But the triploid embryos that arise usually die at later stages of development (Pikó and Bomsel-Helmreich, 1960; Pikó, 1961a; Bomsel-Helmreich, 1966; and Austin in Chapter 10 of this volume).

The block to polyspermy in mammalian eggs operates at two levels: (1) the zona pellucida and (2) the plasma membrane of the egg. In many species the zona is penetrable to but one spermatozoon. But even in cases where two or more spermatozoa traverse the zona (e.g., in rabbits, rats, and mice) only one of these generally becomes incorporated into the egg cytoplasm (Braden et al., 1954; cf. Austin, 1961a). Austin and Braden (1956) postulated that the "zona reaction" and the "vitelline block" (i.e., the block at the level of the plasma membrane) would both have the same underlying mechanism, the breakdown of the cortical granules, elicited by the attachment of the first spermatozoon to the cytoplasmic surface. Recent electron-microscopic studies lend support to this hypothesis.

Cortical granules (vesicles) appear to be of general occurrence in the eggs of mammals (see Section III,D). In the early stages of sperm attachment to the egg plasma membrane the cortical vesicles are intact (see Fig. 27). The cortical reaction, with concomitant release of the contents of the vesicles into the perivitelline space (Figs. 36 and 37), seems to be triggered at the time when membrane fusion and breakdown occur in the postnuclear cap area, that is, 10 minutes or so after the initial attachment of the spermatozoon in the rat egg (cf. preceding section). The mechanism of cortical reaction has been studied in rat and hamster

---

Fig. 31. Frontal section of a rat sperm head being engulfed by the egg cytoplasm (E). The middle and posterior portions of the nucleus (N), shown in the picture, lie free in the cytoplasm and show transformation. Anteriorly, double-membranous structures, composed of the remaining portion of the inner acrosomal membrane and of egg plasma membrane, are seen on each side of the nucleus (arrows). The perforatorium (p) is detached. Posteriorly, the egg plasma membrane and cytoplasm have spread around part of the sperm tail. (For interpretation, see also Fig. 25C.) A section of the anterior portion of the same sperm head is shown in Fig. 32. Magnification: × 25,000.

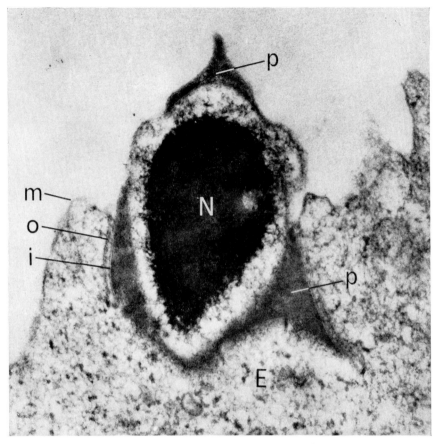

Fig. 32. Transverse section of the anterior portion of the sperm head shown in Fig. 31. The egg cytoplasm (E) only surrounds the ventral part of the sperm head. The inner acrosomal membrane (i) is intact, and the remaining portion of the outer acrosomal membrane (o) is now continuous with the egg plasma membrane (m). The nucleus (N) is compact. (p) Performatorium. (For interpretation, see also Fig. 26B.) Magnification: × 68,000.

eggs (Szollosi, 1967). The cortical granules open up through a fusion of their membrane with the egg plasma membrane and their contents are expelled into the perivitelline space.

It seems probable that the breakdown of the cortical granules has a role in the prevention of polyspermy and that it is responsible for the zona reaction, for the following reasons: (1) the cortical reaction is coincidental with sperm entry; (2) there is no other fine-structural change that could be connected with the reduction, and loss, of the penetrability of the zona by spermatozoa at this time, e.g., the retraction of the microvilli from the zona seems to occur preceding sperm entry (cf. Section

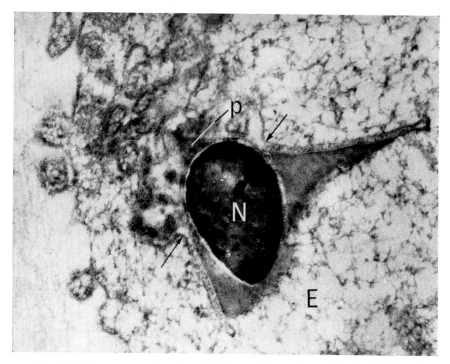

FIG. 33. Transverse section through the anterior portion of the head of another sperm at a level similar to that shown in Fig. 32. The egg plasma membrane and cytoplasm completely surround the sperm head. Note continuity of the egg plasma membrane with the remaining portion of the outer acrosomal membrane (arrows). Many egg microvilli become entrapped during the engulfment process but coalesce later into a continuous cytoplasmic layer. (For interpretation, see also Fig. 26C.) (N) Nucleus; (E) egg cytoplasm; (p) perforatorium. Magnification: × 56,000.

III,C); (3) the structure and properties of the zona pellucida make it unlikely that it would have a propagated reaction of its own; on the other hand, chemical interaction of the contents of the cortical vesicles with the zona is possible, although the nature of this interaction is obscure at present; (4) spontaneous breakdown of the cortical granules in golden hamster eggs seems to be associated with the loss of fertilizability of these eggs (cf. Section III,D); (5) rat and rabbit eggs that have been artificially activated still possess cortical granules (Thibault, 1964) and can undergo fertilization which is monospermic (e.g., Chalmel, 1962).

At the level of the egg plasma membrane (in eggs where more than one sperm penetrates through the zona) the contents of the cortical vesicles may exert a protective action by coating the plasma membrane of the spermatozoa, and that of the egg, thus preventing sperm attachment to the egg surface. That the egg plasma membrane per se has not completely lost its receptivity to spermatozoa is shown by the observation that fertilized rat eggs can be "refertilized" after treatment with the chelating agent, ethylenediaminetetraacetate (EDTA; Versene). Presumably the coating action is dependent on the presence of calcium ions. After Versene treatment the supplementary spermatozoa, which normally

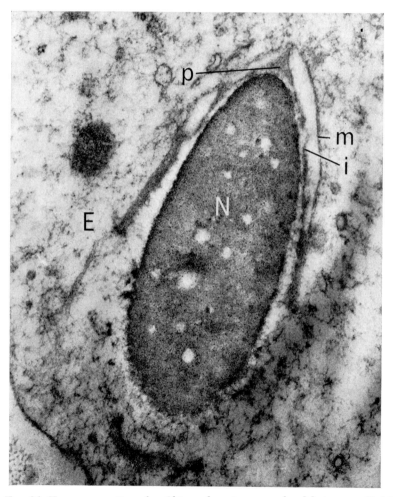

Fig. 34. Tranverse section of a Chinese hamster sperm head being engulfed by egg cytoplasm (E). There are no membranes in the postnuclear cap area. The vesiclelike structure is composed of the inner acrosomal membrane (i) and of egg plasma membrane (m). The nucleus (N) is slightly changed. (p) Perforatorium. (Compare with Fig. 7.) Magnification: × 50,000.

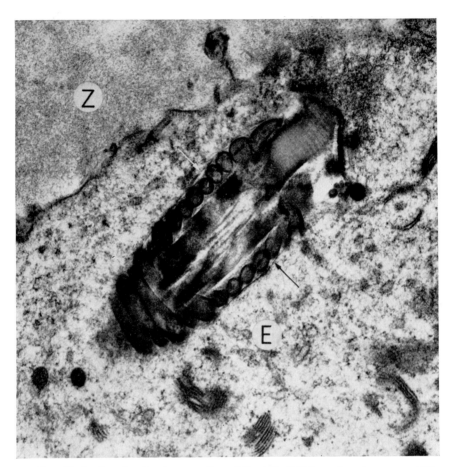

FIG. 35. Section showing incorporation of the tail of Chinese hamster sperm. (E) Egg cytoplasm; (Z) zona pellucida. Where the section angle is favorable, two membranes can be discerned (arrows): sperm plasma membrane and overlying egg plasma membrane. Magnification: × 24,000.

remain free in the perivitelline space, may become incorporated into the egg cytoplasm; a secondary penetration can occur even close to the time of the first cleavage (Pikó, 1961b; Fig. 38). von der Borch (1967) has observed refertilization of rat eggs by supplementary spermatozoa after the injection of a variety of substances, including EDTA, into the fallopian tube; the effect appeared to be nonspecific but might have resulted from a disturbance of the balance of calcium ions. Refertilization has been obtained at various times after fertilization in echinoderm and fish eggs (cf. Ginsburg, 1961; Monroy, 1965).

V. Summary and Conclusions

The plasma membrane of the freshly ovulated mammalian egg is covered by two extracellular coats; the zona pellucida surrounded by the cumulus oophorus. To fulfill their biological function—the activation of the egg and the transfer of paternal genetic material—the spermatozoa must reach and recognize the egg, penetrate the egg's outer coats, and become incorporated into the egg cytoplasm. The specialized sperm organelles that are involved in the cellular mechanisms of sperm entry into the egg (leaving aside the problem of sperm motility which, as is well known, is a prerequisite of sperm penetration) are the plasma membrane of the sperm head, the acrosome, the perforatorium, and the postnuclear cap. Their fine structure and chemical properties have been

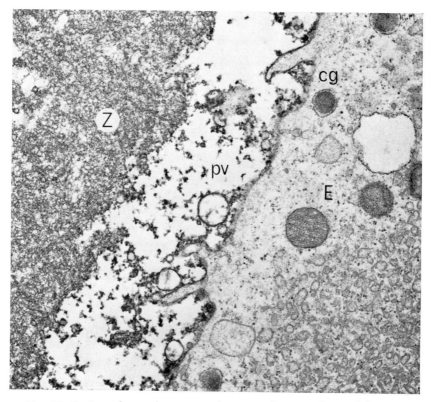

Fig. 36. Section of recently penetrated rat egg showing release of the contents of the cortical vesicles into the perivitelline space (pv). Some deeper located cortical granules (cg) remain in the egg cytoplasm (E). Ruthenium red staining has resulted in strong contrast of the zona pellucida (Z) and of the cortical granule material. (Courtesy of Dr. Daniel Szollosi.) Magnification: × 17,000.

reviewed from a point of view of their function in the sperm-entry process.

It may be concluded from the evidence discussed that the plasma membrane of the sperm head is endowed with special properties which play a critical role in the major events of sperm entry: (1) the recognition of the egg (as distinct from somatic cells and tissues); (2) the initiation of the acrosome reaction (which releases enzymes enabling the spermatozoon to pass through the extraneous coats of the egg); and (3) the attachment of the sperm head to the plasma membrane of the egg (a necessary preliminary step to the incorporation of the spermatozoon into the egg cytoplasm). The evidence lends strong support to the notion that the mechanism of these reactions involves a specific interaction between receptor sites (antifertilizin) on the sperm plasma membrane and complementary reactive groups (fertilizin) in the immediate environment and plasma membrane of the egg. There is evidence indicating that the plasma membrane of the head of epididymal and ejaculated spermatozoa is covered by a protective coating deriving, in part, from secretions of

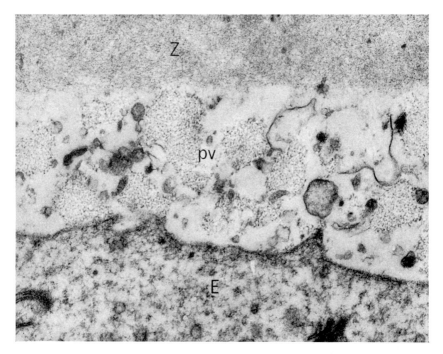

Fig. 37. Release of the contents of the cortical vesicles into the perivitelline space (pv) in recently penetrated Chinese hamster egg. (E) Egg cytoplasm; (Z) zona pellucida. Magnification: × 25,000.

the male genital tract. Sperm capacitation in the female tract probably consists, at least in part, in the gradual removal of coating material from the sperm surface thus exposing the specific sperm receptors.

The extraneous coats of the egg (the cumulus oophorus and the zona pellucida) have an important functional role during the growth and differentiation, and eventual release, of the ovocyte. Their role in the ovulated egg seems to consist in (1) mechanical protection and (2) the maintenance of a suitable microenvironment in the immediate vicinity of the egg. The presence of the cumulus oophorus may increase the chance of sperm penetration and prolong the fertilizable life of the egg. The zone pellucida consists, at least in some species, of several layers of differing chemical properties; an outer mucopolysaccharide layer shows a certain analogy with the gelatinous coat of the echinoderm egg and may be the source of the soluble fertilizin that has been reported to occur in the eggs of several mammalian species. In many species the zona pellucida also functions as an effective barrier against the entry of

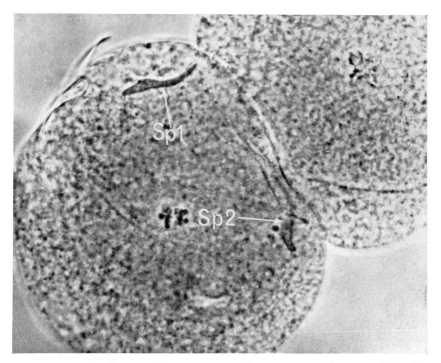

Fig. 38. Refertilization of a Versene-treated rat egg near the time of the first cleavage. Two one-cell eggs are seen, in contact with each other, both in metaphase of the first cleavage division. One of them contains two swollen sperm heads in the cytoplasm (Sp1 and Sp2). Acetocarmine staining. Magnification: × 480.

supernumerary spermatozoa in a manner analogous to the fertilization membrane of sea urchin and other invertebrate eggs. The present evidence makes the above assigned roles the most probable. At the same time it should be noted that they represent areas of gamete physiology in which further investigations would be most desirable and important.

The mechanism of sperm entry through the outer coats of the egg involves the acrosome which probably contains two enzymes: hyaluronidase and zona lysin. The following model of sperm entry is proposed. Contact of capacitated spermatozoa with the egg coats triggers the acrosome reaction consisting in the opening up of the acrosomal sac through a process of membrane fusion and vesiculation. First the hyaluronidase is released, this enzyme probably being located in the apical region of the acrosome, followed by the zona lysin, presumably located mainly in the lateral portion of the acrosome. The perforatorium (subacrosomal material) does not show any definitely visible changes during sperm entry; it appears likely that its role is to provide support for the acrosome and serve as an attachment device between the acrosomal sac and the nuclear envelope. Neither of the acrosomal enzymes is species-specific in its action on the egg coats. Therefore, the species specificity of fertilization can be expected to reside primarily in the mechanism ensuring the release of the acrosomal enzymes during sperm entry, and in the interaction between the plasma membranes of the two gametes.

The spermatozoon is incorporated into the egg cytoplasm by a process akin to phagocytosis, which can be divided into three phases: (1) specific attachment between the plasma membrane of the sperm head and the microvilli of the egg; (2) membrane fusion and breakdown, which occurs first in the postnuclear cap area, exposing the posterior part of the nucleus to the egg cytoplasm; (3) progressive engulfment of the anterior part of the nucleus and of the sperm tail. It is not clear whether the postnuclear cap has a role in the rapid breakdown of surrounding membranes or whether it serves as a purely structural device providing firm attachment between the sperm plasma membrane and the nuclear envelope in this area. The initial fusion and breakdown of sperm and egg plasma membranes appear to trigger the early events of fertilization: (1) the cortical reaction (breakdown of the cortical granules) which probably forms the basis of the block to polyspermy in mammals; (2) the transformation of the sperm nucleus; and (3) initial processes leading to the activation of the egg.

ACKNOWLEDGMENTS

I wish to express my sincere thanks and appreciation to Professor Albert Tyler for his unfailing help and advice in my research work over a period of years and for providing laboratory facilities. I am indebted to Professor C. R. Austin for reading the manuscript and to my colleagues Drs. Joram Piatigorsky and A. E. S. Smith for valuable discussions. The original investigations reported in this chapter were initiated under the tenure of a Fellowship from the Lalor Foundation, and have been further supported by grants from the National Science Foundation (GB-28) and from the National Institutes of Health (GM-12777) to Professor A. Tyler.

REFERENCES

Abraham, K. A., and Bhargava, P. M. (1963). The uptake of radioactive amino acids by spermatozoa. The intracellular site of incorporation into proteins. *J. Biochem.* **86**, 308–313.

Adams, C. E., and Chang, M. C. (1962). Capacitation of rabbit spermatozoa in the Fallopian tube and in the uterus. *J. Exptl. Zool.* **151**, 159–165.

Adams, E. C., and Hertig, A. T. (1964). Studies on guinea pig oocytes. I. Electron microscopic observations on the development of cytoplasmic organelles in oocytes of primordial and primary follicles. *J. Cell Biol.* **21**, 397–427.

Afzelius, B. A. (1955). The fine structure of sea urchin spermatozoa as revealed by the electron microscope. *Z. Zellforsch.* **42**, 134–148.

Afzelius, B. A. (1956). The acrosomal reaction of the sea urchin spermatozoon. *In* "Electron Microscopy—Proceedings of the Stockholm Conference" (F. S. Sjöstrand and J. Rhodin, eds.), pp. 167–169. Academic Press, New York.

Afzelius, B. A., and Murray, A. (1957). The acrosomal reaction of spermatozoa during fertilization or treatment with egg water. *Exptl. Cell Res.* **12**, 325–337.

Afzelius, B. A., and Rosén, B. (1965). Nutritive phagocytosis in animal cells. An electron microscopical study of the gastroderm of the hydroid *Clava squamata* Müll. *Z. Zellforsch.* **67**, 24–33.

Alsterberg, G. (1941). Über Vorkommen und Physiologie der Phosphatide in tierischen Zellen, besonders im Nervensystem. *Z. Zellforsch.* **31**, 364–407.

Ånberg, A. (1957). The ultrastructure of the human spermatozoon. *Acta Obstet. Gynecol. Scand.* **36** (Suppl. 2), 1–133.

Anderson, E., and Beams, H. W. (1960). Cytological observations on the fine structure of the guinea pig ovary with special reference to the oogonium, primary oocyte and associated follicle cells. *J. Ultrastruct. Res.* **3**, 432–446.

André, J. (1962). Contribution à la connaissance du chondriome: étude de ses modifications ultrastructurales pendant la spermatogenèse. *J. Ultrastruct. Res. Suppl.* **3**, 1–185.

André, J. (1963). Some aspects of specialization in sperm. *In* "General Physiology of Cell Specialization" (D. Mazia and A. Tyler, eds.), pp. 91–115. McGraw-Hill, New York.

Austin, C. R. (1948). Function of hyaluronidase in fertilization. *Nature* **162**, 63.

Austin, C. R. (1951a). Observations on the penetration of the sperm into the mammalian egg. *Australian J. Sci. Res.* **B4**, 581–596.

Austin, C. R. (1951b). The formation, growth, and conjugation of the pronuclei in the rat egg. *J. Roy. Microscop. Soc.* **71**, 295–306.

Austin, C. R. (1956). Cortical granules in hamster eggs. *Exptl. Cell Res.* **10**, 533–540.

Austin, C. R. (1960). Capacitation and the release of hyaluronidase from spermatozoa. *J. Reprod. Fertility* **3**, 310–311.

Austin, C. R. (1961a). "The Mammalian Egg." Blackwell, Oxford, England.

Austin, C. R. (1961b). Fertilization of mammalian eggs *in vitro. Intern. Rev. Cytol.* **12**, 337–359.

Austin, C. R. (1961c). Significance of sperm capacitation. *Proc. 4th Intern. Congr. Animal Reprod., Hague,* Vol. 4, pp. 723–725.

Austin, C. R. (1963a). Fertilization and transport of the ovum. *In* "Mechanisms Concerned with Conception" (C. G. Hartman, ed.), pp. 285–320. Macmillan (Pergamon), New York.

Austin, C. R. (1963b). Acrosome loss from the rabbit spermatozoon in relation to entry into the egg. *J. Reprod. Fertility* 6, 313–314.

Austin, C. R. (1964). Behaviour of spermatozoa in the female genital tract and in fertilization. *Proc. 5th Intern. Congr. Animal Reprod., Trento,* Vol. 3, pp. 7–22.

Austin, C. R. (1965). "Fertilization." Prentice-Hall, Englewood Cliffs, New Jersey.

Austin, C. R., and Bishop, M. W. H. (1957a). Preliminaries to fertilization in mammals. *In* "The Beginnings of Embryonic Development" (A. Tyler, R. C. von Borstel, and C. B. Metz, eds.), pp. 71–107. Am. Assoc. Advan. Sci., Washington, D. C.

Austin, C. R., and Bishop, M. W. H. (1957b). Fertilization in mammals. *Biol. Rev.* 32, 296–349.

Austin, C. R., and Bishop, M. W. H. (1958a). Some features of the acrosome and perforatorium in mammalian spermatozoa. *Proc. Roy. Soc.* B149, 234–240.

Austin, C. R., and Bishop, M. W. H. (1958b). Role of the rodent acrosome and perforatorium in fertilization. *Proc. Roy. Soc.* B149, 241–248.

Austin, C. R., and Braden, A. W. H. (1954). Time relations and their significance in the ovulation and penetration of eggs in rats and rabbits. *Australian J. Biol. Sci.* 7, 179–194.

Austin, C. R., and Braden, A. W. H. (1956). Early reactions of the rodent egg to spermatozoon penetration. *J. Exptl. Biol.* 33, 358–365.

Austin, C. R., and Lovelock, J. E. (1958). Permeability of rabbit, rat and hamster egg membranes. *Exptl. Cell Res.* 15, 260–261.

Austin, C. R., and Walton, A. (1960). Fertilisation. *In* "Marshall's Physiology of Reproduction" (A. S. Parkes, ed.), Vol. I (2), pp. 310–416. Longmans, Green, New York.

Balazs, E. A. (1958). Physical chemistry of hyaluronic acid. *Federation Proc.* 17, 1086–1093.

Bane, A., and Nicander, L. (1963). The structure and formation of the perforatorium in mammalian sperm. *Intern. J. Fertility* 8, 865–866 (abstract).

Bangham, A. D. (1961). Electrophoretic characteristics of ram and rabbit spermatozoa. *Proc. Roy. Soc.* B155, 292–305.

Barros, C., and Franklin, L. E. (1968). Behavior of the gamete membranes during sperm entry into the mammalian egg. *J. Cell Biol.* 37, C13–C18.

Barros, C., Bedford, J. M., Franklin, L. E., and Austin, C. R. (1967). Membrane vesiculation as a feature of the mammalian acrosome reaction. *J. Cell Biol.* 34, C1–C5.

Bayer, M. E. (1964). An electron microscope examination of urinary mucoprotein and its interaction with influenza virus. *J. Cell Biol.* 21, 265–274.

Beams, H. W. (1964). Cellular membranes in oogenesis. *In* "Cellular Membranes in Development" (M. Locke, ed.), pp. 175–219. Academic Press, New York.

Beatty, R. A. (1961). Genetics of mammalian gametes. *Animal Breeding Abstr.* 29, 243–256.

Bedford, J. M. (1963a). Morphological changes in rabbit spermatozoa during passage through the epididymis. *J. Reprod. Fertility* 5, 169–177.

Bedford, J. M. (1963b). Changes in the electrophoretic properties of rabbit spermatozoa during passage through the epididymis. *Nature* **200**, 1178–1180.

Bedford, J. M. (1963c). Morphological reaction of spermatozoa in the female reproductive tract of the rabbit. *J. Reprod. Fertility* **6**, 245–255.

Bedford, J. M. (1964a). Fine structure of the sperm head in ejaculate and uterine spermatozoa of the rabbit. *J. Reprod. Fertility* **7**, 221–228.

Bedford, J. M. (1964b). Changes in fine structure of the rabbit sperm head during passage through the epididymis. *Proc. 5th Intern. Congr. Animal Reprod.*, *Trento*, Vol. 3, pp. 397–402.

Bedford, J. M. (1964c). Evidence of change in the sperm head plasma membrane of rabbit uterine spermatozoa. *Proc. 5th Intern. Congr. Animal Reprod.*, *Trento*, Vol. 7, pp. 286–288.

Bedford, J. M. (1965). Effect of environment on phagocytosis of rabbit spermatozoa. *J. Reprod. Fertility* **9**, 249–256.

Bedford, J. M. (1967). Observations on the fine structure of spermatozoa of the bush baby (*Galago senegalensis*), the African green monkey (*Cercopithecus aethiops*) and man. *Am. J. Anat.* **121**, 443–460.

Bedford, J. M. (1968). Ultrastructural changes in the sperm head during fertilization in the rabbit. *Am. J. Anat.* **123**, 329–357.

Bedford, J. M., and Chang, M. C. (1962). Removal of decapacitation factor from seminal plasma by high-speed centrifugation. *Am. J. Physiol.* **202**, 179–181.

Bennett, H. S. (1963). Morphological aspects of extracellular polysaccharides. *J. Histochem. Cytochem.* **11**, 14–23.

Biggers, J. D., and Creed, R. F. S. (1962). Conjugate spermatozoa of the North American opossum. *Nature* **196**, 1112–1113.

Bishop, D. W. (1961). Biology of spermatozoa. *In* "Sex and Internal Secretions" (W. C. Young, ed.), Vol. II, pp. 706–796. Williams & Wilkins, Baltimore, Maryland.

Bishop, D. W. (1962a). Sperm motility. *Physiol. Rev.* **42**, 1–59.

Bishop, D. W., ed. (1962b). "Spermatozoan Motility." Am. Assoc. Advan. Sci., Washington, D. C.

Bishop, D. W., and Tyler, A. (1956). Fertilizin of mammalian eggs. *J. Exptl. Zool.* **132**, 575–602.

Bishop, M. W. H., and Austin, C. R. (1957). Mammalian spermatozoa. *Endeavour* **16**, 137–150.

Bishop, M. W. H., and Walton, A. (1960). Spermatogenesis and the structure of mammalian spermatozoa. *In* "Marshall's Physiology of Reproduction" (A. S. Parkes, ed.), Vol. I (2), pp. 1–129. Longmans, Green, New York.

Björkman, N. (1962). A study of the ultrastructure of the granulosa cells of the rat ovary. *Acta Anat.* **51**, 125–147.

Blanchette, E. J. (1961). A study of the fine structure of the rabbit primary oocyte. *J. Ultrastruct. Res.* **5**, 349–363.

Blandau, R. J. (1961). Biology of eggs and implantation. *In* "Sex and Internal Secretions" (W. C. Young, ed.), Vol. II, pp. 797–882. Williams & Wilkins, Baltimore, Maryland.

Blandau, R. J. (1965). Egg transport from the ovary into oviduct in rodents. Film presented at *Conf. on Gamete Transport, Fertilization, and Preimplantation Mechanisms, Nashville.*

Blandau, R. J., and Odor, D. L. (1952). Observations on sperm penetration into the ooplasm and changes in the cytoplasmic components of the fertilizing spermatozoon in rat ova. *Fertility Sterility* **3**, 13–26.

Blandau, R. J., and Rumery, R. (1962). Observations on cultured granulosa cells from ovarian follicles and ovulated ova of the rat. *Fertility Sterility* **13**, 335–345.

Blandau, R. J., and Rumery, R. E. (1964). The relationship of swimming movements of epididymal spermatozoa to their fertilizing capacity. *Fertility Sterility* **15**, 571–579.

Blom, E., and Birch-Andersen, A. (1965). The ultrastructure of the bull sperm. II. The sperm head. *Nord. Veterinarmed.* **17**, 193–212.

Boettcher, B. (1965). Human ABO blood group antigens on spermatozoa from secretors and non-secretors. *J. Reprod. Fertility* **9**, 267–268.

Bollet, A. J., Bonner, W. M., and Nance, J. L. (1963). The presence of hyaluronidase in various mammalian tissues. *J. Biol. Chem.* **238**, 3522–3527.

Bomsel-Helmreich, O. (1966). Heteroploidy and embryonic death. *Ciba Found. Symp. Preimplantation Stages Pregnancy*, pp. 246–267.

Bowen, R. H. (1922). On the idiosome, Golgi apparatus, and acrosome in the male germ cells. *Anat. Record* **27**, 159–180.

Bowen, R. H. (1924). On the acrosome of the animal sperm. *Anat. Record* **28**, 1–13.

Braden, A. W. H. (1952). Properties of the membranes of rat and rabbit eggs. *Australian J. Sci. Res.* **B5**, 460–471.

Braden, A. W. H. (1955). The reactions of isolated mucopolysaccharides to several histochemical tests. *Stain Technol.* **30**, 19–26.

Braden, A. W. H. (1958). Strain differences in the morphology of the gametes of the mouse. *Australian J. Biol. Sci.* **12**, 65–71.

Braden, A. W. H. (1960). Genetic influences on the morphology and function of the gametes. *J. Cellular Comp. Physiol.* **56** (Suppl. 1), 17–29.

Braden, A. W. H., and Austin, C. R. (1954a). Fertilization of the mouse egg and the effect of delayed coitus and of hot-shock treatment. *Australian J. Biol. Sci.* **7**, 552–565.

Braden, A. W. H., and Austin, C. R. (1954b). The number of sperms about the eggs in mammals and its significance for normal fertilization. *Australian J. Biol. Sci.* **7**, 543–551.

Braden, A. W. H., and Weiler, H. (1964). Transmission ratios at the T-locus in the mouse: inter- and intra-male heterogeneity. *Australian J. Biol. Sci.* **17**, 921–934.

Braden, A. W. H., Austin, C. R., and David, H. A. (1954). The reaction of the zona pellucida to sperm penetration. *Australian J. Biol. Sci.* **7**, 391–409.

Brambell, F. W. R. (1956). Ovarian changes. *In* "Marshall's Physiology of Reproduction" (A. S. Parkes, ed.), Vol. I (1), pp. 397–542. Longmans, Green, New York.

Brökelmann, J. (1963). Fine structure of germ cells and Sertoli cells during the cycle of the seminiferous epithelium in the rat. *Z. Zellforsch.* **59**, 820–850.

Brunish, R., and Högberg, B. (1960). Some of the properties of purified testicular hyaluronidase. *Compt. Rend. Trav. Lab. Carlsberg* **32**, 35–47.

Burgos, M. H., and Fawcett, D. W. (1955). Studies on the fine structure of the mammalian testis. I. Differentiation of the spermatids in the cat (*Felis domestica*). *J. Biophys. Biochem. Cytol.* **1**, 287–300.

Burgos, M. H., and Fawcett, D. W. (1956). An electron microscope study of spermatid differentiation in the toad, *Bufo arenarum* Hensel. *J. Biophys. Biochem. Cytol.* **2**, 223–240.

Buzzell, A., and Hanig, M. (1958). The mechanism of hemagglutination by influenza virus. *Advan. Virus Res.* **5**, 289–346.

Chalmel, M.-C. (1962). Possibilité de fécondation des oeufs de Lapine activés parthénogénétiquement. *Ann. Biol. Animale Biochim. Biophys.* **2**, 279–297.

Chang, M. C. (1951). Fertilizing capacity of spermatozoa deposited into the Fallopian tubes. *Nature* **168**, 697–698.

Chang, M. C. (1955). Development of fertilizing capacity of rabbit spermatozoa in the uterus. *Nature* **175**, 1036–1037.

Chang, M. C. (1957). A detrimental effect of seminal plasma on the fertilizing capacity of sperm. *Nature* **179**, 258–259.

Chang, M. C., and Bedford, J. M. (1962). Fertilizability of rabbit ova after removal of the corona radiata. *Fertility Sterility* **13**, 421–425.

Chang, M. C., and Sheaffer, D. (1957). Number of spermatozoa ejaculated at copulation, transported into the female tract, and present in the male tract of the golden hamster. *J. Heredity* **48**, 107–109.

Chiquoine, A. D. (1960). The development of the zona pellucida of the mammalian ovum. *Am. J. Anat.* **106**, 149–169.

Clermont, Y., and Leblond, C. P. (1955). Spermiogenesis of man, monkey, ram and other mammals as shown by the periodic acid–Schiff technique. *Am. J. Anat.* **96**, 229–250.

Clermont, Y., Glegg, R. E., and Leblond, C. P. (1955a). Presence of carbohydrates in the acrosome of the guinea pig spermatozoon. *Exptl. Cell Res.* **8**, 453–458.

Clermont, Y., Einberg, E., Leblond, C. P., and Wagner, S. (1955b). The perforatorium—an extension of the nuclear membrane of the rat spermatozoon. *Anat. Record* **121**, 1–12.

Colwin, A. L., and Colwin, L. H. (1957). Morphology of fertilization: acrosome filament formation and sperm entry. *In* "The Beginnings of Embryonic Development" (A. Tyler, R. C. von Borstel, and C. B. Metz, eds.), pp. 135–168. Am. Assoc. Advan. Sci., Washington, D. C.

Colwin, A. L., and Colwin, L. H. (1961). Changes in the spermatozoon during fertilization in *Hydroides hexagonus* (Annelida). II. Incorporation with the egg. *J. Biophys. Biochem. Cytol.* **10**, 255–274.

Colwin, A. L., and Colwin, L. H. (1963a). Role of the gamete membranes in fertilization in *Saccoglossus kowalevskii* (Enteropneusta). I. The acrosomal region and its changes in early stages of fertilization. *J. Cell Biol.* **19**, 477–500.

Colwin, L. H., and Colwin, A. L. (1963b). Role of the gamete membranes in fertilization in *Saccoglossus kowalevskii* (Enteropneusta). II. Zygote formation by gamete membrane fusion. *J. Cell Biol.* **19**, 501–518.

Dallam, R. D., and Thomas, L. E. (1953). Chemical studies on mammalian sperm. *Biochim. Biophys. Acta* **11**, 79–89.

Dan, J. C. (1956). The acrosome reaction. *Intern. Rev. Cytol.* **5**, 365–393.

Dan, J. C. (1962). The vitelline coat of the *Mytilus* egg. I. Normal structure and effect of acrosomal lysin. *Biol. Bull.* **123**, 531–541.

Daoust, R., and Clermont, Y. (1955). Distribution of nucleic acids in germ cells during the cycle of the seminiferous epithelium in the rat. *Am. J. Anat.* **96**, 255–283.

Das, C. M. S. (1962). Ultrastructure of the postnuclear cap in the developing sperms of *Microtus pennsylvanicus* (Ord.). *Proc. Zool. Soc. Calcutta* **15**, 75–81.

Dauzier, L. (1958). Physiologie du déplacement des spermatozoïdes dans les voies génitales femelles chez la brebis et la vache. D.Sc. Dissertation, Sorbonne, Paris.

De Salegui, M., Plonska, H., and Pigman, W. (1967). A comparison of serum and testicular hyaluronidase. *Arch. Biochem. Biophys.* **121**, 548–554.

Dickmann, Z. (1963a). Denudation of the rabbit egg: time-sequence and mechanism. *Am. J. Anat.* **113**, 303–335.

Dickmann, Z. (1963b). The zona pellucida of the rabbit egg. *Fertility Sterility* **14**, 490–493.

Dickmann, Z. (1964a). Fertilization and development of rabbit eggs following the removal of the cumulus oophorus. *J. Anat.* **98**, 397–402.

Dickmann, Z. (1964b). The passage of spermatozoa through and into the zona pellucida of the rabbit egg. *J. Exptl. Biol.* **41**, 177–182.

Dickmann, Z., and Dziuk, P. J. (1964). Sperm penetration of the zona pellucida of the pig egg. *J. Exptl. Biol.* **41**, 603–608.

Dukelow, W. R., Chernoff, H. N., and Williams, W. L. (1966a). Enzymatic characterization of decapacitation factor. *Proc. Soc. Exptl. Biol. Med.* **121**, 396–398.

Dukelow, W. R., Chernoff, H. N., and Williams, W. L. (1966b). Stability of spermatozoan decapacitation factor. *Am. J. Physiol.* **211**, 826–828.

Edwards, R. G., Ferguson, L. C., and Coombs, R. R. A. (1964). Blood group antigens on human spermatozoa. *J. Reprod. Fertility* **7**, 153–161.

Farquhar, M. G., and Palade, G. E. (1963). Junctional complexes in various epithelia. *J. Cell Biol.* **17**, 375–412.

Fawcett, D. W. (1958). The structure of the mammalian spermatozoon. *Intern. Rev. Cytol.* **7**, 195–234.

Fawcett, D. W. (1961). Cilia and flagella. *In* "The Cell" (J. Brachet and A. E. Mirsky, eds.), Vol. II, pp. 217–297. Academic Press, New York.

Fawcett, D. W. (1965). The anatomy of the mammalian spermatozoon with particular reference to the guinea pig. *Z. Zellforsch.* **67**, 279–296.

Fawcett, D. W., and Burgos, M. H. (1956). Observations on the cytomorphosis of the germinal and interstitial cells of the human testis. *Ciba Found. Colloq. Aging*, **2**, 86–96.

Fawcett, D. W., and Hollenberg, R. D. (1963). Changes in the acrosome of guinea pig spermatozoa during passage through the epididymis. *Z. Zellforsch.* **60**, 276–292.

Fawcett, D. W., and Ito, S. (1965). The fine structure of bat spermatozoa. *Am. J. Anat.* **116**, 567–609.

Fawcett, D. W., and Phillips, D. M. (1967). Further observations on mammalian spermiogenesis. *J. Cell Biol.* **35**, 152A (Abstr.).

Fazekas de St. Groth, S. (1962). The neutralization of viruses. *Advan. Virus Res.* **9**, 1–125.

Fekete, E., and Duran-Reynals, F. (1943). Hyaluronidase in the fertilization of mammalian ova. *Proc. Soc. Exptl. Biol. Med.* **52**, 119–121.

Fessler, J. H. (1960). Mode of action of testicular hyaluronidase. *Biochem. J.* **76**, 132–135.

Fessler, J. H., and Fessler, L. I. (1966). Electron microscopic visualization of the polysaccharide hyaluronic acid. *Proc. Natl. Acad. Sci. U. S.* **56**, 141–147.

Fitton Jackson, S. (1964). Connective tissue cells. *In* "The Cell" (J. Brachet and A. E. Mirsky, eds.), Vol. VI, pp. 387–520. Academic Press, New York.

Franchi, L. L. (1960). Electron microscopy of oocyte–follicle cell relationship in the rat ovary. *J. Biophys. Biochem. Cytol.* **7**, 397–398.

Franchi, L. L., Mandl, A. M., and Zuckerman, S. (1962). The development of the ovary and the process of oogenesis. *In* "The Ovary" (S. Zuckerman, ed.), Vol. I, pp. 1–88. Academic Press, New York.

Franklin, L. E. (1965). Morphology of gamete membrane fusion and of sperm entry into oocytes of the sea urchin. *J. Cell Biol.* **25**, (Pt. 2), 81–100.

Fuhrmann, G. F., Granzer, E., Bey, E., and Ruthenstroth-Bauer, G. (1963). Die elektrophoretische Beweglichkeit der Spermien des Rindes. *Z. Naturforsch.* **18b**, 236–242.

Gaddum, P., and Glover, T. D. (1965). Some reactions of rabbit spermatozoa to ligation of the epididymis. *J. Reprod. Fertility* **9**, 119–130.

Gatenby, J. B., and Beams, H. W. (1935). The cytoplasmic inclusions in the spermatogenesis of man. *Quart. J. Microscop. Sci.* **78**, 1–29.

Gatenby, J. B., and Wigoder, S. B. (1929). The post-nuclear body in the spermatogenesis of *Cavia cobaya*, and other animals. *Proc. Roy. Soc.* **B104**, 471–480.

Ginsburg, A. S. (1961). The block to polyspermy in sturgeon and trout with special reference to the role of cortical granules (alveoli). *J. Embryol. Exptl. Morphol.* **9**, 173–190.

Glass, L. E. (1963). Transfer of native and foreign serum antigens to oviducal mouse eggs. *Am. Zoologist* **3**, 135–156.

Gothié, S. (1958). Contribution à l'étude de la membrane pellucide de l'oeuf de lapine à l'aide du ^{35}S. *J. Physiol.* (*Paris*) **50**, 293–294.

Green, C. L. (1957). Identification of alpha-amylase as a secretion of the human Fallopian tube and "tubelike" epithelium of Müllerian and mesonephric duct origin. *Am. J. Obstet. Gynecol.* **73**, 402–408.

Gregoire, A. T., Rankin, J., Johnson, W. D., Rakoff, A. E., and Adams, A. (1967). α-Amylase content in cervical mucus of females receiving sequential, nonsequential, or no contraceptive therapy. *Fertility Sterility* **18**, 836–839.

Gresson, R. A. R., and Zlotnik, I. (1945). A comparative study of the cytoplasmic components of the male germ-cells of certain mammals. *Proc. Roy. Soc. Edinburgh* **B62**, 137–170.

Gwatkin, R. B. L. (1963a). Studies on the zona pellucida of the mouse egg. *J. Reprod. Fertility* **6**, 325.

Gwatkin, R. B. L. (1963b). Effect of viruses on early mammalian development. I. Action of Mengo encephalitis virus on mouse ova cultivated *in vitro*. *Proc. Natl. Acad. Sci. U. S.* **50**, 576–581.

Gwatkin, R. B. L. (1964). Effect of enzymes and acidity on the zona pellucida of the mouse egg before and after fertilization. *J. Reprod. Fertility* **7**, 99–105.

Hadek, R. (1963a). Study on the fine structure of rabbit sperm head. *J. Ultrastruct. Res.* **9**, 110–122.

Hadek, R. (1963b). Submicroscopic changes in the penetrating spermatozoon of the rabbit. *J. Ultrastruct. Res.* **8**, 161–169.

Hadek, R. (1963c). Electron microscope study on primary liquor folliculi secretion in the mouse ovary. *J. Ultrastruct. Res.* **9**, 445–458.

Hadek, R. (1963d). Submicroscopic study on the cortical granules in the rabbit ovum. *J. Ultrastruct. Res.* **8**, 170–175.

Hadek, R. (1963e). Submicroscopic study on the sperm-induced cortical reaction in the rabbit ovum. *J. Ultrastruct. Res.* **9**, 99–109.

Hadek, R. (1964). Submicroscopic study on the cortical villi in the rabbit ovum. *J. Ultrastruct. Res.* **10**, 58–65.

Hadek, R. (1965). The structure of the mammalian egg. *Intern. Rev. Cytol.* **18**, 29–71.

Hamner, C. E., and Williams, W. L. (1965). Composition of rabbit oviduct secretions. *Fertility Sterility* **16**, 170–176.

Hancock, J. L. (1957). The morphology of boar spermatozoa. *J. Roy. Microscop. Soc.* **76**, 84–97.

Hancock, J. L. (1966). The ultra-structure of mammalian spermatozoa. *Advan. Reprod. Physiol.* **1**, 125–154.

Hancock, J. L., and Trevan, D. J. (1957). The acrosome and post-nuclear cap of bull spermatozoa. *J. Roy. Microscop. Soc.* **76**, 77–83.

Harrison, R. J. (1962). The structure of the ovary. C. Mammals. *In* "The Ovary" (S. Zuckerman, ed.), Vol. I, pp. 143–187. Academic Press, New York.

Hartman, C. G. (1962). "Science and the Safe Period." Williams & Wilkins, Baltimore, Maryland.

Hartree, E. F., and Srivastava, P. N. (1965). Chemical composition of the acrosomes of ram spermatozoa. *J. Reprod. Fertility* **9**, 47–60.

Hathaway, R. R., and Hartree, E. F. (1963). Observations on the mammalian acrosome: experimental removal of acrosomes from ram and bull spermatozoa. *J. Reprod. Fertility* **5**, 225–232.

Hekman, A., and Rümke, P. (1969). The antigens of human seminal plasma, with special reference to lactoferrin as a spermatozoa-coating antigen. *Fertility Sterility* **20**, 312–323.

Henle, W., Henle, G., and Chambers, L. A. (1938). Studies on the antigenic structure of some mammalian spermatozoa. *J. Exptl. Med.* **68**, 335–352.

Henricks, D. M., and Mayer, D. T. (1965). Isolation and characterization of a basic keratin-like protein from mammalian spermatozoa. *Exptl. Cell Res.* **40**, 402–412.

Hertig, A. T., and Adams, E. C. (1967). Studies on the human oocyte and its follicle. I. Ultrastructural and histochemical observations on the primordial follicle stage. *J. Cell Biol.* **34**, 647–675.

Holstein, A.-F. (1965). Elektronenmikroskopische Untersuchungen am Spermatozoon des Opossum (*Didelphys virginiana* Kerr). *Z. Zellforsch.* **65**, 904–914.

Hope, J. (1965). The fine structure of the developing follicle of the Rhesus ovary. *J. Ultrastruct. Res.* **12**, 592–610.

Hopsu, V. K., and Arstila, A. V. (1965). Development of the acrosomic system of the spermatozoon in the Norwegian lemming (*Lemmus lemmus*). *Z. Zellforsch.* **65**, 562–572.

Horstmann, E. (1961). Elektronenmikroskopische Untersuchungen zur Spermiohistogenese beim Menschen. *Z. Zellforsch.* **54**, 68–89.

Hughes, R. L. (1965). Comparative morphology of spermatozoa from five marsupial families. *Australian J. Zool.* **13**, 533–543.

Hunter, A. G., and Hafs, H. D. (1964). Antigenicity and cross-reactions of bovine spermatozoa. *J. Reprod. Fertility* **7**, 357–365.

Iype, P. T., Abraham, K. A., and Bhargava, P. M. (1963). Further evidence for a positive role of acrosome in the uptake of labelled amino acids by bovine and avian spermatozoa. *J. Reprod. Fertility* **5**, 151–158.

Jacoby, F. (1962). Ovarian histochemistry. *In* "The Ovary" (S. Zuckerman, ed.), Vol. I, pp. 189–245. Academic Press, New York.

Kaye, G. I., Pappas, G. D., Yasuzumi, G., and Yamamoto, H. (1961). The distribution and form of the endoplasmic reticulum during spermatogenesis in the crayfish, *Cambaroides japonicus*. *Z. Zellforsch.* **53**, 159–171.

Kaye, J. S. (1962). Acrosome formation in the house cricket. *J. Cell Biol.* **12**, 411–431.

Kellenberger, E., Bolle, A., Boy de la Tour, E., Epstein, R. H., Franklin, N. C., Jerne, N. K., Reale-Scafati, A., Séchaud, J., Bendet, I., Goldstein, D., and

Lauffer, M. A. (1965). Functions and properties related to the tail fibers of bacteriophage T4. *Virology* **26**, 419–440.

Kirton, K. T., and Hafs, H. D. (1965). Sperm capacitation by uterine fluid or beta-amylase *in vitro. Science* **150**, 618–619.

Kojima, Y. (1962). Electron microscopic study of the bull spermatozoon. *Japan J. Vet. Res.* **10**, 72–74.

Kojima, Y. (1966). Electron microscopic study of the bull spermatozoon. *Japan. J. Vet. Res.* **14**, 1–62.

Korolev, V. A., and Petrov, G. N. (1964). Denudation of animal and human ova during fertilization. *Zh. Obshch. Biol.* **25**, 68–74 (in Russian).

Kühn, K., Grassmann, W., and Hofmann, U. (1958). Die elektronenmikroskopische "Anfärbung" des Kollagens und die Ausbildung einer hochunterteilten Querstreifung. *Z. Naturforsch.* **13b**, 154–160.

Lanzavecchia, G., and Mangioni, C. (1964). Etude de la structure et des constituants du follicle humain dans l'ovaire foetal. I. Le follicle primordial. *J. Microscopie* **3**, 447–464.

Leblond, C. P. (1950). Distribution of periodic acid-reactive carbohydrates in the adult rat. *Am. J. Anat.* **86**, 1–49.

Leblond, C. P., and Clermont, Y. (1952). Spermiogenesis of rat, mouse, hamster and guinea pig as revealed by the periodic acid-fuchsin sulfurous acid technique. *Am. J. Anat.* **90**, 167–215.

Leblond, C. P., Steinberger, E., and Roosen-Runge, E. C. (1963). Spermatogenesis. *In* "Mechanisms Concerned with Conception" (C. G. Hartman, ed.), pp. 1–72. Macmillan (Pergamon), New York.

Leuchtenberger, C., and Schrader, F. (1950). The chemical nature of the acrosome in the male germ cells. *Proc. Natl. Acad. Sci. U. S.* **36**, 677–683.

Lindahl, P. E. (1960). Some factors influencing the biological activity of sperm antagglutins. *J. Reprod. Fertility* **1**, 3–22.

Lindahl, P. E., and Brattsand, R. (1962). Particles in bull seminal plasma abolishing agglutination in washed spermatozoa. *Zool. Bidr. Uppsala* **35**, 504–515.

McClean, D., and Rowlands, I. W. (1942). Role of hyaluronidase in fertilization. *Nature* **150**, 627–628.

McGeachin, R. L., Hargan, L. A., Potter, B. A., and Daus, A. T., Jr. (1958). Amylase in Fallopian tubes. *Proc. Soc. Exptl. Biol. Med.* **99**, 130–131.

Machlis, L., and Rawitscher-Kunkel, E. (1963). Mechanisms of gametic approach in plants. *Intern. Rev. Cytol.* **15**, 97–138.

Mancini, R. E., Alonso, A., Barquet, J., Alvarez, B., and Nemirovsky, M. (1964). Histo-immunological localization of hyaluronidase in the bull testis. *J. Reprod. Fertility* **8**, 325–330.

Mann, T. (1949). Metabolism of semen. *Advan. Enzymol.* **9**, 329–390.

Mann, T. (1964). "The Biochemistry of Semen and of the Male Reproductive Tract." Wiley, New York.

Martan, J., and Allen, J. M. (1964). Morphological and cytochemical properties of the holocrine cells in the epdidymis of the mouse. *J. Histochem. Cytochem.* **12**, 628–640.

Martan, J., and Risley, P. L. (1963). Holocrine secretory cells of the rat epididymis. *Anat. Record* **146**, 173–189.

Mastroianni, L. (1962). The structure and function of the Fallopian tube. *Clin. Obstet. Gynecol.* **5**, 781–790.

Matousek, J. (1964). Antigenic characteristics of spermatozoa from bulls, rams and boars. III. Absorption analysis, precipitins and fructolysis in relation to the antigenicity of bull spermatozoa. *J. Reprod. Fertility* **8**, 13–21.

Mattner, P. E. (1963). Capacitation of ram spermatozoa and penetration of the ovine egg. *Nature* **199**, 772–773.

Merker, H.-J. (1961). Elektronenmikroskopische Untersuchungen über die Bildung der *zona pellucida* in den Follikeln des Kaninchenovars. *Z. Zellforsch.* **54**, 677–688.

Metz, C. B. (1957). Specific egg and sperm substances and activation of the egg. In "The Beginnings of Embryonic Development" (A. Tyler, R. C. von Borstel, and C. B. Metz, eds.), pp. 23–69. Am. Assoc. Advan. Sci., Washington, D. C.

Meyer, K. (1958). Chemical structure of hyaluronic acid. *Federation Proc.* **17**, 1075–1077.

Meyer, K., and Rapport, M. M. (1952). Hyaluronidases. *Advan. Enzymol.* **13**, 199–236.

Miller, R. L. (1966). Chemotaxis during fertilization in the hydroid *Campanularia*. *J. Exptl. Zool.* **161**, 23–44.

Mintz, B. (1962). Experimental study of the developing mammalian egg: removal of the zona pellucida. *Science* **138**, 594–595.

Monroy, A. (1965). "Chemistry and Physiology of Fertilization." Holt, Rinehart and Winston, New York.

Moricard, R. (1960). Observations de microscopie électronique sur des modifications acrosomiques lors de la pénétration spermatique dans l'oeuf des mammifères. *Compt. Rend. Soc. Biol.* **154**, 2187–2189.

Moricard, R. (1961). Des structures de l'acrosome des spermatozoïdes contenus dans l'utérus chez la lapine. *Compt. Rend. Soc. Biol.* **154**, 2187–2189.

Moricard, R. (1964). Fécondation et différentiations ultrastructurales intratubaires des spermatozoïdes provenant du canal déférent. *Proc. 5th Intern. Congr. Animal Reprod., Trento*, Vol. 7, pp. 295–300.

Moricard, R., and Gothié, S. (1955). Etude de la répartition en S³⁵ dans les cellules folliculaires périovocytaires au cours de l'ovogenèse et de la terminaison de la première mitose de maturation chez la lapine adulte. *Compt. Rend. Soc. Biol.* **149**, 1918–1922.

Moricard, R., Guillon, G., and Guerrier, M. (1960). Observations de microscopie électronique sur la pénétration spermatique dans la membrane pellucida de l'ovule des mammifères. *Bull. Gynécol. Obstet.* **12**, 542–552.

Nagano, T. (1962). Observations on the fine structure of the developing spermatid in the domestic chicken. *J. Cell Biol.* **14**, 193–205.

Nath, V. (1956). Cytology of spermatogenesis. *Intern. Rev. Cytol.* **5**, 395–453.

Nath, V. (1966). "Animal Gametes (Male)." Asia Publ. House, New York.

Nevo, A. C., Michaeli, I., and Schindler, H. (1961). Electrophoretic properties of bull and of rabbit spermatozoa. *Exptl. Cell Res.* **23**, 69–83.

Nicander, L., and Bane, A. (1962). Fine structure of boar spermatozoa. *Z. Zellforsch.* **57**, 390–405.

Nicander, L., and Bane, A. (1966). Fine structure of the sperm head in some mammals, with particular reference to the acrosome and the subacrosomal substance. *Z. Zellforsch.* **72**, 496–515.

Noyes, R. W. (1959). The capacitation of spermatozoa. *Obstet. Gynecol. Survey* **14**, 785–797.

Noyes, R. W., Walton, A., and Adams, C. E. (1958). Capacitation of rabbit spermatozoa. *J. Endocrinol.* **17**, 374–380.

Odor, D. L. (1960). Electron microscopic studies on ovarian oocytes and unfertilized tubal ova in the rat. *J. Biophys. Biochem. Cytol.* **7**, 567–574.

Odor, D. L. (1965). The ultrastructure of unilaminar follicles of the hamster ovary. *Am. J. Anat.* **116**, 493–521.

Odor, D. L., and Blandau, R. J. (1949). The frequency of occurrence of supernumerary sperm in rat ova. *Anat. Record* **104**, 1–10.

Odor, D. L., and Blandau, R. J. (1951). Observations on fertilization and the first segmentation division in rat ova. *Am. J. Anat.* **89**, 29–62.

Onuma, H. (1963). Studies on the acrosomic system of spermatozoa of domestic animals. V. The acrosomic system of ejaculated spermatozoa of bulls, stallions and boars. *Bull. Natl. Inst. Animal Ind. Chiba*, No. 2, 243–248.

Onuma, H., and Nishikawa, Y. (1963). Studies on the acrosomic system of spermatozoa of domestic animals. I. Cytochemical nature of the PAS positive material in the acrosomic system of boar spermatids. *Bull. Natl. Inst. Animal Ind. Chiba*, No. 1, 125–134.

Orgebin-Crist, M. C. (1967). Sperm maturation in rabbit epididymis. *Nature* **216**, 816–818.

Parat, M. (1933a). Nomenclature, genèse, structure et fonction de quelques éléments cytoplasmiques des cellules sexuelles males. *Compt. Rend. Soc. Biol.* **112**, 1131–1134.

Parat, M. (1933b). L'acrosome du spermatozoïde dans la fécondation et la parthénogenèse expérimentale. *Compt. Rend. Soc. Biol.* **112**, 1134–1137.

Parkes, A. S. (1960). The biology of spermatozoa and artificial insemination. *In* "Marshall's Physiology of Reproduction" (A. S. Parkes, ed.), Vol. I (2), pp. 161–263. Longmans, Green, New York.

Pasteels, J. J. (1961). La réaction corticale de fécondation ou d'activation (revue comparative). *Bull. Soc. Zool. Fr.* **86**, 600–629.

Pasteels, J. J. (1965a). Étude au microscope électronique de la réaction corticale. I. La réaction corticale de fécondation chez *Paracentrotus* et sa chronologie. II. La réaction corticale de l'oeuf vierge de *Sabellaria alveolata*. *J. Embryol. Exptl. Morphol.* **13**, 327–339.

Pasteels, J. J. (1965b). Aspects structuraux de la fécondation vus au microscope électronique. *Arch. Biol.* (*Liège*) **76**, 463–509.

Pasteels, J. J. (1965c). La fécondation étudiée au microscope électronique. Etude comparative. *Bull. Soc. Zool. Fr.* **90**, 195–224.

Picheral, B. (1967). Structure et organisation du spermatozoïde de *Pleurodeles waltlii* Michah. (Amphibian Urodèle.) *Arch. Biol.* (*Liège*) **78**, 193–221.

Pikó, L. (1961a). La polyspermie chez les animaux. *Ann. Biol. Animale Biochim. Biophys.* **1**, 323–383.

Pikó, L. (1961b). Repeated fertilization of fertilized rat eggs after treatment with Versene (EDTA). *Am. Zoologist* **1**, 467 (Abstract).

Pikó, L. (1964). Mechanism of sperm penetration in the rat and the Chinese hamster based on fine structural studies. *Proc. 5th Intern. Congr. Animal Reprod., Trento*, Vol. 7, pp. 301–302.

Pikó, L. (1967). Immunological phenomena in the reproductive process. *Intern. J. Fertility* **12**, 377–383.

Pikó, L., and Bomsel-Helmreich, O. (1960). Triploid rat embryos and other chromosomal deviants after colchicine treatment and polyspermy. *Nature* **186**, 737–739.

Pikó, L., and Tyler, A. (1962). Antigenic analysis of Chinese hamster sperm. *Am. Zoologist* **2**, 548 (Abstr.).

Pikó, L., and Tyler, A. (1964a). Ultrastructure of the acrosome and the early events of sperm penetration in the rat. *Am. Zoologist* **4**, 287 (Abstr.).

Pikó, L., and Tyler, A. (1964b). Fine structural studies of sperm penetration in the rat. *Proc. 5th Intern. Congr. Animal Reprod., Trento*, Vol. 2, pp. 372–377.

Pincus, G. (1936). "The Eggs of Mammals." Macmillan, New York.

Pinsker, M. C., and Williams, W. L. (1967). Properties of a spermatozoan antifertility factor. *Arch. Biochem. Biophys.* **122**, 111–117.

Prokofieva-Belgovskaya, A. A., and Chai, C. H. (1961). Electron microscopic investigation of spermiogenesis in the mouse. *Biophysics (USSR) (Engl. Transl.)* (6), **1**, 43–51.

Randall, J. T., and Friedlaender, M. H. G. (1950). The microstructure of ram spermatozoa. *Exptl. Cell Res.* **1**, 1–32.

Raven, C. P. (1961). "Oogenesis: The Storage of Developmental Information." Macmillan (Pergamon), New York.

Restall, B. J. (1967). The biochemical and the physiological relationships between the gametes and the female reproductive tract. *Advan. Reprod. Physiol.* **2**, 181–212.

Risley, P. L. (1963). Physiology of the male accessory organs. *In* "Mechanisms Concerned with Conception" (C. G. Hartman, ed.), pp. 73–133. Macmillan (Pergamon), New York.

Roosen-Runge, E. C. (1962). The process of spermatogenesis in mammals. *Biol. Rev.* **37**, 343–377.

Rothschild, Lord (1956a). "Fertilization." Wiley, New York.

Rothschild, Lord (1956b). The fertilizing spermatozoon. *Discovery* **18**, 64–65.

Rothschild, Lord (1962). Spermatozoa. *Brit. Med. J.* **ii**, 743–749 and 812–817.

Saacke, R. G., and Almquist, J. O. (1964). Ultrastructure of bovine spermatozoa. I. The head of normal, ejaculated sperm. *Am. J. Anat.* **115**, 143–162.

Schenk, S. L. (1878). Das Säugethierei künstlich befruchtet ausserhalb des Mutterthieres. *Mitt. Embryol. Inst. K. K. Univ. Wien* **1**, 107–118.

Schrader, F., and Leuchtenberger, C. (1951). The cytology and chemical nature of some constituents of the developing sperm. *Chromosoma* **4**, 404–428.

Searcy, R. L., Craig, R. G., and Bergquist, L. M. (1964). Immunochemical properties of normal and pathologic seminal plasma. *Fertility Sterility* **15**, 1–8.

Seshachar, B. R., and Bagga, S. (1963). Cytochemistry of the oocyte of *Loris tardigradus lydekkerianus* (Cabr.) and *Macaca mulatta mulatta* (Zimmerman). *J. Morphol.* **113**, 119–137.

Sjöstrand, F. S. (1963). A comparison of plasma membrane, cytomembranes, and mitochondrial membrane elements with respect to ultrastructural features. *J. Ultrastruct. Res.* **9**, 561–580.

Sotelo, J. R., and Porter, K. R. (1959). An electron microscope study of the rat ovum. *J. Biophys. Biochem. Cytol.* **5**, 327–342.

Soupart, P., and Clewe, T. H. (1965). Sperm penetration of rabbit zona pellucida inhibited by treatment of ova with neuraminidase. *Fertility Sterility* **16**, 677–689.

Soupart, P., and Noyes, R. W. (1964). Sialic acid as a component of the zona pellucida of the mammalian ovum. *J. Reprod. Fertility* **8**, 251–253.

Srivastava, M. D. L. (1965). Cytoplasmic inclusions in oogenesis. *Intern. Rev. Cytol.* **18**, 73–98.

Srivastava, P. N., Adams, C. E., and Hartree, E. F. (1965a). Enzymatic action of lipoglycoprotein preparations from sperm-acrosomes on rabbit ova. *Nature* **205**, 498.

Srivastava, P. N., Adams, C. E., and Hartree, E. F. (1965b). Enzymic action of acrosomal preparations on the rabbit ovum *in vitro*. *J. Reprod. Fertility* **10**, 61–67.

Stambaugh, R., and Buckley, J. (1968). Zona pellucida dissolution enzymes of the rabbit sperm head. *Science* **161**, 585–586.

Stegner, H. E., and Wartenberg, H. (1961). Elektronenmikroskopische und histotopochemische Untersuchungen über Struktur und Bildung der Zona pellucida menschlicher Eizellen. *Z. Zellforsch.* **53**, 702–713.

Stent, G. S. (1963). "Molecular Biology of Bacterial Viruses." Freeman, San Francisco, California.

Sternberger, L. A. (1967). Electron microscopic immunochemistry: A review. *J. Histochem. Cytochem.* **15**, 139–159.

Stone, W. H. (1964). The significance of immunologic phenomena in animal reproduction. *Proc. 5th Intern. Congr. Animal Reprod., Trento*, Vol. 2, pp. 89–109.

Swyer, G. I. M. (1947a). The release of hyaluronidase from spermatozoa. *Biochem. J.* **41**, 413–417.

Swyer, G. I. M. (1947b). A tubal factor concerned in the denudation of rabbit ova. *Nature* **159**, 873–874.

Szollosi, D. (1962). Cortical granules: a general feature of mammalian eggs? *J. Reprod. Fertility* **4**, 223–224.

Szollosi, D. (1967). Development of cortical granules and the cortical reaction in rat and hamster eggs. *Anat. Record* **159**, 431–446.

Szollosi, D. G., and Ris, H. (1961). Observations on sperm penetration in the rat. *J. Biophys. Biochem. Cytol.* **10**, 275–283.

Takashima, R., and Takashima, Y. (1960). Electron microscopical observations on the fertilization phenomenon of sea urchin with special reference to the acrosome filament. *Tokushima J. Exptl. Med.* **6**, 334–339.

Tardini, A., Vitali-Mazza, L., and Mansani, F. E. (1960). Ultrastruttura dell'ovocita umano maturo. I. Rapporti fra cellule della corona radiata, pellucida ed ovoplasma. *Arch. "de Vecchi" Anat. Patol. Med. Clin.* **33**, 281–305.

Tardini, A., Vitali-Mazza, L., and Mansani, F. E. (1961). Ultrastruttura dell'ovocita umano maturo. II. Nucleo e citoplasma ovulare. *Arch. "de Vecchi" Anat. Patol. Med. Clin.* **35**, 25–71.

Thibault, C. (1959). Analyse de la fécondation de l'oeuf de la truie après accouplement ou insémination artificielle. *Ann. Zootech. Suppl.* **8**, 165–177.

Thibault, C. (1964). Presence d'une spermalysine dans le spermatozoïde maturé (capacité) du lapin. *Proc. 5th Intern. Congr. Animal Reprod., Trento*, Vol. 7, pp. 294–295.

Thibault, C., and Dauzier, L. (1960). "Fertilisines" et fécondation *in vitro* de l'oeuf de lapine. *Compt. Rend. Acad. Sci.* **250**, 1358–1359.

Thibault, C., and Dauzier, L. (1961). Analyse des conditions de la fécondation *in vitro* de l'oeuf de la lapine. *Ann. Biol. Animale Biochim. Biophys.* **1**, 277–294.

Tice, L. W., and Barrnett, R. J. (1963). Fine structural localization of some testicular phosphatases. *Anat. Record* **147**, 43–63.

Trujillo-Cenóz, O., and Sotelo, J. R. (1959). Relationships of the ovular surface with follicle cells and origin of the zona pellucida in rabbit oocytes. *J. Biophys. Biochem. Cytol.* **5**, 347–350.

Tyler, A. (1948). Fertilization and immunity. *Physiol. Rev.* **28**, 180–219.

Tyler, A. (1949). Properties of fertilizin and related substances of eggs and sperm of marine animals. *Am. Naturalist* **83**, 195–219.

Tyler, A. (1956). Physico-chemical properties of the fertilizins of the sea urchin

Arbacia punctulata and the sand dollar *Echinarachnius parma*. *Exptl. Cell Res.* **10**, 377–386.

Tyler, A. (1959). Some immunological experiments of fertilization and early development in sea urchins. *Exptl. Cell Res. Suppl.* **7**, 183–199.

Tyler, A. (1960). Introductory remarks on theories of fertilization. In "Symposium on Germ Cells and Development" (S. Ranzi, ed.), pp. 155–174, Fondazione Baselli, Milano, Italy.

Tyler, A. (1961). Approaches to the control of fertility based on immunological phenomena. *J. Reprod. Fertility* **2**, 473–506.

Tyler, A. (1965). The biology and chemistry of fertilization. *Am. Naturalist* **99**, 193–219.

Tyler, A., and Bishop, D. W. (1963). Immunological phenomena. In "Mechanisms Concerned with Conception" (C. G. Hartman, ed.), pp. 397–482. Macmillan (Pergamon), New York.

Tyler, A., and Tyler, B. S. (1965). Physiology of fertilization and early development. In "Physiology of Echinodermata" (R. A. Boolootian, ed.), pp. 683–741. Wiley, New York.

von der Borch, S. (1967). Abnormal fertilization of rat eggs after injection of substances into the ampullae of the fallopian tubes. *J. Reprod. Fertility* **14**, 465–468.

Von Lenhossék, M. (1898). Untersuchungen über Spermatogenese. *Arch. Mikroskop. Anat.* **51**, 215–318.

Waldeyer, W. (1906). Die Geschlechtszellen. In "Handbuch der vergleichenden und experimentellen Entwickelungslehre der Wirbeltiere" (O. Hertwig, ed.), Vol. I, pp. 86–476. Fischer, Jena, Germany.

Wartenberg, H., and Stegner, H. E. (1960). Über die elektronenmikroskopische Feinstructur des menschlichen Ovarioleies. *Z. Zellforsch.* **52**, 450–474.

Weakley, B. S. (1966). Electron microscopy of the oocyte and granulosa cells in the developing ovarian follicles of the golden hamster (*Mesocricetus auratus*). *J. Anat.* **100**, 503–534.

Weakley, B. S. (1967). Light and electron microscopy of developing germ cells and follicle cells in the ovary of the golden hamster: twenty-four hours before birth to eight days post partum. *J. Anat.* **101**, 435–459.

Weil, A. J. (1960). Immunological differentiation of epididymal and seminal spermatozoa of the rabbit. *Science* **131**, 1040–1041.

Weil, A. J. (1965). The spermatozoa-coating antigen (SCA) of the seminal vesicle. *Ann. N. Y. Acad. Sci.* **124**, 267–269.

Weil, A. J., and Rodenburg, J. M. (1960). Immunological differentiation of human testicular (spermatocele) and seminal spermatozoa. *Proc. Soc. Exptl. Biol. Med.* **105**, 43–45.

Weil, A. J., and Rodenburg, J. M. (1962). The seminal vesicle as the source of the spermatozoa-coating antigen of seminal plasma. *Proc. Soc. Exptl. Biol. Med.* **109**, 567–570.

Weissmann, B. (1955). The transglycosylative action of testicular hyaluronidase. *J. Biol. Chem.* **216**, 793–794.

Weissmann, B., Hadjiioannou, S., and Tornheim, J. (1964). Oligosaccharase activity of β-N-acetyl-D-glucosaminidase of beef liver. *J. Biol. Chem.* **239**, 59–63.

Willson, J. T., and Katsh, S. (1965). Cyto-immunological studies of guinea pig sperm antigens. I. Testicular versus epididymal spermatozoa. *Z. Zellforsch.* **65**, 16–26.

Wilson, E. B. (1928). "The Cell in Development and Heredity." Macmillan, New York.

Wintenberger-Torrès, S. (1961). Motilité des trompes et progression des oeufs chez la brebis. *Ann. Biol. Animale Biochim. Biophys.* **1,** 121–133.

Wislocki, G. B. (1949). Seasonal changes in the testis, epididymides and seminal vesicles of deer investigated by histochemical methods. *Endocrinology* **44,** 167–189.

Yamada, E., Muta, T., Motomura, A., and Koga, H. (1957). The fine structure of the oocyte in the mouse ovary studied with electron microscope. *Kurume Med. J.* **4,** 148–171.

Yamane, I. (1930). The proteolytic action of mammalian spermatozoa and its bearing upon the second maturation division of ova. *Cytologia* (*Tokyo*) **1,** 394–403.

Yamane, I. (1935). Kausal analytische Studien über die Befruchtung des Kanincheneies. I. Die Dispersion der Follikelzellen und die Ablösung der Zellen der Corona radiata des Eies durch Spermatozoen. *Cytologia* (*Tokyo*) **6,** 233–255.

Yanagimachi, R., and Chang, M. C. (1961). Fertilizable life of golden hamster ova and their morphological changes at the time of losing fertilizability. *J. Exptl. Zool.* **148,** 185–204.

Yanagimachi, R., and Chang, M. C. (1964). *In vitro* fertilization of golden hamster ova. *J. Exptl. Zool.* **156,** 361–375.

Young, W. C. (1931). A study of the function of the epididymis. III. Functional changes undergone by spermatozoa during their passage through the epididymis and vas deferens in the guinea pig. *J. Exptl. Biol.* **8,** 151–162.

Zachariae, F. (1959). Acid mucopolysaccharides in the female genital system and their rôle in the mechanism of ovulation. *Acta Endocrinol.* 33 (Suppl. 47), 1–62.

Zamboni, L., and Mastroianni, L., Jr. (1966a). Electron microscopic studies on rabbit ova. I. The follicular oocyte. *J. Ultrastruct. Res.* **14,** 95–117.

Zamboni, L., and Mastroianni, L., Jr. (1966b). Electron microscope studies on rabbit ova. II. The penetrated tubal ovum. *J. Ultrastruct. Res.* **14,** 118–132.

Zamboni, L., Hongsanand, C., and Mastroianni, L., Jr. (1965). Influence of tubal secretions on rabbit tubal ova. *Fertility Sterility* **16,** 177–184.

Zamboni, L., Mishell, D. R., Jr., Bell, J. H., and Baca, M. (1966). Fine structure of the human ovum in the pronuclear stage. *J. Cell Biol.* **30,** 579–600.

Zlotnik, I. (1943). A nuclear ring in the developing male germ cells of dog and cat. *Nature* **151,** 670.

CHAPTER 9

In Vitro Fertilization
of the Mammalian Egg

C. Thibault

DÉPARTMENT DE PHYSIOLOGIE ANIMALE, INSTITUT NATIONAL DE LA
RECHERCHE AGRONOMIQUE, JOUY-EN-JOSAS, FRANCE

I. Introduction

Fertilization in mammals is not fundamentally different from that in other species, even those which are zoologically very distant. As early as 1875, Van Beneden studied almost simultaneously fertilization in the rabbit and in *Ascaris*.

Yet, the site of fertilization and the chemical specificity of the medium in which fertilization takes place in mammals have always presented a considerable obstacle to experimental research.

Until recently our knowledge of the mechanisms by which mammalian gametes meet and fuse was very limited in comparison with species whose fertilization occurs naturally outside the animal's body. This explains why for nearly a century numerous efforts have been made to try and fertilize a mammalian egg outside the maternal body and why most of our knowledge on mammalian fertilization results from experiments carried out recently *in vitro* or partly *in vitro*.

II. Attempts to Induce *in Vitro* Fertilization

A. CRITERIA OF FERTILIZATION

The success claimed by some authors has generally been criticized by others, who questioned the results because they considered that proof of *in vitro* fertilization was insufficient.

It is only by adopting precise criteria that it is possible to select the cases from the literature that really constitute fertilization and to establish a reference base for further work.

The most convincing criterion of successful *in vitro* fertilization is the birth of young resulting from eggs fertilized and cultured *in vitro* and then transplanted into foster mothers. However, this method cannot be generally used, for the means and the materials required are too costly, and furthermore, the delay in waiting for the results is too long (Plate II, Fig. 8).

The criterion of *in vitro* cleavage of eggs is much more efficient as the results of the experiment are known within 24–28 hours. However, it presents the disadvantage that cleavage can follow parthenogenetic activation and can be as regular as after normal fertilization in the sheep, the hamster, and especially in the rabbit (Thibault, 1949; Chang, 1954; Austin, 1956).

The first polar body may divide or disappear soon after ovulation or fertilization. Thus, the number of polar bodies is not a reliable criterion.

By using the precise stages of pronuclei evolution as a basis, one can quickly determine whether or not fertilization is really occurring.

Six stages in the fertilization of the mammalian egg may be considered (Thibault, 1967) (Plate I):

Stage 0—The capacitated spermatozoa pass through the corona radiata and the zona pellucida.

Stage 1—The spermatozoa adhere to the microvilli of the vitelline membrane, and the sperm and egg membranes begin to fuse. At the end of this stage, the cortical granules disappear and the telophase of the second maturation division is achieved.

Stage 2—The sperm head swells, and the midpiece usually penetrates; the second polar body forms.

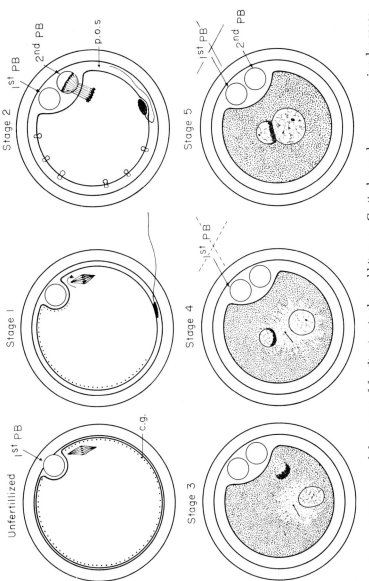

PLATE I. Diagram of the stages of fertilization in the rabbit. c.g.—Cortical granules; p.o.s.—periovular space; PB—polar body.

Stage 3—The pronuclei begin to form. The male pronucleus is often recognized by the proximity of the midpiece, with the sperm aster radiating from its anterior end. This is particularly evident in the rabbit (Plate II, Fig. 3) and the guinea pig, but is somewhat less obvious in the ewe and sow.

At the end of stage 3, the pronuclei are easily identified, mainly by the asymmetric distribution of the deoxyribonucleic acid (DNA) of the female pronucleus [rabbit (Plate II, Fig. 4), cat, sow, sheep, and cow], but also by their size (the male being generally larger) or the number of their nucleoli.

Stage 4—Increase in the size of the pronuclei and their subcentral migration; DNA duplication preparing the first cleavage mitosis (Oprescu and Thibault, 1965). During this stage, the DNA remains asymmetrically distributed in the female pronucleus, thus allowing the nuclei to be identified (Plate II, Fig. 4).

Thus, observation of eggs in stages 3 or 4 provides unquestionable proof that fertilization has taken place.

Stage 5—The pronuclei in subcentral position reach their maximum size and are in contact. The first polar body is not necessarily seen and the midpiece is distinctly visible in some species (woman, sow, rat, and rabbit) but not in others (cow).

In many species the two pronuclei become so similar at stage 5 that they are indistinguishable from the result of parthenogenetic activation. In stage 2, the second polar body is always distinctly visible, but swelling of the sperm head sometimes requires careful examination before it can be seen (Plate II, Fig. 1).

B. ANALYSIS OF RESULTS

Up to 1953 capacitated sperms were not used in *in vitro* fertilization attempts. Capacitation of the spermatozoa was discovered in the rabbit and rat by Chang (1951) and Austin (1951) working independently. However, in spite of this, Shettles (1953, 1955) and Petrov (1958) still used freshly ejaculated sperm.

During the period extending from Shenk's first attempt (1878) to Shettles' experiments on the human egg (1953, 1955), only Smith (1951) avoided cooling of the egg during manipulation in order to prevent the possibility of parthenogenetic activation. The following investigators used ovarian oocytes: Shenk (1878), Pincus and Enzmann (1935), Pincus (1939) working on the rabbit, Menkin and Rock (1948), Rock and Menkin (1944), Shettles (1953), and Petrov (1958) working on man. However, it is probably impossible to fertilize ovarian oocytes until they have achieved maturation.

With the exception of Smith's results, it appears from all the data summarized in Table I, that no *in vitro* fertilization had been obtained. Many causes for failure can be named: use of noncapacitated sperm, no precautions against temperature drop, frequent use of ovarian primary oocytes, and inadequacy of histological controls.

The more recent work of the following investigators has been successful as demonstrated by histological control of the eggs, normal *in vitro* cleavage rate, or the birth of young (Table II): Dauzier and Thibault (1956, 1959; Dauzier *et al.*, 1954), Chang (1959), Thibault and Dauzier (1961), Brackett and Williams (1965), Suzuki and Mastroianni (1965), and Brackett and Williams (1967–1968) working on rabbit eggs, and Yanagimachi and Chang (1964), Yanagimachi (1964), and Barros and Austin (1967) working on hamster eggs.

Finally, attempts to fertilize eggs maintained *in vitro* in the fallopian tubes have been partially successful (Table III).

III. *In Vitro* Fertilization of the Rabbit Egg

A. SYNTHETIC MEDIA. ANALYSIS OF THE FERTILIZATION PROCESS

1. *Passage through the Corona Radiata*

The spermatozoa, whether obtained directly from the male tract (epididymal or ejaculated) or from the female tract (capacitated), pass through the corona radiata in the same manner. The presence of hyaluronidase in sperm has distracted attention from the first action of the sperm on the corona cells. This first reaction is revealed by a slow-motion microcinematographic study of the arrival of the spermatozoa. It consists of a contraction of the cytoplasmic processes that join the corona cells to the oocytes and to each other. The egg is then surrounded by a compact tissue consisting of cells adhering together. Their dissociation is a secondary phenomenon (Thibault, unpublished observations).

2. *Passage through the Zona Pellucida*

The noncapacitated spermatozoa adhere to the zona pellucida in great numbers, but do not pass through it, irrespective of the duration of the experiment *in vitro*. On the other hand, fertilization is possible if freshly ovulated oocytes are introduced immediately after recovery into a suspension of capacitated sperm flushed from the uterine horn of rabbit mated 12 hours previously (Experiment A, Thibault and Dauzier, 1961).

Fertilization rate is slightly higher if oocytes are introduced into the sperm suspension after a short passage (5–15 minutes) in a Locke solution (Experiment B, Thibault and Dauzier, 1961). The average fertiliza-

TABLE I

Authors	Species	Egg[a]	Sperm[b]	Culture media
Shenk, 1878	Rabbit, guinea pig	O.E.	EPI	Genital mucus and uterine mucosa
Onanoff, 1893	Rabbit, guinea pig	—	—	—
Long, 1912	Mouse, rat	T.E.	EPI	Ringer or blood serum
Krasovskaya, 1935	Rabbit	T.E.	EPI (from rat, guinea pig, and dog)	Ringer, 20 min, then rabbit blood plasma
Pincus, 1930	Rabbit	T.E.	VD	Egg recovered in Pannett-Compton solution culture: serum + chicken embryo extract
Pincus and Enzmann, 1934	Rabbit	T.E.	VD	Sperm and oocytes in Ringer for 20 min, then, eggs transferred to a rabbit doe
Frommolt, 1934	Rabbit	T.E.	—	—
Pincus and Enzmann, 1935	Rabbit	O.E.	VD	In Ringer-Locke for 2 hr (pH, 7.3–7.5)
Pincus and Enzmann, 1936	Rabbit	T.E.	VD	
Pincus, 1939	Rabbit	T.E. O.E.	VD (rabbit, ferret, guinea pig, bull, ram, . . .)	Tyrode or Pannett-Compton solution, 20–30 min then rabbit blood serum
Rock and Menkin, 1944; Menkin and Rock, 1948	Man	O.E.	EJ	27 hr of culture in homologous blood serum Locke during in vitro insemination; culture in heterologous serum
Smith, 1951	Rabbit	T.E.	EJ	Blood serum + Sim + tubal mucosa + streptomycin

Summary of Early Investigations of *in Vitro* Fertilization

Temperature during manipulation	Results	Comments
Not controlled	Cleavage in 10 to 12 hr	—
—	—	Not *in vitro* fertilization, but first successful attempt to inseminate by intraperitoneal injection
Not controlled	Mouse: none Rat: formation of the 2nd polar body	Parthenogenetic activation (?)
Not controlled	Formation of the 2nd polar body; eggs with 2 pronuclei and normal blastomeres. With rat sperm, no penetration	Authentic cleavages. Parthenogenesis (?)
Not controlled	Formation of the 2nd polar body and cleavage	Also cleavage in controls. One photograph of a sperm head inside an egg is an artifact
Not controlled	Birth of young	*In vivo* fertilization after egg transfer
—	Cleavage	—
Not controlled	Formation of the 2nd polar body: sperm head or male pronucleus. Close relationship between number of sperms and fertilization rate	Noncapacitated sperm. Doubtful photographs of pronuclei
Not controlled	Activation and cleavage with rabbit, rat, ram, and bull sperm	Noncapacitated sperm. Figs. 22 and 25—sperm heads swept away by the blade; Figs. 15, 23, and 24— "fertilized" eggs are often haploid even with rabbit sperm[c]
At least 1 hr at room temp.	3 out of 138 eggs are cleaved	Fragmentation: many nuclei in each "blastomere"
37°C	1 male pronucleus; 1 out of 24 eggs is cleaved	1 Egg probably fertilized

(Continued)

TABLE I

Authors	Species	Egg[a]	Sperm[b]	Culture media
Shettles, 1953	Man	O.E.	EJ	In follicular fluid and fragments of tubal mucosa
Shettles, 1955	Man	O.E.	EJ	—
Petrov, 1958	Pig	O.E.	EJ	Ringer and follicular fluid
	Rabbit	O.E.		
	Man	O.E.		

[a] O.E.— ovarian egg; T.E.—tubal egg.
[b] EPI—epididymis; VD—vas deferens; EJ—ejaculated.
[c] Figures referred to are from Pincus, 1939.

tion rate is then 13.7% and fertilized oocytes are present in 32% of the trials (Dauzier and Thibault, 1956; Thibault and Dauzier, 1960). Such fertilization rate is low, and results are not constant from one trial to another. Chang (1959), Yanagimachi (1964), and Brackett and Williams (1965) have confirmed this.

The average number of supernumerary spermatozoa found under the zona pellucida of eggs fertilized *in vitro* is less than 1, whereas the average number is 8 at the same stage in eggs fertilized normally *in vivo* (Thibault, 1967).

Although there are more spermatozoa present in the culture medium than in the tubal fluid, very few spermatozoa penetrate under the zona pellucida. This seems to be due to a repelling effect of the zona pellucida on the capacitated spermatozoa, as is shown after fixation, sectioning, and staining of the eggs (Plate II, Fig. 5).

By using aged oocytes (*1*) recovered only 4–5 hours after ovulation (Experiment B2, Thibault and Dauzier, 1961), (*2*) recovered as soon as possible after ovulation (11½–12 hours after mating) but cultured *in vitro* for 1 to 3 hours in a saline solution (Experiment C, Thibault and Dauzier, 1961), and (*3*) recovered as in the preceding experiment and mechanically washed for 30 to 120 minutes in a saline medium (Experiment D), the percentage of fertilization was increased from 13.7% (Experiment B) to 24.5% (Experiment B2) to 32% (Experiment C), and to 66% (Experiment D) (Thibault and Dauzier, 1961).

In Experiment D, the results are reproducible in 94% of the trials. The number of sperm under the zona pellucida reaches 4. Spermatozoa adhere to the zona pellucida as in *in vivo* fertilization (Plate II, Fig. 6). Fertilization (stage 1) occurs approximately 90 minutes after introduc-

Continued

Temperature during manipulation	Results	Comments
Not controlled	2nd polar body; 1 penetration; 1 egg with 2 pronuclei	The spermatozoon deeply penetrating in the egg is a typical example of penetration in a degenerated egg
	Morula	
Heated room	Formation of the 2nd polar body and pronuclei	Parthenogenesis ?

tion of the capacitated sperm. Such a delay is of the same order as that observed during natural fertilization in the fallopian tubes (Thibault, 1967).

Bedford and Chang (1962) have failed to corroborate the favorable results obtained by previous washing of the eggs for 1 to 3 hours in Ringer's lactate solution. On the contrary, they state that in the majority of their experiments, fertilization was higher when freshly ovulated eggs were used. Unfortunately, they related a large number of experiments, the results of which vary between 0 and 97% fertilization. Moreover, no chronological analysis of sperm penetration is mentioned. Thus, the necessary comparison between treatments is not possible. Since different delays in fertilization may have occurred according to the type of experiment.

Brackett and Williams (1968) obtained high fertilization rates (between 50 and 75%) using a large volume (<5 ml) of a modified Krebs-Ringer bicarbonate solution plus bovine plasma albumin and glucose. The authors insisted on the necessity for the following conditions during manipulation: absence of oxygen ($N + 5\%$ CO_2), maintenance of a constant temperature, high humidity (under paraffin oil), and pH stability of about 7.8. This medium preserves capacitated spermatozoa at 37°C remarkably well and facilitates thinning of the zona pellucida; sperm are numerous in and under the zona pellucida (Thibault, 1968, unpublished).

3. Fertilizin-Like Substance

Thibault and Dauzier (1960, 1961) postulated that rabbit oocytes liberate a substance similar to a fertilizin. The function of this substance is to attach capacitated spermatozoa to the oocyte. This substance, dif-

TABLE II

SUMMARY OF RECENT INVESTIGATIONS OF *in Vitro* FERTILIZATION

Authors	Species[a]	Sperm type[b]	Culture media	Temperature (°C) during manipulation	Results	Comments
Dauzier et al., 1954; Dauzier and Thibault, 1956; Thibault and Dauzier, 1960	Rabbit T.E.	CA, (12 hr) perfusion from uterus	Locke and Locke serum	37–38	19/121: 16% cleaved 217/1587: 13.7% fertilized	Typical fertilization stages
Moricard, 1954	Rabbit T.E.	CA, pipetted out of uterus 10 hr after mating	Tubal or uterine fluids (anaerobic)	—	7/21 penetrated; no cleavage	Evident penetration in the periovular space; no evidence of sperm pronuclei on the photographs
Dauzier and Thibault, 1959	Rabbit T.E.	CA, pipetted out of uterus 12 hr after mating (as Moricard's)	Locke	37–38	28/91: 30% fertilized	Typical fertilization stages
Chang, 1957	Rabbit T.E.	CA	Aerobic or anaerobic conditions with or without Fallopian tube	—	No fertilization	—
Chang, 1959	Rabbit T.E.	CA, perfusion	Locke serum	37	77/266: 29% fertilized 55/266: 20% cleaved	Using Dauzier and Thibault's techniques, birth of 13 young after the transfer of 30 eggs
			Homologous serum Heterologous serum	37 37	50/102: 50% fertilized 18/109: 16.5% fertilized	(Not clear whether fertilization occurred in serum or if eggs were transferred before fertilization.)

				Temp. (°C)	Results	Remarks
Thibault and Dauzier, 1961	Rabbit T.E. washed 30 min to 2 hr in Locke medium	CA, perfusion 12 hr after mating	Locke, discontinuous rotary glass tube rack	37	305/593: 67% fertilized	Typical fertilization stages: birth of a litter of 2 young
Suzuki and Mastroianni, 1965	Rabbit T.E.	CA, pipetted out of uterus 12 hr after mating	Tubal and uterine fluids	—	30.9 and 63.7% in 2 trials	Pronuclei and cleavage (59.4%; Suzuki, 1966)
Brackett and Williams, 1965	Rabbit T.E.	CA	Uterine fluid	39	53/73: 72.5% fertilized	Pronuclei and cleavage (64%)
Brackett and Williams, 1967–1968	Rabbit T.E.	CA	Modified Tyrode + albumin and glucose, pH 7.8, no oxygen ($N_2 + CO_2$)	39	35/54: 65% fertilized	Cleavage rate *in vitro*
Yanagimachi and Chang, 1963, 1964	Hamster T.E.	CA EPI	Tubal fluid + Tyrode or TC 199, Gey, etc.	30–38	From 9.5 to 64.7%, depending on the medium	In some experiments high incidence of polyspermy
Barros and Austin, 1967	Hamster T.E.	EPI	Tubal fluid + Eagle Eagle Follicular fluid + Eagle	37	123/262: 47% fertilized 5/109: 4.6% 94/141: 66%	Typical pronuclei Tubal or follicular fluid is necessary

[a] T.E.—tubal egg.
[b] CA—capacitated; EPI—epididymis.

TABLE III
SEMI-*in Vitro* FERTILIZATION

Author	Species[a]	Sperm[b]	Culture medium	Temperature (°C)	Results	Comments
Moricard, 1950	Rabbit T.E.	From the vagina	Fallopian tube *in vitro*	—	Penetrated eggs, 15/21; 5 with sperm head in cytoplasm	Photographs show typical artifacts
Thibault and Dauzier, 1961	Rabbit T.E.	CA	Fallopian tube in oxygenated Locke medium	37	40/219—penetrated; 31/219—14.1% fertilized	♀ and ♂ pronuclei (stages 3–4)
Brinster and Biggers, 1965	Mouse T.E.	EPI	Fallopian tube in Brinster medium	32–37	111/447 cleaved; 67/447 morula and blastocysts—15%	Normal morula and blastocysts
Pavlok, 1967	Mouse T.E.	EPI	Fallopian tube over TC 199 + glucose + 5% homologous serum	37.5	101/132 cleaved—76%	Pronuclei cleavage and embryos

[a] T.E.—tubal egg.
[b] CA—capacitated; EPI—epididymis.

fusing from the zona pellucida of the freshly ovulated oocytes, becomes too concentrated around the oocytes when they are neither agitated nor washed. This may be true particularly in a small amount of medium where the sperm receptors become saturated, and the capacitated spermatozoa, thus, lose their property to adhere to the egg. If, on the other hand, the oocyte is previously washed or simply moved continuously during the fertilization process (especially in a large volume of medium), the fertilizin may diffuse in the medium, and the sperm receptors are not saturated in the narrow zone of high fertilizin concentration around the oocyte.

This hypothesis is supported by the following:

1. The longer that the recovered, freshly ovulated oocytes remain in the saline medium before introduction of capacitated sperm, the higher is the fertilization rate.

2. In a very small volume (0.1 ml), although the density of the spermatozoa is sufficiently great to completely separate the corona cells in 2 to 3 hours, no spermatozoa are found to adhere to the oocytes, and the total percentage of fertilization is very low—8.2% (Experiment E, Thibault and Dauzier, 1961).

3. When freshly ovulated oocytes are immediately mixed with capacitated spermatozoa (as in Experiment B), but placed in the discontinuous rotary washing system (as in Experiment D), the fertilization rate is as high as in Experiment D (65 vs 66%). Sperm penetration, however, is delayed by about 1½ hours (Experiment F; Table IV) (Thibault and Dauzier, 1961).

It should be noted that *in vivo* the eggs are continuously moved by the action of currents due to ciliary beating and to muscular contractions of the tubes (Wintenberger-Torres, 1961).

Under Brackett and Williams' experimental conditions (1968), penetration of the fertilizing sperm is also delayed, and fertilization rate is very high. It appears that fertilizin dissipation may result from a large volume of medium, high spermatozoa survival which helps to agitate the eggs, and thinning of the zona pellucida (Thibault, 1968, unpublished).

The question arises, "Is this fertilizin similar to the one described by Bishop and Tyler (1956) which produced agglutination of noncapacitated spermatozoa?" Rapid agglutination of the spermatozoa does not seem to occur in the medium, and it is difficult to state that spermatozoa which do not adhere to the eggs after incubation specifically agglutinate. We have found them attached to dispersed corona cells and to diverse cellular fragments. Nevertheless, some nonagglutinating fertilizins are known to exist (Tyler, 1961).

4. Penetration into the Cytoplasm

About 10% of the unfertilized eggs have spermatozoa under the zona pellucida: 76 vs 66% (Experiment D, Thibault and Dauzier, 1961). This has also been observed in the hamster by Yanagimachi and Chang (1964). This difference may be due to the conditions affecting *in vitro* fertilization itself which produce a dilution of the spermatozoa in an

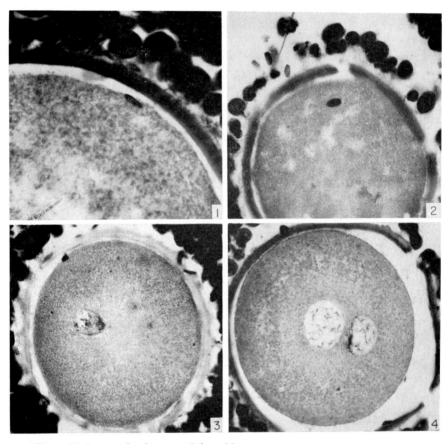

PLATE II. *In vitro* fertilization of the rabbit egg.

FIG. 1. The beginning of the sperm head penetration into the cytoplasm.

FIG. 2. A typical artifact: the sperm head on the cytoplasm has been displaced by the microtome blade.

FIG. 3. The male pronucleus during stage III and the sperm aster radiating from the centrosome of the midpiece (the black point in front of the pronucleus).

FIG. 4. The male pronucleus in central position and the female pronucleus with asymmetrical DNA distribution, at the end of the stage 4.

artificial medium outside the female tract and which greatly reduce their viability.

The frequency of polyspermy and digyny remains the same as *in vivo* when freshly ovulated eggs are used (Thibault and Dauzier, unpublished).

5. *Spermalysine-Like Substances*

When spermatozoa cannot pass through the zona pellucida, for example, when unwashed oocytes are placed in a small volume (Experiment

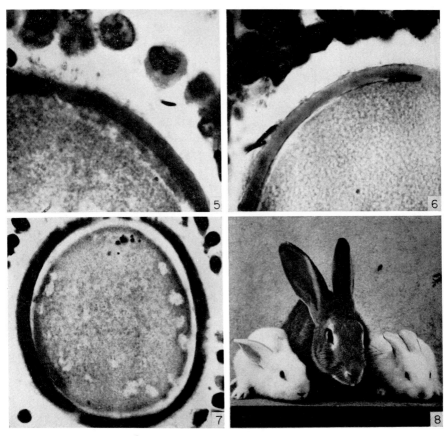

PLATE II. *Continued.*

FIG. 5. Unwashed egg inseminated with capacitated sperm: three sperms are in the vicinity of, but do not adhere to the zona pellucida.

FIG. 6. Egg washed during 1 hour before insemination with capacitated sperm. Sperm heads are stuck into the zona pellucida.

FIG. 7. The cytolytic action of "spermalysine" from capacitated sperms.

FIG. 8. Young rabbits from *in vitro* fertilization, with their foster mother.

TABLE IV

In Vitro Fertilization of the Rabbit Egg[a]

Technique	No. of does	Successful experiments No.	Successful experiments %	Total No. eggs	Fertilized eggs No.	Fertilized eggs %	Penetrated[b] No.	Penetrated[b] %	No. of sperms On or in zona pellucida	No. of sperms Under zona pellucida	Time of culture	Previous washing of eggs[c]
"Washing"[c] during all the culture time (Locke medium)	8	8	100	63	41	65	54	86	13.5	3	3 hr 58 min–4 hr 12 min	No
	8	5	62	42	17	40	21	50	6	2.5	2 hr 58 min–3 hr 10 min	No
	16	9	56	109	31	28	38	35	6	0.7	2 hr 18 min–2 hr 36 min	No
	10	9	90	67	42	63	46	69	12.1	0.9	2 hr 28 min–2 hr 36 min	1 hr
	51	48	94	289	191	66	22	76	17.0	4.0	4 hr–7 hr	0.5–2.0 hr
No washing (Locke medium)	132	43	32.5	922	121	13.1	162	17.6	1	0	4 hr–7 hr	No

[a] From Thibault and Dauzier, 1961.

[b] Eggs not fertilized, but with sperm (generally 1) in the periovular space.

[c] Discontinuous rotation of the glass tubes through 180° once every 30 seconds, causes the eggs to fall continuously through the medium in a diametrically opposite position.

E, Thibault and Dauzier, 1961), the egg is cytolyzed at a distance by the spermatozoa in 2 to 3 hours (Plate II, Fig. 7). From the results of several similar observations Thibault (1964) has postulated the existence of a spermalysine having properties analogous to those of the spermalysines described in other species. This substance has no effect on the fertilized egg and this may explain why the general appearance of fertilized eggs is better than that of unfertilized eggs in the same experiment (Chang, 1959; Thibault and Dauzier, 1961). The spermalysine seems to be released, in particular, by capacitated spermatozoa.

B. IN BIOLOGICAL FLUIDS

In vitro fertilization is more easily obtained in uterine and tubal fluids than in saline media. Moricard (1950, 1954) was the first to recover spermatozoa in uterine fluid, and to put eggs and spermatozoa together in this medium. When tubouterine fluid containing spermatozoa is pipetted from a living rabbit, the percentage of fertilization of unwashed oocytes is higher (30%) than when spermatozoa are obtained by perfusion of the uterus with a saline solution [Experiment B—13.7% (Thibault and Dauzier, 1961)]. Dauzier and Thibault (1959) concluded that an antifertilizin-like substance may exist in the living female tract.

Maintaining strict temperature and relative humidity conditions and using fresh uterine fluid as a culture medium, the fertilization rate is high (73%). It is lower (30%) when uterine fluid has been stored before use (Brackett and Williams, 1965). These authors accepted the conclusions of Dauzier and Thibault (1959).

When the fluid from ligated fallopian tubes is used as a culture medium, and sperms are pipetted from the uterine horn, 63.7% of the oocytes are fertilized (Suzuki and Mastroianni, 1965). If eggs are maintained in the same medium, the cleavage rate is 59.4% (Suzuki, 1966).

Even though these tubal and uterine fluids give excellent results, they do not have the same advantages as an artificial medium. They may contain unknown substances (capacitation factors, antifertilizins, metabolites, etc.) which could obscure the fundamental mechanisms of fertilization. Their use represents a "semi-*in vitro*" stage.

Other biological media have also been employed. For the collection of uterine sperm and as a fertilization medium, Chang (1959) and Bedford and Chang (1962) used 5% heated serum in acidic saline containing either 0.1% of sodium lactate or 0.25% of glucose. They obtained a 45–57% fertilization rate, but in pure serum the percentage was more variable, from 17 to 49% (Chang, 1959). Fertilizin-like substance does not appear to diffuse as easily from the eggs in pure serum as in a saline medium (Thibault and Dauzier, 1961).

IV. *In Vitro* Fertilization of the Hamster Egg

The physiological mechanisms involved in *in vitro* fertilization of the hamster egg (Table V) appear simpler than those observed in the rabbit egg. Yanagimachi and Chang (1963, 1964) obtained fertilization *in vitro* by mixing freshly shed eggs and either capacitated or epididymal sperm in tubal fluid.

From the results of these authors the following conclusions may be drawn:

1. The medium in which the sperms are suspended is of great importance. With Gey's medium the percentage of fertilization is poor (8.3%). This is also the case with the more complex medium, TC 199, where the rate is 9.5%. Better results were obtained with Krebs-Ringer's bicarbonate solution (19.5%), and the highest percentage with Tyrode's solution (64.7%).

2. Glycine increases spermatozoan motility and the percentage of fertilization. When three parts of an $M/8$ solution of glycine were added to ten parts of TC 199, the percentage of fertilization rose from 9.5 to 59.5%. Curiously enough, however, in other media, glycine may diminish the percentage of fertilized eggs (Dulbecco's medium) or have no effect (Tyrode's solution).

3. Polyspermy occurs more often than in normal fertilization. It varies from 10 to 33% depending on the experiment, whereas the frequency of *in vivo* polyspermy is low (2.1%) and does not increase with the age of the egg (1.8%) even though the hamster egg is very sensitive to aging after ovulation (Chang and Fernandez-Cano, 1958).

4. Sperm removed directly from the epididymis can pass through the zona pellucida and fertilize the egg. Such sperm is initially noncapaci-

TABLE V

In Vitro FERTILIZATION OF THE HAMSTER EGG[a]

Sperm type	No. eggs[b]	Fertilized eggs		No. of dispermic eggs	Time lapse between insemination and penetration into zona pellucida
		No.	%		
Uterine sperm					
4–5 hr after mating	207	134	64.7	20	
1.5–3 hr after mating	89	66	74.0	—	2–5 hr
0.5 hr after mating	100	52	52	2	
Epididymal sperm	180	80	44.4	10	longer?

[a] From Yanagimachi and Chang, 1964, and Yanagimachi, 1964.

[b] Culture media—1 part of tubal fluid + 1–3 parts of Tyrode's solution.

tated. The percentage of fertilized eggs is lower, however, and epididymal sperm penetrates the egg much more slowly than uterine sperm. It tends less to adhere to the zona pellucida. This conclusion is puzzling since capacitation appears to be necessary in this species even though it may be achieved within 3 to 4 hours. However, this is much more rapid than in the rabbit (Chang and Sheaffer, 1957). The delay time of 3 to 4 hours could correspond to the delay observed in *in vitro* fertilization with epididymal sperm. Consequently, it is possible that epididymal sperms are capacitated by tubal secretions in the fertilization system since the eggs are not obtained by perfusion but by dissection of the fallopian tubes under paraffin oil. In fact, if eggs are washed so that tubal or follicular fluid is completely removed, the percentage of fertilization is very low (4.6 vs 49%) (Barros and Austin, 1967).

5. Even with capacitated sperm remaining between 1½ and 3 hours in the uterus, the penetration of the sperm through the zona pellucida occurs generally 2–5 hours after insemination (Yanagimachi, 1964). This delay implies modification either of the eggs or of the spermatozoa, and one may wonder if this delay can be compared to that observed in unwashed rabbit eggs. The variability of experimental results (Yanagimachi, 1964, personal communication) suggests that there are many facts yet to be clarified.

V. *In Vitro* Fertilization in Other Species

All attempts to fertilize rat eggs have been unsuccessful, and in the mouse, Brinster and Biggers (1965) and Pavlok (1967) only obtained semi-*in vitro* fertilization.

Fallopian tubes containing freshly ovulated oocytes from virgin mice were placed in a suspension of sperm for 10 to 30 minutes, and they were then cultured on agar sheets in TC 199 with glucose and 5% homologous serum added. The fertilization rate reached 73% (Pavlok, 1967).

Freshly ovulated sheep oocytes were not fertilized when mixed with freshly ejaculated sperm (0 out of 41 eggs), in spite of the considerable number of spermatozoa that adhered to the eggs.

The use of sperm capacitated for 11 to 24 hours in the uterine horn or cervix of the ewe resulted in few fertilizations (4 out of 78), and the spermatozoa had penetrated into the zona pellucida in only 7 of the 78 eggs. Kraemer (1966) obtained similar results. He reported 1 definitely fertilized egg out of 23 when the eggs were mixed with spermatozoa previously incubated for 2 to 7 hours in the fallopian tubes or uterine horns. The same phenomenon of exclusion of capacitated spermatozoa occurred here as was observed in the rabbit. However, washing the eggs by the same procedure as used for the rabbit did not improve sperm attachment (Thibault and Dauzier, 1961).

In the sow, only 1 fertilization in 56 unwashed eggs resulted when capacitated sperm was used for insemination. However, in spite of the constant temperature maintained during the experiment, 6 of the 41 eggs underwent parthenogenetic activation as attested by the presence of only one pronucleus and two polar bodies. Since cytological examination showed that many eggs had no spermatozoa either on or in the zona pellucida, it may be assumed that mature spermatozoa do not attach themselves in large numbers to the zona.

In these species, the mechanisms of fertilization are still to be discovered.

VI. Environmental Conditions

During *in vitro* fertilization one or more spermatozoa pass through the zona pellucida without penetrating the cytoplasm (Table IV) in about 10% of the oocytes. This suggests that the culture medium is incompatible with good sperm or egg survival. Therefore, unsuccessful *in vitro* fertilization in some species may be due to lack of knowledge of a suitable culture medium.

A. Egg Culture Media

1. *Rabbit Egg Culture*

Fertilized rabbit eggs do not cleave in pure saline media. The addition of glucose and bovine plasma albumin to Krebs-Ringer bicarbonate is without effect. However, 2–16-cell eggs can cleave in this medium (Adams, 1956).

The presence of some amino acids may permit cleavage from the 1-cell stage to the morula, for example, in Eagle's basal medium (Purshottam and Pincus, 1961) and in simplified Ham's medium consisting only of salts, glucose, and five amino acids (Daniel, 1963).

If rabbit blood serum is added to Locke's medium (Chang, 1949) or to Eagle's medium (Purshottam and Pincus, 1961), fertilized eggs cleave regularly from the 1-cell stage to the blastocyst. Rabbit serum is more effective than horse serum, and the latter is better than human serum.

There may be good egg survival, however, in a saline medium in which regular cleavage fails to occur. Adams (1956) obtained 7 embryos when he transferred seven 16-cell eggs which had been stored for 21.5 hours in a Krebs-Ringer's bicarbonate, glucose, albumin medium.

Unfertilized oocytes survive in a Locke medium at 10°C for 24 hours because parthenogenetic activation begins when they are put back in a serum medium at 37.5°C. However, an abnormal pronucleus is formed. A normal pronucleus is always found if the oocyte is stored in blood serum (Chalmel, 1962).

2. Rodent Egg Culture

Attempts to culture 1-cell rat egg were unsuccessful (Pincus, 1936; Washburn, 1951). Occasionally, a division was observed, but the blastomeres immediately became abnormal (Defrise, 1933; Washburn, 1951). The hamster egg, cultured in TC 199 and Dulbecco's medium plus bovine serum, did not develop beyond the 2-cell stage, although the first cleavage sometimes occurred (Yanagimachi and Chang, 1964).

In contrast to this, the culture of the mouse egg was progressively successful. Using a saline medium slightly different from the Krebs-Ringer bicarbonate which contained albumin, and later on, egg yolk, Hammond, Jr. (1949) obtained cleavage from the 8-cell to the blastocyst stage. Whitten (1956) confirmed these results using an essentially similar saline medium, but containing glucose and bovine serum albumin. Such blastocysts obtained *in vitro* were able to develop *in vivo* (MacLaren and Biggers, 1958). The substitution of calcium lactate for calcium chloride permitted division from the 2-cell stage (Whitten, 1957). The same result was obtained by Mulnard (1964), who first mixed three parts of Whitten's saline solution with one part of mouse serum, and then tried rat serum with greater success.

Brinster (1963, 1965a,b,c) clearly defined the essential requirements for the development of the egg from the 2-cell to the blastocyst stage: (*a*) pH 6.82; (*b*) osmotic pressure, 0.2760 osmoles; (*c*) source of energy—lactate, pyruvate, oxalacetate, or phosphoenolpyruvate (the intermediate stages of the Krebs cycle or sugar cannot be used to provide energy); and (*d*) source of nitrogen—bovine serum albumin or its amino acids. The removal of one amino acid from the medium (with the exception of cystine) does not disturb normal blastocyst development. When no amino acids are present in the saline solution a few cleavages may occur (Brinster, 1965c).

Unfertilized oocytes as well as fertilized eggs can only metabolize pyruvate and oxalacetate. Nevertheless, if corona cells are present, lactate, phosphoenolpyruvate, and glucose can be utilized, and the first division of fertilized eggs takes place (Biggers et al., 1967). In this medium, 2-cell eggs recovered from mated mice can divide regularly up to the blastocyst stage. However, 1-cell eggs cleave only once, and blastocyst stages fail to occur. Two-cell eggs must be replaced in the ampulla to survive the critical period from the 2- to the 4-cell stage (Whittingham, 1966). Thus, tubal secretions provide a specific factor in development from the 1- to the 4-cell stage (Whittingham, 1966; Biggers et al., 1962).

3. Culture of Sheep, Goat, Cow, and Sow Eggs

The fertilized sheep egg divides *in vitro* when it is cultured in homologous serum (Robinson, 1950). However, Wintenberger-Torres *et al.* (1953) have shown that undivided fertilized sheep and goat eggs do not develop further than the 6–8-cell stage. Only eggs at the 16-cell stage reach the blastocyst stage. The eggs cannot pass from the 8- to 16-cell stage when *in vitro*. The secretions of the female genital tract evidently contribute an essential factor for this period of the egg's life.

Hancock (1963) and Kraemer (1966) have never observed, respectively, more than one or two sheep egg divisions in Whitten's medium. If this medium is not favorable for division, it is, however, favorable for egg survival, for Hancock obtained three pregnancies from ova that had been cultured for 48 hours in the medium. Kardymowicz *et al.* (1963) obtained a pregnancy after 72 hours of storage at 8° to 11°C in a medium consisting partly of serum and partly of Locke's solution. Only one survival has been reported in a pure saline solution—lambing occurred after transplantation of an egg preserved in 0.9% sodium chloride for 24 hours at 0°C (Loginova, 1961).

The cow egg has been successfully cultured from the 1-cell to the 12–24-cell stage in follicular fluid, but it dies rapidly in other media (Thibault, 1966).

No division has been observed in sow eggs cultured in homologous serum or in a mixture of Locke's solution and serum (Thibault, unpublished observations).

4. Summary

Mouse eggs are the only eggs that have survived and developed *in vitro* in a nonbiological medium. This includes oocyte maturation as well as cleavage of fertilized eggs. These eggs have been fertilized *in vitro*.

Cleavage of rabbit eggs is a very rare occurrence in a saline medium without the addition of blood serum. However, the eggs survive long enough so that *in vitro* fertilization may be accomplished.

The first division of the hamster egg may occur *in vitro* in a medium containing bovine serum, but egg survival is short. Fertilization, however, may occur *in vitro* in synthetic media.

Ewe and goat eggs divide regularly in homologous serum. The ewe egg survives in a saline medium, and may go through the two first cleavages in a synthetic medium. However, *in vitro* fertilization has almost never been obtained. The cow egg cleaves in follicular fluid. In the other media so far tested it rapidly dies.

Sow and rat eggs not only fail to cleave but also survive poorly in the media mentioned. Thus there is less chance of *in vitro* fertilization than in the other species.

B. SPERMATOZOA SUSPENSION MEDIA

Yanagimachi and Chang (1964) working on the hamster demonstrated the influence of the sperm suspension medium on the results of fertilization.

When the classic egg yolk–citrate diluent was used for egg recovery and perfusion in the recovery of capacitated sperm in the rabbit, no fertilization was noted (0 out of 44 eggs in six trials). After storage for 2 hours in Locke's solution at 37.5°C, the capacitated sperm removed from the uterus had lost its fertilizing ability (1 fertilization out of 81 eggs; Thibault and Dauzier, 1961). Contrary to this, Brackett and Williams' medium is able to preserve capacitated sperm for at least 5 hours.

The sperm diluent is, thus, an important factor in the success or failure of fertilization. Saline media permit sperms to survive for only a few hours, but complex biological media may contain ovicidal, spermicidal, or decapacitating factors.

Phillips' (1939) discovery that hen's egg yolk protects freshly ejaculated bull spermatozoa resulted in considerable progress, and, thus, reduced interest in the use of saline and strictly synthetic diluents. The beneficial action of egg yolk on spermatozoa is related to lipoproteins and phospholipids (Kampschmidt *et al.*, 1953; Miller and Mayer, 1960). Almost at the same time as Philips' discovery, milk (a biological medium containing phosphoproteins) was shown to have the same properties (Mockel, 1937; Underbjerg *et al.*, 1942; Mihailov, 1949; and, especially, Thacker and Almquist, 1951, 1953).

The problems of capacitated sperm storage at 37°C for a time long enough to permit fertilization to take place are different from those of preserving ejaculated sperm fertilizing capacity at low temperatures. Thus systematic research using either saline or synthetic sperm diluents must be carried out before studying the *in vitro* fertilization of a species.

VII. Experimental Approaches Using *in Vitro* Fertilization

A. GAMETE AGING

The duration of the fertile life of the egg and the fertilizing ability of the sperm varies from one species to another. Gamete aging results in a decrease of fertilization rate and an increase of abnormal fertilizations which are essentially digynic or polyspermic (Thibault, 1967).

Unfertilized eggs that have aged *in vivo* do not have any spermatozoa on, in, or under the zona pellucida. In addition, the number of fertilized eggs which do not contain supernumerary spermatozoa increases with the age of the eggs. This reduction in the number of spermatozoa surrounding the eggs may be due either to genital tract changes (secretions and muscular contractions) after estrus or to loss by the egg of substances necessary for the attachment of the spermatozoa. Only *in vitro* fertilization experiments can establish whether spermatozoa reduction is due to changes in the genital tract or in the oocyte itself.

In vitro fertilization experiments also may aid in studying fertilization abnormalities in relation to egg and sperm aging (Table VI). In the rabbit, when eggs are recovered 9 hours after ovulation, washed for 1 hour, and inseminated with spermatozoa that had remained for 20 hours in the genital tract, a lower percentage of fertilization than normal results (36 vs 66%). Most of the fertilizations are digynic or dispermic. In this case, the ages of the eggs and the sperm could be equally responsible since aged spermatozoa are capable of producing digynic fertilizations in freshly ovulated eggs and vice versa (Table VI).

The part played by tubal secretions in egg aging is completely unknown. In the rabbit, the deposit of an albuminous layer and the dispersion of the corona radiata demonstrate the action of tubal secretions. They have been considered as the cause of egg fertility loss (Hammond, 1934; Chang, 1955).

Comparison of the percentage of *in vitro* fertilization of eggs remaining for 9 hours after ovulation either in the fallopian tubes or in a homologous serum, shows that aging also occurs *in vitro*. There is a decrease in the percentage of fertilization and an increase in abnormal fertilizations (Table VI).

B. ACTION OF VARIOUS COMPOUNDS AT THE TIME OF FERTILIZATION

The effects of various compounds on fertilization can be easily studied *in vitro*, especially in the case of fertilization inhibitors.

A study using colchicine and Colcemid as inhibitors of mitosis has shown the possibilities of this technique. These mitotic poisons have been used to produce triploid eggs by retaining the second polar body (digyny). Previously, colchicine was mixed with spermatozoa (rabbit—Haggvist and Bane, 1950a; sow—Melander, 1951; goat—Propaczy *et al.*, 1959) without any definite results. At the time of fertilization, an intraperitoneal injection of colchicine was also given (rabbit—Venge, 1954; mouse—Waldo and Wimsatt, 1945; Edwards, 1954, 1958a,b; rat—Pikó and Bomsel-Helmreich, 1960).

Edwards obtained mouse embryos which were tri- and tetraploid, but Edwards and Sirlin (1959) stated that numerous anomalies, such as com-

TABLE VI

In Vitro Fertilization with Aged Gametes

Experiment	No. of rabbits	No. of eggs	Penetrated eggs		Fertilized eggs		Abnormal fertilization					
							Dispermy		Digyny		Abnormal sperm head swelling	
			No.	%	No.	%	No.	%	No.	%	No.	%
Controls[a]	7	56	42	75	36	64	0		0		0	0
Aged sperms[b] and aged eggs[c]	6	57	28	50	23	40	5	21	12	52	10	43
Aged sperms[b] and aged eggs[d]	6	38	11	30	15	40	0	0	6	40	1	6
Aged sperms[b] and control eggs	7	60	27	45	24	40	0		6	25	3	12.5
Aged eggs[d] and control sperms	14	92	13	14	10	11	0		5	50	0	0
Aged eggs fertilized *in vivo* (delayed fertile mating + 12 hr)	10	109	39	36	36	33	2	5.5	3	8	0	0

[a] Controls—ova and sperms are recovered from females mated either with a vasectomized buck or with a normal buck 11.5 hours previously.

[b] Aged sperms are recovered 20 hours after mating; they are 10 hours older than in normal fertilization.

[c] Aged eggs are eggs recovered 11.5 hours after sterile mating and maintained 8 hours *in vitro* before insemination.

[d] Aged eggs are eggs recovered 10 hours after ovulation.

plete rejection of the male pronucleus and of all or part of the female chromatin and subnuclei, were produced by colchicine or by Colcemid (Edwards, 1961). Variations have been observed in the proportion of digynic eggs from one experiment to another. Anomalies are due to an extended action of mitotic poisons before fertilization, and absence of these effects is observed if fertilization takes place sooner than expected.

Introducing colchicine *in vitro* about 30 minutes before cytoplasmic penetration of the spermatozoa (which usually occurs 1½ hours after insemination of the eggs *in vitro*), Bomsel-Helmreich and Thibault (1962) obtained digyny in 97% of the eggs (60 out of 62) with hardly any anomalies (1.5%). This experiment permitted a complete study to be made of the evolution of triploid embryos during gestation (Bomsel-Helmreich, 1965).

C. Other Possibilities

An experimental study of egg and spermatozoa interaction can only be carried out by combining *in vitro* fertilization with microcinemato-graphic analysis. This is also the undisputable method of studying the phenomenon of capacitation.

Finally, mention should be made of the study of the action of radiations which may be used at a precise time during fertilization or pronuclear evolution.

VIII. Conclusions

In vitro fertilization has been achieved with the egg of the rabbit, hamster, mouse, and human (see addendum).

The study of the rabbit egg has provided an understanding of the preliminary step essential to egg penetration, that is, the attachment of the spermatozoa to the egg.

In vitro fertilization has now become a routine technique resulting in several thousands of fertilizations which have been used for various experiments.

In relation to this we have been able to define the relative effects of egg and spermatozoa aging on the frequency of abnormal fertilization.

In vitro fertilization has also been an aid in studying the action of mitotic inhibitors. Eggs treated during *in vitro* fertilization with Colcemid develop triploid embryos. In the same way it may be possible to study specific fertilization inhibitors.

However, complete or partial failures in other species show that one cannot generalize on the basis of the facts found for the rabbit and the hamster. We do not yet know enough about the existence of specific mechanisms or special media which allow eggs and sperms to survive and the sperm-capacitating ability to be conserved.

REFERENCES

Adams, C. E. (1956). Egg transfer and fertility in the rabbit. *Proc. 3rd Intern. Congr. Animal Reprod., Cambridge* Sect. 3, pp. 5–6.

Austin, C. R. (1951). Observations on the penetration of the sperm into the mammalian egg. *Australian J. Sci. Res.* 4, 581–596.

Austin, C. R. (1956). Activation of eggs by hypothermia in rats and hamsters. *J. Exptl. Biol.* 33, 338–347.

Barros, C., and Austin, C. R. (1967). *In vitro* fertilization of golden hamster ova (Abst.). *Anat. Record* 157, 209–210.

Bedford, J. M., and Chang, M. C. (1962). Fertilization of rabbit ova *in vitro*. *Nature* 193, 898–899.

Biggers, J. D., Gwatkin, R. B. L., and Brinster, R. L. (1962). Development of mouse embryos in organ cultures of fallopian tubes on a chemically defined medium. *Nature* 194, 747–749.

Biggers, J. D., Whittingham, D. G., and Donahue, R. P. (1967). The pattern of energy metabolism in the mouse oocyte and zygote. *Proc. Natl. Acad. Sci. U. S.* 58, 560–567.

Bishop, D. W., and Tyler, A. (1956). Fertilizin of mammalian eggs. *J. Exptl. Zool.* 132, 575–602.

Bomsel-Helmreich, O. (1965). Heteroploidy and embryonic death. *Ciba Found. Symp., Preimplantation Stages Pregnancy*, pp. 246–269.

Bomsel-Helmreich, O., and Thibault, C. (1962). Fécondation *in vitro* en présence de colchicine et polyploïdie expérimentale chez le lapin. *Ann. Biol. Animale Biochim. Biophys.* 2, 13–16.

Brackett, B. G., and Williams, W. L. (1965). *In vitro* fertilization of rabbit ova. *J. Exptl. Zool.* 160, 271–281.

Brackett, B. G., and Williams, W. L. (1968). Fertilization of rabbit ova in a defined medium. *Fertility Sterility* 19, 144–155.

Brinster, R. L. (1963). A method for *in vitro* cultivation of mouse ova from two-cell to blastocyst. *Exptl. Cell Res.* 32, 205–208.

Brinster, R. L. (1965a). Studies on the development of mouse embryos *in vitro*. I. The effect of osmolarity and hydrogen ion concentration. *J. Exptl. Zool.* 158, 49–58.

Brinster, R. L. (1965b). Studies on the development of mouse embryos *in vitro*. II. The effect of energy source. *J. Exptl. Zool.* 158, 59–68.

Brinster, R. L. (1965c). Studies on the development of mouse embryos *in vitro*. III. The effect of fixed-nitrogen source. *J. Exptl. Zool.* 158, 69–78.

Brinster, R. L., and Biggers, J. D. (1965). *In vitro* fertilization of mouse ova within the explanted fallopian tubes. *J. Reprod. Fertility* 10, 277–280.

Chalmel, M.-C. (1962). Possibilités de fécondation des oeufs de lapine activés parthénogénétiquement. *Ann. Animale Biol. Biochim. Biophys.* 2, 279–297.

Chang, M. C. (1949). Effects of heterologous sera on fertilized rabbit ova. *J. Gen. Physiol.* 32, 291.

Chang, M. C. (1951). Fertilizing capacity of spermatozoa deposited into the fallopian tubes. *Nature* 168, 697–698.

Chang, M. C. (1954). Development of parthenogenetic rabbit blastocysts induced by low temperature storage of unfertilized ova. *J. Exptl. Zool.* 125, 127–150.

Chang, M. C. (1955). Développement de la capacité fertilisante des spermatozoïdes du lapin à l'intérieur du tractus génital femelle et fécondabilité des oeufs de lapine. In "La Fonction tubaire et ses troubles," pages 40–52. Masson, Paris.

Chang, M. C. (1957). Some aspects of mammalian fertilization. In "The Beginning of Embryonic Development," pp. 109–134. Am. Assoc. Advan. Sci.,Washington, D. C.

Chang, M. C. (1959). Fertilization of rabbit ova in vitro. Nature 184, 466–467.

Chang, M. C., and Fernandez-Cano, L. (1958). Effects of delayed fertilization on the development of pronucleus and the segmentation of hamster ova. Anat. Record 132, 307–320.

Chang, M. C., and Sheaffer, D. (1957). Number of spermatozoa ejaculated at copulation transported into the female tract and present in the male tract of the golden hamster. J. Heredity 48, 107.

Daniel, J. C., Jr. (1963). Cleavage of rabbit ova in protein-free media, Am. Zoologist 3, 526.

Dauzier, L., and Thibault, C. (1956). Recherche expérimentale sur la maturation des gamètes mâles chez les Mammifères, par l'étude de la fécondation in vitro de l'oeuf lapine. Proc. 3rd Intern. Congr. Animal Reprod., Cambridge, Vol. 1, pp. 58–61.

Dauzier, L., and Thibault, C. (1959). Données nouvelles sur la fécondation in vitro de l'oeuf de la lapine et de la brebis. Compt. Rend. Acad. Sci. 248, 2655–2656.

Dauzier, L., Thibault, C., and Wintenberger, S. (1954). La fécondation in vitro de l'oeuf de lapine. Compt. Rend Acad. Sci. 238, 844–845.

Defrise, A. (1933). Some observations on living eggs and blastulae of the albino rat. Anat. Record 57, 239–250.

Edwards, R. G. (1954). Colchicine-induced heteroploidy in early mouse embryos. Nature 174, 276.

Edwards, R. G. (1958a). Colchicine-induced heteroploidy in the mouse. I. The induction of triploidy by treatment of the gametes. J. Exptl. Zool. 137, 317–348.

Edwards, R. G. (1958b). Colchicine-induced heteroploidy in the mouse. II. The induction of tetraploidy and other types of heteroploidy. J. Exptl. Zool. 137, 349–362.

Edwards, R. G. (1961). Induced heteroploidy in mice: effect of deacetylmethyl-colchicine on eggs at fertilization. Exptl. Cell Res. 24, 615–617.

Edwards, R. G., and Sirlin, J. L. (1959). Identification of C¹¹-labelled male chromatin at fertilization in colchicine-treated mouse eggs. J. Exptl. Zool. 140, 19–27.

Frommolt, G. (1934). Die Befruchtung und Furchung des Kanincheneies im Film. Zentr. Gynäkol. 58, 7–12.

Haggvist, G., and Bane, A. (1950a). Chemical induction of polyploid breeds of mammals. Svenska Vet. Akad. Handl. 1, 1–12.

Haggvist, G., and Bane, A. (1950b). Studies in triploid rabbits produced by colchicine. Hereditas 36, 329–334.

Hammond, J. (1934). The fertilization of rabbit ova in relation to time. A method of controlling the litter size, the duration of pregnancy, and the weight of the young at birth. J. Exptl. Zool. 11, 140–161.

Hammond, J., Jr. (1949). Recovery and culture of tubal mouse ova. Nature 163, 28.

Hancock, J. L. (1963). Survival in vitro of sheep eggs. Animal Prod. 5, 237–243.

Kampschmidt, R. F., Mayer, D. T., Herman, H. A., and Dickerson, G. E. (1953). Lipid and lipoprotein constituants of egg yolk in the resistance and storage of bull spermatozoa. J. Dairy Sci. 36, 733–742.

Kardymowicz, M., Kardymowicz, O., Kuhl, W., and Lada, A. (1963). Storage of fertilized sheep ova at low temperature. *Acta Biol. Cracoviensia* **6**, 31–37.

Kraemer, D. C. (1966). A study of *in vitro* fertilization and culture of ovarian ova. Ph.D. Thesis, Texas Univ., 75 pp.

Krasovskaya, O. V. (1935). Cytological study of the heterogeneous fertilization of the egg of rabbit outside the organism. *Acta Zool.* **16**, 449–459.

Loginova, N. V. (1961). La conservation des embryons de brebis en dehors de l'organisme maternel. *Ovtsevodstvo* **8**, 18–20 (in Russian).

Long, J. A. (1912). The living eggs of rats and mice with a description of apparatus for obtaining and observing them. *Univ. Calif. (Berkeley) Publ. Zool.* **9**, 105–136.

McLaren, A., and Biggers, J. D. (1958). Successful development and birth of mice cultivated *in vitro* as early embryos. *Nature* **182**, 877–878.

Melander, Y. (1951). Polyploidy after colchicine treatment of pigs. *Hereditas* **37**, 288–289.

Menkin, M. F., and Rock, J. (1948). *In vitro* fertilization and cleavage of human ovarian eggs. *Am. J. Obstet. Gynecol.* **55**, 440–452.

Mihailov, N. N. (1949). Milk as a diluent for the semen of animals. *Kunevodstvo* **6**, 14–16; (1950). *Ann. Breeding* **18**, 43 (Abstr.).

Miller, L. D., and Mayer, D. T. (1960). Survival of bovine spermatozoa in media containing a lipoprotein complex isolated from spermatozoa and other sources. *Missouri Univ. Agr. Expt. Sta. Res. Bull.*, No. 742, 16pp.

Mockel, H. (1937). Zur Physiologie des Ziegenbockspermas (ein Hinblick auf die künstliche Besammung). Dissertation Univ. of Leipzig, 47 pp.

Moricard, R. (1950). Premières observations de la pénétration du spermatozoïde dans le membrane pellucide d'ovocytes de lapine fécondés *in vitro*. Niveau de potentiel d'oxydo-réduction de la sécrétion tubaire. *Compt. Rend. Assoc. Anat.*, 37e Réunion, Louvain, pp. 337, 349.

Moricard, R. (1954). Etude cinématographique de la fécondation réussie *in vitro* de l'ovule de lapine. *Bull. Fédération Soc. Gynécol. Obstet. Langue Franc.* **6**, 271–275.

Moricard, R. (1955). La fonction fertilisatrice des sécrétions utéro-tubaires (étude microcinématographique de la fécondation *in vitro* de l'ovocyte de lapine). "La Fonction tubaire et ses troubles," pp. 60–90. Masson, Paris.

Mulnard, J. (1964). Obtention *in vitro* du développement continu de l'oeuf de souris du stade II au stade du blastocyste. *Compt. Rend. Acad. Sci.* **258**, 6228–6229.

Onanoff, J. (1893). Recherches sur la fécondation et la gestation des mammifères (conclusions). *Compt. Rend. Soc. Biol.* **45**, 719.

Oprescu, S., and Thibault, C. (1965). Duplication de l'ADN dans les oeufs de lapine avant la segmentation. *Ann. Biol. Animale Biochim. Biophys.* **5**, 151–156.

Pavlok, A. (1967). Development of mouse ova in explanted oviducts—fertilization, cultivation and transplantation. *Science* **157**, 1457–1458.

Petrov, G. N. (1958). Fécondation et phases primaires de division de l'oeuf humain hors de l'organisme. *Arch. Anat. Histol. Embryol.* **35**, 88–91. (in Russian).

Phillips, P. H. (1939). Preservation of bull semen. *J. Biol. Chem.* **130**, 415.

Pikó, L., and Bomsel-Helmreich, O. (1960). Triploid rat embryos and other chromosomal deviants after colchicine and polyspermy. *Nature* **186**, 737–739.

Pincus, G. (1930). Observations on the living eggs of the rabbit. *Proc. Roy. Soc.* **B107**, 132–167.

Pincus, G. (1936). "The Eggs of Mammals," Vol. 1. Exptl. Biol. Monographs, Macmillan, New York.

Pincus, G. (1939). The comparative behaviour of mammalian eggs *in vivo* and *in vitro*. IV. Development of fertilized and artificially activated rabbit eggs. *J. Exptl. Zool.* **82**, 85–129.

Pincus, G., and Enzmann, E. V. (1934). Can mammalian eggs undergo normal development *in vitro*. *Proc. Natl. Acad. Sci. U. S.* **20**, 121–122.

Pincus, G., and Enzmann, E. V. (1935). The comparative behaviour of mammalian eggs *in vivo* and *in vitro*. *J. Exptl. Med.*, **62**, 665–675.

Pincus, G., and Enzmann, E. V. (1936). The comparative behaviour of mammalian eggs *in vivo* and *in vitro*. II. The activation of tubal eggs of the rabbit. *J. Exptl. Zool.* **73**, 195–208.

Porpaczy, A., Gyorffy, B., Farago, M., and Vaskutti, L. (1959). Experiments on treatment of sheep with colchicine *Kyserletuggi Kozlemerujek* **4**, 91–100 (in Hungarian).

Purshottam, N., and Pincus, G. (1961). *In vitro* cultivation of mammalian eggs. *Anat. Record* **140**, 51–55.

Robinson, T. J. (1950). The control of fertility in sheep. II. The augmentation of fertility by gonadotrophin treatment of the ewe in the normal breeding season. *J. Agr. Sci.* **41**, 63.

Rock, J., and Menkin, M. F. (1944). *In vitro* fertilization and cleavage of human ovarian eggs. *Science* **100**, 105–106.

Shenk, S. L. (1878). Das Saugetierei Künstlich befruchtet ausserhalb der Muttertieres. *Mitt. Embryol. Inst. Univ. Wien, 1877–1885.*

Shettles, L. B. (1953). Observations on human follicular and tubal ova. *Am. J. Obstet. Gynecol.* **66**, 235–247.

Shettles, L. B. (1955). A morula stage of human ovum developed *in vitro*. *Fertility Sterility* **6**, 287–289.

Smith, A. U. (1951). Fertilization *in vitro* of the mammalian egg. *Biochem. Soc. Symp.* (Cambridge, Engl.) **7**, 3–10.

Suzuki, S. (1966). *In vitro* cultivation of rabbit ova following *in vitro* fertilization in tubal fluid. *Cytologia* **31**, 416–421.

Suzuki, S., and Mastroianni, L. (1965). *In vitro* fertilization of rabbit ova in tubal fluid. *Am. J. Obstet. Gynecol.* **93**, 465–471.

Thacker, D. L., and Almquist, J. O. (1951). Milk and milk products as diluters for bovine semen. *J. Animal Sci.* **10**, 1082 (Abstr.).

Thacker, D. L., and Almquist, J. O. (1953). Diluter for bovine semen. I. Fertility and motility of bovine spermatozoa in boiled milk. *J. Dairy Sci.* **36**, 173–180.

Thibault, C. (1949). L'oeuf des mammifères. Son développement parthénogénétique. *Ann. Sci. Nat. Zool. Biol. Animale* **11**, 136–219.

Thibault, C. (1964). Présence d'une spermalysine dans le spermatozoïde maturé (capacitated) du lapin. *Proc. 5th Intern. Congr. Animal Reprod. and Artificial Insemmination, Trento,* **7**, 294–295.

Thibault, C. (1966). La culture *in vitro* de l'oeuf de vache. *Ann. Biol. Animale Biophys.* **6**, 159–164.

Thibault, C. (1967). Analyse comparée de la fécondation chez la Brebis, la Vache et le Lapine. *Ann. Biol. Animale Biochim. Biophys.* **7**, 5–23.

Thibault, C., and Dauzier, L. (1960). Fertilisines et fécondation *in vitro* de l'oeuf de lapine. *Compt. Rend. Acad. Sci.* **250**, 1358–1359.

Thibault, C., and Dauzier, L. (1961). Analyse des conditions de la fécondation *in vitro* de l'oeuf de la lapine. *Ann. Biol. Animale Biochim. Biophys.* **1**, 277–294.

Tyler, A. (1961). The fertilization process. *In* "Sterility" (E. T. Tyler, ed.), pp. 25–56. McGraw-Hill, New York.

Underbjerg, G. K. L., Davis, H. P., and Spangler, R. E. (1942). Effects of diluters and storage upon fecondity of bovine semen. *J. Animal Sci.* 1, 149–154.

Van Beneden, E. (1875). La maturation de l'oeuf, la fécondation et les premières phases du développement embryonnaire des mammifères d'après les recherches faites chez le lapin. *Bull. Acad. Roy. Med. Belg.* [2] 40, 53 pp.

Venge, O. (1954). Experiments on polyploidy in the rabbit. *Kgl. Lantbrukshögskol. Ann.* 21, 417–444.

Waldo, C. M., and Wimsatt, W. H. (1945). The effect of colchicine on early cleavage of mouse ova. *Anat. Record* 93, 363–373.

Washburn, W. W., Jr. (1951). A study of the modifications in rat eggs observed *in vitro* and following tubal retention. *Arch. Biol.* 62, 439–458.

Whitten, W. K. (1956). Culture of tubal mouse ova. *Nature* 177, 96.

Whitten, W. K. (1957). Culture of tubal ova. *Nature* 179, 1081.

Whittingham, D. G. (1966). A critical phase in the cultivation of mouse ova *in vitro*. *J. Cell Biol.* 31, 123 (Abstr.).

Whittingham, D. G., and Biggers, J. D. (1967). Fallopian tube and early cleavage in the mouse. *Nature* 213, 942–943.

Wintenberger-Torres, S. (1961). Mouvements des trompes et progression des oeufs chez la brebis. *Ann. Biol. Animale Biochim. Biophys.* 1, 121–133.

Wintenberger-Torres, S., Dauzier, L., and Thibault, C. (1953). Le développement *in vitro* de l'oeuf de brebis et de celui de la chèvre. *Compt. Rend. Soc. Biol.* 147, 1971–1973.

Yanagimachi, R. (1964a). The behavior of hamster sperm to the hamster and mouse ova *in vitro*. *Proc. 5th Intern. Congr. Animal Reprod. and Artificial Insemmination, Trento*, 7, 292–294.

Yanagimachi, R., and Chang, M. C. (1963). Fertilization of hamster eggs *in vitro*. *Nature* 200, 281–282.

Yanagimachi, R., and Chang, M. C. (1964). *In vitro* fertilization of golden hamster ova. *J. Exptl. Zool.* 156, 361–376.

Addendum

Mouse and human oocytes were fertilized *in vitro* after the preparation of this chapter.

Superovulated mouse oocytes were released into 0.5 ml of Biggers medium by incising the wall of the ampulla under paraffin oil. Spermatozoa were collected from the uterus of a mouse mated 1–2 hr previously. A sample of 50 μl of sperm suspension was added to the medium containing the oocytes.

In seven experiments, between 10.4 and 40.9% of the eggs proceeded through the first cleavage division (Whittingham, 1968).

The experimental procedure was very similar to that used for hamster fertilization, and its success must be related to the presence of follicular fluid in the tubal medium. This fluid was also the major constituent of the medium used by Edwards *et al.* (1969). At least 7 human eggs were fertilized out of 36 inseminated after *in vitro* maturation. The follicular fluid seems to contain the elements necessary for fertilization in the hamster, mouse, and man.

REFERENCES TO ADDENDUM

Whittingham, D. G. (1968). Fertilization of mouse eggs *in vitro*. *Nature* 220, 592–593.

Edwards, R. G., Bavister, B. D., and Steptoe, P. C. (1969). Early stages of fertilization *in vitro* of human oocytes matured *in vitro*. *Nature* 221, 632–635.

CHAPTER 10

Variations and Anomalies in Fertilization

C. R. Austin[*]

PHYSIOLOGICAL LABORATORY, CAMBRIDGE, ENGLAND

I. Introduction

The unusual and abnormal forms of fertilization in plants and animals are treated in this chapter, with attention given primarily to mammals. It is sometimes difficult to decide whether or not a particular process

[*] The author gratefully acknowledges support by U. S. Public Health Service grants FR00164 and GM13226 while he was Head of the Genetic and Developmental Disorders Research Program at the Delta Regional Primate Research Center, Covington, Louisiana, during which period this Chapter was prepared.

should be regarded as abnormal, since a wide range of fertilization patterns exists in the plant and animal kingdoms, and events that are abnormal and deleterious in some organisms are normal or at least innocuous in others. Consideration is given first to variations that appear to be normal features of several different patterns of fertilization; later, the frank anomalies are described. Steps in gametogenesis subject to variation and error are also included if their modification is evident at, or influences the course of, fertilization.

Previous reviews dealing with aspects of this general topic include: Gustafsson (1948), Rothschild (1954), Pasteels (1956), Thibault (1959), Austin (1960), Pikó (1961a), Beatty (1964a,b), Fofanova (1965). Books with relevant information include: Wilson (1928), McLean and Ivimey-Cook (1951, 1956), White (1954), Beatty (1957), Swanson (1957), Austin (1961, 1965).

II. Variations in Fertilization

A. GAMETOGENESIS

Mammalian gametogenesis is typically one by which specialized heterogamous gametes are formed. The gonial cells are diploid in chromosome constitution and they undergo a succession of divisions, forming at first more gonial cells; later a proportion of the divisions leads to production by mitosis of primary spermatocytes or oocytes. Division of primary spermatocytes and oocytes occurs in the first meiotic division and gives rise to secondary spermatocytes and oocytes; these, passing through the second meiotic division, form haploid spermatids and ootids. Centromeres divide in the first meiotic division and chromatids separate in the second. Each primary spermatocyte ideally yields two secondary spermatocytes, and division of these, in turn, yields four spermatids. The spermatids differentiate into spermatozoa in the course of spermateleosis. Within a narrow range of variation, all spermatozoa produced by any one mammalian species are normally the same in size and shape. Division of the primary oocyte yields a single secondary oocyte and the first polar body, and division of the secondary oocyte in its turn gives the ootid and the second polar body. Occasionally, the first polar body divides so that a total of three polar bodies is formed. It is more common, however, in mammals to find that the first polar body either fails to divide or else breaks up and disappears by the time the egg is ovulated, so that two or often only one polar body persists into the cleavage phase. The polar bodies of the mammalian egg are comparatively large, approaching one-twentieth of the volume of the vitellus. In the majority of mammals, eggs are released from the ovary as secondary oocytes, the meiotic divisions being suspended at the second metaphase; exceptions are found in the dog, fox, and probably the horse, in which the egg is ovulated as a primary oocyte.

Meiosis, although with numerous variations in detail, is almost a universal preliminary to syngamy in the plant and animal kingdoms. Exceptions are found in a few organisms, such as the green alga *Oedogonium*, in which meiosis takes the place of mitosis at the beginning of cleavage and is referred to as zygotic meiosis. Meiosis is nearly always involved in the formation of gametes, though in many protozoa cytoplasmic fusion precedes visible nuclear changes so that initial cell union is essentially between two adult organisms. Among the plants (except the simpler forms, such as the algae), meiosis occurs in mega- and microsporogenesis; the spores give rise to haploid gametophytes and these, in turn, produce the specialized gametes; the gametes are thus haploid without there being any immediately preceding meiosis. In different organisms, gametes may be of much the same size and structure as the adult individuals (hologamy) or somewhat smaller than the adult cell and showing also certain differences in structure (merogamy). Gametes can be very similar to one another (isogamy), or a little different (anisogamy), or strikingly different from each other (heterogamy). Where heterogamy exists the smaller microgamete is nearly always motile, whereas the larger macrogamete is nonmotile and is usually termed an egg cell.

Microgametes are generally uniform in size and shape in any one species, but this is not always so, and dimorphism and polymorphism are known in certain organisms (Wilson, 1928). This is thought to reflect some abnormality of production, but it is found quite frequently in certain species, and under apparently normal circumstances, so that its pathological nature is by no means certain. Unusually small spermatozoa lacking nuclear material and known as apyrene spermatozoa are found in some Lepidoptera. In other forms, such as the gastropod *Paludina*, a remarkable giant spermatozoon is produced in addition to the normal type; the large form appears as though compounded of a number of normal forms and is referred to as an oligopyrene spermatozoon. It is not known whether apyrene or oligopyrene spermatozoa can function in fertilization. In the pentatomid bug *Arvelius*, the testis contains lobes with spermatocytes of three different sizes. Meiosis, however, is essentially normal and of the same kind in all three lobes, so that three different types of spermatozoa are formed differing greatly in size; it has yet to be determined whether all types function genetically, or whether larger or smaller spermatozoa play a subsidiary role of some kind in fertilization.

Patterns of meiosis involving chromosome reduction to less than haploid, or exhibiting irregular forms of cytoplasmic division, are known in several insects (White, 1954; Swanson, 1957). In the gall midge *Miastor*, the spermatogonia and primary spermatocytes contain 48 chromosomes. At the first anaphase, the chromosomes become segregated into two groups—one containing 42 and the other only 6 chromosomes. The larger group is then gathered into a nucleus which soon becomes pycnotic and degenerates; the smaller group of 6 chromosomes enters the second meiotic division, from which are derived two spermatids, each containing 6 chromosomes. Consequently, the mature spermatozoa have only 6 chromosomes and require the addition of 42 chromosomes by the egg to restore the normal somatic number of the species. Just how the egg through its meiotic divisions comes to contain 42 chromosomes is obscure. In the human louse *Pediculus*, the primary spermatogonium contains 12 chromosomes; this undergoes a reduction division to give secondary spermatogonia, each containing 6 chromosomes. After several generations of secondary spermatogonia, spermatocytes are formed still containing the 6 chromosomes, and these then undergo a meiotic division involving unequal division of the cytoplasm: a small bud containing 6 chromosomes is extruded from the cell and degenerates, while the main body of the cell, also containing 6 chromosomes, develops into a spermatozoon. The spermatozoon containing 6 chromosomes enters an egg with the same number, its chromosome complement having been reduced by two normal meiotic divisions.

In parthenogenetic development, meiosis may be entirely lacking in oogenesis; this situation is believed to exist in the mollusk *Campeloma*, the crustaceans *Trichoniscus* and *Daphnia*, and the insects *Saga* and *Tetraneura*. Alternatively, there is either inhibition of the first meiotic division (as in the moth *Solenobia lichenella*) or of the second meiotic division (as in another moth *Apterona helix*).

Polar bodies in many nonmammalian eggs are relatively small; this especially is so in the large yolk eggs of birds and reptiles wherein the bulk of the egg may be several thousand times that of the polar body. In many insect and crustacean eggs, polar bodies are not formed but groups of chromosomes (the polar nuclei) are relegated to the peripheral cytoplasm. The first polar nucleus may divide, just as the first polar body does in some animals, and, thus, a group of three polar nuclei is commonly found in the peripheral cytoplasm of the egg.

B. Gamete Union—Cytoplasm

The time of gamete conjugation in relation to the progress of maturation varies widely among animals. As already noted, in protozoa such as *Paramecium*, cytoplasmic union occurs between individuals with unreduced nuclear status. Sperm entry into the primary oocyte is seen in some mesozoans, sponges, annelids, platyhelminths, mollusks, and in *Peripatopsis*. In other annelids and mollusks, and in many insects, the maturation of the egg tends to stop at the first metaphase. Comparatively few lower animals resemble the majority of mammals where sperm entry is halted at the second metaphase. *Amphioxus* and *Siredon* are two animals for which definite information is available. [Exceptions among the mammals includes the dog, in which sperm entry may occur before breakdown of the germinal vesicle or more commonly during the course of the first meiotic division (Van der Stricht, 1923).] In echinoids and coelenterates, fertilization begins after maturation is complete and when the egg chromosomes are enclosed in a well-formed female pronucleus.

Commonly, only one spermatozoon enters and initiates the process of fertilization in mammals, surplus spermatozoa being excluded by the zona reaction or the vitelline block to polyspermy.[*] The relative importance of the two reactions varies in different mammals. In the course of entering the egg, the spermatozoon follows a curved path through the zona pellucida and comes to rest with its head lying flat upon the surface of the vitellus. The sperm head passes into the vitellus by a process involving fusion of plasma membranes and, in the majority of mammals studied, is followed successively by the midpiece and main piece of the tail. In some mammals, the tail may not be drawn into the vitellus; it is left in the perivitelline space in about 40% of the eggs of the field vole (Austin, 1957), and in most eggs of the Chinese hamster (Austin and Walton, 1960; Pikó, 1961b).

Failure of entry of the sperm tail is seen also in some nonmammals, for example, as a regular feature in *Nereis*, and, apparently, as an occasional occurrence in *Arbacia* (Wilson, 1928).

With gametes, cytoplasmic fusion is complete and permanent; with conjugants, it is partial and temporary. Thus, in bacteria, a cytoplasmic bridge has been observed connecting supposedly conjugating cells; the bridge is believed to provide the means for chromosome transfer. Chromosome passage is a protracted affair, and the cells may not remain connected long enough for its completion; the time at which they separate determines the proportion of chromosome that is passed from one cell to the other (see Stanier *et al.*, 1963). In ciliates, such as *Paramecium*, the area over which fusion takes place is greater in proportion than that in bacteria (Elliott and Tremor, 1958; André and Vivier, 1962; Schneider, 1963). The connecting neck of cytoplasm is sufficient to permit the migration of nuclei between the two cells. Some cytoplasmic elements

---

[*] See L. Pikó, Chapter 8 of this volume, for these and other matters relating to sperm penetration in mammals.

also get through, as is shown by the transfer of kappa particles between "killer" and "sensitive" strains. Occasionally, through an anomaly, the region of fusion between two paramecia becomes larger than normal, and, apparently, because of this the cells subsequently fail to separate ("nonseps"). Each may divide so as to give rise to one or more daughter cells, though still connected with its partner (see Metz, 1954).

Fertilization involving the entry of only one spermatozoon into the egg cytoplasm is the normal course of events not only in mammals but also in sea urchins, most insects, fish, some angiosperms, and in frogs. The entry of two or more spermatozoa in these forms ("pathological polyspermy") leads to abnormalities of development, which will be described later on. By contrast, there are a number of animals in which entry of several to many spermatozoa at fertilization is normal ("physiological polyspermy"); these include some angiosperms, the bryozoans, the elasmobranchs, the urodeles, the reptiles, and the birds. (For references to work on both forms of polyspermy, the reader should consult the review by Pikó, 1961a.) In the urodeles, bryozoans, and angiosperms, only the first male pronucleus to become associated with the female pronucleus is able to advance in its development, and the supernumerary male pronuclei arising from the extra spermatozoa are suppressed. In the other organisms listed, the supernumerary male pronuclei are repelled to the periphery of the blastodisc. This process has been studied particularly in the bird egg, and it has been noted that the extra pronuclei set up a certain amount of cleavage in this location. Possibly the extra cells that form have some role in embryonic development; there is evidence that some kinds of mosaicism in the fowl are attributable to the activity of the extra pronuclei. It is to be noted that the animals in which physiological polyspermy occurs are typically those with large yolk eggs. Presumably the difficulty of excluding extra spermatozoa from large eggs has resulted in the development of other mechanisms of preventing polyandrous syngamy, which leads either to chaotic cleavage or to ultimate degeneration of the embryo.

C. Gamete Union—Nucleus

Fertilization in mammals ends with the union of chromosomes from spermatozoon and egg, without the occurrence of actual pronuclear fusion. Maternal and paternal chromosome complements unite after pronuclear breakdown and become arranged as components of the metaphase plate of the first cleavage spindle.

Syngamy between maternal and paternal nuclei in nonmammals does not invariably follow cytoplasmic fusion of gametes, even under normal circumstances. The male pronucleus may not become involved at all in further development of the egg, and the spermatozoon is then reduced to that merely of activation; this is known as a natural phenomenon in strains of the nematode *Rhabditis pellio* (P. Hertwig, 1920), the planarians *Polycelis nigra* (Lepori, 1949) and *Dugesia benazzi*, (Benazzi, 1952), and the lepidopteran *Luffia* (Narbel-Hofstetter, 1957). It is, in fact, a type of parthenogenesis; when it takes place in nematodes it is referred to as nematode pseudogamy, but is more generally and specifically designated gynogenesis. Since the sperm chromosomes are not involved in this method of propagation, the species homology of the gametes is no longer important, and a few animals have, in fact, come to depend entirely upon the males of related species for the activation of their eggs. This is seen, for example, in the spider beetle *Ptinus latro*— here males are lacking, and the eggs can only begin developing if they are penetrated by spermatozoa of *Ptinus hirtellus* (Moore, *et al.*, 1956). The same type of activation is known in the worm *Lumbricillus lineatus* (Christensen and O'Connor, 1958) and even in a fish, the Amazon molly, *Mollienesia* (*Poecilia*) *formosa*, a species that consists only of females and whose eggs are activated by spermatozoa of related species *Mollienesia sphenops*, *Mollienesia latipinna*, and others) (Kallman, 1962a,b; see also Schultz, 1961).

Where maternal and paternal chromosomes do become incorporated into the embryo, there still exist different degrees of union between the two sets of chromosomes. Thus in some forms, such as the sea urchin, the two pronuclei appear to undergo complete fusion and the chromosome complements evidently become intermingled. In *Ascaris*, on the other hand, though the pronuclei come into close contact, fusion does not result; the egg and sperm chromosomes condense within the persisting nuclear membranes, and, even after the membranes have disappeared, remain isolated from each other as they lie on the first cleavage spindle. In the crustacean *Cyclops*, separation is even more persistent; two distinct cleavage spindles are formed, and the resulting two nuclei in each of the blastomeres of the two-cell egg remain separate though in close apposition. The same arrangement is seen in subsequent cleavage stages. In the ameba *Sappinea diploidea*, the two nuclei brought together by cytoplasmic fusion remain side-by-side until the early phases of the next conjugation and only then do they fuse.

Nuclear fusion in fertilization need not involve only the nuclei of spermatozoon and egg. Thus in the angiosperms, two nuclei are released by the pollen tube; one of these unites with the egg cell nucleus and the other with the two polar nuclei or with their fusion product. The result of the latter union is a triploid nucleus, which gives rise to the endosperm nuclei.

In some insects and crustaceans, the second polar nucleus fuses with one of the products of division of the first; alternatively, all three polar nuclei may unite. During the course of embryonic development the compound nucleus may undergo a large number of divisions, eventually causing subdivision of the cytoplasm so that a mass of cells finally comes to surround part of the embryo. This is seen, for example, in the parasitic hymenopteran *Litomastix*. The polar mass consists, of course, of either diploid or triploid cells.

In the scale insect *Pseudococcus citri* the fusion product of polar nuclei is triploid and undergoes a succession of divisions which are all endomitotic; that is to say, there is no separation of the chromosomes after division, and as a result, the nuclei become highly polyploid. When the nuclei have reached the 50–70 chromosome stage they take in a fungal symbiont and so constitute the primary mycetocytes. These now undergo further divisions and give rise to the mycetome organ of the adult animal (see Swanson, 1957).

III. Errors of Gametogenesis

A. MULTINUCLEAR AND "GIANT" GAMETES

Binuclear primary oocytes have been reported in a number of different mammals [various species (Hartman, 1926), mouse (Engle, 1927), cat (Dederer, 1934), man (Pankratz, 1938; Bacsich, 1949), rat (Lane, 1938), goat (Harrison, 1948), and hamster (Skowron, 1956)]. The fate of these oocytes—whether they can give rise to secondary oocytes and fertilizable ootids—is not known; they seem likely to have twice the normal chromosome complement and may possibly be ovulated as giant eggs.

Giant gametes are occasionally encountered in mammals as well as in other animals. These are most commonly cells of about twice the normal volume and, in the case of eggs, with one or two nuclei, they may arise through either (*a*) failure of cytoplasmic division in gonial cells, with or without failure of anaphase separation in the nuclear division or (*b*) fusion of two cells with or without fusion of nuclei. Giant-cell formation, often involving higher multiples of cells, is well known in monocytes and muscle cells cultured *in vitro*, with fusion of separate cells as the underlying mechanism, and also in tissues affected by neoplasia or infection. Giant-cell formation in apparently normal testis tissue has been described on several occasions; the cells may be either spermatocytes or spermatids, and may contain as many as sixteen or more nuclei (Bishop and Walton, 1960). The phenomenon in the testis has been attributed to such factors as high environmental temperatures, ingestion of heavy metals, and vitamin A deficiency; it is not certain by which mechanism the giant cells arose. The evidence of Fawcett *et al.* (1959) concerning delayed separation of secondary spermatocytes and spermatids (persistence of cytoplasmic bridges between the cells) supports the idea that nonseparation is, in this case, more likely than fusion; consistently, giant spermatocytes and spermatids nearly always contain an even number of nuclei, often multiples of 4. It is not known whether these giant spermatogenic cells are destined for degeneration or whether they subsequently split up and form normal spermatozoa.

Mono- and binuclear eggs of twice the normal volume and spermatozoa of 2 or 3 times the normal size—both gametes being apparently normal in other respects—are not common but have been reported on several occasions in invertebrates (see Wilson, 1928) and in some mammals [rat, cotton rat, and rabbit eggs and cat spermatozoa (Austin and Braden, 1954a; Austin and Bishop, 1957; Austin, 1960)].

There is little information on the functional aspects of the anomalous gametes. Giant rat eggs have been observed in meiosis and in the pronuclear stage of fertilization (Austin and Braden, 1954a), and a fertilized 2-cell giant cotton rat egg has been reported (Austin and Amoroso, 1959). Giant sea urchin eggs on fertilization gave rise to apparently normal embryos during the early stages of development, though these were, in fact, triploid in their chromosomal constitution and did not develop to maturity (Wilson, 1928).

B. Chromosome Anomalies

Well-recognized chromosome anomalies in mammals, which occur in the course of meiosis and can lead to formation of abnormal gametes, are scattering, nondisjunction, lagging, and reciprocal translocation.

By "scattering" is implied dispersal of chromosomes, either individually or in groups, following degeneration of the spindle, usually the second metaphase spindle. The term "nondisjunction" refers to failure of chromosome homologs to separate, and passage of both to the same pole of the spindle. "Lagging" implies the slower movement (or lack of movement) of a chromosome during anaphase resulting in its absence from either pole of the spindle at telophase. "Reciprocal translocation" implies mutual exchange of parts between two chromosomes.

Both scattering and lagging (if the lagging chromosome is retained within the egg) can lead to formation of subnuclei, the number and size of which will depend on the number of chromosomes involved and on whether or not the chromosomes remain in groups. The common experience is to find one or perhaps more small subnuclei in eggs undergoing fertilization, with two apparently normal pronuclei. The incidence of

subnuclei in eggs was observed by Braden (1957) to be genetically determined in the mouse: the mean figure for most strains was not more
than 0.2%, but in certain strains and crosses between strains this figure
was significantly augmented. Thus, the incidence in the V strain was
5.4% and in the progeny of the cross between V strain females and CBA
males, 8.6%. The incidence at which eggs with subnuclei are seen can
also be greatly increased by experimental procedures: the employment of
artificial insemination late in estrus in the rat was a factor noted by
Blandau (1952); and the treatment of eggs *in situ* with heat or the administration of colchicine to the animal has been observed to increase the
incidence in rats and mice (Austin and Braden, 1954b; Edwards, 1958a,b;
Pikó and Bomsel-Helmreich, 1960). Irradiation of spermatozoa with
ultraviolet or X-radiations, or treatment of the animal with radiomimetic
drugs were also effective (Edwards, 1957a,b, 1958a,b; Russell, 1965).
The incidence with delayed mating rose to as high as 18%, and when heat
treatment was superimposed, to as high as 34%.

The significance of subnuclei for subsequent development is not
known for certain. If the chromosomes in the subnuclei fail to assemble
with the main chromosome complement on the first cleavage spindle or
if they take up an abnormal location at this stage, the resulting embryo
could show hypodiploidy, or mosaicism, or both. The ultimate outcome
would depend on the number and kind of chromosomes involved. Loss of
an autosome seems to be inconsistent with extensive embryonic development, and, as yet, no case of survival to birth is known in which no X
chromosome was present. The consequence of loss of the Y chromosome
is mentioned below.

Nondisjunction, lagging, and reciprocal translocation have been in-
voked as likely modes of origin of certain human chromosomal disorders:
(a) the sex chromosome status XXY (Klinefelter's syndrome), and XXX,
in both of which it is generally presumed that one X chromosome failed
to be discarded in a polar body during maturation of the egg;[*] (b) the
status XO (Turner's syndrome) in which it is presumed that both maternal X chromosomes were discharged in the polar body and the X present
derives from the fertilizing spermatozoon, or else that the sex chromosome supplied by the spermatozoon has failed to take its place in the
karyotype of the embryo; (c) trisomy-21 (Down's syndrome or "mongol-

---

[*] By the use of color-vision testing and Xg typing in selected cases of Klinefelter's syndrome, it has been possible to show that one of the X chromosomes
sometimes is paternal in origin. Nondisjunction may, therefore, take place in either
of the gametes. Maternal nondisjunction, which may also occur at the first cleavage,
seems more likely with advancing age (Ferguson-Smith *et al.*, 1964; see also de la
Chapelle *et al.*, 1964).

in the form of a dense precursor material located in a number of cortical alveoli; on sperm penetration, the alveoli break down, the jelly precursor is released, passes through pores in the vitelline membrane through which microvilli project, and becoming hydrated, forms the broad jelly coat (Fallon and Austin, 1967).

In fish eggs, protection takes a different form. The eggs in the sturgeon, for example, are provided with micropyles, namely, small holes through the tough outer membrane (chorion). In the unpenetrated condition, the vitellus nearly fills the space within the chorion, and a tongue of cytoplasm projects a short distance into each micropyle. Spermatozoa appear to be attracted to the micropyles and soon enter these. The first spermatozoon to make contact with and become attached to the cytoplasmic projection causes retraction of the cytoplasmic process which seemingly draws the spermatozoon through the remainder of the micropyle. The spermatozoon then becomes incorporated in the cytoplasm of the egg. Following attachment of the spermatozoon to the cytoplasmic process, a jellylike substance is emitted from the vitellus which fills the perivitelline space; it is apparently impermeable to spermatozoa (Ginsburg, 1957, 1961).

Multiple union can also take place in ciliates, despite the fact that normal conjugation involves a precise succession of steps between two specific participants. In *Paramecium*, attachment of a third individual involves fusion of plasma membranes and establishment of cytoplasmic continuity between its anterior pole and the posterior pole of one of the normal conjugants. This form of union does not lead to nuclear transfer.

C. Incidence of Multiple Union

The incidence of polyspermy in such animals as sea urchins under natural conditions is probably of a low order. Under experimental conditions in the laboratory, the incidence depends largely upon the concentration of spermatozoa used for seminating the eggs. The frequency can also be augmented by treating eggs with various toxic substances, such as chloral hydrate and nicotine.

Among mammals, the presence of three pronuclei in the eggs at fertilization has been recorded in a number of species; however, in many instances it was not known whether the condition was due to polyspermy or to suppression of a polar body (polygyny). Identification depends upon the finding of two sperm tails within the vitellus in addition to three pronuclei, and this has proved possible in only a few species owing to the difficulty of finding the comparatively small sperm tails in yolk-laden eggs. The incidence of polyspermy as thus diagnosed has been reported as follows: outbred albino rat, 1.2% (Austin and Braden, 1953); outbred hooded rat, 1.8% (Austin, 1955); Wistar albino rat, 0.3% (Odor and Blandau, 1956); Jouy rat, 0% (Pikó, 1958); crossbred Jouy X WAG rat, 3.2% (Pikó, 1958); Jouy rat crossed with Wistar CF, 0.9% (Pikó, 1958); outbred albino mouse, 0.3% (Braden and Austin, 1954b); mice of various stocks, 0.9% (Braden, 1957); golden hamster, 1.4 and 3.8% (Braden and Austin, 1956, and Chang and Fernandez-Cano, 1958, respectively); field vole, 2% (Austin, 1957); pig, 1.8% (Thibault, 1959). The incidence of polyspermy can often be augmented by various experimental devices, but whether there is an increase at all, and the degree of increase, depends very much on the strain or stock of animal being tested. Thus in the outbred albino rats investigated by Austin and Braden (1953), delayed mating caused an increase from 1.2 to 8.2%, whereas in the outbred hooded rats used by Austin (1955) the increase from 1.8% was only to 3.3%. The Wistar albino rats of Odor and Blandau (1956) responded to delayed mating with an increase from 0.3 to 3.3%, and the three stocks of rats investigated by Pikó (1958) gave increases, respectively, from 0, 3.2, and 0.9 to 4.5, 6.6, and 7.1%. In the mice and hamsters no increase was detectable as the result of a delay in time of mating. Heat treatment, whether applied locally to the fallopian tube or through an increase in the general body temperature of the animal, produced an incidence of up to 16% with the outbred albino rats, and 34% with the outbred hooded rats. Heat treatment was also effective in mice, causing an increase in the incidence of polyspermy from 0.3 to 3.8%. Delayed artificial insemination in the pig increased the incidence from 1.8 to 12% (Thibault, 1959). A high rate of polyspermy has also been observed in the pig when fertilization occurred during the luteal phase of the estrous cycle, after ovulation had been provoked by hormone injection (Hunter, 1967).

V. Errors of Nuclear Union between Gametes

A. FAILURE OF PRONUCLEAR DEVELOPMENT

Either the male or the female pronucleus may fail to develop in an egg following sperm penetration, and, under these circumstances, fertilization may proceed with a single nucleus; if this were the male pronucleus, development would be known as androgenesis, and if this were female, gynogenesis. The resulting zygote in animals is generally found to have very limited viability; it is haploid unless, through a mechanism such as meiosis inhibition, there is regulation to diploidy.

Androgenesis is comparatively uncommon in animals, though it can be produced experimentally. G. Hertwig (1913) subjected frog and toad eggs to radium emanations, and some of the embryos developing after fertilization were found to be androgenetic, owing to the early degeneration of the egg nucleus. Packard (1918) obtained essentially the same results with *Chaetopterus*. Haploid androgenetic embryos have been produced in axolotls by cold-shock treatment of fertilized eggs (Humphrey and Fankhauser, 1957), and diploid androgenetic hybrids in moths (*Bombyx mandarina* and *Bombyx mori*) were seen after temperature and X-ray treatment of fertilized eggs (the temperature change inhibiting the second meiotic division) (Astaurov and Ostriakova-Varshaver, 1957). When eggs of the newt *Toricha rivularis* were fertilized with spermatozoa of *Toricha torosa* and then heat-shocked, some haploid (hybrid) androgenetic larvae developed to the stage where species characters could be distinguished; skin pigmentation and certain other features clearly showed the dominant influence of the paternal nucleus (Brandom, 1962). Edwards (1954a, 1958b) injected colchicine solution into the uteri of mice before artificial insemination. Subsequently he recovered several eggs at the time of fertilization, each of which had a single, apparently male, pronucleus; at 3½ days gestation, he found some haploid blastocysts which were considered to be androgenetic in origin.

Instances of androgenesis in plants have been known for some years and are of special interest because they show an overriding influence of the nucleus on the development of cytoplasmic characters. One of the earliest reports was that of Kostoff (1929) who pollinated a triploid *Nicotiniana tabacum* var. *macrophylla* with *Nicotiniana Langsdorffii* and obtained a single hybrid plant which showed no *N. tabacum* features at all and was haploid. A paternal haploid was also described by Clausen and Lammerts (1929) from a *Nicotiniana digluta* × *N. tabacum* cross. Haploid progeny with entirely paternal features resulted also from similar breeding trials with maize (Goodsell, 1961; Chase, 1963), and Chase points out that androgenesis provides geneticists with the means for transferring cytoplasm directly from one strain to another without disturbing the chromosomal complement of the pollen donor.

The classic work on the artificial induction of gynogenesis is that of O. Hertwig (1911), who irradiated frog spermatozoa with radium emanations, and added them to eggs. With low doses of irradiation, the resulting embryos died during development, but when the dose was raised above a certain point, apparently normal young were produced. Hertwig inferred that at low doses the spermatozoa took part in syngamy and contributed dominant lethals to the embryo genotype, but at high doses the spermatozoa were too severely damaged to take part in syngamy, and merely stimulated the eggs to gynogenetic development. This work was extended by G. Hertwig (1913) who treated the eggs of *Bufo vulgaris* with the spermatozoa of *Rana temporaria*. The hybrid embryo fails to develop to gastrulation, but if the spermatozoa are first irradiated at high dosage the resulting embryos, developing now gynogenetically, can reach at least the larval stage. Rugh and Exner (1940) confirmed these findings, using leopard frog eggs and bullfrog spermatozoa; here again the hybrid embryo does not continue beyond the gastrula stage, but tadpoles are formed if the spermatozoa are first X-irradiated at 66,000 r.

More recently, it has been found possible to produce gynogenetic diploids in the leopard frog *Rana pipiens* by seminating the eggs with spermatozoa from the spadefoot toad *Scaphiopus holbrooki* and then subjecting the eggs to heat shock. The heterologous spermatozoa activate the eggs without taking part in development, and the heat shock inhibits the second meiotic division. The diploid embryos complete gastrulation and neurulation in a normal manner, but afterward developmental anomalies appear. A few individuals, nevertheless, survive through metamorphosis (Volpe and Dasgupta, 1962).

Experiments on mammals concerned with the effects of ionizing radiations on sperm function show that the first part of the "Hertwig effect" can be produced without difficulty. Irradiation of the whole male animal or of the testes or epididymides at relatively low dose levels resulted in increased prenatal mortality and number of young born dead, in the rabbit, guinea pig, mouse, and rat (Regaud and Dubreuil, 1908; Strandskov, 1932; Snell, 1933; Brenneke, 1937; Hensen, 1942). There was, however, no recognizable evidence of gynogenesis in any of these experiments. Higher dose levels were then applied to mice, treatment being restricted to the genital region (Bruce and Austin, 1956). Some eggs were found with single nuclei (probably female pronuclei) at the time of fertilization, but there was no evidence of continuing gynogenetic development. Treatment with irradiation (ultraviolet and X-rays) and with radiomimetic drugs was also applied to semen before its use for artificial insemination in rabbits and mice (Amoroso and Parkes, 1947; Edwards, 1954a,b, 1957a,b, 1958a; Chang et al., 1958). Again no instances of fargoing gynogenetic development were encountered; in mice, the gynogenetic haploids degenerated at about 3½ days gestation.

The most frequent occurrence of spontaneous though rudimentary gynogenesis in mammals is seen in the hamster, in which about 4% of the eggs have been found to form only a single large pronucleus, though penetrated by a spermatozoon; close examination of the eggs with a phase-contrast microscope revealed that the sperm head had failed to complete its metamorphosis into a pronucleus, and consequently these eggs were in the early stage of gynogenesis (Austin, 1956c; Chang and Fernandez-Cano, 1958). Their subsequent fate is as yet unknown. Further information on gynogenesis is given by Beatty (1964a).

B. SUPERNUMERARY PRONUCLEI

When fertilization involves the presence of supernumerary pronuclei, the condition may be attributable either to the development of extra female pronuclei (polygyny) or of extra male pronuclei (polyandry), or occasionally of both. The commonest forms of the multinuclear state in mammalian eggs are those in which there are three pronuclei: polygyny with one male and two female pronuclei, or polyandry with one female and two male pronuclei. Available evidence indicates that, in both instances, all three pronuclei come together at syngamy and the three sets of chromosomes take their place on a single bipolar first cleavage spindle. The first and subsequent cleavage divisions take place in a normal manner and thus a uniformly triploid embryo develops (Austin and Braden, 1953; Pikó and Bomsel-Helmreich, 1960).

The outcome is similar in the urodele egg in which polygyny has been induced by heat shock (Fankhauser and Godwin, 1948), and in the oosphere of some angiosperms when syngamy is effected with two male nuclei—a single bipolar spindle forms and the initially normal-looking embryo displays triploidy.

Differences appear in other species. In the frog, polyandrous syngamy rarely follows polyspermy, but development is abnormal because cleavage spindles appear in association with each of the supernumerary male pronuclei. The first cleavage of the zygote is, therefore, into four or more cells; only two contain a diploid chromosome complement, the others being haploid. Some of the embryos advance to the tadpole stage but none reach maturity. In the sea urchin egg, polyandrous syngamy follows polyspermy and is succeeded by the formation of a tripolar, tetrapolar, or pentapolar spindle on which the separation of chromosomes appears to be random. Thus the first cleavage produces an embryo of three or more cells which all contain different numbers of chromosomes. In addition some of the supernumerary male pronuclei in the egg induce extra cleavage spindles. Development soon becomes chaotic, therefore, and the embryo has a low viability and rarely survives the blastula stage.

Eggs of the kind that exhibit physiological polyspermy (see Section IV,B) rarely suffer polyandrous syngamy or disturbance of cleavage, the supernumerary pronuclei being eliminated in various ways.

C. SOURCES AND CONSEQUENCES OF TRIPLOIDY

The main sources of triploidy are polygyny (with one male and two female pronuclei) and polyandry (with one female and two male pronuclei), and the less common include fertilization of a normal egg with a diploid, probably giant spermatozoon, and fertilization by a normal spermatozoon of a diploid egg (deriving from a tetraploid, often giant, oocyte). In addition, eggs developing a diploid female pronucleus, through anaphase inhibition in the first or second polar division, would give rise to a triploid zygote on fertilization.

Triploid embryos in rats and mice develop in an apparently normal fashion through approximately the first half of gestation, but then undergo regression and death (Fischberg and Beatty, 1952; Pikó and Bomsel-Helmreich, 1960; Bomsel-Helmreich, 1965). The reason for the inviability is not understood. In man, development of triploids (or more correctly of diploid/triploid mosaics) very occasionally proceeds to survival beyond birth (e.g., Böök and Santesson, 1960; Böök et al., 1961; Ellis et al., 1963); the subjects were mentally retarded and had numerous physical defects. Penrose and Delhanty (1961) found triploid cells in a macerated fetus obtained by curettage at about 4 months of pregnancy, and expressed the view that triploidy may well be a common cause of early interruption of pregnancy in man. Delhanty et al. (1961) reported almost only triploid cells (XXY + 3A) in an apparently normal embryo aborted at 9 weeks.

Triploid individuals are known to show protracted development in nonmammalian vertebrates. Ohno et al. (1963) described a fully grown hen with a cock's comb and hackles, and spurs. The gonad was an ovotestis containing early oocytes and apparently spermatozoa. Chromosome analysis showed that the bird was probably derived from a ZZO + 3A zygote. Many instances exist of experimentally produced triploid newts, toads, and frogs that developed through metamorphosis (see Beatty, 1964b), and some *Triturus* triploids have survived for as long as 3 years (Fischberg, 1945). A triploid *Triturus* individual (male) was also found in the wild (Böök, 1940).

Fully functional adult triploids (and other polyploids) are known in a number of parthenogenetic species or races of crustaceans and insects (see White, 1954; Swanson, 1957). In plants, triploidy is not a bar to full development.

REFERENCES

Amoroso, E. C., and Parkes, A. S. (1947). Effects on embryonic development of X-irradiation of rabbit spermatozoa *in vitro*. *Proc. Roy. Soc. (London)*, **B134**, 57.

André, J., and Vivier, E. (1962). Quelques aspects ultrastructuraux de l'échange micronucleaire lors de la conjugaison chez *Paramecium caudatum*. *J. Ultrastruct. Res.* **6**, 390.

Astaurov, B. L., and Ostriakova-Varshaver, V. P. (1957). Complete heterospermic androgenesis in silkworms as a means for experimental analysis of the nucleus-cytoplasm problem. *J. Embryol. Exptl. Morphol.* **5**, 449.

Austin, C. R. (1955). Polyspermy after induced hyperthermia in rats. *Nature* **175**, 1038.

Austin, C. R. (1956a). Ovulation, fertilization and early cleavage in the hamster (*Mesocricetus auratus*). *J. Roy. Microscop. Soc.* **75**, 141.

Austin, C. R. (1956b). Cortical granules in hamster eggs. *Exptl. Cell Res.,* **10**, 533.

Austin, C. R. (1956c). Activation of eggs by hypothermia in rats and hamsters. *J. Exptl. Biol.* **33**, 338.

Austin, C. R. (1957). Fertilization, early cleavage and associated phenomena in the field vole (*Microtus agrestis*). *J. Anat.* **91**, 1.

Austin, C. R. (1960). Anomalies of fertilization leading to triploidy. *J. Cellular Comp. Physiol.* **56** (Suppl. 1), 1.

Austin, C. R. (1961). "The Mammalian Egg." Blackwell, Oxford, England.

Austin, C. R. (1963). Fertilization and transport of the ovum. *In* "Mechanisms Concerned with Conception" (C. G. Hartmen, ed.), p. 285. Macmillan (Pergamon), New York.

Austin, C. R. (1965). "Fertilization." Prentice-Hall, Englewood Cliffs, New Jersey.

Austin, C. R. (1967). Chromosome deterioration in aging eggs of the rabbit. *Nature* **213**, 1018.

Austin, C. R., and Amoroso, E. C. (1959). The mammalian egg. *Endeavour* **18**, 130.

Austin, C. R., and Bishop, M. W. H. (1957). Fertilization in mammals. *Biol. Rev.* **32**, 296.

Austin, C. R., and Braden, A. W. H. (1953). An investigation of polyspermy in the rat and rabbit. *Australian J. Biol. Sci.* **6**, 674.

Austin, C. R., and Braden, A. W. H. (1954a). Anomalies in rat, mouse and rabbit eggs. *Australian J. Biol. Sci.* **7**, 537.

Austin, C. R., and Braden, A. W. H. (1954b). Induction and inhibition of the second polar division in the rat egg and subsequent fertilization. *Australian J. Biol. Sci.* **7**, 195.

Austin, C. R., and Braden, A. W. H. (1956). Early reactions of the rodent egg to spermatozoon penetration. *J. Exptl. Biol.* **33**, 358.

Austin, C. R., and Walton, A. (1960). Fertilisation. *In* "Marshall's Physiology of Reproduction" (A. S. Parkes, ed.), Vol. 1, Part 2, p. 310. Longmans, Green, New York.

Bacsich, P. (1949). Multinuclear ova and multiovular follicles in the young human ovary and their probable endocrinological significance. *J. Endocrinol.* **6**, i.

Beatty, R. A. (1957). "Parthenogenesis and Polyploidy in Mammalian Development." Cambridge Univ. Press, London and New York.

Beatty, R. A. (1964a). Gynogenesis in vertebrates: fertilization by genetically inactivated spermatozoa. *Proc. Intern. Symp. Effects Ionizing Radiation on the Reproductive System, Fort Collins, Colorado, 1962,* p. 229.

Beatty, R. A. (1964b). Chromosome deviations and sex in vertebrates. *In* "Intersexuality in Vertebrates Including Man" (C. N. Armstrong and A. J. Marshall, eds.). Academic Press, New York.

Beatty, R. A., and Fischberg, M. (1949). Spontaneous and induced triploidy in preimplantation mouse eggs. *Nature* **163**, 807.

Benazzi, L. G. (1952). Nuove ricerche sulla riproduzione pseudogamica e sul ciclo cromosomico in biotipi poliploidi di *Dugesia benazzi* (*Tricladida paludicola*). *Caryologia* **5**, 223.

Bishop, M. W. H., and Walton, A. (1960). Spermatogenesis and the structure of mammalian spermatozoa. *In* "Marshall's Physiology of Reproduction" (A. S. Parkes, ed.), Vol. I, Part 2, p. 1. Longmans, Green, New York.

Blandau, R. J. (1952). The female factor in fertility and infertility. 1. Effects of delayed fertilization on the development of the pronuclei in rat ova. *Fertility Sterility* **3**, 349.

Bomsel-Helmreich, O. (1965). Heteroploidy and embryonic death. *Ciba Found. Symp. Preimplantation Stages Pregnancy*, pp. 246–267.

Bomsel-Helmreich, O., and Thibault, C. (1962). Fertilization *in vitro* in the presence of colchicine and experimental polyploidy in the rabbit. *Ann. Biol. Animale Biochim. Biophys.* **2**, 13.

Böök, J. A. (1940). Triploidy in *Triton taeniatus* Laur. *Hereditas* **26**, 107.

Böök, J. A., and Santesson, B. (1960). Malformation syndrome in man associated with triploidy (69 chromosomes). *Lancet* **i**, 858.

Böök, J. A., Santesson, B., and Zetterqvist, P. (1961). Association between congenital heart malformation and chromosomal variations. *Acta Paediat.* **50**, 216.

Braden, A. W. H. (1957). Variation between strains in the incidence of various abnormalities of egg maturation and fertilization in the mouse. *J. Genet.* **55**, 476.

Braden, A. W. H. (1958). Strain differences in the incidence of polyspermia in rats after delayed mating. *Fertility Sterility* **9**, 243.

Braden, A. W. H., and Austin, C. R. (1954a). Reaction of unfertilized mouse eggs to some experimental stimuli. *Exptl. Cell Res.* **7**, 277.

Braden, A. W. H., and Austin, C. R. (1954b). Fertilization of the mouse egg and the effect of delayed coitus and of hot-shock treatment. *Australian J. Biol. Sci.* **7**, 552.

Braden, A. W. H., and Austin, C. R. (1956). Early reactions of the rodent egg to spermatozoon penetration. *J. Exptl. Biol.* **33**, 358.

Brandom, W. F. (1962). Karyoplasmic studies in haploid, androgenetic hybrids of California newts. *Biol. Bull.* **123**, 253.

Brenneke, H. (1937). Strahlenschadigung von Mause- und Rattensperma, beobachtet an der Fruhentwicklung der Eier. *Strahlentherapie* **60**, 214.

Bruce, H. M., and Austin, C. R. (1956). An attempt to produce the Hertwig effect by X-irradiation of male mice. *Proc. Soc. Study Fertility* **8**, 121.

Butcher, R. L., and Fugo, N. W. (1967). Overripeness and the mammalian ova. II. Delayed ovulation and chromosome anomalies. *Fertility Sterility* **18**, 297.

Cattanach, B. M., and Edwards, R. G. (1958). The effects of triethylenemelamine on the fertility of male mice. *Proc. Roy. Soc. Edinburgh* **57**, 54.

Chang, M. C. (1954). Development of parthenogenetic rabbit blastocysts induced by low temperature storage of unfertilized ova. *J. Exptl. Zool.* **125**, 127.

Chang, M. C., and Fernandez-Cano, L. (1958). Effects of delayed fertilization on the development of pronucleus and the segmentation of hamster ova. *Anat. Record* **132**, 307.

Chang, M. C., Hunt, D. M., and Romanoff, E. B. (1958). Effects of radiocobalt irradiation of unfertilized or fertilized rabbit ova *in vitro* on subsequent fertilization and development *in vivo*. *Anat. Record* **132**, 161.

Chase, S. (1963). Androgenesis—its use for transfer of maize cytoplasm. *J. Heredity* **54**, 152.

Christensen, B., and O'Connor, F. B. (1958). Pseudofertilization in the genus *Lumbricillus* (Enchytraeidae). *Nature* **181**, 1085.

Clausen, R. E., and Lammerts, W. E. (1929). Interspecific hybridization in *Nicotiana*. X. Haploid and diploid merogony. *Am. Naturalist* **63**, 279.

Clement, A. C. (1935). The formation of giant polar bodies in centrifuged eggs of *Ilyanassa*. *Biol. Bull.* **69**, 403.

Conklin, E. G. (1917). Effects of centrifugal force on the structure and development of the eggs of *Crepidula*. *J. Exptl. Zool.* **22**, 311.

Dederer, P. H. (1934). Polyovular follicles in the cat. *Anat. Record* **60**, 391.

de la Chapelle, A., Hortling, H., Sanger, R., and Race, R. R. (1964). Successive non-disjunction at first and second meiotic division of spermatogenesis: evidence of chromosomes and Xg. *Cytogenetics* **3**, 334.

Delhanty, J. D. A., Ellis, J. R., and Rowley, P. T. (1961). Triploid cells in a human embryo. *Lancet* **i**, 1286.

Dziuk, P., and Polge, C. (1965). Fertility in gilts following induced ovulation. *Vet. Record* **77**, 236.

Edwards, R. G. (1954a). Colchicine-induced heteroploidy in early mouse embryos. *Nature* **174**, 276.

Edwards, R. G. (1954b). The experimental induction of pseudogamy in early mouse embryos. *Experientia* **10**, 499.

Edwards, R. G. (1957a). The experimental induction of gynogenesis in the mouse. I. Irradiation of the sperm by X-rays. *Proc. Roy. Soc. (London)* **B146**, 469.

Edwards, R. G. (1957b). The experimental induction of gynogenesis in the mouse. II. Ultra-violet irradiation of the sperm. *Proc. Roy. Soc. (London)* **B146**, 488.

Edwards, R. G. (1958a). Colchicine-induced heteroploidy in the mouse. I. *J. Exptl. Zool.* **137**, 317.

Edwards, R. G. (1958b). The experimental induction of gynogenesis in the mouse. III. *Proc. Roy. Soc. (London)* **B149**, 117.

Elliott, A. M., and Tremor, J. M. (1958). The fine structure of the pellicle in the contact area of conjugating *Tetrahymena pyriformis*. *J. Biophys. Biochem. Cytol.* **4**, 839.

Ellis, J. R., Marshall, R., Normand, I. C. S., and Penrose, L. S. (1963). A girl with triploid cells. *Nature* **198**, 411.

Engle, E. T. (1927). Polyovular follicles and polynuclear ova in the mouse. *Anat. Record* **35**, 341.

Fallon, J. F., and Austin, C. R. (1967). Fine structure of the gametes of *Nereis limbata*, before and after interaction. *J. Exptl. Zool.* **166**, 225.

Fankhauser, G., and Godwin, D. (1948). The cytological mechanism of the triploidy-inducing effect of heat on eggs on the newt, *Triturus viridescens*. *Proc. Natl. Acad. Sci. U. S.* **34**, 544.

Fawcett, D. W., Ito, S., and Slautterbach, D. (1959). The occurrence of intercellular bridges in groups of cells exhibiting synchronous differentiation. *J. Biophys. Biochem. Cytol.* **5**, 453.

Ferguson-Smith, M. A., Mack, W. S., Ellis, P. M., Dickson, M., Sanger, R., and Race, R. R. (1964). Parental age and the source of the X chromosomes in XXL Klinefelter's syndrome. *Lancet* **i**, 46.

Fischberg, M. (1945). Uber die Ausbildung des Geschlechts bei triploiden und ein haploiden *Triton alpestris*. *Rev. Suisse Zool.* **52**, 407.

Fischberg, M., and Beatty, R. A. (1952). Heteroploidy in mammals. II. Induction of triploidy in pre-implantation mouse eggs. *J. Genet.* **50**, 455.

Fofanova, K. A. (1965). Morphologic data on polyspermy in chickens. *Federation Proc.* **24** (Transl. Suppl.) No. 2, Part II, T239.

Gartler, M., Waxman, H., and Giblett, E. (1962). An XX/XY human hermaphrodite resulting from double fertilization. *Proc. Natl. Acad. Sci. U. S.* **48**, 332.

Ginsburg, A. S. (1957). Monospermy in sturgeons in normal fertilization and the consequences of penetration into the egg of supernumerary spermatozoa (trans. title). *Dokl. Akad. Nauk. S.S.S.R.* **114**, 445.

Ginsburg, A. S. (1961). The block to polyspermy in sturgeon and trout with special reference to the role of cortical granules (alveoli). *J. Embryol. Exptl. Morphol.* **9**, 173.

Goodsell, S. F. (1961). Male sterility in corn by androgenesis. *Crop. Sci.* **1**, 227.

Grosser, O. (1927). Fruhentwicklung, Eihautbildung und Placentation des Menschen und der Saugetiere. *Deut. Frauenheilk.* **5**, 454.

Gustafsson, A. (1948). "Apomixis in Higher Plants. Pt. 1. The Mechanism of Apomixis." Gleerup, Lund and Harrassowitz, Leipzig.

Hagstrom, B. (1956). Studies on polyspermy in sea urchins. *Arch. Zool.* **10**, 307.

Hagstrom, B., and Hagstrom, Br. (1954). Refertilization of the sea-urchin egg. *Exptl. Cell Res.* **6**, 491.

Hancock, J. L. (1961). Fertilization in the pig. *J. Reprod. Fertility* **2**, 307.

Harrison, R. J. (1948). The changes occurring in the ovary of the goat during the estrous cycle and in early pregnancy. *J. Anat.* **82**, 21.

Hartman, C. G. (1926). Polynuclear ova and polyovular follicles in the oppossum and other mammals, with special reference to the problem of fecundity. *Am. J. Anat.* **37**, 1.

Hensen, M. (1942). The effect of roentgen irradiation of sperm upon the embryonic development of the albino rat (*Mus norvegicus albinus*). *J. Exptl. Biol.* **91**, 405.

Hertwig, G. (1913). Parthenogenesis bei Wirbeltieren, hervorgerufen durch artfremden radiumbestrahlten Samen. *Arch. Mikroskop. Anat. Entwicklungsmech.* **81**, 87.

Hertwig, O. (1911). Die Radiumkrankheit tierischer Keimzellen. *Arch. Mikroskop. Anat. Entwicklungsmech.* **77**, 97.

Hertwig, P. (1920). Haploid and diploid parthenogenesis. *Biol. Zentr.* **40**, 145.

Humphrey, R. R., and Fankhauser, G. (1957). The origin of spontaneous and experimental haploids in the Mexican axolotl (*Siredon* or *Ambystoma mexicanum*). *J. Exptl. Zool.* **134**, 427.

Hunter, R. H. F. (1964). Superovulation and fertility in the pig. *Animal Prod.* **6**, 189.

Hunter, R. H. F. (1967). Polyspermic fertilization in pigs during the luteal phase of the estrous cycle. *J. Exptl. Zool.* **165**, 451.

Josso, N., de Grouchy, J., Auvert, J., Nezelof, C., Jayle, M. F., Moullec, J., Frezal, J., de Casaubon, A., and Lamy, M. (1965). True hermaphroditism with XX/XY mosaicism, probably due to double fertilization of the ovum. *J. Clin. Endocrinol. Metab.* **25**, 114.

Kallman, K. D. (1962a). Population genetics of the gynogenetic teleost, *Mollienesia formosa* (*Girard*). *Evolution* **16**, 497.

Kallman, K. D. (1962b). Gynogenesis in the teleost, *Mollienesia formosa* (Girard), with a discussion of the detection of parthenogenesis in vertebrates by tissue transplantation. *J. Genet.* **58**, 7.

Kostoff, D. (1929). An androgenic *Nicotiana* haploid. *Z. Zellforsch. Mikroskop. Anat.* **9**, 640.

Krzanowska, H. (1960). Studies on heterosis. II. Fertilization rate in inbred lines of mice, and their crosses. *Folia Biol.* **8**, 269.

Lane, C. E. (1938). Aberrant ovarian follicles in the immature rat. *Anat. Record* **71**, 243.

Lepori, N. G. (1949). Ricerche sulla ovogenesi e sulla fecondazione nella planaria

Polycelis nigra Ehrenberg con particolare riguardo all'ufficio del nucleo spermatico. *Caryologia* **1**, 280.

Longley, W. H. (1911). The maturation of the egg and ovulation in the domestic cat. *Am. J. Anat.* **12**, 139.

McLean, R. C., and Ivimey-Cook, W. R. (1951). "Textbook of Theoretical Botany," Vol. 1. Longmans, Green, New York.

McLean, R. C., and Ivimey-Cook, W. R. (1956). "Textbook of Theoretical Botany," Vol. 2. Longmans, Green, New York.

Metz, C. B. (1954). Specific egg and sperm substances and activation of the egg. *In* "The Beginnings of Embryonic Development" (A. Tyler, R. C. von Borstel, and C. B. Metz, eds.), Publ. No. 48, p. 23. Am. Assoc. Advance. Sci., Washington, D. C.

Moore, B. P., Woodroffe, G. E., and Sanderson, A. R. (1956). Polymorphism and parthenogenesis in a ptinid beetle. *Nature* **177**, 847.

Narbel-Hofstetter, M. (1957). Thélytoquie et pseudogamie chez *Luffia* (Lépidoptère Psychide). *Arch. Vererbungsforsch. Sozial. Rassenhyg.* **32**, 3.

Odor, D. L., and Blandau, R. S. (1956). Incidence of polyspermy in normal and delayed matings in rats of the Wistar strain. *Fertility Sterility* **7**, 456.

Ohno, S., Kittrell, W. A., Christian, L. C., Stenius, C., and Witt, G. A. (1963). An adult triploid chicken (*Gallus domesticus*) with a left ovotestis. *Cytogenetics* **2**, 42.

Packard, C. (1918). The effect of radium radiations on the development of *Chaetopterus*. *Biol. Bull.* **35**, 50.

Pankratz, D. S. (1938). Some observations on the graafian follicles in an adult human ovary. *Anat. Record* **71**, 211.

Parkes, A. S., Rogers, H. J., and Spensley, P. C. (1954). Biological and biochemical aspects of the prevention of fertilization by enzyme inhibitors. *Proc. Soc. Study Fertility* **6**, 65.

Pasteels, J. J. (1956). La polyspermie chez les Lacertiliens. *Arch. Biol.* **67**, 513.

Penrose, L. S., and Delhanty, J. D. A. (1961). Triploid cell cultures from a macerated foetus. *Lancet* **i**, 1261.

Pesonen, S. (1946a). Abortive egg cells in the mouse. *Hereditas* **32**, 93.

Pesonen, S. (1946b). Uber Abortiveier. I. *Acta Obstet. Gynecol. Scand. Suppl.* **2**, 152.

Pikó, L. (1958). Etude de la polyspermie chez la rat. *Compt. Rend. Soc. Biol.* **152**, 1356.

Pikó, L. (1961a). La polyspermie chez les animaux. *Ann. Biol. Animale Biochim. Biophys.* **1**, 323.

Pikó, L. (1961b). Fertilization in the Chinese hamster after normal matings. *Am. Zoologist* **1**, 467.

Pikó, L. (1961c). Repeated fertilization of fertilized rat eggs after treatment with Versene. *Am. Zoologist* **1**, 467.

Pikó, L., and Bomsel-Helmreich, O. (1960). Triploid rat embryos and other chromosomal deviants after colchicine treatment and polyspermy. *Nature* **186**, 737.

Polani, P. E. (1963). Chromosome aberrations and birth defects. *In* "Birth Defects" (M. Fishbein, ed.), p. 136. Lippincott, Philadelphia, Pennsylvania.

Polge, C., and Dziuk, P. (1965). Recovery of immature eggs penetrated by spermatozoa following induced ovulation in the pig. *J. Reprod. Fertility* **9**, 357.

Regaud, C., and Dubreuil, G. (1908). Perturbations dans le developpement des oeufs fécondes par des spermatozoides roentgenises chez le lapin. *Compt. Rend. Soc. Biol.* **64**, 1014.

Rothschild, Lord. (1954). Polyspermy. *Quart. Rev. Biol.* **29**, 332.

Rothschild, Lord, and Swann, M. M. (1949). The fertilization reaction in the sea-urchin egg. A propagated response to sperm attachment. *J. Exptl. Biol.* **26**, 164.

Rothschild, Lord, and Swann, M. M. (1950). The fertilization reaction in the sea-urchin egg. The effect of nicotine. *J. Exptl. Biol.* **27**, 400.

Rothschild, Lord, and Swann, M. M. (1951). The conduction time of the block to polyspermy in the sea-urchin egg. *Exptl. Cell Res.* **2**, 137.

Rothschild, Lord, and Swann, M. M. (1952). The fertilization reaction in the sea-urchin. The block to polyspermy. *J. Exptl. Biol.* **29**, 469.

Rugh, R., and Exner, F. (1940). Developmental effects resulting from exposure to X-rays. II. Development of leopard frog eggs activated by bull-frog sperm. *Proc. Am. Phil. Soc.* **83**, 607.

Russell, L. B. (1965). Radiation sensitivity of phases of fertilization in the mouse. *Ciba Found. Symp. Preimplantation Stages Pregnancy,* pp. 217–241.

Schneider, L. (1963). Electronenmikroskopische Untersuchungen der Konjugation von *Paramecium. Protoplasma* **61**, 109.

Schultz, R. J. (1961). Reproductive mechanism of unisexual and bisexual strains of the viviparous fish *Poeciliopsis. Evolution* **15**, 302.

Skowron, S. (1956). The development of the oocytes in Graafian follicles of the golden hamster *Mesocricetus auratus* (trans. title). *Folia Biol.* (*Warsaw*) **4**, 23.

Snell, G. D. (1933). X-ray sterility in the male house mouse. *J. Exptl. Zool.* **65**, 421.

Spalding, J. F., Berry, R. O., and Moffitt, J. G. (1955). The maturation process of the ovum of swine during normal and induced ovulations. *J. Animal Sci.* **14**, 609.

Stanier, R. Y., Doudoroff, M., and Adelberg, E. A. (1963). "The Microbial World," 2nd Ed. Prentice-Hall, Englewood Cliffs, New Jersey.

Strandskov, H. H. (1932). Effects of X-rays in an inbred strain of guinea-pigs. *J. Exptl. Zool.* **63**, 175.

Sugiyama, M. (1951). Re-fertilization of the fertilized eggs of the sea-urchin. *Biol. Bull.* **101**, 335.

Swanson, C. P. (1957). "Cytology and Cytogenetics." Prentice-Hall, Englewood Cliffs, New Jersey.

Szollosi, D. G. (1962). Cortical granules: a general feature of mammalian eggs? *J. Reprod. Fertility* **4**, 223.

Szollosi, D. G. (1967). Development of cortical granules and the cortical reaction in rat and hamster eggs. *Anat. Rec.* **159**, 431.

Thibault, C. (1949). L'oeuf des mammifères. Son développement parthénogénétique. *Ann. Sci. Nat. Zool. Biol. Animale* **11**, 136.

Thibault, C. (1959). Analyse de la fécondation de l'oeuf de la truie après accouplement ou insemination artificielle. *Ann. Zootech. Suppl.* **8**, 165.

Tyler, A. (1932). Changes in volume and surface of *Urechis* eggs upon fertilization. *J. Exptl. Zool.* **63**, 155.

Tyler, A., Monroy, A., and Metz, C. B. (1956). Fertilization of fertilized sea-urchin eggs. *Biol. Bull.* **110**, 184.

Van der Stricht, O. (1923). Etude comparée des ovules des mammifères aux différentes périodes de l'ovogenèsis, d'après les travaux du Laboratoire d'Histologie et d'Embryologie de l'Université de Gand. *Arch. Biol.* (*Paris*) **33**, 229.

Volpe, E., and Dasgupta, S. (1962). Gynogenetic diploids of mutant leopard frogs. *J. Exptl. Zool.* **151**, 287.

White, M. J. D. (1954). "Animal Cytology and Evolution," 2nd Ed. Cambridge Univ. Press, London and New York.

Wilson, E. B. (1928). "The Cell in Development and Heredity." Macmillan, New York.

Witschi, E., and Laguens, R. (1963). Chromosomal aberrations in embryos from overripe eggs. *Develop. Biol.* **7**, 605.

Woodiel, F. N., and Russell, L. B. (1963). A mosaic formed from fertilization of two meiotic products of oogenesis. *Genetics* **48**, 917 (Abstr.).

Zuelzer, W. W., Beattie, K. M., and Reisman, L. E. (1964). Generalized unbalanced mosaicism attributable to dispermy and probable fertilization of a polar body. *Am. J. Human Genet.* **16**, 38.

Addendum

Since this chapter was written, four reviews and a book of particular relevance have appeared: Carr (1967a), Ford (1969), Ginburg (1968), Polani (1969), and Race and Sanger 1969).

A good deal of interest has been aroused recently in meiotic chromosomal configurations by the observations of Henderson and Edwards (1968) who found that the chiasma frequency in mouse oocytes declined with advancing maternal age, and this trend was associated with increase in the numbers of univalent chromosomes. Univalents often become separated prematurely, and the hypothesis was put forward that the lower chiasma frequency in the oocytes of older mothers was responsible for the greater likelihood of nondisjunction during meiosis and of nondiploidy among the resulting embryos. Similar anomalies were found in human oocytes (Edwards and Henderson, 1968), providing a possible explanation for the origin of trisomy-21 (Down's syndrome), and its association with maternal age. An alternative explanation, relating nondisjunction to delayed fertilization, was advanced by German (1968).

Intrafollicular overripeness in *Xenopus* oocytes was found to lead to nondisjunction of chromosomes and to high rates of embryonic malformation (Mikamo, 1968a,b).

The effect of delayed fertilization was investigated also by Shaver and Carr (1967) who found that 6-day blastocysts recovered from rabbits mated near to the time of ovulation often showed triploidy, but this was the only anomaly that could be associated with the delay. In pigs, too, the incidence of polyspermy was found to be increased (to 15.4%) when eggs were penetrated as late as 16 hours after ovulation (Hunter, 1967). Progesterone treatment in pigs had a similar effect: administered 24 or 36 hours before ovulation it produced an increase in the incidence of polyspermy to 40 and 36%, respectively (Day and Polge, 1968). On the other hand, progesterone treatment in hamsters caused only a reduction in the fertilization rate (Hunter, 1968). Delayed fertilization in the eggs of some strains of mice was associated with a high incidence (nearly 20%) of eggs undergoing total cleavage with the second polar spindle; about 5% of these eggs exhibited stages of fertilization in both "blastomeres" (J. H. Marston, personal communication, 1969).

Women becoming pregnant after discontinuance of the contraceptive pill have been found more frequently to have fetuses that are triploid or XO in constitution (Carr, 1967b).

"Refertilization" of rat eggs was found to occur after injection into the ampullae of various substances in solution including isosmotic sodium chloride (von der Borch, 1967).

REFERENCES TO ADDENDUM

Carr, D. H. (1967a). Cytogenetics of abortions. *In* "Comparative Aspects of Reproductive Failure" (K. Benirschke, ed.), pp. 96–117. Springer-Verlag, New York.

Carr, D. H. (1967b). Chromosomes after oral contraceptives. *Lancet* ii, 830.

Day, B. N., and Polge, C. (1968). Effects of progesterone on fertilization and egg transport in the pig. *J. Reprod. Fertility* 17, 227.

Edwards, R. G., and Henderson, S. A. (1968). Unpublished data. Cited by R. G. Edwards, *Proc. VI World Congr. Fertility Sterility Tel Aviv, 1968.*

Ford, C. E. (1969). Mosaics and chimaeras. *Brit. Med. Bull.* 25, 104.

German, J. (1968). Mongolism, delayed fertilization and human sexual behaviour. *Nature* 217, 516.

Ginsburg, A. S. (1968). "Fertilization in Fish and the Problem of Polyspermy." Nauka, Moscow.

Henderson, S. A., and Edwards, R. G. (1968). Chiasma frequency and maternal age in mammals. *Nature* 218, 22.

Hunter, R. H. F. (1967). The effects of delayed insemination on fertilization and early cleavage in the pig. *J. Reprod. Fertility* 13, 133.

Hunter, R. H. F. (1968). Effect of progesterone on fertilization in the golden hamster. *J. Reprod. Fertility* 16, 499.

Mikamo, K. (1968a). Mechanism of nondisjunction of meiotic chromosomes and of degeneration of maturation spindles in eggs affected by intrafollicular overripeness. *Experientia* 24, 75.

Mikamo, K. (1968b). Intrafollicular overripeness and teratological development. *Cytogenetics* 7, 212.

Polani, P. E. (1969). Autosomal imbalance and its syndromes excluding Down's. *Brit. Med. Bull.* 25, 81.

Race, R. R., and Sanger, R. (1969). Xg and sex chromosome abnormalities. *Brit. Med. Bull.* 25, 99.

Shaver, E. L., and Carr, D. H. (1967). Chromosome abnormalities in rabbit blastocysts following delayed fertilization. *J. Reprod. Fertility* 14, 415.

von der Borch, S. M. (1967). Abnormal fertilization of rat eggs after injection of substances into the ampullae of the fallopian tubes. *J. Reprod. Fertility* 14, 465.

CHAPTER 11

Control of Fertility Mechanisms Affecting Gametogenesis

Harold Jackson

PATERSON LABORATORIES, CHRISTIE HOSPITAL, MANCHESTER, ENGLAND

I. Introduction

Extensive histological studies of the gonads after exposure to radiation have been a feature of radiobiological research for many years. Today, for obvious reasons, such work is proceeding at a greatly enhanced pace with particular reference to chronic dosage and genetic effects. However, studies of drug action on reproductive processes, with all the potential of specific action envisaged by the modern pharmacological outlook, have been greatly neglected. Thus, until little more than 10 years ago, reproductive pharmacology was virtually confined to investigations of steroid sex hormones, i.e., the topic was essentially endocrinological in outlook. Damage induced to the gonads by a variety of other chemicals was regarded as a nonspecific process due to general toxicity and high susceptibility of the germinal epithelium—a view now known to be substantially

467

unbalanced. The only relatively selective type of action appeared to correlate with the special requirement of vitamin E in reproduction in the rat.

An appreciation of the potential importance of drug effects on reproductive cells is developing, although much of the emphasis has remained closely associated with outstanding developments in steroid research, in relation to fertility control in the female.

As a whole, reproductive pharmacology is still in its early stages, particularly in the male. It presents problems of great complexity since so little detailed knowledge is available of underlying basic physiological and biochemical mechanisms. There is common ground with cancer research where knowledge is required of the mechanisms controlling cell proliferation and differentiation.

Important advances may be anticipated from investigations in the largely unexplored field of genetic damage by chemicals in experimental animals, which might be produced in highly selective fashion. For human society the results of exposure of the gametes to chemicals during developmental stages *in utero,* prepubertal, or adult life may present serious hazard.

In the mammal, the stages in development of the mature male and female gamete, fertilization, the transport and nidation of the zygote, and postimplantation development comprise a natural sequence of events and a convenient basis for discussion of chemical interference. Overall control is exercised by the hypothalamus and pituitary gonadotrophins.

II. Pituitary Control of Reproductive Processes

The functional activity of the reproductive system in either sex is governed primarily by circulating gonadotrophins as the efferent pathway. Considerable evidence supports the view that the afferent mechanism involves the blood level of appropriate steroid sex hormone (androgen in the male; estrogen and/or progesterone in the female, according to the circumstances prevailing), operating on hypothalamic centers—presumably via specific chemoreceptor cells.

Accordingly, as yet unidentified humoral agents from hypothalamic neurosecretory cells pass through the vascular network linking hypothalamus to the anterior pituitary, controlling the secretory activity of the gonadotrophs. Administered sex steroids can, therefore, function as pituitary inhibitors, indirectly affecting gametogenesis and fertility. Excessive quantities, however, may affect the gonad directly, causing stimulatory effects, e.g., on spermatogenesis in the male, or impairing follicular development in the female. Interpretations of the role of gonadotrophins based upon hypophysectomy and replacement therapy are relatively crude, particularly in the female, because of the cyclic nature of events involving the three gonadotrophins [follicle-stimulating hormone (FSH), luteinizing hormone (LH), and luteotropic hormone (LTH)] successively and no doubt with overlap. The discovery of nonsteroidal pituitary inhibitors, e.g., reserpine and hydrazine derivatives, opens up prospects of greater knowledge of the influence of gonadotrophins on gametogenesis, which will be referred to later.

III. Spermatogenesis

Mammalian spermatogenesis as in the rat, is now visualized as a highly dynamic process proceeding with clocklike precision. Successive generations of cells enter the proliferative stages from stem cell division at precise intervals. After a series of divisions and morphological changes culminating in meiosis, spermatids are formed. Then follows the long and complex metamorphosis into spermatozoa, with the essential participation of Sertoli cells and relative movement of the metamorphosing spermatids occasioned by the concurrent development of succeeding generations of spermatogenic cells. The net result is the continued shedding of morphologically mature sperm into the lumen of the seminiferous tubule from which they are swept rapidly into the epididymis through which they pass as packed masses of nonmotile cells, as a result of combined ciliary action of the lining epithelial cells and fluid movement—secretion by the head of the epididymis and reabsorption in the distal segment. The duration of the testicular phase in the mouse has been estimated by cell population analysis following radiation damage (Oakberg, 1956) and in the rat and man using tritiated thymidine labeling (Clermont *et al.*, 1959; Heller and Clermont, 1964). The overall time of spermatogenesis to sperm emission can also be inferred from the onset of sterility or aspermia after administration of drugs blocking spermatogenesis in its early stages.

An interesting feature of the spermatogenic process is the manner in which the fluid dynamic mechanisms of the testis and epididymis continue to operate, apparently normally, after cell proliferation has been blocked at different stages of the process. This fact, and the steady progression of spermatogenic cells through various stages of development, are important factors in relation to selective drug actions and the production and maintenance of sterility.

The influence of gonadotropins and androgen in the control of spermatogenesis seems far from resolved. The general view is that testosterone from the intertubular Leydig cells plays a direct controlling role in postmeiotic events; male hormone secretion is, in turn, controlled by pituitary interstitial cell-stimulating hormone (ICSH or LH), but the role of FSH is not clearly established. Knowledge of these interrelationships has been derived from the results of hypophysectomy followed by gonado-

trophin or androgen therapy. The purity of the gonadotrophin preparations used is obviously of the utmost importance. The success of this
experimental procedure, so far as the structural integrity of the seminiferous epithelium is concerned, depends both upon the time allowed to
elapse before commencing treatment and persistence with the latter; the
contributory importance of somatotrophin and thyroid hormone has been
discussed (Boccabella, 1963). Although spermatogenesis in hypophy-
sectomized rats could remain or be restored to normal by hormonal
therapy, the criterion of fertility was below normal. In man, using
thymidine labeling, it has been shown that neither injection of gonadotrophin nor completely blocking later stages of spermatogenesis by a
progestational steroid, norethandrolone (Nilevar), altered the rate con-
stancy of the proliferative stages (Heller and Clermont, 1964). Similar
information had been obtained in the rat. Since hypophysectomy is a
relatively drastic procedure, the use of nonsteroidal compounds to produce pituitary inhibition could be more informative (see below). Pituitary
gonadotrophins constitute the efferent pathway for the control of testicular
function, but the afferent mechanism is not well-defined. In addition to
information reaching the pituitary via the blood androgen level, there is
evidence of more direct information from the seminiferous epithelium. In
fact, a suggestion is that gonadotrophin production primarily depends
upon tubular function in the testis, a hypophyseal inhibitor being normally
liberated during late stages of spermatid development (Johnsen, 1964).
Here again, studies with direct chemical inhibitors of spermatogenesis
may help to clarify the situation.

There is little detailed knowledge of the *mechanism* of hormonal control of spermatogenesis and epididymal function. This system, of precise
order and continuity, possesses unique features which imply the possibility
of selective chemical interference at various points, which is supported by
the limited studies so far made.

Two special nutritional requirements of the seminiferous epithelium
are known, in the form of vitamins E and A. Although recognized many
years ago, vitamin E work in the testis (Mason, 1941; Bryan and Mason,
1941) appears to have been neglected and might repay further studies
using modern techniques. Whereas a vitamin E-deficient diet rapidly produced irreversible changes in the seminiferous epithelium of the male rat,
that of the mouse was unaffected. In the hamster, a progressive degeneration occurred which, however, gradually responded to the inclusion of
α-tocopheryl acetate in the diet (Mason and Mauer, 1957). Mode of
action and reason for the species difference are unknown. Recent studies
have revealed a specific dependence of rat spermatogenesis upon vitamin
A alcohol, since severe degenerative changes of the seminiferous epi-

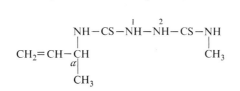

1-(α-methylallyl thiocarbamoyl)-2-(methyl thiocarbamoyl) hydrazine
(compound 33,828)
(I)

Fig. 1

thelium occurred when vitamin A acid was provided in amount adequate
to support normal growth (Howell *et al.*, 1963).

Although the antimetabolite approach has not so far achieved notable
success in the selective impairment of spermatogenesis and fertility in
experimental animals, the development of antagonists to substances with
specific and essential functions in sperm development is an attractive
possibility.

IV. Chemical Interference with Spermatogenesis

A. Pituitary Inhibition

Selective inhibition of gonadotropin secretion from the pituitary could
provide more precise information concerning the hormonal control of
spermatogenesis. Methallibure (compound 33,828), a hydrazine derivative (Fig. 1, I) (Walpole, 1965) produces reversible suppression of spermatogenesis in rats. For this action, persistent daily treatment is required.
Cell counts showed that the proliferative stages were normal until late
spermatocytes (in Stage VII), which were reduced to about 30% of normal. Spermatids were also seen at this stage but no later (Hemsworth
et al., 1968). Study of associated changes in fertility are complicated by
the loss in libido (due to the androgen deficiency), which in the rat, is
not readily distinguished from aspermic copulation. However, full fertility was not restored until about 10 weeks post-treatment, i.e., the overall duration of spermatogenesis, although mating capacity had reappeared
several weeks previously. The implication is that pituitary influence also
extends over early phases of the proliferative stages of spermatogenesis.
The ability of a few other nonsteroidal, centrally acting compounds, e.g.,
reserpine, 5-hydroxytryptamine, iproniazide, and other amine oxidase inhibitors, to suppress spermatogenesis and affect male fertility, has not
been studied systematically. Observations of the effects of steroidal compounds in experimental animals, e.g., estrogen and progestational steroids,
are also inadequate and complicated by the loss of libido. A number of

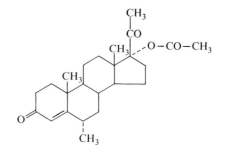

Provera (6α-methyl-17α-hydroxyprogesterone acetate)
(II)

FIG. 2

publications report testicular effects with oligospermia and aspermia following administration of progestational steroids in men (Heller *et al.*, 1958, 1959). These steroids either contained estrogenic material as contaminant or the major progestational component was partially metabolized to female hormone. The antispermatogenic effects of two depot-type, long-acting steroid preparations have been described (MacLeod, 1965). Testosterone oenanthate suppressed spermatogenesis in man without affecting libido and medroxyprogesterone acetate (Provera) (Fig. 2, II) caused severe histological changes in the testis by a weekly dose (250 mg) or single large injection (1 gm). The effect appears to be due to inhibition of gonadotrophin secretion but a direct mechanism upon the gonad must be considered for the progestational steroid, which is not metabolized to estrogen. In a small number of subjects, the sperm count approached 0 about 70 days from the first or only dose. Two months after the last injection of testosterone oenanthate, the count rose quite rapidly to normal levels. With medroxyprogesterone acetate (1 gm single injection) the near-aspermic state was more protracted, recovery occurred slowly, and normal levels were not reached 1 year later. Aberrations in sperm morphology occurred after both steroids. These were of a minor character with the testosterone ester but were severe following the medroxyprogesterone acetate.

The varied physiological role of androgen in the control of spermatogenesis, accessory sexual structures, and secretions, as well as its actions on the central nervous system (CNS) shown by effects on libido and gonadotrophin secretion, suggests different end-organ mechanisms, for which pharmacological dissociation may be possible. Perhaps the greatly varied range of compounds that show antiandrogen properties (Dorfman, 1962, 1963) are a manifestation of the complexity of the end-organ mechanisms. So far as the present author is aware, studies of the antispermatogenic and antifertility actions of such compounds have not been reported.

Antifertility potential in the male should not be judged solely on ability to produce obvious impairment of spermatogenesis. It may also result from less perceptible or even imperceptible damage to spermatogenic cells during development or perhaps from modifications in normal or essential constituents of epididymal fluid or other accessory secretions.

B. ANTISPERMATOGENIC COMPOUNDS

Direct interference with spermatogenesis at the testicular level is a more obvious approach to male fertility control, since the consequence of interference with sexual activity may be avoided. Substances that suppress spermatogenesis are conveniently referred to as antispermatogenic compounds, although general mode of action is not necessarily clear. There are three principal categories to consider: (*1*) alkylating chemicals; (*2*) nitroaromatic compounds; and (*3*) diamines.

1. *Alkylating Chemicals*

Any chemical that can react with and donate an alkyl radical to one of a variety of chemical groupings (e.g., —OH, —SH, —COOH, —NH$_2$) widely occurring in substances present in blood, tissue fluids, and cells is designated an alkylating agent. Many types of such chemicals are known and the alkylating group may be very simple (e.g., methyl, ethyl, propyl) or consist of substituted alkyl groups, of which there are numerous possibilities. The nature of the carrier molecule bearing the alkyl grouping(s) is particularly important in relation to biological activity. For a comprehensive discussion of these widely studied substances and their biological activities, the reader is referred elsewhere (Ross, 1962; Wheeler, 1962). The general tendency to regard the biological actions of alkylating chemicals as presenting a similar type of activity spectrum seems no longer acceptable. Studies on the antifertility effects of these substances in rodents (H. Jackson, 1959, 1964a,b) have provided evidence of distinctive actions on reproductive cells according to their stage of development. Two general types of action can be discerned—the "antispermatogenic" and the "functional" effects. The former is associated with histological changes in the testis, whereas the latter is detectable only by fertility tests, since there is no evidence of cellular damage.

Derivatives of ethyleneimine (Fig. 3, III) and esters of methanesulfonic acid have provided the most interesting results so far. Chronologically the antifertility effect in male rats of the ethyleneimine tumor inhibitor, Tretamine (Fig. 3, IV), first aroused interest (Bock and Jackson, 1955, 1957) and has since been widely investigated (for references, see

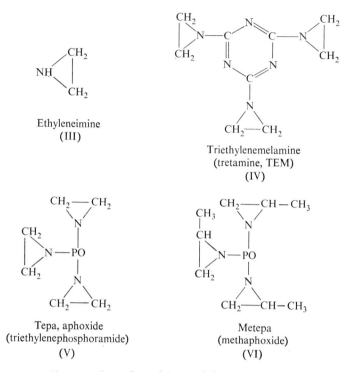

Ethyleneimine
(III)

Triethylenemelamine
(tretamine, TEM)
(IV)

Tepa, aphoxide
(triethylenephosphoramide)
(V)

Metepa
(methaphoxide)
(VI)

FIG. 3. Chemosterilants derived from ethyleneimine (aziridine).

H. Jackson, 1964b). Other ethyleneimine derivatives such as TEPA (Fig. 3, V) and Metepa (Fig. 3, VI) produce actions resembling those of TEM. Metepa is an effective chemosterilant for male flies and mosquitoes and its acute and chronic effects on rat fertility have been reported (Gaines and Kimbrough, 1964). The predominant effect of tretamine in mammalian and insect spermatogenesis is a postmeiotic action on spermatids, which leads to incompetent but motile spermatozoa—the functional antifertility action. An antispermatogenic action against spermatogonia can also be recognized by quantitative cell population analysis in the rat testis (Steinberger, 1962) but only accentuated sufficiently to impair fertility in rats by chronic administration (Bock and Jackson, 1957; Gaines and Kimbrough, 1964). By structural modifications with the introduction of more ethyleneimine groups, it is possible to enhance the antispermatogenic action of ethyleneimine derivatives (Table I) and biphasic antifertility episodes appear when the serial mating technique is applied to treated males (H. Jackson, 1964a). This latter technique, in effect, samples cell stages present at the time of treatment as they later mature (or should mature) and emerge as spermatozoa, according to the dynamic nature of

TABLE I

Enhancement of Antispermatogonial Effect in Rats by Ethyleneimine Derivatives

Drug	No. of ethyleneimine groups in molecule	Intra-peritoneal dose (mg/kg)	Average litter size (at weeks)													
			Epididymal spermatozoa[a]		Testicular sperm and spermatids[a]			Spermatocytes[a]			Spermatogonia[a]				
			1	2	3	4	5	6	7	8	9	10	11	12	13
Tretamine	3	1	0	0	0	0	0	4	5	10	6	4	8	9	—
Tetraethylene-pyrophosphoramide	4	5	1	0	0	1	5	10	8	7	9	9	8	9	—
		10	0	0	1	2	0	0	0	4	9	9	11	6	—
PN6[b]	6	10	4	0	0	0	0	3	5	2	0	0	0	0	5

[a] Treated cell type represented at mating.
[b] PN6, 2,2,4,4,6,6,hexa(1-aziridinyl)2,4,6-triphospha-1,3,5-triazine.

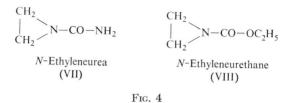

N-Ethyleneurea   N-Ethyleneurethane
(VII)   (VIII)

FIG. 4

the spermatogenic process. The antispermatogenic action is also apparent as episodes of oligospermia or aspermia in serial samples of sperm from rabbits (Fox *et al.*, 1963).

Simple ethyleneimine derivatives such as ethyleneurea (Fig. 4, VII) and ethyleneurethane (Fig. 4, VIII) can produce destruction of the seminiferous epithelium in rats with protracted episodes of sterility (H. Jackson *et al.*, 1959; H. Jackson, 1964b). In all cases, the antifertility effects develop without apparent loss of sexual activity, so that the action appears to be directly on the seminiferous epithelium. This aspect of the antifertility mechanism has been applied with advantage in the development of insect chemosterilants, where the sexual competitiveness of treated male with normal insects is an essential factor in the method (Smith *et al.*, 1964). Overall, the antifertility effects of ethyleneimine derivatives in rodents present a rather uniform pattern, predominantly affecting spermatids and spermatogonia but capable of extending over the entire seminiferous epithelium. The dose–response action is generally steep. Whereas the effect of tretamine was shown to be cumulative, other compounds have not been adequately studied in this respect, except for a recent study of Metepa (Gaines and Kimbrough, 1964).

In the methanesulfonic ester series $(CH_3 \cdot SO_2 \cdot OR)$, the acid radical $(CH_3 \cdot SO_2O\text{---})$ is the donor system for the alkylating radical R. So far, only work with simple di- and monoesters of methanesulfonic acid (Table II) in rodents has been reported. The diesters have the general formula shown and form a homologous series $n = 1, 2, 3, 4$ and so forth. The antispermatogenic action of busulfan (Myleran, $n = 4$) appears as a selective impairment of early spermatogonial development in the rat, mouse, and rabbit (H. Jackson *et al.*, 1959, 1961; Fox *et al.*, 1963, 1964). The time of onset of sterility with aspermia after minimal treatment provides a pharmacological measure of the duration of spermatogenesis. After minimal dosage the effect is transient, but, with larger quantities sterility is prolonged and with earlier onset due to spread of action of spermatocytes. Compounds $n = 1$ and $n = 3$ show a similar action. Methylenedimethanesulfonate $(n = 1)$ also shows a remarkable effect on epididymal sperm, producing sterility for 1 to 2 weeks after a dose of the compound (Fox and Jackson, 1965). This action appears to be antispermatozoal in

TABLE II. STERILE PHASES IN THE MALE RAT AFTER SULFONIC ESTERS

Compound	Intraperitoneal dose (mg/kg)	Sperm[a]		Spermatids[a]			Spermatocytes[a]			Spermatogonia[a]		
		1	2	3	4	5	6	7	8	9	10	11
Methylenedimethanesulfonate $CH_2\!<\!^{OSO_2CH_3}_{OSO_2CH_3}$ $n=1$	15	▓										
Ethylenedimethanesulfonate $(CH_2)_2\!<\!^{O-SO_2CH_3}_{O-SO_2CH_3}$ $n=2$	100		▓	▓	▓	▓	▓	▓	▓	▓		
Butylenedimethanesulfonate $(CH_2)_4\!<\!^{O-SO_2CH_3}_{O-SO_2CH_3}$ $n=4$	8									▓	▓	▓
Methylmethanesulfonate $CH_3O-SO_2CH_3$	50			▓								
Isopropylmethanesulfonate $CH_3\!>\!CHO-SO_2CH_3$ CH_3	100			▓	▓	▓	▓	▓	▓	▓	▓	▓

Weeks post-injection

[a] Cell type present at time of treatment.

nature, whereby the male gamete is rendered functionally incompetent. The spermatogonial action of this compound is demonstrable by repeated smaller daily doses. The action of methylene dimethanesulfonate on sperm recalls the marked subfertility noted in early weeks after treatment with the diesters, dimethylmyleran (dimethylbusulfan) and β-chloroethylmethanesulfonate (H. Jackson et al., 1961).

Ethylene-1,2-dimethanesulfonate ($n = 2$) produced a completely different action from the other members of this series. Predominantly it disorganizes or destroys intermediate spermatogenic cells providing another distinct antifertility pattern (Table II). Also, this compound does not produce hemopoietic damage characteristic of other members of this group and is the least toxic member of the series. A common feature of all these esters, however, is their highly cumulative action in small divided doses by mouth, irrespective of chemical stability. The compound $n = 1$ is remarkable in this respect since it has only a short chemical half-life (Fox and Jackson, 1965) yet shows its selective antispermatozoal action at a daily dose level of 1 mg/kg by mouth (H. Jackson, 1965).

Simple monoalkyl esters of methanesulfonic acid (e.g., methyl-, ethyl-, and n-propyl; Table II) are relatively stable, but produce yet another distinctive effect on spermatogenesis (H. Jackson et al., 1961). Unlike the diesters, the action is not antispermatogenic, for gamete production continues unhindered. The resultant functional sterility is apparently due to impairment of the capacity of sperm to sustain embryonic development. The specificity of the effect is indicated in Table III, which presents the detailed experimental results from a group of rats treated with methyl methanesulfonate. Although the potency of the straight-chain esters dwindles from methyl- to ethyl- and n-propyl, the branched-chain isopropyl ester, an unstable substance, showed a remarkable transformation in action (Table II) to that of the diester type, being now antispermatogenic against the premeiotic stages (H. Jackson et al., 1961; Partington et al., 1964; H. Jackson, 1964a). The reason for this is obscure but, again, illustrates the difficulty of predicting structure–activity relationships.

In the production of antifertility effects the methyl-, ethyl-, and isopropyl esters of methanesulfonic acid are also highly cumulative but the n-propyl ester did not show a corresponding activity. The cumulative effect was restricted to cell types affected by the single dose. For example, with methyl methanesulfonate, fertility always returned predictably within 5 weeks post-treatment, the time being dependent upon the daily dose level but independent of the duration of treatment. There was no evidence of cumulative damage to premeiotic cells, even after many months of treatment. Uniform sterility can be produced by single daily doses of 5 mg/kg by mouth or about 8 mg/kg consumed in the drinking water daily.

TABLE III

EFFECT OF METHYLMETHANESULFONATE ON MALE RAT FERTILITY

Week from treatment[a]	No. of females inseminated	Individual litter sizes[b]	Mean
1	7	7, 0, 2, 8, 7, 5(1), 6.	4.9
2	5	0, 0, 0, *0*, 0, 0, *0*.	0.0
3	5	0, 0, 0, 0, *0*, *0*, 0.	0.0
4	6	8(1), 12, 8, 8(1), 8, 0, 9.	8.5
5	7	9(2), 6, 7, 9, 8, 11, 7(1).	10.8
6	5	14, *0*, 8(1), 7, 2, 0, 11(1).	8.0
7	7	6, 0, 9, 9, 10, 8, 9(1).	7.1
8	7	12, 6, 11, 12, 10, 9, 10.	10.0
9	7	10, 0, 11(1), 8, 10, 9, 11.	8.3
10	7	11, 7, 13(7), 10, 13, —, 11.	8.3
11	2	10, 10(2), 11, 12, 10, 6(2), 8.	9.0
12	7	12, 8(1), 5, 10, 10, 8, 12(1)	9.0

[a] One dose of 50 mg/kg, intraperitoneally.

[b] Numbers in parentheses refer to young born dead; italic zeros refer to pairings in which insemination was not observed. Normal average litter size is about 8.

The production of infertility appears to require exposure of postmeiotic cells to a critical total dose of drug (H. Jackson, 1964a, 1965). Whether an effect is produced depends upon the dose–rate in relation to the time spent by cells moving through susceptible stages of spermatogenesis. There is no indication of the development of resistance to the antifertility effects during prolonged courses of treatment. Male rats may be maintained in a state of reversible infertility throughout much of their life span by this type of pharmacological action. With compounds affecting the premeiotic stages the predictability is less certain, although the cumulative potential of the esters is similar.

The relative contribution of the number of alkylating groups and chemical reactivity to the antifertility action is difficult to evaluate, and physical properties such as water or lipoid solubility do not appear to provide a decisive contribution. There is still no real clue as to the mode of action of these compounds. Explanations must take into account the varied types of effect—particularly the susceptibility of some cells and the insusceptibility of others within the same environment—to compounds of varying stability, chemical reactivity, and physical properties. A uniform theory of alkylation of cell nuclear material seems unacceptable. The nature of the alkylating group clearly plays an essential role in determining the type of cell affected. There is no certainty that the actions are exerted directly by the chemicals on the seminiferous epithelium. The

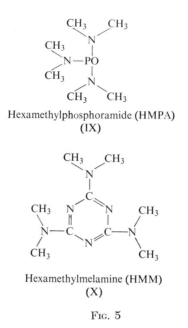

Hexamethylphosphoramide (HMPA)
(IX)

Hexamethylmelamine (HMM)
(X)

Fig. 5

postmeiotic functional antifertility actions on spermatids and sperm are associated with damage to genetic material as indicated by the production of "dominant lethal mutations" (Cattanach and Edwards, 1958; Partington and Jackson, 1963; Partington and Bateman, 1964) and heritable partial sterility (Cattanach, 1959; H. Jackson et al., 1964), but in the present state of knowledge no suggestion can be accepted unreservedly. The toxicology of alkylating substances with particular reference to chemosterilants has been recently surveyed (Hayes, 1964).

The major outcome of work with alkylating compounds is the demonstration that selective attack is possible at various stages of spermatogenesis, with the production of short or prolonged inhibition of fertility, apparently without loss of libido. The time relationships and predictability of antifertility actions are governed by the dynamics of spermatogenesis and the cell type(s) affected. There seems little doubt that other focal points will be susceptible to chemical interference. A great deal remains to be learned about the pharmacology of this unique, proliferating cell system, particularly with regard to species variations and the risk of production of heritable damage.

The important application of alkylating chemicals and antimetabolites to the problem of insect chemosterilization has been surveyed elsewhere (Smith et al., 1964; H. Jackson, 1965). It has recently been suggested that alkylation is not necessarily a requisite for sterilization of postmeiotic cells, on the basis of results with hexamethylphosphoramide (HMPA) (Fig. 5, IX) and hexamethylmelamine (HMM) (Fig. 5, X) in insects. In these analogs of TEPA and TEM, the aziridinyl rings (i.e., the alkylating groups), have been replaced by dimethylamino residues. Like TEM and TEPA, these two compounds also possess the ability to produce chemosterilization of house flies with retention of sexual activity (Chang *et al.*, 1964) but relatively high doses were required. In male rats the TEPA analog (hexamethylphosphoramide) in maximum tolerated dose (5 daily doses, 500 mg/kg), produced a rapid and prolonged loss of fertility associated with aspermia. At lower dose levels, the action became rapidly less intense unless prolonged courses of treatment were given (H. Jackson and Craig, 1966). So far, it appears that, in rodents, there is some analogy with the effects of various ethyleneimine derivatives but only at greatly increased dose levels. It would be a remarkable coincidence if the dimethylamino analogs of TEM and TEPA also produced antispermatogenic action without loss of sexual activity, and the possibility of common factors needs further investigation. Perhaps the action may be due to some metabolite, and the possibility of an alkylating mechanism has not been definitely excluded.

Assuming the pharmacological effects of alkylating substances are the result of direct chemical reaction with cellular components, there is no clear demarcation between this type of control of cell population growth and the classic antimetabolite method. An alkylating chemical might react *in vivo* with some essential substance forming a metabolic antagonist. This mechanism could also halt spermatogenesis, cause cell death, or at a more subtle level, result in the production of functionally incompetent spermatozoa. In general there have been few developments in this approach to fertility control. Ethionine, the methionine antagonist, was reported to cause an interesting retrospective loss of spermatogenic cells, i.e., sperm first and then spermatids, the effect spreading to the proliferating stages so that all germ cells were lost by 24 days (Goldberg *et al.*, 1959). An interesting feature was the proliferation of the Sertoli cells, also said to occur in hypo-vitamin A-induced degeneration of the seminiferous epithelium (Howell *et al.*, 1963; Howell, 1965).

2. *Nitroaromatic Compounds and Dichloroacetyl Diamines*

Studies of the antispermatogenic effects of such compounds seem to have originated from chemotherapeutic research based upon the structure of chloramphenicol, since active substances contain chemical groupings found in this antibiotic. Antispermatogenic activity has been described among nitroderivatives of furane, pyrrole, and thiazole, in which the ben-

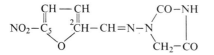

Nitrofurantoin, Furadantin
N-(5-nitro-2-furfurylidene)-1-aminohydantoinate)
(XI)

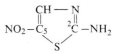

Enheptin
2-amino-5-nitrothiazole
(XII)

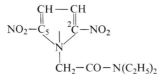

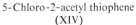

1(N, N-Diethylcarbamoyl methyl)-2, 5-dinitropyrrole
(XIII)

5-Chloro-2-acetyl thiophene
(XIV)

Fig. 6. Heterocyclic antispermatogenic compounds.

zene ring present in chloramphenicol corresponds to the five-membered
heterocyclic structures (Fig. 6, XI to XIV).

These nitro compounds are active by mouth and predominantly suppress the meiotic stage of spermatogenesis. Administered to rats in the
diet, Furacin (5-nitro-2-furfuraldehyde semicarbazone) and two related
compounds (Furadroxyl and Furadantin) cause gradual inhibition of
spermatogenesis reaching a peak after several weeks (Nelson and Steinberger, 1953). A similar action occurred in male mice at all ages (Cranston, 1961). Even after prolonged treatment the action remained fully reversible and, although originally considered to be directed specifically
against the meiotic division (Nelson and Steinberger, 1957) another view
maintained that spermatogonia were involved (Nissim, 1957). Gonadotrophins failed to antagonize the action, whereas effects on the prostate
and seminal vesicles were consistent with increased output of androgen
from the interstitial cells (Montemurro, 1960). Furadroxyl did not affect
reproductive function in the female rat, suggesting a specific link with the
role of androgen in spermatogenesis. Available experimental data does
not permit the formulation of a satisfactory explanation of the mode of
action.

An antispermatogenic dinitropyrrole (Fig. 6, XIII) gradually inhibited
rat spermatogenesis when given in the diet (20 mg/kg/day), producing
sterility in 6 weeks. A large single dose caused infertility after 3 weeks,
which could be maintained by comparable treatment at monthly intervals
(Blye and Berliner, 1963). Analysis of the effects of single doses revealed
changes in primary spermatocytes a few hours after treatment, spreading

to involve spermatids and Type A spermatogonia (Patanelli and Nelson, 1963). This is consistent with breeding studies, where rats were sterile after 3 weeks and remained infertile for 28 days, although full restoration of the spermatogenic epithelium required 75 days, i.e., the entire duration of spermatogenesis. As with nitrofurans, the action was said to be dependent on the presence of pituitary gonadotrophin.

Although aromatic nitro compounds are not therapeutically attractive, further data on structure–activity relationships and mode of action are desirable; possibly the active materials are metabolites. The fact that nitrofurans have achieved a therapeutic role in chronic urinary infections provides a further incentive for such investigations.

3. *Diamines*

Bisdichloroacetyldiamines (Fig. 7, XVI) produce spermatogenic arrest in the rat, dog, monkey, and man. The dichloroacetyl group is also a feature of the chloramphenicol molecule. Although persistent administration of rather high doses was required, the compounds were of low toxicity (Coulston *et al.*, 1960; Beyler *et al.*, 1962; Drobeck and Coulston, 1962). The selective inhibition of spermatogenesis was reversible and, in the rat, the onset of sterility was delayed in accordance with inhibition of the spermatocyte stage. Histologically, the epithelium could be denuded of cells apart from Sertoli cells so that evidently spermatogonia could be affected. The antispermatogenic changes were similar in the rat, monkey, and dog, although there were marked differences in the minimal effective dose. Full restoration of spermatogenic function occurred after the antispermatogenic action had been maintained for prolonged periods. The effect on the seminiferous epithelium was highly specific and no other

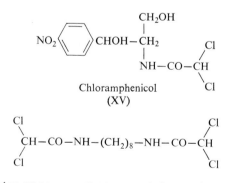

Chloramphenicol
(XV)

N,N^1-Bis(dichloroacetyl)-1,8-octamethylenediamine (Win 18446)
(XVI)

Fig. 7

tissues were affected (Drobeck and Coulston, 1962). The mode of action was undecided and injection of gonadotrophin (mainly FSH) or testosterone, apparently enhanced the action of the diamines. The ovaries and uterus of female rats were unaffected after prolonged treatment with large doses of the compounds, again indicating close association between the drug action and male germinal cells. A variety of vitamins and amino acids did not interfere with the pharmacological effect.

In man, these dichloroacetyldiamines reduced the sperm count to very low levels within 8 to 11 weeks from the commencement of treatment, and the drugs were well tolerated (Heller et al., 1961; MacLeod, 1965). The testicular changes induced by the most potent of these compounds (the octamethylene compound) have been described (Heller et al., 1963). However, an antabuse type of reaction occurred, and highly abnormal forms of sperm were found in semen specimens. The use of such compounds in controlling human fertility would necessitate the production of aspermia to insure sterility. Possible hazards from fertilization by morphologically abnormal or functionally impaired sperm during the phase of declining sperm count and in the post-treatment recovery would have to be adequately explored in laboratory animals. Again, the need for a high dose rate suggests possible conversion to pharmacologically active metabolites, but no studies of this kind appear to have been published.

C. MISCELLANEOUS AGENTS

Small amounts of the cadmium ion are highly destructive to the seminiferous epithelium and interstitial cells of the rat (Parizek, 1960; Allanson and Deansley, 1962) and monkey (Kar, 1961), but the hen is insusceptible (Erickson and Pincus, 1964). Whether repopulation of the rat testis tubules occurs after cadmium is said to be dose-dependent (Allanson and Deansley, 1962). Pituitary gonadotrophs reflect the damage to spermatogenic and interstitial cells by a hyperactive response typical of the castration pituitary. With appropriate doses of cadmium the tubules remained atrophic but interstitial cells reappeared under the tunica albuginea, and androgenic function was fully restored judged by the state of the prostate and seminal vesicles. This was associated with return of centrally disposed pituitary gonadotrophs to a normal state. Peripheral gonadotrophs appeared to be related to the integrity of the germinal epithelium, being restored to normal levels of activity as repopulation by spermatogenic cells occurred. This provides further indication of a direct homeostatic relationship between the germ cell population and pituitary cells.

Administration of zinc acetate before or after cadmium treatment can prevent the toxic action of the latter, and selenium oxide has been reported to exert a greater protective action than zinc. Zinc-deficient diet in the rat results in irreversible atrophy of the seminiferous epithelium (Miller *et al.*, 1958), but apparently no data for the rooster have been published. A suggestion was that the action of cadmium may involve an enzyme peculiar to the pampiniform plexus and testicular vessels (Gunn *et al.*, 1963). Thus the mechanism might involve interference with the sensitive thermoregulation of the testicular environment, and changes in the vascular endothelium have been shown to precede tubular damage (Gunn *et al.*, 1963; Chiquoine, 1963). Although nonscrotal animals (grass frog, pigeon, rooster, and armadillo) were insusceptible to cadmium, the testis of the opossum and ferret were also unaffected by subcutaneous injection of the element (Chiquoine, 1963; Chiquoine and Suntzeff, 1965). All species are sensitive to intratesticular cadmium treatment, so that the insusceptibility shown by some species is presumably related to the ability of the metal ion to reach an effective level in the gonad after subcutaneous administration.

In summary, therefore, the pharmacological actions of chemical agents on spermatogenesis may involve cell destruction, inhibition of proliferation at different stages, and injury to metamorphosing cells or to the mature gametes. Spermatozoa from such damaged, postmeiotic stages may still fertilize, but result in death of the embryo in pre- or early postimplantation development. Investigation of genetic damage by chemical agents in the experimental mammal has received little attention so far. Evidence continues to accumulate of selective interference with various regions of spermatogenic development, which is encouraging for future studies. So far as human fertility control is concerned, the complexity of the problem presents formidable problems requiring a great deal more research effort than is at present available. Other potential rewards are, nevertheless, substantial in terms of rodent and pest control, besides the prospect of considerable advances in fundamental knowledge of physiological processes concerned with cell proliferation and differentiation.

V. Ovogenesis

Development of the embryonic sex cells proceeds more rapidly in the female than in the male. By day 14 of pregnancy in the rat, the majority of such cells in the female fetus have entered meiotic prophase, so that these cells are the counterpart of premeiotic spermatocytes in the male, the first generation of which only appear some days after birth. The oocytes remain within primordial follicles, a finite population of cells of some hundreds of thousands of which only a few hundred to a few thousand, depending upon the species, will proceed to ovulation. There is no mechanism for replenishment in the eventuality of damage or destruction, in striking contrast to the extravagant, continual production of spermatozoa in the male. Approaches to fertility control in the female are necessarily very different.

The population of germ cell follicles undergoes a continuous partial development and atresia within the ovary throughout prepubertal and adult reproductive life. This is the fate of the vast majority and only a few progress to ovulation under the appropriate hormonal stimulus from the pituitary. The basic integrity of the ovary is dependent upon gonadotrophin, whereas ovulation follows additional cyclic surges of pituitary hormones. To the latter, only follicles in the appropriate stage of development within the endogenous ovarian cycle, are presumably able to respond.

The gonadotrophin stimulus to ovulation can be modified or inhibited by a variety of natural stimuli in rodents, as well as by pregnancy, which forms the basis of one major approach to fertility control. Pharmacological agents selectively affecting germ cell development in fetal, neonatal, and adult life could help to unravel the complexities of these obscure phases of reproductive physiology. For example, the fetal gonads show a differential sensitivity to certain alkylating chemicals (Hemsworth and Jackson, 1963a,b, 1965) which is related to the varying rates of development of the fetal sex cells in the two sexes. The gonads of both sexes during the first few days postpartum in the rat are damaged by exogenous androgen with the production of sterility (Barraclough, 1961), apparently due to interference with hypothalamic controlling mechanisms (Barraclough and Gorski, 1962).

The egg follicle when stimulated to develop for ovulation, undergoes a complex period of growth in which the germ cell does not divide. Surrounding follicular cells proliferate, fluid is secreted with cyst formation as the surface of the ovary is approached, and rupture finally occurs. Pituitary FSH apparently promotes follicular development, whereas the ovulation process and estrogen secretion by interstitial and thecal cells is believed to be due to the synergistic stimulus of FSH and LH. Progesterone secretion by the corpus luteum is dependent upon the stimulus of LTH. The latter hormone is produced spontaneously during the menstrual cycle of primates but requires the stimulus of mating in rodents, where estrous cycles normally prevail and the corpus luteum is otherwise nonfunctional. After sterile matings in rats (e.g., using vasectomized males) the activated corpora lutea remain functional for about 15 days, thus inducing a pseudopregnant state, before degeneration occurs with resumption of estrous cycles. Following a fertile mating, progesterone secretion is essential for the transport and implantation of the zygote, besides subsequent postimplantation development, where it is implemented by hormones from the placental trophoblast.

The precise roles of gonadotrophins on ovarian function are extremely difficult if not impossible to disentangle (Carter *et al.*, 1962), and the dual possibility of actions of ovarian hormones on the hypothalamic–pituitary mechanism, as well as directly on the ovary, add to the complexity. The fundamental effects of natural steroid hormones on pituitary and ovarian function are very relevant to the possible antifertility mechanisms of a variety of steroidal derivatives now available. Much experimental evidence exists (Shipley, 1962) of the ability of sex hormones, natural and synthetic, to inhibit overall gonadotrophin secretion. Although this is an indication of possible antifertility activity, details of the actions on individual gonadotrophins is not yet forthcoming. This may be particularly important in relation to the efficiency with which antigonadotrophic substances interfere with fertility.

The complexity of the situation has been increased by growth of knowledge of the controlling influence of hypothalamic centers upon pituitary activity, involving neurosecretory hormone(s) from hypothalamic cells (Harris, 1962). The existence of a "hypophysiotrophic area" (Halasz *et al.*, 1962) governing the functional gonadotrophic activity of the pituitary and a "mating center" located in the anterior hypothalamus (Barraclough and Gorski, 1962) have been suggested. Other experiments in the rat suggest that two hypothalamic regions control gonadotrophin release, one being responsible for the cyclic secretion of LH concerned with ovulation and the other with continuous LH release (Desclin *et al.*, 1962).

Stilbestrol implants in the posterior pituitary inhibit gonadotrophin secretion (Harris and Michael, 1958), and evidence has also been produced for a direct action of estradiol upon hypothalamic neurons (Lisk and Newlon, 1963). Blockade of ovulation also resulted from the application to the hypothalamic area in rabbits of the progestational steroid, norethindrone [19-norethisterone (Fig. 8, XIX); Kanematsu and Sawyer, 1965]. The specific region involved, the posterior median eminence, is known to control the release of ovulating hormone by the pituitary. Appropriate destructive lesions of the area cause irreversible atrophy of the ovaries. Implants of norethindrone were not associated with atrophic changes in ovary or uterus. Their blocking effect to ovulation later disappeared, presumably as the steroid was absorbed and metabolized. The likelihood is that chemoreceptive areas exist in the hypothalamus, which respond selectively to the natural steroid hormones. In addition, the susceptibility of central reproductive mechanisms to a wide variety of afferent impulses—visual, olfactory, and tactile, presents a situation of extraordinary complexity, the investigation of which is hampered by considerable species variation. These intricate pathways offer opportunities of interference by a variety of mechanisms. The chemical control of fertility over prolonged periods of time could involve these central pathways, the elucidation of which is of considerable importance.

The greater proportion of antifertility research has, so far, been devoted to exploring the potential of modifying the structure of sex hormones and their synthetic counterparts, on the basis of their ability to suppress gonadotrophin secretion and so inhibit ovulation. Relatively small changes in the chemical structure of steroid hormones can greatly alter potency and type of biological action. The same applies to various nonsteroidal, basically estrogenic compounds. In the development of antifertility substances along these lines, arises the possibility of multiple effects—antigonadotrophic, estrogenic, antiestrogenic, progestational, and antiprogestational—which may or may not contribute to a desirable result. The ability of various nonsteroidal, nonhormonal types of chemical to inhibit gonadotrophin secretion, e.g., dithiocarbamoylhydrazines and reserpine, are important indications toward the delineation of the effects of selective chemical pituitary inhibition, uncomplicated by peripheral actions of a steroidal nature.

VI. Steroids and Inhibition of Ovulation

Before ovulation, the peripheral role of estrogen is to condition the oviducts and the uterine mucosa for the subsequent action of progesterone, in preparation for transport of the egg and nidation. Centrally, this estrogen possibly induces the secretion of LH which actually triggers off ovulation. However, various estrogens, especially by repeated doses, are highly effective in preventing ovulation in both experimental animals and human subjects. In women, synthetic compounds such as ethinyl estradiol and stilbestrol effectively lower gonadotrophin output. The natural sex hormones, progesterone and testosterone (Fig. 8, XVII) are not particularly effective in preventing ovulation or inhibiting gonadotrophin secretion, although ovulation in the rabbit can be prevented by progesterone for many months using a subcutaneous pellet (Pincus, 1958).

The specific physiological role of progesterone upon the female reproductive pathway insures the safe transport, implantation, and development of the fertilized ovum. In passing, however, it is interesting to note that progesterone is toxic in low concentration (of the order of $10^{-5} M$) to cleaving mouse and rabbit ova *in vitro* (Daniel and Levy, 1964; Whitten, 1957). Much experimental evidence is available regarding other possible functions of progesterone in relation to mating and ovulation induction (see Kotz and Herrmann, 1961). In constant estrous rats, progesterone can induce ovulation and corpus luteum development, whereas evidence for both induction and inhibition of ovulation was found in the rabbit (Sawyer and Everett, 1959). During pregnancy (and pseudopregnancy), progesterone secretion presumably suppresses cyclic gonadotrophin secretion and, consequently, ovulation. The safety of progesterone coupled with the fact that, through repeated pregnancies, ovulation can be prevented for a considerable proportion of reproductive life, provided a physiological basis for the steroid approach via this mechanism, with orally effective progestational compounds. There has been a good deal of controversy over the mode of action, for, besides suppression of ovulation via pituitary mechanisms, there arise questions of direct action at the ovarian level or upon the zygote, or by the production of an unsuitable endometrial environment. Evidence concerning these questions has been discussed elsewhere (H. Jackson, 1964b; Mears, 1965b).

The broad aim of this field of research has been the enhancement of the gonadotrophin inhibitory facet of progestational activity by the oral route, with retention of the requisite peripheral activity. It seems, however, that the latter can be greatly increased without notable intensification of antigonadotrophic activity; there is little indication that the reverse is possible. The successful introduction some 10 years ago of progestational compounds derived from 19-nortestosterone for fertility control in women, involved the presence of traces of estrogen contaminant, augmented by metabolic conversion of the nortestosterone to estrogen. The cyclic courses of treatment used are thus combined estrogen–progestagen therapy, producing inhibition of ovulation and, after discontinuation of treatment, a withdrawal bleeding simulating the natural menstrual shedding. Further variations on this theme include continued treatment with the steroid combination to eventual breakthrough bleeding, perhaps after many weeks (Flowers, 1964) and the introduction of sequential treatment. In the latter, estrogen is first administered to suppress ovulation and "condition" the endometrium, followed by combined estrogen–progestagen for a short time to produce a secretory-type endometrium and to insure that withdrawal bleeding will occur. This procedure possesses the advantage of a more physiological approach as well as economy with the expensive progestagen. Its efficiency is rated high but more experience is required for an accurate overall assessment (Goldzieher, 1964; Mears, 1965b). Whatever the contraceptive procedure used, the general view seems to be that breakthrough ovulation occurs in about 10% of cycles, but the state of the endometrium is unreceptive to any blastocyst which might develop. As mentioned below, progestational compounds have now been developed which neither contain contaminant estrogen nor undergo metabolic conversion to this material. Nevertheless, they are relatively poor ovulation

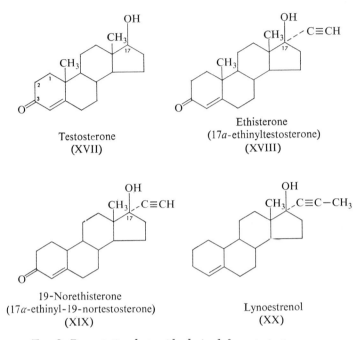

Fig. 8. Progestational steroids derived from testosterone.

inhibitors and concurrent or prior administration of estrogen is required to produce the desired therapeutic effect. Unlike estrogen, progestational steroids are expensive preparations, but estrogen alone is unsatisfactory as a contraceptive substance.

Chemical trends in the development of antifertility steroids over the years, first utilized as a basis the long-known progestational activity of 17α-ethinyltestosterone (ethisterone) (Fig. 8, XVIII). The essential step was the enhanced progestational activity with dissociation from androgenicity, which accompanied elimination of the angular methyl group designated C19 (i.e., the 17α-ethinyl-19-nor series of testosterone analogs). It is also established that the oxygen atom in position 3 of 17α-ethinyl-19-nortestosterone (Fig. 8, XIX) can be eliminated with retention of progestational potency, as in lynoestrenol (Fig. 8, XX).

Later followed the rational approach of stabilizing the progesterone molecule, in view of its very rapid metabolic conversion to pregnanediol. The enhanced progestational activity which accompanied elimination of the methyl group in position 19 of the ethinyl testosterone molecule was duly paralleled in the progesterone series, with the synthesis of 19-norprogesterone (Fig. 9, XXI). Possible antifertility application of the latter has not found favor, perhaps for economic reasons and its demonstrable

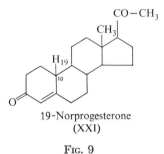

19-Norprogesterone
(XXI)

FIG. 9

carcinogenicity in mice (Lipschutz *et al.*, 1962). Instead, 17α-hydroxy-progesterone (Fig. 10, XXIII) derivatives have been developed (Fig. 10, XXIV and XXV). This compound (XXIII) lies on the biosynthetic pathway in which testosterone and estrogen are produced from progesterone (Fig. 10, XXII) with cholesterol as starting material. Although 17-hydroxyprogesterone is inactive, certain of its esters inhibit fertility in ani-

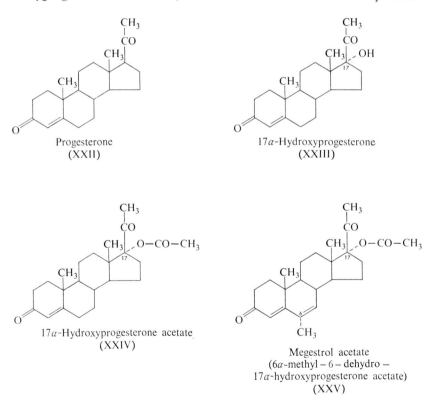

Progesterone
(XXII)

17α-Hydroxyprogesterone
(XXIII)

17α-Hydroxyprogesterone acetate
(XXIV)

Megestrol acetate
(6α-methyl – 6 – dehydro –
17α-hydroxyprogesterone acetate)
(XXV)

FIG. 10. Progestational steroids derived from progesterone.

mals when given by mouth (H. S. Bryan, 1960; Harris and Wolchuk, 1963). Further modifications have enhanced both stability and potency, especially with the introduction of methyl in position 6α (i.e., spatially above the plane of the rings), and additional unsaturated linkages in rings A and B. It has been concluded from experimental studies with progestational substances that there is little or no relation between progestational and antigonadotrophic activity (Ercoli and Falconi, 1961; Falconi and Bruni, 1962). It may be that progestational agents have differential effects upon FSH and LH release or show varying degrees of antagonism peripherally to estrogen and perhaps to progesterone. Even toward the uterine mucosa, differential effects have been designated by use of the terms "progestagen" and "gestagen" (Overbeek et al., 1962). Thus compounds such as norethynodrel, norethisterone, and lynoestrenol are progestational by the Clauberg assay (i.e., progestagenic) but unable to support pregnancy in the rat or the decidual cell response in the mouse.

Numerous modifications in the steroid structure are possible, and results from tests for ovulation inhibition in the rabbit with 187 analogs of progesterone and 19-nortestosterone gave activity in over 30% (Pincus and Merrill, 1961). In this series, certain antifertility steroids in common use (norethisterone and norethynodrel) exhibited enhanced activity by mouth, suggesting possible conversion to more potent material. Surprisingly, the 3-methyl ether of ethinylestradiol (a potent estrogen) produced no notable effect by this test, although there is general agreement on the efficiency with which this substance and ethinyl estradiol inhibit ovulation in women. This may well be a reflection of the importance of timing of treatment with estrogen in relation to preovulatory events. In human treatment, compounds are administered from early days of the menstrual cycle and the high cumulative potency of estrogen has been referred to in this respect.

In another study of the antiovulatory, progestational, and antigonadotrophic activity of thirty-six steroids (Kincl and Dorfman, 1963), 19-norprogesterone was the most potent antiovulatory compound, being 150 times as active as subcutaneous progesterone in the rabbit. Although a number of other 17α-hydroxyprogesterone derivatives were highly active antiovulatory agents, it was noted that this property was not necessarily related to progestational effectiveness nor to antigonadotrophic activity by the rat parabiosis test. Thus, 19-norprogesterone was 25 times more active in preventing ovulation than as a progestational agent; perhaps the explanation lies in a combination of mechanisms operative at both pituitary and ovarian levels. By the oral route, the progestationally potent 6-chloro-6-dehydro-17α-acetoxyprogesterone (Chlormadinone) was the most active antiovulatory steroid of the group, averaging about 35 times the activity of norethisterone. A number of nonprogestational steroids showed antiovulatory activity in the rabbit, including 2-hydroxymethyl-17α-methyl-17β-hydroxyandrostan-3-one, which was twice as active as norethisterone. Ethinylestradiol-3-methyl ether (mestranol) showed less than one-third of the antiovulatory activity of norethisterone, whereas 17α-acetoxyprogesterone was inactive by this test. The lack of correlation between antigonadotrophic and antiovulatory activities was notable, particularly with mestranol and 6α-chloro-6-dehydro-17α-acetoxyprogesterone. Whereas a few micrograms of the estrogen inhibited pituitary gonadotrophin in the rat test, 1 mg failed to inhibit ovulation in the rabbit. On the other hand, the progesterone derivative did not inhibit gonadotrophin in a 20-mg dose, but 0.3 mg was antiovulatory. In general, the compounds were less active by mouth with the exception of the nonprogestational compound, 2-hydroxymethyl-17α-methyl-17β-hydroxyandrostan-3-one, which was 2.5 times as effective by gavage.

Another facet of progestational activity, which is used as an index of antifertility potential, is the ability of a compound to "support" the endometrium and postpone normal menstrual shedding (Greenblatt *et al.*, 1958; Swyer *et al.*, 1960; Mears, 1961). For example, norethisterone may thus be effective for many months although, in general, there is a tendency for the action of the steroid to be overcome and breakthrough bleeding to occur. This varies with different compounds and from one subject to another. Certain antifertility progesterone derivatives are ineffective by this test (Greenblatt *et al.*, 1958) even with added estrogen, and the relative potency of progesterone and testosterone derivatives has been compared (Swyer and Little, 1962a,b). Enhancement of activity by added estrogen in this test, as well as in oral contraception, is striking. It is generally agreed that the dose of estrogen alone is sufficient to inhibit ovulation, and the role of the progestational compound is more of a subsidiary nature, enhancing ovulation inhibition and providing progestational support to the endometrium.

In women, therefore, the contraceptive mechanism may involve, besides suppression of the actual ovulatory process, interference with earlier stages of development of the preovulatory egg follicle. It is also recognized that ovulation may occur in spite of steroid administration, in which event reception of the ovum is prevented by an inadequate endometrium due to exposure to the drugs. It has also been suggested that the administration of steroid combinations blockades formation and/or release of LH and renders the cervical mucus hostile to spermatozoa (Diczfalusy, 1965) and the consecutive treatment suppresses release of FSH and LH, thereby inhibiting ovulation.

The present contraceptive procedure thus relies on cyclic treatment with estrogen and progestational steroid (concurrent or sequential administration of the two hormonal agents). The choice of progestational agent is a question of clinical trial and error, mainly dependent upon the natural hormone balance of the subject and the corresponding estrogenic or progestational tendencies of the steroid preparation (Mears, 1965a). A few patients pass into a state of amenorrhea, which circumstance may remain perfectly acceptable (M. C. N. Jackson, 1963). Perhaps this group represents the nearest approach to the idea of a contraceptive mechanism involving sustained inhibition of ovulation and menstruation, as in pregnancy, by an oral progestational compound.

Nevertheless, the present steroid method has been an outstanding success in spite of limitations imposed by technique and economics. Anxieties have been expressed (e.g., Dodds, 1961) concerning possible hazards from the long-term administration of these potent agents and the sustained disturbance of endocrine mechanisms. Although the results from careful clinical supervision over 10 years are reassuring and the antifertility effect remains rapidly reversible, crucial years lie ahead. It is also possible that the steroid contraceptive method may be of prophylactic value regarding the development of cancer of the reproductive organs and accessory tissues. Unlike most other chemical agents, these steroids are remarkable in their lack of acute systemic toxicity and lethality.

A considerable literature has formed on the various facets of steroidal agents in relation to oral contraception and the reader is referred to recent publications for further details and discussion (Pincus, 1965; Jackson, 1964b; Mears, 1965b).

VII. Interference with Postovulatory Events

Experimental data in laboratory rodents have demonstrated the existence of timed mechanisms in relation to transport, development, and implantation of the fertilized egg, suggesting that these phases provide approaches to fertility control. It must be borne in mind, however, that there are wide species variations in the timing of postovulatory events reflecting considerable quantitative variations in the operative mechanisms—presuming these latter are qualitatively similar and dependent upon the same hormones in mammalian species. Conclusions drawn from experimental animals may have limited validity in the human subject. The role of estrogen in contraceptive preparations prompted further research into the effects of substances of this nature upon postovulatory events. Subsequently the possibility arose of fertility control by the use of nonsteroidal antiestrogens and now possible antiprogestational steroids are also being sought—there are as yet no nonsteroidal chemicals operating by this latter mechanism.

After ovulation and fertilization, the ovum is actively transported and nurtured through the oviducts into the uterus, developing as it proceeds. In the rat and mouse this period occupies about one-fifth of the gestation time. The relationship between progression of the egg and endometrial development under the influence of the now functional corpus luteum, has been investigated using the technique of ovum transfer. Donor eggs are transferred into the reproductive pathway of recipients previously stimulated (e.g., by mating with vasectomized males) in order to insure functional corpora lutea. The age of the eggs at transfer and that of the conditioned endometrium of the recipient are dated by precise knowledge of the time of mating in each case. In such experiments, transfer is successful in high percentage when the ages correlate. In the mouse, ova transferred 3½ days postinsemination into females 2½ days after copulation gave a high yield of pregnancies (MacLaren and Michie, 1956); the reverse gave a low yield. According to others, transfers into the rat uterus of ova immature relative to the age of the uterine mucosa, were strikingly less successful than those where the developmental ages were similar (synchronous transfers) or where the eggs were a day "older" than the uterus (Noyes *et al.*, 1963). Similar qualitative results prevailed in the mouse, with transfer of eggs less developed than the uterus yielding few term fetuses. These authors believe the processes controlling specific stages of ovular and endometrial development to be independent phenomena. Thus the hormonal environment of the uterus controls its development to a point where appropriately developed or overdeveloped ova are stimulated to become attached to the endometrium and, thus, provoke the decidual cell response (see Section VII,B). Beyond this time (day 5 in the rat), ova and endometrium rapidly lose the ability to take part in an implantation process. In particular, the relatively overdeveloped endometrium is inhospitable to rat and mouse zygotes. Relevant to this subject is the survival of eggs at the blastocyst stage in the uterine cavity of rodents ovariectomized 3 days after mating and maintained on progesterone. The viability of such zygotes with time is readily assessed through implantation and development induced by a small dose of estrogen. In fact, the blastocyst can survive for considerable periods under these artificial conditions, comparable with the physiological state of delayed implantation which occurs in a variety of species.

Cleavage of mouse and rabbit ova *in vitro* was inhibited by concentrations of estradiol or progesterone of the order of 10^{-4} to $10^{-5} M$, respectively, although after a critical time, mitosis occurred in spite of addition of hormone (Daniel, 1964; Daniel and Levy, 1964). The inhibitory

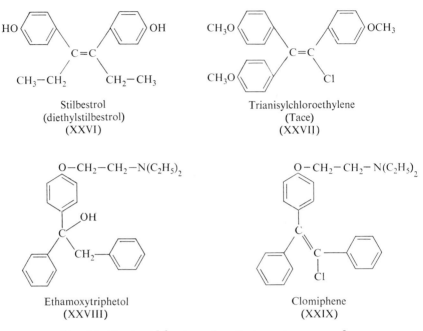

FIG. 11. Nonsteroidal estrogenic-antiestrogenic compounds.

action was prevented by addition of amino acids or raising the serum content of the media to over 30%. Blastocysts in the mouse still exhibited sensitivity to progesterone but apparently not those of the rabbit. Cleavage of the zygote of the sea urchin, *Arbacia punctulata,* was inhibited by estradiol $(3 \times 10^{-3} M)$, and the effect was associated with reduction in deoxyribonucleic acid (DNA) but little change in ribonucleic acid (RNA) synthesis (Jolley *et al.,* 1962). Cleaved ova were destroyed by stilbestrol (Fig. 11, XXVI) or hexestrol $(10^{-5}-10^{-4} M)$ but not by trianisylchloroethylene (Fig. 11, XXVII) (Segal and Tyler, 1958). This latter substance is believed to be activated by metabolism in the mammalian liver. A nonsteroidal antiestrogen, ethamoxytriphetol (Fig. 11, XXVIII) $(2 \times 10^{-4}\%)$, added 5 minutes after fertilization also completely inhibited division, but the effect was reversed by washing. The antifertility action in the rat of the antiestrogen, Clomiphene (Fig. 11, XXIX) (a similar antiestrogen) (Segal and Davidson, 1962) is also considered a direct antizygotic action (Nelson *et al.,* 1963; Schlough and Meyer, 1965).

In vivo, the oviductal stage is readily disturbed by administration of estrogen, but sensitivity to this hormone has waned once the uterus is reached (Dreisbach, 1959; Edgren and Shipley, 1961; Emmens, 1962; Emmens and Finn, 1962). Eggs may be greatly hurried or hindered in their progress according to the dose of hormone and the species (Green-wald, 1961a,b). The accelerating mechanism alone can inhibit fertility for reasons referred to above, with the possibility of antagonism to some other aspect of progesterone action. The antifertility effects of estrogen and estrogenic compounds as exemplified in rodents seem obvious and attractive possibilities. According to a recent study, estradiol-17β or mestranol (0.45 μg daily for 7 days from the day of proestrus) completely inhibited pregnancy and implantation sites were not found (Kincl and Dorfman, 1965). However, the absence of reports on the successful use of estrogens as postcoital agents is perhaps indicative of a failure to produce analogous results in the human subject.

A. Antiestrogenic and Antiprogestational Compounds

The combined role of estrogen and progesterone in the nutrition, transport, and implantation of the developing zygote also provides prospects of fertility inhibition via antagonism to these hormones. Details of the target cell mechanism of hormone action in relation to these aspects of physiological control have to be unravelled. Experiments have implicated estrogen in specific mechanisms controlling the activities of endometrial cells by influencing the synthesis of messenger RNA (Talwar and Segal, 1963; Segal *et al.*, 1965). A macromolecular cell fraction was isolated from the rat uterus which inhibits RNA polymerase. This fraction also binds estradiol as a result of which the inhibition is reversed. On the evidence presented it is suggested that the stimulatory role of estrogen in the uterus may be limited to this primary releasing mechanism. The estrogen activation apparently produces both qualitative and quantitative changes in uterine RNA synthesis. Estrogen is also believed to be an essential factor in the postulated histamine mechanism concerned in nidation (Marcus *et al.*, 1964). In rodents, various nonsteroidal, antiestrogenic substances possess remarkable antifertility actions especially related to the oviductal phase. Basically these compounds are related to the triphenylethylene series, of which trianisylchloroethylene (Tace) (Fig. 11, XXVII) is a well-known but feeble synthetic estrogen. The first notable antagonist along these lines was ethamoxytriphetol (MER-25) (Fig. 11, XXVIII) (Segal and Nelson, 1958), later followed by Clomiphene (Fig. 11, XXIX) (Segal and Nelson, 1961; Barnes and Meyer, 1962). More recent chemical modifications have involved ring closures, substantially modifying the underlying triphenylethylene pattern but producing substances of high potency in the present context. The antifertility effectiveness of these compounds in the rat is mainly limited to the first 4 days postinsemination, corresponding to tubal transport of the fertilized ovum. After day 4, when the blastocysts are within the uterine cavity, sensitivity

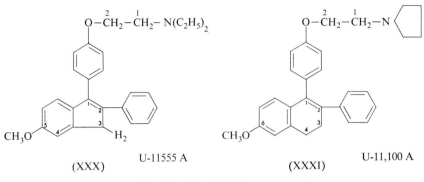

FIG. 12. "Cyclized" triphenylethylene postcoital agents.

is largely lost, recalling the diminishing susceptibility to estrogen. In rats, ethamoxytriphetol was feebly antigonadotrophic (Lerner *et al.*, 1958) although the pituitary hypertrophy induced by estradiol valerate could be blocked by concurrent treatment with the antiestrogen (Cutler *et al.*, 1961), with resulting increase in pituitary content of gonadotrophin. The major action of Clomiphene in the same species is considered to be antigonadotrophic with suppression of ovulation (Holtkamp *et al.*, 1961). This compound did not antagonize the action of exogenous gonadotrophin nor produce increase in pituitary weight or gonadotrophin content. There are also indications that ova are destroyed by the drug during passage through the oviduct (Nelson *et al.*, 1963; Schlough and Meyer, 1965; Prasad *et al.*, 1965). The antifertility effect of Clomiphene could not be explained in terms of its antigonadotrophic and antiestrogenic effects.

Chronic administration of these antiestrogens to rats produced diestrus smears, although occasional cycles occurred, and reduced ovarian weight (Barnes and Meyer, 1962); the effects were rapidly reversible. In women, Clomiphene unexpectedly stimulated ovulation (Greenblatt, 1961; Greenblatt *et al.*, 1961). This action draws attention to the importance of species variation, although in this instance such an effect does not necessarily mitigate against a postovulatory application. Administered after mating, the "cyclized" triphenylethylenes (Fig. 12, XXX and XXXI) were remarkably potent fertility inhibitors, being effective in the mouse, rat, rabbit, and guinea pig although not in the hamster (Duncan *et al.*, 1962, 1963). In minimal antifertility dosage the compounds did not show significant antigonadotropic, estrogenic, androgenic, or antiprogestational activity. Throughout chronic treatment, mating activity occurred whereas estrous cycles and fertility were rapidly restored when administration ceased. It has been further shown that these compounds, in antifertility doses, did not interfere with the survival of rat blastocysts in Provera-treated animals. These could still be induced to implant by estrogen or be transferred successfully to new hosts (Duncan and Forbes, 1965). Administered concomitantly with estrogen, these estrogen antagonists could prevent implantation, which process is much more sensitive to these agents than other estrogen-dependent functions associated with the estrous cycle. The antifertility mechanism favored is interference with the estrogen-sensitivity phase of implantation rather than a direct effect on the blastocyst or changes in tubal transport.

These developments emphasize the possibility of producing anti-fertility compounds which could be free from the hazard of significant and persistent disturbance of the overall endocrine balance. They might also avoid the necessity of regular cyclic administration. Species variations provide a familiar obstacle to progress, as manifest, for example, by the resistance of the guinea pig to these cyclized triphenylethylenes and the fact that ovulation inhibition in one species becomes ovulation stimulation in another.

The availability of relatively simple synthetic estrogens, manifesting all the pharmacological actions of the natural steroid hormones, facilitates investigations of structural modifications on a scale impracticable with the complex steroid molecule. Since there are as yet no nonsteroidal progestational substances, the possibility of interrupting postovulatory events by a mechanism of target end-organ competition must rely upon modifications of the steroid structure. Antiprogestational activity has been assessed by ability to prevent implantation (Banik and Pincus, 1962). Besides conventional steroid derivatives, compounds produced by deletion of a carbon atom in ring A (A-nor steroids) or the introduction of heteroatoms (e.g., N or S) at appropriate points in the steroid structure are being explored. Steroid derivatives inhibiting early pregnancy included a 19-nor steroid diacetate and an A-nor steroid (Fig. 13, XXXIII and XXXII). Blastocysts recovered from the rat on day 4 postinsemination appeared normal and the antifertility action was not reversed by progesterone. Mode of action will have to be carefully defined before the term "antiprogestational" may properly be applied. Postovulatory events move smoothly and synchronously for both zygote and endometrium under precise hormonal control, so that delineation of the action of chemical agents is very difficult. Timed administration is no guarantee that interference bears a direct relation to reproductive events occurring about the same period since the pharmacological effect may be delayed.

Antifertility studies in the rat with fifty-three steroids administered from the day of proestrus for 7 days (Kincl and Dorfman, 1965) showed that activity frequently correlated with estrogenicity. However, several

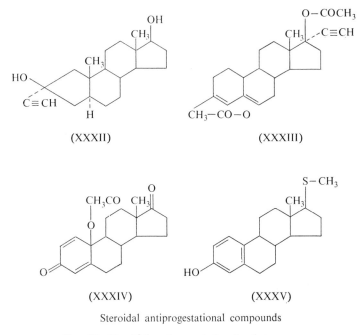

Steroidal antiprogestational compounds

FIG. 13. Steroidal antiprogestational substances.

tetrahydropyranyl ethers derived from estradiol lacking either a 3 or 17 oxygen atom showed a separation of these two parameters. The most active oral compound compared with mestranol was 3-methoxy-17β-cyanethoxyestra-1,3,5(10)-triene which showed a sixtyfold increase in antifertility activity compared with estrogenicity. The results are considered to indicate the possibility of producing a relatively effective steroidal antifertility compound of low estrogenicity.

B. COMPOUNDS AFFECTING IMPLANTATION

The series of complex events following the encounter between blastocyst and uterine wall at the correct time and site offers opportunities for pharmacological action. Such mechanism would be more properly defined as "interference with implantation" than impairment of the development of the zygote or endometrium at earlier stages, which also results in failure to implant. In bare outline the process of nidation proceeds as follows. Contact between the blastocyst and endometrium provokes a reaction in the latter, manifested by the formation of decidual tissue composed of characteristic cells derived by transformation of stromal uterine cells. This constitutes the primary response in the maternal contribution to the placenta. Much of the work on early stages of the implantation

mechanisms utilized the analogous decidual cell response (DCR) produced in pseudopregnant rats. This reaction follows a traumatic stimulus
to the uterus of female rats mated 4 days previously with vasectomized
males (or where the cervix was mechanically or electrically stimulated)
thereby producing functional corpora lutea.

The implantation mechanism is considered to be biphasic in character, the primary decidual reaction being followed by an invasive process
leading to engulfment of the blastocyst (Shelesnyak, 1963). The initial
response of the endometrium is thought to be associated with a surge of
estrogen secretion from the ovary, which sensitizes endometrial cells to
local decidualizing influences, presumably triggered off in some way by
the blastocyst (Shelesnyak, 1960). As would be expected, therefore, the
DCR can be prevented by early ovariectomy or injection of the estrogen
antagonist, ethamoxytriphetol (Shelesnyak, 1962). The pituitary stimulus
causing the estrogen surge is believed to be discharged on night 3 postinsemination (Alloiteau, 1961) by a process which is clocklike in precision. The mechanism is presumed to be "set" by the act of mating (or
other cervical stimulus) but only operates after the requisite delay, when
the blastocyst is expected to reach the endometrium. It will be recalled
that estrogen as well as progesterone need to be administered to mated
rats ovariectomized 4 days postinsemination, in order for implantation to
occur. If ovariectomy is delayed beyond this time, then progesterone only
is required for implantation to proceed.

Daily administration of tranquillizing substances (reserpine, chlor-
promazine, Stelazine) after mating will delay implantation in rats and
the effect can be terminated by a minute dose (0.1 μg) of estradiol
(Psychoyos, 1963), but not if treatment is delayed to day 4. The drug
interference is presumed to operate at central levels and only be effective
if given before the hypothalamic–pituitary interaction induces a discharge of gonadotrophin. A similar action was produced by surgical interruption of connections between hypothalamus and pituitary, reported
to result in LH and LTH release (Everett, 1954). The suggestion is,
therefore, that an estrogen surge occurs consequent on pituitary hormone
stimulation of the ovary, about 18 hours prior to the time of maximum
sensitivity of the uterus to decidualization; however, evidence against this
theory has been presented. Thus, in the rat, the time of maximum sensitivity was unaffected by pentobarbitone administration, known to block
LH release. Either the barbiturate treatment was inadequate or a CNS
surge mechanism affecting uterine sensitivity did not occur (DeFeo,
1963). Furthermore, in castrated rats (without cervical stimulation), the
events of maximal sensitivity to decidualization and its subsequent loss
could be produced by a constant ratio of estrogen–progesterone admin-

istration (Yochim and DeFeo, 1963). The uterine sensitivity to decidualization was dated to the previous ovulation rather than to cervical stimulation.

Histamine release from mast cells within the endometrium has been demonstrated (Spaziani and Szego, 1958; Shelesnyak, 1959). Depletion of mast cells from the uterine wall is said to prevent decidualization and, therefore, nidation (Kraicer et al., 1963). It may be questioned how close the analogy is between the events of pseudopregnancy and gestation. According to a recent study the hormonal influences are not identical in the two circumstances (Marcus et al., 1964). No release of histamine occurred after the estrogen surge of pseudopregnancy, even when supplemented by additional estrogen (estradiol or estriol, up to 40 μg). The absence of seminal fluid, spermatozoa, and, of course, the blastocyst are obvious differences between the two states, which might play some part.

The view that histamine or some similar substance provides a final link in the chain of events has also been questioned. The histamine hypothesis grew around the idea that tissue injury was involved in implantation and supported by the observation that atraumatic introduction of histamine into the "conditioned" uterus induced the DCR (Shelesnyak, 1957). Other work has failed to show that histamine produced any greater DCR response than saline alone. The failure of the antihistamine drug, mepyramine, to influence the response suggested the need for further work into the postulated role of histamine (Finn and Keen, 1962). In other experiments, pharmacological reduction of the mast cell population in the rat uterus by a chemical designated 48/80 did not affect the incidence of pregnancy in rats, although effective in mice (Banik et al., 1963).

A range of steroidal hormones (e.g., estrogen, testosterone, cortisone) has long been known to inhibit the DCR, presumably by antagonism to progesterone (Hisaw et al., 1954). Extensive studies of the mode of action of ergotoxine alkaloids, particularly ergocornine, in terminating pregnancy and pseudopregnancy in rats have failed to establish the mechanism, even though it is reversed by progesterone. Earlier data suggested a disturbance of the hormone balance operating via the pituitary; the hypothalamus was excluded from participation in the action of the alkaloids (Carlsen et al., 1961). A proposal now is that the active material may be a metabolic product to ergocornine (Shelesnyak and Barnea, 1963). A marked fall in urinary pregnandiol which followed a small dose of ergocornine to women postovulation, was interpreted as possible interference with progesterone biosynthesis (Shelesnyak et al., 1963). Ergokryptine has recently been assessed the most effective of four ergot alkaloids tested for termination of pregnancy on day 4 postmating in the rat (minimal effective doses, 175 μg per rat) (Kraicer and Shelesnyak, 1965).

Chemical termination of gestation in its early stages could be a useful approach to fertility control and from the point of view of interference with maternal mechanisms, the DCR response forms a useful screening system. Earlier studies on the interruption of pregnancy in experimental animals, mainly by cytotoxic agents and antimetabolites, have been surveyed (Jackson, 1959, 1964b) and little has been added in this direction. Other work has utilized the rationale of interference with central mechanisms along the lines of ovulation inhibition and pregnancy suppression by reserpine (Gaunt *et al.*, 1955; Khazan *et al.*, 1960; Hopkins and Pincus, 1963). The actions of reserpine are associated with depletion of 5-hydroxytryptamine (5-HT) from areas rich in this substance, e.g., brain and intestinal tract. So, the possible effects of 5-HT and amine oxidase inhibitors (AOI), which normally antagonize the enzyme degrading 5-HT were studied.

5-Hydroxytryptamine interrupts pregnancy in the rabbit, rat, and mouse, particularly in the later stages when it causes placental hemorrhage. In the mouse the effect is prevented by administration of progesterone or prolactin only in the early stages of gestation, for the later action is considered to be directly upon the placental circulation (Poulson and Robson, 1963). Also 5-HT caused degeneration of formed decidual tissue in mice, which was prevented by progesterone (Poulson *et al.*, 1960).

Various AOI terminated pregnancy in rats until mid-term (Lindsay *et al.*, 1961), which action was also antagonized by progesterone. Evidence suggested the ovary was necessary for the interfering mechanism (Lindsay *et al.*, 1962). Treatment with methysergamide and cyproheptadine (both histamine and 5-HT antagonists) countered the actions of 5-HT later in pregnancy (Poulson and Robson, 1963). The pituitary inhibitor, methallibure, terminated pregnancy in rats at any stage when administered on three successive days (100 mg/kg) (Paget *et al.*, 1961), but the mechanism is not known. With inadequate treatment, surviving offspring were normal and subsequent fertility unimpaired.

The pharmacology of interference with implantation and early pregnancy is still in early stages and is a difficult field of investigation.

Overall, an occasional brief disorganization of the complex events between ovulation and the early implantation process may be a preferable and more generally practicable approach to fertility control than the present methods. It is clearly necessary to obtain greater information of these intricate and sensitive mechanisms. Far greater effort is required in the study of the effects of chemical agents on reproductive processes in many species, in relation to the control of fertility and, hence, of biological populations.

REFERENCES

Allanson, M., and Deansley, R. (1962). *J. Endocrinol.* **24**, 453.

Alloiteau, J. J. (1961). *Compt. Rend. Acad. Sci.* **253**, 1348.

Banik, U. K., Kobayashi, Y., and Ketchel, M. M. (1963). *J. Reprod. Fertility* **6**, 179.

Banik, U. K., and Pincus, G. (1962). *Proc. Soc. Exptl. Biol. Med.* **111**, 595.

Barnes, L. E., and Meyer, R. K. (1962). *Fertility Sterility* **13**, 472.

Barraclough, C. A. (1961). *Endocrinology* **68**, 62.

Barraclough, C. A., and Gorski, R. A. (1962). *J. Endocrinol.* **25**, 175.

Beyler, A. L., Potts, G. O., Coulston, F., and Surrey, A. R. (1962). *Endocrinology* **25**, 221.

Blye, R. P., and Berliner, V. R. (1963). *Am. Chem. Soc. Abstr.*, 31L, March–April.

Boccabella, A. V. (1963). *Endocrinology* **72**, 787.

Bock, M., and Jackson, H. (1955). *Nature* **175**, 1037.

Bock, M., and Jackson, H. (1957). *Brit. J. Pharmacol.* **12**, 1.

Bryan, H. S. (1960). *Proc. Soc. Exptl. Biol. Med.* **105**, 23.

Bryan, W. L., and Mason, K. E. (1940–1941). *Am. J. Physiol.* **131**, 263.

Carlsen, R. A., Zeilmaker, G. H., and Shelesnyak, M. C. (1961). *J. Reprod. Fertility* **2**, 369.

Carter, F., Woods, M. C., and Simpson, M. E. (1962). *In* "Control of Ovulation" (C. A. Villee, ed.), p. 251. Macmillan (Pergamon), New York.

Cattanach, B. M. (1959). *Intern. J. Radiation Biol.* **1**, 228.

Cattanach, B. M., and Edwards, R. G. (1958). *Proc. Roy. Soc. Edinburgh* **B67**, 54.

Chang, S. C., Terry, P. H., and Borkovec, A. B. (1964). *Science* **144**, 57.

Chiquoine, A. D. (1963). *Anat. Record* **145**, 216.

Chiquoine, A. D., and Suntzeff, V. (1965). *J. Reprod. Fertility* **10**, 455.

Clermont, Y., Leblond, C. P., and Messier, B. (1959). *Arch. Anat. Microscop. Morphol. Exptl.* **48**, 37.

Coulston, F., Beyler, A. L., and Drobeck, H. P. (1960). *Toxicol. Appl. Pharmacol.* **2**, 715.

Cranston, E. M. (1961). *Endocrinology* **69**, 331.

Cutler, A., Ober, W. B., Epstein, T. A., and Kupperman, H. S. (1961). *Endocrinology* **69**, 473.

Daniel, J. C., Jr. (1964). *Endocrinology* **75**, 706.

Daniel, J. C., Jr., and Levy, J. D. (1964). *J. Reprod. Fertility* **7**, 323.

DeFeo, V. J. (1963). *Endocrinology* **72**, 305.

Desclin, L., Flament-Durand, J., and Gepts, W. (1962). *Endocrinology* **70**, 429.

Diczfalusy, E. (1965). *Brit. Med. J.* **2**, 1394.

Dodds, E. C. (1961). *J. Endocrinol.* **23**, i.

Dorfman, R. I. (1962). *In* "Methods in Hormone Research" (R. I. Dorfman, ed.), Vol. II, p. 113. Academic Press, New York.

Dorfman, R. I. (1963). *In* "Perspectives in Biology," p. 42. Elsevier, Amsterdam.

Dreisbach, R. H. (1959). *J. Endocrinol.* **18**, 271.

Drobeck, H. P., and Coulston, F. (1962). *Exptl. Mol. Pathol.* **1**, 251.

Duncan, G. W., and Forbes, A. D. (1965). *J. Reprod. Fertility* **10**, 161.

Duncan, G. W., Stucki, J. C., Lyster, S. C., and Lednicer, D. (1962). *Proc. Soc. Exptl. Biol. Med.* **109**, 163.

Duncan, G. W., Lyster, S. C., Clark, J. J., and Lednicer, D. (1963). *Proc. Soc. Exptl. Biol.* (*N. Y.*) **112**, 439.

Edgren, R. A., and Shipley, G. C. (1961). *Fertility Sterility* **12**, 178.

Emmens, C. W. (1962). *J. Reprod. Fertility* **3**, 246.

Emmens, C. W., and Finn, C. A. (1962). *J. Reprod. Fertility* **3**, 239.

Ercoli, A., and Falconi, G. (1961). *Biochem. Pharmacol.* **8**, 103.

Erickson, A. E., and Pincus, G. (1964). *J. Reprod. Fertility* **7**, 379.

Everett, J. W. (1954). *Endocrinology* **54**, 685.

Falconi, G., and Bruni, G. (1962). *Proc. Soc. Exptl. Biol.* **108**, 3.

Finn, C. A., and Keen, P. M. (1962). *Nature* **194**, 602.

Flowers, C. E. (1964). *J. Am. Med. Assoc.* **188**, 115.

Fox, B. W., and Jackson, H. (1965). *Brit. J. Pharmacol.* **24**, 24.

Fox, B. W., Jackson, H., Craig, A. W., and Glover, T. D. (1963). *J. Reprod. Fertility* **5**, 13.

Fox, B. W., Partington, M., and Jackson, H. (1964). *Exptl. Cell Res.* **33**, 78.

Gaines, T. B., and Kimbrough, R. D. (1964). *Bull. World Health Organ.* **31**, 737.

Gaunt, R. R., Renzi, A. A., and Chase, J. J. (1955). *J. Clin. Endocrinol.* **15**, 621.

Goldberg, G. M., Pfau, A., and Ungar, H. (1959). *Am. J. Pathol.* **35**, 649.

Goldzieher, J. W. (1964). *Med. Clin. N. Am.* **48**, 529.

Greenblatt, R. B. (1961). *Fertility Sterility* **12**, 402.

Greenblatt, R. B., Jungck, E. C., and Barfield, W. E. (1958). *Ann. N. Y. Acad. Sci.* **71**, 717.

Greenblatt, R. B., Barfield, W. G., Jungck, E. C., and Ray, A. W. (1961). *J. Am. Med. Assoc.* **178**, 101.

Greenwald, G. S. (1961a). *Endocrinology* **69**, 1068.

Greenwald, G. S. (1961b). *Fertility Sterility* **12**, 80.

Gunn, S. A., Gould, T. C., and Anderson, W. A. D. (1963). *Am. J. Pathol.* **42**, 685.

Halasz, B., Pupp, L., and Uhlarik, S. (1962). *J. Endocrinol.* **25**, 147.

Harris, G. W. (1962). *In* "Control of Ovulation" (C. A. Villee, ed.), p. 56. Macmillan (Pergamon), New York.

Harris, G. W., and Michael, R. P. (1958). *J. Physiol.* (*London*) **142**, 26.

Harris, T. W., and Wolchuk, N. (1963). *Am. J. Vet. Res.* **24**, 1003.

Hayes, W. J. (1964). *Bull. World Health Organ.* **31**, 721.

Heller, C. G., and Clermont, Y. (1964). *Recent Progr. Hormone Res.* **20**, 545.

Heller, C. G., Laidlaw, W. M., Harvey, H. T., and Nelson, W. O. (1958). *Ann. N. Y. Acad. Sci.* **71**, 649.

Heller, C. G., Moore, D. J., Paulsen, C. A., Nelson, W. O., and Laidlaw, W. M. (1959). *Federation Proc.* **18**, 1057.

Heller, C. G., Moore, D. J., and Paulsen, C. A. (1961). *Toxicol. Appl. Pharmacol.* **36**, 1.

Heller, C. G., Flayolle, B. Y., and Matson, L. J. (1963). *Exptl. Mol. Pathol.* Suppl. **2**, 107.

Hemsworth, B. N., and Jackson, H. (1963a). *J. Reprod. Fertility* **5**, 187.

Hemsworth, B. N., and Jackson, H. (1963b). *J. Reprod. Fertility* **6**, 229.

Hemsworth, B. N., and Jackson, H. (1965). *In* "Biological Council Symposium on Embryopathic Activity of Drugs" (J. M. Robson, F. Sullivan, and R. L. Smith, eds.), pp. 305–319. Churchill, London.

Hemsworth, B. N., Jackson, H., and Walpole, A. L. (1968). *J. Endocrinol.* **40**, 275.

Hisaw, F. L., Velardo, J. T., and Ziel, H. K. (1954). *J. Clin. Endocrinol.* **14**, 763.

Holtkamp, D. E., Davis, R. H., and Rhoads, J. E. (1961). *Federation Proc.* **20**, 419.

Hopkins, T. F., and Pincus, G. (1963). *Endocrinology* **73**, 775.

Howell, J. McC. (1965). *In* "Biological Council Symposium on Agents Affecting Fertility" (C. R. Austin and J. S. Perry, eds.), 319 pp. Churchill, London.

Howell, J. McC., Thompson, J. N., and Pitt, G. A. J. (1963). *J. Reprod. Fertility* **5,** 159.

Jackson, H. (1959). *Pharmacol. Rev.* **11,** 135.

Jackson, H. (1964a). *Brit. Med. Bull.* **20,** 107.

Jackson, H. (1964b). *Fortschr. Arzneimittelforsch.* **7,** 134.

Jackson, H. (1965). *In* "Biological Council Symposium on Agents Affecting Fertility" (C. R. Austin and J. S. Perry, eds.), pp. 319. Churchill, London.

Jackson, H., and Craig, A. W. (1966). *Nature* **212,** 86.

Jackson, H., Fox, B. W., and Craig, A. W. (1959). *Brit. J. Pharmacol.* **14,** 149.

Jackson, H., Fox, B. W., and Craig, A. W. (1961). *J. Reprod. Fertility* **2,** 447.

Jackson, H., Partington, M., and Walpole, A. L. (1964). *Brit. J. Pharmacol.* **23,** 521.

Jackson, M. C. N. (1963). *J. Reprod. Fertility* **6,** 153.

Johnsen, S. G. (1964). *Acta Endocrinol.* Suppl. 90, 99.

Jolley, W. B., Martin, W. E., Bamberger, J. W., and Stearns, L. W. (1962). *J. Endocrinol.* **25,** 183.

Kanematsu, S., and Sawyer, C. H. (1965). *Endocrinology* **76,** 691.

Kar, A. B. (1961). *Endocrinology* **69,** 1116.

Khazan, N., Sulman, F. G., and Winnik, H. Z. (1960). *Proc. Soc. Exptl. Biol. Med.* **105,** 201.

Kincl, F. A., and Dorfman, R. I. (1963). *Acta Endocrinol. (Kbh) Suppl.* **73,** 17.

Kincl, F. A., and Dorfman, R. I. (1965). *J. Reprod. Fertility* **10,** 105.

Kotz, H. L., and Herrmann, W. (1961). *Fertility Sterility* **12,** 202.

Kraicer, P. F., and Shelesnyak, M. C. (1965). *J. Reprod. Fertility* **10,** 221.

Kraicer, P. F., Marcus, G. J., and Shelesnyak, M. C. (1963). *J. Reprod. Fertility* **5,** 417.

Lerner, L. J., Holthaus, F. J., and Thompson, C. R. (1958). *Endocrinology* **63,** 295.

Lindsay, D., Poulson, E., and Robson, J. M. (1961). *J. Endocrinol.* **23,** 209.

Lindsay, D., Poulson, E., Robson, J. M. (1962). *J. Endocrinol.* **25,** 53.

Lipschutz, A., Iglesias, R., and Salinas, S. (1962). *Nature* **196,** 946.

Lisk, R. D., and Newlon, M. (1963). *Science* **139,** 223.

McLaren, A., and Mitchie, D. (1956). *J. Exptl. Biol.* **33,** 394.

MacLeod, J. (1965). *In* "Biological Council Symposium on Agents Affecting Fertility" (C. R. Austin and J. S. Perry, eds.), 319 pp. Churchill, London.

Marcus, G. J., Shelesnyak, M. C., and Kraicer, P. F. (1964). *Acta Endocrinol.* **47,** 255.

Mason, K. E. (1940–1941). *Am. J. Physiol.* **131,** 268.

Mason, K. E., and Mauer, S. I. (1957). *Anat. Record* **127,** 329.

Mears, E. (1961). *Brit. Med. J.,* p. 1179.

Mears, E. (1965a). *In* "Handbook of Oral Contraception" (E. Mears, ed.), pp. 107. Churchill, London.

Mears, E. (1965b). *In* "Biological Council Symposium on Agents Affecting Fertility" (C. R. Austin and J. S. Perry, eds.), 319 pp. Churchill, London.

Miller, M. J., Fischer, M. P., Elcoate, P. V., and Mawson, C. A. (1958). *Can. J. Biochem.* **36,** 557.

Montemurro, D. G. (1960). *Brit. J. Cancer* **14,** 319.

Nelson, W. O., and Steinberger, E. (1953). *Federation Proc.* **12,** 103.

Nelson, W. O., and Steinberger, E. (1957). *Endocrinology* **60,** 105.

Nelson, W. O., Davidson, O. W., and Wada, K. (1963). *In* "Delayed Implantation" (A. C. Enders, ed.), p. 183. Univ. of Chicago Press, Chicago, Illinois.

Nissim, J. A. (1957). *Lancet* i, 304.
Noyes, R. W., Dickman, Z., Doyle, L. L., and Gates, A. H. (1963). *In* "Delayed Implantation" (A. C. Enders, ed.), p. 197. Univ. of Chicago Press, Chicago, Illinois.
Oakberg, E. F. (1956). *Am. J. Anat.* **99**, 507.
Overbeek, G. A., Madjerek, Z., and de Visser, J. (1962). *Acta Endocrinol.* **41**, 351.
Paget, G. E., Walpole, A. L., and Richardson, D. N. (1961). *Nature* **192**, 1191.
Parizek, J. (1960). *J. Reprod. Fertility* **1**, 294.
Partington, M., and Bateman, A. J. (1964). *Heredity* **19**, 191.
Partington, M., and Jackson, H. (1963). *Genet. Res.* (*Cambridge*) **4**, 333.
Partington, M., Fox, B. W., and Jackson, H. (1964). *Exptl. Cell Res.* **33**, 78.
Patanelli, D. J., and Nelson, W. O. (1963). *Am. Chem. Soc. Abstr.*, 32L, March–April.
Pincus, G. (1958). *Ann. N. Y. Acad. Sci.* **71**, 531.
Pincus, G. (1965). *In* "Control of Fertility," pp. 360. Academic Press, New York.
Pincus, G., and Merrill, A. P. (1961). *In* "Control of Ovulation" (C. A. Villee, ed.), p. 251. Macmillan (Pergamon), New York.
Poulson, E., and Robson, J. M. (1963). *Brit. J. Pharmacol.* **21**, 150.
Poulson, E., Botros, M., and Robson, J. M. (1960). *Science* **131**, 1101.
Prasad, M. R. N., Kalra, S. P., and Segal, S. J. (1965). *Fertility Sterility* **16**, 101.
Psychoyos, A. (1963). *J. Endocrinol.* **27**, 337.
Ross, W. C. J. (1962). "Biological Alkylating Agents," pp. 232. Butterworths, London.
Sawyer, C. H., and Everett, T. W. (1959). *Endocrinology* **65**, 644.
Schlough, J. S., and Meyer, R. K. (1965). *Fertility Sterility* **16**, 106.
Segal, S. J., and Davidson, O. W. (1962). *Anat. Record* **142**, 278.
Segal, S. J., and Nelson, W. O. (1958). *Proc. Soc. Exptl. Biol. Med.* **98**, 431.
Segal, S. J., and Nelson, W. O. (1961). *Anat. Record* **139**, 273.
Segal, S. J., and Tyler, A. (1958). *Biol. Bull.* **115**, 364.
Segal, S. J., Davidson, O. W., and Wada, K. (1965). *Proc. Natl. Acad. Sci. U.S.* **54**, 782.
Shelesnyak, M. C. (1957). *Recent Progr. Hormone Res.* **13**, 269.
Shelesnyak, M. C. (1959). *Proc. Soc. Exptl. Biol. Med.* **100**, 739.
Shelesnyak, M. C. (1960). *Proc. 1st Intern. Congr. Endocrinol.* (*Copenhagen*), p. 677.
Shelesnyak, M. C. (1962). *Perspectives Biol. Med.* **5**, 503.
Shelesnyak, M. C. (1963). *J. Reprod. Fertility* **5**, 295.
Shelesnyak, M. C., and Barnea, A. (1963). *Acta Endocrinol* (*Kbh.*) **43**, 469.
Shelesnyak, M. C., Lunenfield, B., and Honig, B. (1963). *Life Sci.* **1**, 73.
Shipley, E. G. (1962). *In* "Methods in Hormone Research" (R. I. Dorfman, ed.), Vol. II, p. 180. Academic Press, New York.
Smith, C. N., LaBrecque, G. C., and Borkovec, A. B. (1964). *Ann. Rev. Entomol.* **9**, 269.
Spaziani, E., and Szego, C. M. (1958). *Endocrinology* **63**, 669.
Steinberger, E. (1962). *J. Reprod. Fertility* **3**, 250.
Swyer, G. I. M., and Little, V. (1962a). *J. Endocrinol.* **24**, xxii.
Swyer, G. I. M., and Little, V. (1962b). *Proc. Roy. Soc. Med.* **55**, 861.
Swyer, G. I. M., Sebok, L., and Barnes, D. F. (1960). *Proc. Roy. Soc. Med.* **53**, 435.
Talwar, G. P., and Segal, S. J. (1963). *Proc. Natl. Acad. Sci. U. S.* **50**, 226.
Walpole, A. L. (1965). *In* "Biological Council Symposium on Agents Affecting Fertility" (C. R. Austin and J. S. Perry, eds.), 319 pp. Churchill, London.
Wheeler, G. P. (1962). *Cancer Res.* **22**, 651.

Whitten, W. K. (1957). *J. Endocrinol.* **16**, 80.
Yochim, J. M., and DeFeo, V. J. (1963). *Endocrinology* **72**, 317.

Addendum

Research in the fields of reproductive physiology and pharmacology is developing so rapidly in both breadth and depth that it is necessary to provide the reader with more recent references and brief indication of more significant developments. The monograph by Pincus (1965) deals mainly with fertility control in the female mammal via steroidal chemicals while that by Jackson (1966) discusses the subject of antifertility chemicals on a broader basis. A number of articles have appeared in annual publications, with three reviews in consecutive years in *Annual Reviews of Pharmacology* by Pincus and Fridhandler (1964), Tyler (1967), and Jackson and Schnieden (1968). Harper (1968) and Fox and Fox (1967), have surveyed the literature on antifertility steroidal chemicals and biochemical aspects of drug action on spermatogenesis, respectively. Studies of the neuroendocrine processes involved in reproduction have been greatly intensified in recent years. In the hypothalamus and perhaps adjacent areas, a number of centers have been implicated in mechanisms controlling sexual processes; these included regions chemoreceptively sensitive to circulating sex hormones. The physiological activity of pituitary gonadotrophs is then controlled via specific releasing factors transmitted from the hypothalamus to anterior pituitary by means of its portal blood system. The situation has been admirably reviewed in a recent article by McCann and collaborators (1968).

Reports of a number of symposia are recommended, namely by the Biological Council, on agents affecting fertility (1965), and a Ciba Foundation Colloqium concerned with the endocrine function of the testis (1967). A series of supplements to the *Journal of Reproduction and Fertility* are of special interest, dealing, respectively, with ovarian regulatory mechanisms (1966), sperm capacitation and endocrine control of the testis (1967), and the effects of pharmacologically active substances on sexual function (1968).

The search for more potent progestational compounds has led to "Norgestrol," a 13-alkyl gonane produced by total synthesis of the steroid molecule, the uses of which form the topic of a further symposium and supplement (1968). Modifications in the technique of application of contraceptive steroids have yielded the sequential administration of estrogen and progestational steroid—seemingly not quite so reliable as the concurrent presentation of the two hormones but a more economical and perhaps more physiological approach to the problem. The discovery of a practicable approach to oral contraception by low-dosage progestational compound only (using chlormadinone, Megestrol acetate or Norgestrol) may avoid the hazards of long-term interference with hypothalamic and pituitary function (see supplement on Norgestrol for references). This method is compatible with normal ovarian function and menstrual cycles and the mechanism of action may involve minimal disturbances of endometrial function unfavorable to implantation, although modification of the cervical mucus rendering it "hostile" to sperm penetration is a favored explanation. Postcoital agents of an antiestrogenic nature are at the experimental level and antiprogestational compounds remain a potential for the future.

In the male, intensified efforts are being made to explore physiological mechanisms concerned with the spermatogenic process, sperm transport, and maturation in the epididymis as well as sperm capacitation within the uterus. The discovery of two further simple antifertility chemicals comprises a significant advance in male reproductive pharmacology. In the rat, trimethylphosphate produces a functional type of sterility involving spermatids and sperm (Jackson and Jones, 1968), while administration of 3-chloro-1,2-propanediol renders only epididymal sperm incompetent in the same species (Coppola, 1969). Consequently, the sterilant action rapidly disappears in the latter instance, although trimethylphosphate also shows antispermatogenic properties at higher dose levels.

Experimental chemosterilization of the male is progressing along a wide front toward the control of insect and rodent populations, but human contraception via the male remains in the future. Studies of chemicals affecting male fertility may have implications in the control of widespread parasitic worm vectors of disease, as exemplified by Bilharzia (Jackson *et al.*, 1968). Information on insect sterilants has been assembled by Borkovec (1966).

REFERENCES TO ADDENDUM

Aspects of the chemistry, pharmacology and clinical use of a new progestagen— Norgestrol (WY 3707). (1968). *J. Reprod. Fertility Suppl.* **5**, 177 pp.

Austin, C. R., and Perry, J. S. (eds.). (1965). *Biological Council Symp. Agents Affecting Fertility*, 319 pp. Churchill, London.

Borkovec, A. B. (1966). "Advances in Pest Control Research," Vol. VII: Insect Chemosterilants, pp. 143. Wiley (Interscience), New York.

Capacitation of spermatozoa and endocrine control of spermatogenesis. (1967). *J. Reprod. Fertility Suppl.* **2**, 155 pp.

Ciba Foundation Colloquium. (1967). *Endocrinology* **16**, 352 pp.

Coppola, J. A. (1969). *Life Sci.* **8**, 43.

Effects of pharmacologically active substances on sexual function. (1968). *J. Reprod. Fertility Suppl.* **4**, 114 pp.

Fridhandler, L., and Pincus, G. (1964). *Ann. Rev. Pharmacol.* **4**, 177.

Fox, B. W., and Fox, M. (1967). *Pharmacol. Rev.* **19**, 21.

Harper, M. J. K. (1968). *Arzneimittel-forsch.* **12**, 47.

Jackson, H. (1966). *Antifertility Compounds in the Male and Female, Am. Lecture Ser. No.* **631**, 214.

Jackson, H., and Jones, A. R. (1968). *Nature* **220**, 511.

Jackson, H., and Schnieden, H. (1968). *Ann. Rev. Pharmacol.* **8**, 467.

Jackson, H., Davies, P., and Bock, M. (1968). *Nature* **218**, 977.

McCann, S. McD., Dhariwal, A. P. S., and Porter, J. C. (1968). *Ann. Rev. Physiol.* **30**, 589.

Ovarian regulatory mechanisms. (1966). *J. Reprod. Fertility Suppl.* **1**, 136 pp.

Pincus, G. (1965). "The Control of Fertility," 360 pp. Academic Press, New York.

Tyler, E. T. (1967). *Ann. Rev. Pharmacol.* **7**, 381.

Author Index

Numbers in italics refer to the pages on which the complete references are listed.

511

Index to Genus and Species Names

529

Subject Index

A

Acetamide, conjugation and, 281, 282, 284

Acetylene, chemotaxis and, 150

Acid, fish egg activation by, 306

Acipenserids
egg
fertilization impulse in, 304
membrane formation in, 305–306
sperm of, 310

Acridine orange
chromosome transfer and, 74
colicinogeny and, 68
fertility episome and, 27
pili and, 55
plasmids and, 64
resistance transfer factor and, 69

Acridines, bacterial fertility factors and, 51, 52, 67

Acriflavine
induction of conjugation and, 282, 284, 286
uptake, phage infection and, 33–34

Acrosome
fish sperm and, 309–310
hyaluronidase and, 364–367
membrane of, 335
structure of, 329
zona lysin in, 368–370

Acrosomin, nature of, 338

Actinomycin D, mating reactivity and, 278

Adenosine diphosphate (ADP), fish egg and, 313–314

Adenosine triphosphate (ATP)
boron effects and, 207
fish egg and, 313, 323

Age, meiotic chromosome abnormalities, and, 465

Agglutination, specificity of, 151–158

Agglutinin, sperm, 350

Aldehydes, egg membrane and, 305

Algae
chemotaxis, chemotropism and chemotactica in, 140–150

fertilization
manifestation forms, 135–138
induction effects and substances, 139–140
gametangiogamy in, 165–178
gametes, fusion of, 158–165
monoecious, sexual differentiation in, 157–158
sexual incompatibility in, 178–181

Alkylating agents, spermatogenesis and, 473–481

Allogamy, occurrence of, 173–178

Aluminum, conjugation and, 282

Amberlite IR-120, mating-type and, 275–277

Amebas, cytonucleoproteins in, 124

Amenorrhea, production of, 494

Amine oxidase inhibitors, fertility and, 471, 503

Amino acids
egg culture and, 424, 425
micropylar fluid and, 232
oogenesis and, 298
phosphoprotein and, 315
pollen tube metabolism and, 211
selective fertilization and, 240
starvation, competence and, 76–77
trout egg, 316
uptake by oocytes, 323
zygote cleavage and, 496

Ammonium ions, conjugation and, 281

Ammonium sulfate, mating reaction and, 268

Ampulla, fertilization and, 362–363

Amylases, pollen germination and, 210

β-Amylase, sperm capacitation and, 352

Androgen(s)
spermatogenesis and, 469, 470, 482
sterility and, 486

Androgenesis, occurrence of, 454–455

Androspores, development of, 143–144

Angiosperms
fertilization in, 223–231
macrosporogenesis in, 190–192

unequal, intermediate males and, 71–73

Cruciferae, incompatibility in, 242

Cumulus oophorus
fertilization and, 356
nature of, 355–356
sperm passage through, 364–367

Cutin
pollen germination and, 210
self-incompatibility and, 242

Cutinase, self-incompatibility and, 242

Cyanide
superinfection breakdown and, 34
transformation and, 77

Cyanophyta, sexuality in, 137

Cycadales, fertilization in, 232, 233

Cyclohexane, sperm attraction by, 149

Cyproheptadine, pregnancy and, 503

Cysteine, egg membrane and, 305

Cytochrome oxidase
egg development and, 311
ovary and, 193–194, 196
pollen germination and, 210
self-incompatibility and, 243

Cytogamy, conjugation and, 263

Cytoplasm
ascomycete plasmogamy and, 106–107
differentiation and, 122–124
errors of union of gametes and, 451–454
fish oocyte, 296
penetration, *in vitro* fertilization and, 421
variations in gamete union and, 440–442

D

Decidual cell response, prevention of, 501–503

Deer, sperm of, 347

Dehydrogenase, ovary and, 194, 196

Deoxyribonuclease
sperm and, 328
superinfection breakdown and, 34

Deoxyribonucleic acid (DNA)
Allomyces gametes and, 102
colicins and, 68
cytoplasmic, 299
estradiol and, 496
exogenous, 84–85
autonomous replication of, 85

insertion and release of, 85
linear inheritance and reciprocal recombination of, 85–86
fertility factor and, 29
infection with, 12, 14, 16, 30, 32
myxomycete gametes and, 99, 100, 130–131
noncompatible pollen and, 215
oogenesis and, 299
plant egg nucleus and, 230–231
pollen formation and, 202–203
pronuclei and, 408
release, lipocarbohydrate and, 26
replication, fertilization and, 93
resistance transfer factor and, 69
sex factor and, 65, 67
sperm, 327–328
transduction and, 81, 82
transformation and, 75, 77

Desmidiaceae, fertilization in, 165–167

Desmids, chemotropism in, 148

Detergents, acrosome and, 338

Dextrose, *see* Glucose

Diamines, spermatogenesis and, 483–484

Diastase
cortical granules and, 361
integumentary tapetum and, 196

Diatoms
chemotropism in, 148
fertilization in, 161–163

Diazonium compounds, mating-type substance and, 273

1(N,N-Diethylcarbamoyl methyl)-2,5-dinitropyrrole, fertility and, 482–483

Differentiation, nucleocytoplasmic interactions and, 122–125

Digyny, colchicine and, 428, 430

Dikaryons, glucan degradation by, 132

Dikaryosis
in ascomycetes, 104, 107, 108
in basidiomycetes, 108–110

Dimethylbusulfan, fertility and, 478

Dimethylmyleran, fertility and, 478

Dinitrofluorobenzene, mating-type substance and, 273

2,4-Dinitrophenol, lysis from without and, 33

Dinophyta, sexuality in, 137

Diphenylamine, fungal gametes and, 102, 103